Lecture Notes in Computational Science and Engineering 114

Editors:

More information about this series at http://www.springer.com/series/3527

Gabriel R. Barrenechea • Franco Brezzi •
Andrea Cangiani • Emmanuil H. Georgoulis
Editors

Building Bridges: Connections and Challenges in Modern Approaches to Numerical Partial Differential Equations

Springer

Editors
Gabriel R. Barrenechea
Department of Mathematics & Statistics
University of Strathclyde
Glasgow, United Kingdom

Franco Brezzi
Istituto di Matematica Applicata e
Tecnologie Informatiche - Pavia
Consiglio Nazionale delle Ricerche
Pavia, Italy

Andrea Cangiani
Department of Mathematics
University of Leicester
Leicester, United Kingdom

Emmanuil H. Georgoulis
Department of Mathematics
University of Leicester
Leicester, United Kingdom
and
Department of Mathematics
National Technical University of Athens
Zografou, Greece

ISSN 1439-7358 ISSN 2197-7100 (electronic)
Lecture Notes in Computational Science and Engineering
ISBN 978-3-319-82402-4 ISBN 978-3-319-41640-3 (eBook)
DOI 10.1007/978-3-319-41640-3

Mathematics Subject Classification (2010): 65M06, 65M08, 65M12, 65M60, 65M99

Printed on acid-free paper

This Springer imprint is published by Springer Nature
The registered company is Springer International Publishing AG Switzerland

Preface

This volume contains survey articles from the topics of a number of the plenary talks of the 101st LMS-EPSRC Symposium entitled *Building Bridges: Connections and Challenges in Modern Approaches to Numerical Partial Differential Equations* held at Durham University between 8 and 16 July 2014.

The symposium was devoted to recent advances in numerical methods for partial differential equations (PDEs) using non-polynomial basis functions, mimetic finite difference methods, and discontinuous Galerkin (dG) methods. Moreover, since recent works in these directions have highlighted unforeseen pairwise connections between the different approaches, a particular emphasis in this symposium has been to explore further links and to exchange ideas and techniques between them.

The scientific program included six short courses given by Tom Hughes (Austin, Texas) on isogeometric analysis, Ralf Hiptmair (ETH, Zurich) on plane wave discontinuous Galerkin methods, Leszek Demkowicz (Austin, Texas) on discontinuous Petrov-Galerkin method, Bernardo Cockburn (Minnesota) on hybridized DG methods, Chi-Wang Shu (Brown) on discontinuous Galerkin methods for hyperbolic equations with delta-singularities, and Konstantin Lipnikov (Los Alamos) on mimetic finite difference methods. These were complemented by plenary lectures by Assyr Abdulle (EPFL, Lausanne), Mark Ainsworth (Brown), Lourenco Beirão Da Veiga (Milan), Pavel Bochev (Sandia), Annalisa Buffa (Pavia), Erik Burman (UCL), Alexandre Ern (EPC, Paris), Oleg Davydov (Giessen), Charles Elliott (Warwick), Alexandre Ern (Paris), Ivan Graham (Bath), Paul Houston (Nottingham), Charalambos Makridakis (Sussex), Gianmarco Manzini (Pavia), Peter Monk (Delaware), Alessandro Russo (Milan), Robert Scheichl (Bath), and Frédéric Valentin (Petropolis). Further, a *session with speakers from industry* was organized with invited speakers Paul Childs (Schlumberger Gould Research, Cambridge) and Halvor Nilsen (SINTEF, Norway) who presented very interesting lectures closely related to the symposium's topics.

During the symposium, a number of extremely interesting discussions took place, in particular, on the connections, similarities, and differences of related numerical methods, especially in the context of variational methods on polytopic meshes, such as mimetic finite difference methods, virtual element methods, polygonal discontinuous Galerkin methods, and hybrid-type discontinuous Galerkin approaches. Further, a number of stimulating discussions and talks were given on abstraction of numerical methods through different frameworks (such as the HDG and DPG frameworks of discontinuous variational methods). As an example of an outcome from the aforementioned discussions, two of the main speakers in the symposium, B. Cockburn and A. Ern, recently wrote a research article entitled "Bridging the hybrid high-order and hybridizable discontinuous Galerkin methods".

This volume collects 13 contributions, by several of the main speakers. The type of contributions ranges from new applications of some of the emerging techniques, to new powerful frameworks in which many of the new techniques can be inserted and better understood, to careful analyses of the differences and of the similarities of wide ranges of new methods that have been proposed independently, by various groups all over the planet, in the very last few years.

Only an intensive comparison of all these new ideas on realistic problems of interest in applications will help in understanding which method is more suited for each class of problems. However, we believe that this book could be an excellent guide for young (and less young) researchers who are willing to get closer, to familiarize, and possibly to start working, on some of these new methods, trying to dig into their inner mathematical nature and/or testing them on new problems.

If you have been curious about all these new instruments, this book is an ideal help to start learning more about them, their main features, their reciprocal relationships, and their possibilities in applications.

We would like to express our gratitude to Mark Ainsworth (Brown) and Endre Süli (Oxford), for their help and support as scientific advisors to the meeting, and to the LMS and EPRSC for giving us the opportunity to organize this event. Moreover, we would also like to thank the Numerical Algorithms and Intelligent Software (NAIS) consortium for additional financial support. Finally, we would like to extend our thanks to the Durham Symposia administration staff for all their help.

Glasgow, UK — Gabriel R. Barrenechea
Pavia, Italy — Franco Brezzi
Leicester, UK — Andrea Cangiani
Leicester, UK & Athens, Greece — Emmanuil H. Georgoulis

Contents

Numerical Homogenization Methods for Parabolic Monotone Problems

Assyr Abdulle

Abstract In this paper we review various numerical homogenization methods for monotone parabolic problems with multiple scales. The spatial discretisation is based on finite element methods and the multiscale strategy relies on the heterogeneous multiscale method. The time discretization is performed by several classes of Runge-Kutta methods (strongly A-stable or explicit stabilized methods). We discuss the construction and the analysis of such methods for a range of problems, from linear parabolic problems to nonlinear monotone parabolic problems in the very general $L^p(W^{1,p})$ setting. We also show that under appropriate assumptions, a computationally attractive linearized method can be constructed for nonlinear problems.

1 Introduction

Parabolic problems with multiple scales enter in the modelling of a wide range of problems, e.g., thermal diffusion in composite materials, flow problems in heterogeneous medium, etc. We are interested in problems in which the microscopic heterogeneities occur at a much smaller scale than the macroscopic length scale of interest that describes the physical phenomenon of interest. For such problems mathematical homogenization [18, 40] gives the adequate theoretical framework to describe an effective solution originating from the limit of the fine scale solution when the size of the small scales tends to zero. An effective equation for this effective solution can also be established. However, except for special cases, there are no explicit expressions for the effective coefficients (diffusion tensor) of the upscaled equation. The aim of numerical homogenization is to construct computational strategy to compute an approximation of these effective equations and sometimes to capture fine scale oscillations of the multiscale solution. The

A. Abdulle (✉)
ANMC, Mathematics Section, École Polytechnique Fédérale de Lausanne, CH-1015 Lausanne, Switzerland
e-mail: assyr.abdulle@epfl.ch

G.R. Barrenechea et al. (eds.), *Building Bridges: Connections and Challenges in Modern Approaches to Numerical Partial Differential Equations*, Lecture Notes in Computational Science and Engineering 114,
DOI 10.1007/978-3-319-41640-3_1

theory of homogenization is at the root of two classes of numerical methods that we briefly discuss

- methods based on oscillatory basis functions built into a coarse FE space: this idea goes back to Babuška and Osborn [16] and is based on solving local fine scale problems within each macroscopic element of the coarse FE space. Elaboration and generalization have been developed within the multiscale finite element method (MsFEM) [15, 38];
- methods supplementing upscaled data for resolving the effective equation: this idea has been widely used by engineer (see e.g., the references in [32]) and turned into a general framework in the heterogeneous multiscale method (HMM) [3, 4, 55]. In the finite element context, this latter method is called the finite element heterogeneous multiscale method (FE-HMM) and is based on a macroscopic finite element method with input data given by microscopic sampling of the original fine scale problem in patches of size proportional to the fine scale oscillation.

These two classes of methods use either in their formulation or in their analysis the theory of homogenization in an essential way. Further related to homogenization theory we mention the sparse tensor product FEM based on the two-scale convergence theory and its generalization [14, 48] and the projection based numerical homogenization [20, 31] using successive projection of a fine scale discretization of the multiscale equation into a lower dimensional space and iteratively eliminating the fine scale component of the numerical solution.

We also mention multiscale methods that share some similarities with numerical homogenization methods and have been used for homogenization problems. We start with the variational multiscale method [39]. In this approach one starts from a coarse finite element space that cannot resolve the multiscale structure of the fine scale problem. This coarse space is supplemented by a fine scale space and one seeks a numerical solution in the form of a coarse and fine scale components. The fine scale component is obtained by solving localized fine scale problems. Once these problems solved one can solve the coarse scale approximation. Using local quasi-interpolation and an orthogonal decomposition of the coarse and fine spaces, exponential decay of the localisation error has been first proved in [46] (see also [37]). This new approach of the variational multiscale method is called Localised Orthogonal Decomposition (LOD). Finally we also mention methods based on harmonic coordinates [49]. The idea of this method is to compute an appropriate change of coordinates (based on the full fine scale problem) so its composition with the fine scale problem is a slowly varying function that can be approximated in a coarse space. This approach share some similarity with the MsFEM proposed in [15].

In this article, we review several numerical homogenization methods based on the HMM for the solution of the following class of monotone parabolic multiscale problems in a finite time interval $(0, T)$

$$\begin{aligned} \partial_t u^\varepsilon(x,t) - \operatorname{div}(\mathcal{A}^\varepsilon(x, \nabla u^\varepsilon(x,t))) &= f(x) \text{ in } \Omega \times (0,T), \\ u^\varepsilon(x,t) = 0 \text{ on } \partial\Omega \times (0,T), \quad u^\varepsilon(x,0) &= g(x) \text{ in } \Omega, \end{aligned} \tag{1}$$

with initial source and initial conditions f and g. The maps $\mathcal{A}^\varepsilon\colon \Omega \times \mathbb{R}^d \to \mathbb{R}^d$ are defined on a domain $\Omega \times \mathbb{R}^d$, where $\Omega \subset \mathbb{R}^d$ $d \leqslant 3$, and $\mathcal{A}^\varepsilon(\cdot, \xi)\colon \Omega \to \mathbb{R}^d$ are Lebesgue measurable for every $\xi \in \mathbb{R}^d$. The indexing by an (abstract) parameter $\varepsilon > 0$ indicates that these maps are subject to rapid variations on a very fine scale relative to the size of the domain Ω. For the finite element method we will assume that Ω is a polygonal domain and we will sometimes assume that it is convex. For simplicity neither time dependent source terms $f(x,t)$ or time-dependent maps of the form $\mathcal{A}^\varepsilon(x, t, \nabla u^\varepsilon(x,t))$ are considered but we note that many of the results presented in this review can be extended for these situations.

Let us briefly review the literature on multiscale methods for the parabolic problems (1). For linear problems, most of the methods described above can be used. We mention [29] for MsFEM type methods, [11, 47] for HMM type methods, [45] for LOD type methods. While most of the numerical method have been analysed for the Euler explicit or implicit time discretization, a fully discrete a priori error analysis in space and time for several classes of implicit and explicit Runge-Kutta methods has been given in [9]. For nonlinear monotone parabolic problems, the literature is much more scarce and only methods supplementing upscaled data for resolving the effective equation have been analyzed. In [30] monotone problems with stochastic heterogeneities have been analysed however without convergence rates and for non-discretized micro-problems. In [6, 13] a priori error analysis (in space and time) for two different types of HMM is established under general assumption on the nonlinearity. We close this review by mentioning that for elliptic problems, a posteriori error estimates have been obtained for an HMM type method in the strongly monotone and Lipschitz case in [36] and a priori error estimates for general numerical quadrature methods have been derived in [5]. Finally in [33] numerical homogenization methods (both of HMM and MsFEM types) for monotone PDEs associated to minimization problems have been studied. We note in contrast that for the class of problems (1) discussed in this review, we make no assumptions of an associated scalar potential for $\mathcal{A}^\varepsilon$.

In this paper we aim at reviewing the numerical homogenization methods based on the HMM that have been developed in [6, 9, 13] for parabolic problems (1). We aim at giving a unified description of various error estimates and numerical discretization variant of the FE-HMM

- for linear problem the spatial discretisation based on the FE-HMM is coupled with general classes of Runge-Kutta methods (strongly A-stable and explicit stabilized methods), and fully-discrete space-time analysis is proposed for this family of space-time multiscale solvers [9];
- for nonlinear monotone problems a fully discrete space-time method that couples the FE-HMM in space with the backward Euler method in time is shown to

converge in the $L^p(W^{1,p})$ and $C^0(L^2)$ norms towards the homogenized solution u^0 for Problem 1 under the general assumptions. Space-time convergence rates are established for strongly monotone and Lipschitz maps [6];
- for strongly monotone and Lipschitz maps $\mathcal{A}^\varepsilon$ a new linearized scheme that relies only on linear micro and macro finite element (FE) solvers is proposed and analyzed. A fully discrete space time analysis is also provided for this scheme [13].

We briefly sketch the type of convergence rates that we aim at deriving in this paper: under appropriate assumptions on the tensor $\mathcal{A}^\varepsilon$, the family of solutions u^ε converges, up to a subsequence, to a homogenized solution u^0 solution of a homogenized equation similar to (1) but with $\mathcal{A}^\varepsilon$ replaced by an effective map $\mathcal{A}^0$ that is unknown explicitly (see Sect. 2). In the context of an FE-HMM method, the goal is to derive an error estimate of the type

$$\max_{1\leqslant n\leqslant N} \left\|u^0(\cdot,t_n) - u_n^H\right\|_{L^2(\Omega)} + \left(\sum_{n=1}^{N} \Delta t \left\|\nabla u^0(\cdot,t_n) - \nabla u_n^H\right\|_{L^2(\Omega)}^2\right)^{1/2} \leqslant C\left[(\Delta t)^r + H^s + \left(\frac{h}{\varepsilon}\right)^q + r_{mod} + \left\|g - u_0^H\right\|_{L^2(\Omega)}\right], \tag{2}$$

where C is independent of $\Delta t, H, h$ and r_{mod}. Here H is the size of a macroscopic triangulation that is used in the FE-HMM to approximate the effective solution u^0 and h is the mesh size of a microscopic triangulation used on a patch K_δ around macroscopic quadrature points. The diameter of the patch K_δ is of size δ typically $\delta = \mathcal{O}(\varepsilon)$. As h must resolve the fine scale oscillation we have $h < \varepsilon \leqslant \delta$. We notice two important facts

- as $h/\varepsilon = 1/N_{mic}$, where N_{mic} is the number of points per oscillation length and the quantity h/ε in the estimate (2) is thus independent of ε and measure the degrees of freedom used to resolve the oscillation; if $\varepsilon \to 0$, so does the patch K_δ hence we solve the fine scale only on small fraction of the macroscopic computational domain and the overall computational cost is independent of ε;
- the quantities, $\Delta t, H, h$ are discretisation parameters while r_{mod} quantifies the error due to the upscaling procedure, i.e., by replacing the true homogenized map $\mathcal{A}^0$ by a map computed from some microscopic models. The coupling condition (periodic, Dirichlet), the size of the sampling domain enter in this modelling error that is not influenced by the macro or micro discretisation parameter H, h. In the most favourable case (e.g., locally periodic homogenization), r_{mod} can be shown to vanish.

In view of the above prototypical error estimate in this paper we will speak about fully discrete spatial error estimates when we have an estimate in terms of both the macroscopic and microscopic spatial mesh H, h and a fully discrete space-time error estimate when we derive an estimate in terms of H, h and Δt.

Several difficulties arise when analyzing a numerical homogenization method: first as the effective data are only available at quadrature points, we necessarily rely on a FEM with numerical quadrature on the macroscale and have to deal with variational crimes. Second, as the upscaled data are obtained from micro solvers (FEM) one has to precisely quantify the propagation of the errors across scales. Finally the modelling error that originates from the averaging procedure designed to recover the effective data need also to be quantified. To close this introduction, we review several important contributions concerning FE methods for single scale nonlinear monotone problems and contrast these results with the numerical homogenization literature. Using quasi-norm techniques, convergence rates have been derived in [17, 27] in the $L^p(W^{1,p})$ setting for single scale parabolic monotone problems with the following p-structure

$$
\begin{aligned}
&|\mathcal{A}(\xi) - \mathcal{A}(\eta)| \leqslant L(\kappa_1 + |\xi| + |\eta|)^{p-2}|\xi - \eta|,\\
&(\mathcal{A}(\xi) - \mathcal{A}(\eta)) \cdot (\xi - \eta) \geqslant \lambda(\kappa_2 + |\xi| + |\eta|)^{p-2}|\xi - \eta|^2, \ \forall\, \xi, \eta \in \mathbb{R}^d
\end{aligned}
$$

including for example the p-Laplacian. Note however that under the most general assumptions on the map $\mathcal{A}^\varepsilon$ under which homogenization results are proved (see e.g. [50]) and under which we can show convergence of an FE-HMM method [6], we have a p-structure if and only if the map $\mathcal{A}^\varepsilon$ is strongly monotone and Lipschitz.

This review is organized as follows. In Sect. 2 we briefly review the homogenization theory for the class of parabolic problems considered and introduce the numerical homogenization method. In Sect. 3 we review the coupling of the FE-HMM with various families of Runge-Kutta methods and explain the techniques used to derive a fully-discrete space-time error analysis. Convergence of a fully-discrete numerical method for general nonlinear monotone parabolic problems is discussed in Sect. 4 and a linearized method is presented in Sect. 5.

2 Assumptions and Homogenization

We consider Problem 1 and the "evolution triple" $W^{1,p}(\Omega) \subset L^2(\Omega) \subset W^{1,p}(\Omega)'$, $f \in L^{p'}(\Omega)$, $g \in L^2(\Omega)$. Very general hypotheses for the maps $\mathcal{A}^\varepsilon$ under which homogenization for (1) can be established, see [21, 50] are the following assumptions assumed to hold uniformly in $\varepsilon > 0$ for all $\xi_1, \xi_2 \in \mathbb{R}^d$ and almost every $x \in \Omega$. For $1 < p < \infty$ and $p > 2d/(d+2)$ we assume

($\mathcal{A}_0$) there is some $C_0 \geqslant 0$ such that $|\mathcal{A}^\varepsilon(x, 0)| \leqslant C_0$ for almost every (a.e.) $x \in \Omega$;

($\mathcal{A}_1$) there exist $\kappa_1 \geqslant 0$, $L > 0$ and $0 < \alpha \leqslant \min\{p-1, 1\}$ such that

$$
|\mathcal{A}^\varepsilon(x, \xi_1) - \mathcal{A}^\varepsilon(x, \xi_2)| \leqslant L(\kappa_1 + |\xi_1| + |\xi_2|)^{p-1-\alpha}|\xi_1 - \xi_2|^\alpha;
$$

$(\mathcal{A}_2)$ there exist $\kappa_2 \geqslant 0$, $\lambda > 0$ and $\max\{2,p\} \leqslant \beta < \infty$ such that

$$(\mathcal{A}^\varepsilon(x,\xi_1) - \mathcal{A}^\varepsilon(x,\xi_2)) \cdot (\xi_1 - \xi_2) \geqslant \lambda(\kappa_2 + |\xi_1| + |\xi_2|)^{p-\beta}|\xi_1 - \xi_2|^\beta.$$

Then under the assumptions $(\mathcal{A}_{0-2})$ the problem (1) has a unique solution $u^\varepsilon \in E$ for any $\varepsilon > 0$

$$E = \{v \in L^p(0,T; W_0^{1,p}(\Omega)) \mid \partial_t v \in L^{p'}(0,T; (W_0^{1,p}(\Omega))')\}, \tag{3}$$

endowed with the norm $\|v\|_E = \|v\|_{L^p(0,T;W_0^{1,p}(\Omega))} + \|\partial_t v\|_{L^{p'}(0,T;(W_0^{1,p}(\Omega))')}$ (see e.g., [57, Theorem 30.A]).

The aim of homogenization is to find a limiting effective solution for the family of oscillatory solutions $\{u^\varepsilon\}$ and an equation for this effective solution involving a parabolic PDE, where the small scales have been averaged out. We briefly describe this procedure. First, observe that the solution satisfies the bound

$$\begin{aligned}\|u^\varepsilon\|^p_{L^p(0,T;W_0^{1,p}(\Omega))} &+ \|\partial_t u^\varepsilon\|^{p'}_{L^{p'}(0,T;(W_0^{1,p}(\Omega))')} \\ &\leqslant C((L_0 + \kappa_1 + \kappa_2)^p + \|f\|^{p'}_{L^{p'}(\Omega)} + \|g\|^2_{L^2(\Omega)}),\end{aligned}$$

independently of ε and $\{u^\varepsilon\}$ is a bounded sequence in E. By compactness, there exists a subsequence, still denoted by $\{u^\varepsilon\}$, and some $u^0 \in E$, such that

$$u^\varepsilon \rightharpoonup u^0 \text{ in } L^p(0,T; W_0^{1,p}(\Omega)) \quad \text{and} \quad \partial_t u^\varepsilon \rightharpoonup \partial_t u^0 \text{ in } L^{p'}(0,T; (W_0^{1,p}(\Omega))') \tag{4}$$

for $\varepsilon \to 0$.

The question answered in the framework of homogenization theory is that of a limiting equation for u^0. For the above parabolic problems, one refers to the so called G-convergence of parabolic operators, sometimes called PG for strong G-convergence [50, 52].

The following compactness result can be shown: there exists a subsequence of $\{u^\varepsilon\}$ (still denoted by $\{u^\varepsilon\}$) and a map $\mathcal{A}^0: \Omega \times \mathbb{R}^d \to \mathbb{R}^d$, such that u^ε weakly converges to u^0 in the sense of (4) and the corresponding maps $\mathcal{A}^\varepsilon(x, \nabla u^\varepsilon) \rightharpoonup \mathcal{A}^0(x, \nabla u^0)$ weakly converges in $L^{p'}(0,T; (L^{p'}(\Omega))^d)$. The homogenized solution $u^0 \in E$ is the solution of the following homogenized problem

$$\begin{aligned}&\partial_t u^0(x,t) - \operatorname{div}(\mathcal{A}^0(x, \nabla u^0(x,t))) = f(x) \text{ in } \Omega \times (0,T), \\ &u^0(x,t) = 0 \text{ on } \partial\Omega \times (0,T), \quad u^0(x,0) = g(x) \text{ in } \Omega,\end{aligned} \tag{5}$$

where $\mathcal{A}^0$ satisfies $(\mathcal{A}_{0-2})$ (with possibly different constants $C_0, \kappa_1, \kappa_2, \lambda$ and L) with Hölder exponent $\gamma = \alpha/(\beta - \alpha)$ in $(\mathcal{A}_1)$. We note that $\gamma = \alpha$, if and only if $p = 2$, $\alpha = 1$, $\beta = 2$. Convergence of the whole sequence $\{u^\varepsilon\}$ to u^0 can be obtained under additional structure of the maps $\mathcal{A}^\varepsilon$, for example if $\mathcal{A}^\varepsilon(x,\xi) = \mathcal{A}(x/\varepsilon,\xi)$, where $\mathcal{A}(y,\xi)$ is a $Y = (0,1)^d$-periodic function in y. In this case one can also derive a description of $\mathcal{A}^0$ in terms of the solutions of a boundary value problems in the reference domain Y. When the maps $\mathcal{A}^\varepsilon$ depend on both a slow and a fast variable, i.e. $\mathcal{A}(x, x/\varepsilon, \xi)$, the boundary value problems depends on $x \in \Omega$. For completeness we introduce the weak formulation of the homogenized problem, by introducing the map $B^0\colon W_0^{1,p}(\Omega) \times W_0^{1,p}(\Omega) \to \mathbb{R}$ given by

$$B^0(v;w) = \int_\Omega \mathcal{A}^0(x, \nabla v(x)) \cdot \nabla w(x) dx, \qquad v, w \in W_0^{1,p}(\Omega), \tag{6}$$

We will also sometimes need a discrete weak form based on a quadrature formula $\{x_{K_j}, \omega_{K_j}\}_{j=1}^J$ defined in the next section that reads

$$\hat{B}^0(v^H; w^H) = \sum_{K\in\mathcal{T}_H} \sum_{j=1}^J \omega_{K_j} \mathcal{A}^0(x_{K_j}, \nabla v^H(x_{K_j})) \cdot \nabla w^H(x_{K_j}), \quad v^H, w^H \in S_0^1(\Omega, \mathcal{T}_H), \tag{7}$$

provided $\mathcal{A}^0(\cdot,\xi)$ has a continuous representative for every $\xi \in \mathbb{R}^d$.

2.1 *Multiscale Methods: The Finite Element Heterogeneous Multiscale Methods*

We give in this section a general formulation of the FE-HMM for parabolic problem. The method relies on

- a macroscopic FE method based on a macroscopic spatial discretization of Ω;
- a microscopic solver defined in sampling domains around sampling points $x \in \Omega$, where an approximation of the map $\mathcal{A}^0(x)$ is required;
- a time discretization method.

Macro Discretization Let $\mathcal{T}_H$ be a family of macro partitions of the polygonal domain Ω consisting of conforming, shape-regular meshes with simplicial elements.[1] The macro elements $K \in \mathcal{T}_H$ are open and such that $\cup_{K\in\mathcal{T}_H} \bar{K} = \overline{\Omega}$. Let $\operatorname{diam} K$ be the diameter of $K \in \mathcal{T}_H$ we define by $H = \max_{K\in\mathcal{T}_H} \operatorname{diam} K$ the macroscopic mesh size and consider the macro finite element space

$$S_0^\ell(\Omega, \mathcal{T}_H) = \{v^H \in W_0^{1,p}(\Omega) \mid v^H|_K \in \mathcal{P}^\ell(K), \forall\, K \in \mathcal{T}_H\}, \tag{8}$$

[1]We concentrate on simplicial elements for simplicity but note that many results presented in this paper can be extended to rectangular elements (see for example [9]).

where $\mathcal{P}^\ell(K)$ is the space of polynomials on $K \in \mathcal{T}_H$ of degree at most ℓ. We also consider within each macro element $K \in \mathcal{T}_H$ quadrature points $x_{K_j} \in K$ and weights ω_{K_j} for $j = 1, \ldots, J$. We assume that $\{x_{K_j}, \omega_{K_j}\}_{j=1}^J$ are obtained from a quadrature formula $\{\hat{x}_j, \hat{\omega}_j\}_{j=1}^J$ by $x_{K_j} = F_K(\hat{x}_j),\ \omega_{K_j} = \hat{\omega}_j|\det(\partial F_K)|, j = 1, \ldots, J$ where F_K is the affine mapping such that $K = F_K(\hat{K})$. We will make the following assumption on the quadrature formula

(Q1) $\int_{\hat{K}} \hat{p}(\hat{x})d\hat{x} = \sum_{j\in J} \hat{\omega}_j \hat{p}(\hat{x}_j),\ \forall \hat{p}(\hat{x}) \in \mathcal{P}^\sigma(\hat{K})$, where $\sigma = \max(2\ell - 2, \ell)$.

These requirements on the quadrature formula ensure that the optimal convergence rates for elliptic FEM hold when using numerical integration [23].

Multiscale Method The FE-HMM method for parabolic problems can be defined as follows. Find $u^H \in [0, T] \times S_0^\ell(\Omega, \mathcal{T}_H) \to \mathbb{R}$ such that

$$\begin{aligned} &(\partial_t u^H, v^H) + B_H(u^H, v^H) = (f, v^H) \ \ \forall v^H \in S_0^\ell(\Omega, \mathcal{T}_H) \\ &u^H = 0 \ \text{ on } \partial\Omega \times (0, T) \\ &u^H(x, 0) = u_0^H, \end{aligned} \tag{9}$$

where

$$B_H(v^H; w^H) = \sum_{K\in\mathcal{T}_H} \sum_{j=1}^{J} \omega_{K_j} \mathcal{A}_{K_j}^{0,h}(\nabla v^H(x_{K_j})) \cdot \nabla w^H(x_{K_j}) \quad v^H, w^H \in S_0^\ell(\Omega, \mathcal{T}_H) \tag{10}$$

and $\mathcal{A}_{K_j}^{0,h}(\cdot)$ is a numerically upscaled tensor defined in (13).

Micro Solver We see that for the map B_H in (9), we need to a procedure to recover the effective data $\mathcal{A}_{K_j}^{0,h}(\nabla v^H(x_{K_j}))$. This rely on micro solvers in each sampling domain K_{δ_j}, $j = 1, \ldots, J$, associated to a macro element $K \in \mathcal{T}_H$. Let $K_{\delta_j} = x_{X_j} + \delta I$, $I = (-1/2, 1/2)^d$, $\delta \geqslant \varepsilon$ be discretized by micro meshes $\mathcal{T}_h$ consisting of simplicial elements $T \in \mathcal{T}_h$, with size h is defined by $h = \max_{T\in\mathcal{T}_h} \operatorname{diam} T$. We then consider the micro finite element spaces

$$S^q(K_{\delta_j}, \mathcal{T}_h) = \{v^h \in W(K_{\delta_j}) \mid v^h|_T \in \mathcal{P}^q(T), \forall\, T \in \mathcal{T}_h\}, \tag{11}$$

where $\mathcal{P}^q(T)$ is the space of linear polynomials on $T \in \mathcal{T}_h$ and $W(K_{\delta_j}) \subset W^{1,p}(K_{\delta_j})$ is some Sobolev space. The choice of the space $W(K_{\delta_j})$ sets the coupling condition between the macro and micro solver, e.g.,

- $W(K_{\delta_j}) = \mathcal{W}_{per}^{1,p}(K_{\delta_j}) = \{v \in W_{per}^{1,p}(K_{\delta_j}) \mid \int_{K_{\delta_j}} v\, dx = 0\}$ (periodic coupling);
- $W(K_{\delta_j}) = W_0^{1,p}(K_{\delta_j})$ (Dirichlet coupling).

For $\xi \in \mathbb{R}^d$ and $K_{\delta_j} \subset K \in \mathcal{T}_H$, we introduce the function $\chi_{K_j}^{\xi,h}$ as the solution to the variational problem: find $\chi_{K_j}^{\xi,h} \in S^q(K_{\delta_j}, \mathcal{T}_h)$ such that

$$\int_{K_{\delta_j}} \mathcal{A}^{\varepsilon}(x, \xi + \nabla \chi_{K_j}^{\xi,h}) \cdot \nabla w^h \, dx = 0, \qquad \forall\, w^h \in S^q(K_{\delta_j}, \mathcal{T}_h). \tag{12}$$

Based on the functions $\chi_{K_j}^{\xi,h}$ we can compute the effective data by

$$\mathcal{A}_{K_j}^{0,h}(\xi) = \frac{1}{|K_{\delta_j}|} \int_{K_{\delta_j}} \mathcal{A}^{\varepsilon}(x, \xi + \nabla \chi_{K_j}^{\xi,h}) dx. \tag{13}$$

We also define an auxiliary flux useful for the analysis

$$\bar{\mathcal{A}}_{K_j}^{0}(\xi) = \frac{1}{|K_{\delta_j}|} \int_{K_{\delta_j}} \mathcal{A}^{\varepsilon}(x, \xi + \nabla \bar{\chi}_{K_j}^{\xi}) dx, \tag{14}$$

where $\bar{\chi}_{K_j}^{\xi} \in W(K_{\delta_j})$ solve (12) in the infinite dimensional space $W(K_{\delta_j})$.

Upscaling Error We define the upscaling error, called r_{HMM}, as the total error made by approximating the effective flux $\mathcal{A}^0$ by the numerics flux $\mathcal{A}_{K_j}^{0,h}$, precisely for any $v^H \in S_0^{\ell}(\Omega, \mathcal{T}_H)$ we define

$$r_{HMM}(\nabla v^H) = \left(\sum_{K\in\mathcal{T}_H} \sum_{j=1}^{J} \omega_{K_j} \left| \mathcal{A}^0(x_{K_j}, \nabla v^H(x_{K_j})) - \mathcal{A}_{K_j}^{0,h}(\nabla v^H(x_{K_j})) \right|^{p'} \right)^{\frac{1}{p'}}, \tag{15}$$

where $p' = p/(p-1)$ is the dual exponent of $1 < p < \infty$. Thanks to the auxiliary flux, we can further decompose r_{HMM} into two components

$$r_{mic}(\nabla v^H) = \left(\sum_{K\in\mathcal{T}_H} \sum_{j=1}^{J} \omega_{K_j} \left| \bar{\mathcal{A}}_{K_j}^{0}(\nabla v^H(x_{K_j})) - \mathcal{A}_{K_j}^{0,h}(\nabla v^H(x_{K_j})) \right|^{p'} \right)^{\frac{1}{p'}}, \tag{16a}$$

$$r_{mod}(\nabla v^H) = \left(\sum_{K\in\mathcal{T}_H} \sum_{j=1}^{J} \omega_{K_j} \left| \mathcal{A}^0(x_{K_j}, \nabla v^H(x_{K_j})) - \bar{\mathcal{A}}_{K_j}^{0}(\nabla v^H(x_{K_j})) \right|^{p'} \right)^{\frac{1}{p'}}. \tag{16b}$$

We observe that using the Minkowski inequality we get $r_{HMM}(\nabla v^H) \leqslant r_{mic}(\nabla v^H) + r_{mod}(\nabla v^H)$ for every $v^H \in S_0^1(\Omega, \mathcal{T}_H)$. The first term $r_{mic}(\nabla v^H)$ quantifies the error made by solving the micro problems (12) in $S^q(K_{\delta_j}, \mathcal{T}_h)$. The second term r_{mod} quantifies the error due to the upscaling procedure used to replace the true homogenized flux $\mathcal{A}^0$ by (14). The coupling condition (periodic, Dirichlet), the size of the sampling domain enter in this modelling error that is not influenced by the macro or micro discretisation parameter H and h. In the most favourable case (e.g., locally periodic homogenization), when $\delta/\varepsilon \in \mathbb{N}^*$ and periodic coupling is used we can have $r_{mod}(\nabla v^H) = 0$ (see [8]).

Existence and Uniqueness of the Micro Nonlinear Problem The assumptions $(\mathcal{A}_0)$–$(\mathcal{A}_2)$ are sufficient to guarantee existence and uniqueness of a solution to the nonlinear problem (12). To treat both the exact and the FE approximation of this nonlinear problem we consider the more general problem: find $z \in X$ such that

$$a^{\xi}_{K_j}(z; w) := \int_{K_{\delta_j}} \mathcal{A}^{\varepsilon}(x, \xi + \nabla z) \cdot \nabla w \, dx = 0, \qquad \forall\, w \in X, \tag{17}$$

where X is any closed linear subspace of the Banach space $W(K_{\delta_j})$.

Lemma 1 *Assume that $\mathcal{A}^{\varepsilon}$ satisfies $(\mathcal{A}_{0-2})$. Then problem* (17) *has a unique solution.*

Sketch of the Proof Unless specified otherwise, all the constants below depend on $\kappa_1, |K_\delta|, \xi, L$ and C_0 (see $(\mathcal{A}_{0-2})$). Using a Hölder inequality and $(\mathcal{A}_0)$ yield for any $z \in X$ the bound $|a^{\xi}_K(z; w)| \leqslant C(z)\|w\|_{L^p(K_\delta)}$ for a constant C depending on z, hence the nonlinear operator $M : X \to X^*$ defined by $\langle Mz, w\rangle = a^{\xi}_K(z; w)$ is well-defined and the problem (17) is equivalent to the problem $Mz = 0$. We next list the properties of the operator M:

1) Using $(\mathcal{A}_1)$ and a Hölder inequality yields

$$\|Mz - Mw\|_{X^*} \leqslant C(\|z\|_{L^p(K_{\delta_j})}, \|w\|_{L^p(K_{\delta_j})})\|z - w\|^{\alpha}_{L^p(K_{\delta_j})},$$

and M is continuous.
2) Thanks to $(\mathcal{A}_2)$ we have $\langle Mz - Mw, z - w\rangle > 0$ for all $z \neq w$ and the operator M is strictly monotone.
3) Finally the bound [26, Lemma 3.1]

$$\|\nabla z - \nabla w\|_{L^p(K_{\delta_j})} \leqslant \left[\kappa \left|K_{\delta_j}\right|^{\frac{1}{p}} + \|\nabla z\|_{L^p(K_{\delta_j})} + \|\nabla w\|_{L^p(K_{\delta_j})}\right]^{\frac{\beta-p}{\beta}}$$
$$\left(\int_{K_{\delta_j}} (\kappa + |\nabla z| + |\nabla w|)^{p-\beta} |\nabla z - \nabla w|^{\beta} dx\right)^{\frac{1}{\beta}},$$

for any $z, w \in X$ that holds for $1 < p < \infty$, $\beta \geqslant p$ and $\kappa \geqslant 0$ together with $(\mathcal{A}_2)$ yields

$$\langle Mz, z\rangle \geqslant C_1\|\nabla z\|^p_{L^p(K_{\delta_j})} - C_2$$

where C_1, C_2 in addition also depends on $\kappa_2, \beta, p, \lambda$ and the operator M is coercive. Hence we can apply the Browder-Minty theorem that ensure that the equation $Mz = 0$ with the operator M that continuous, strictly monotone and coercive, has a unique solution. □

We next list several properties of the map $B_H(v^H; w^H)$ that follows from the assumption $\mathcal{A}^\varepsilon$ (we refer to [6] for a detailed derivation).

Lemma 2 *Assume that $\mathcal{A}^\varepsilon$ satisfies ($\mathcal{A}_{0-2}$). Let $v^H, w^H, z^H \in S_0^\ell(\Omega, \mathcal{T}_H)$ then the nonlinear map B_H defined in* (10) *satisfies the following properties*

$$|B_H(v^H; w^H)| \leqslant C\left[C_d + \|\nabla v^H\|_{L^p(\Omega)}\right]^{p-1} \|\nabla w^H\|_{L^p(\Omega)}, \tag{18}$$

$$|B_H(v^H; z^H) - B_H(w^H; z^H)| \leqslant C\left[C_d + \|\nabla v^H\|_{L^p(\Omega)} + \|\nabla w^H\|_{L^p(\Omega)}\right]^{p-1-\gamma} \|\nabla v^H - \nabla w^H\|^{\gamma}_{L^p(\Omega)} \|\nabla z^H\|_{L^p(\Omega)}, \tag{19}$$

$$B_H(v^H; v^H - w^H) - B^H(w^H; v^H - w^H) > 0 \quad \text{for } v^H \neq w^H \tag{20}$$

$$B_H(v^H; v^H) \geqslant \lambda_c \|\nabla v^H\|^p_{L^p(\Omega)} - C(C_d)^p, \tag{21}$$

where C may depend on p, α, β, λ, L and the measure of Ω, with $\lambda_c > 0$ depending only on p, β, λ and L and $C_d = L_0 + \kappa_1 + \kappa_2$, $\gamma = \alpha/(\beta - \alpha)$.

The above properties are sufficient to guarantee the existence and uniqueness of a solution to the problem (9). We note that while (20) is sufficient to ensure the strict monotonicity of $B_H(\cdot, \cdot)$ for the error estimate we will need the following monotonicity estimate

$$\|\nabla v^H - \nabla w^H\|_{L^p(\Omega)} \leqslant C\left[1 + \|\nabla v^H\|_{L^p(\Omega)} + \|\nabla w^H\|_{L^p(\Omega)}\right]^{\frac{\beta-p}{\beta}} \times (B^H(v^H; v^H - w^H) - B^H(w^H; v^H - w^H))^{\frac{1}{\beta}}, \tag{22}$$

where C depends on $C_d, \lambda_c, |\Omega|, p$ and β.

Theorem 1 *Assume that ($\mathcal{A}_{0-2}$) hold and that $f \in L^{p'}(\Omega)$. Then, for any parameter $H, h, \delta > 0$, there exists a unique numerical solution of* (9) *that satisfies*

$$\|u^H\|_{L^p(W_0^{1,p})} \leqslant C, \qquad \|u^H\|_{C^0(L^2)} \leqslant C, \tag{23}$$

where C is independent of H, h, ε.

Proof The map $B : S_0^\ell(\Omega, \mathcal{T}_H) \to S_0^\ell(\Omega, \mathcal{T}_H)$, defined by $\langle Bv^H, w^H\rangle = B_H(v^H; w^H)$ is (strictly) monotone (20), hemicontinuous (the map $v^H \to \langle Bv^H, w^H\rangle$ is continuous for all $w^H \in S_0^\ell(\Omega, \mathcal{T}_H)$ thanks to (19)), coercive (21) and satisfies a growth condition $\|Bv\|_{(W_0^{1,p})^*} \leqslant c_1 + c_2\|v^h\|^{p-1}_{W_0^{1,p}}$. Hence the ordinary differential equation (9) satisfies the hypothesis of the Caratheodory theorem that guarantees the existence and uniqueness of a solution [57, Lemma 30.4]. The monotonicity and coercivity of B yield the a priori bound. □

3 Fully Discrete Space-Time Error Estimates for Linear Parabolic Problems

In this section we consider linear parabolic multiscale problems for which $\mathcal{A}^\varepsilon(x,\xi) = a^\varepsilon(x)\xi$. We assume $a^\varepsilon(x) \in (L^\infty(\Omega))^{d\times d}$ and for all $\xi \in \mathbb{R}^d$ and a.e. $x \in \Omega, t \in (0,T)$ there exists $\lambda, L > 0$ such that, uniformly for all $\varepsilon > 0$

$$\lambda|\xi|^2 \leqslant a^\varepsilon(x)\xi\cdot\xi,\ |a^\varepsilon(x)\xi| \leqslant L|\xi|. \tag{24}$$

The maps $\mathcal{A}^\varepsilon$ then satisfy ($\mathcal{A}_{0-2}$) for $p = 2$, $\alpha = 1$, $\beta = 2$ and with constants $C_0 = 0$ and λ, L given by the ellipticity and continuity constants. For simplicity we consider tensors $a^\varepsilon(x)$ independent of time but all the results of this section can be generalised for time dependent tensors [9]. The numerical method that we consider is still given by (9) but we have now the following explicit expression for the flux

$$\mathcal{A}^{0,h}_{K_j}(\xi) = \frac{1}{|K_{\delta_j}|}\int_{K_{\delta_j}} a^\varepsilon(x)(\xi + \nabla\chi^{\xi,h}_{K_j})dx.$$

Now since $\nabla v^H = \sum_{i=1}^d \mathbf{e}_i\partial_i v^H$, where $\mathbf{e}_i$ $i = 1,\dots,d$ is the canonical basis of $\mathbb{R}^d$, it is easy to see that $\mathcal{A}^{0,h}_{K_j}(\nabla v^H(x_{K_j})) = a^{0,h}(x_{K_j})\nabla v^H(x_{K_j})$, where the i-th row of the matrix $a^{0,h}(x_{K_j})$ is given by

$$a^{0,h}(x_{K_j}) = \frac{1}{|K_{\delta_j}|}\int_{K_{\delta_j}} a^\varepsilon(x)(I + \nabla\chi^{h}_{K_j})dx. \tag{25}$$

Here I is the $d\times d$ identity matrix and $\chi^h_{K_j}$ is a $d\times d$ matrix with column given by $\nabla\chi^{\mathbf{e}_i,h}_{K_j}$, where $\chi^{\mathbf{e}_i,h}_{K_j}$ are the (linear) solution of (12). We can thus rewrite the bilinear form (10) as

$$B_H(v^H,w^H) = \sum_{K\in\mathcal{T}_H}\sum_{j=1}^J \omega_{K_j} a^{0,h}(x_{K_j})\nabla v^H(x_{K_j})\cdot\nabla w^H(x_{K_j}), \tag{26}$$

for all $v^H, w^H \in S^\ell_0(\Omega,\mathcal{T}_H)$. We will also use below the bilinear form

$$B_{0,H}(v^H,w^H) = \sum_{K\in\mathcal{T}_H}\sum_{j=1}^J \omega_{K_j} a^{0}(x_{K_j})\nabla v^H(x_{K_j})\cdot\nabla w^H(x_{K_j}), \tag{27}$$

where a^0 is the (usually unknown) exact homogenized tensor that is known to satisfy similar bound (24) as a^ε. The solution of the homogenized problem (5) will be denoted by $u^0(t)$ and the corresponding bilinear by

$$B(v, w) = \int_\Omega a^0(x)\nabla v \cdot \nabla w dx. \tag{28}$$

We next mention classical estimates for FEM with numerical quadrature that are needed in the analysis below [23, Theorems 4, 5]. Assuming (**Q1**) and appropriate regularity of the homogenised solution u^0 we have for all $v^H, w^H \in S_0^\ell(\Omega, \mathcal{T}_H)$ (where $\mu = 0$ or 1),

$$|B(v^H, w^H) - B_{0,H}(v^H, w^H)| \leq CH\|v^H\|_{H^1(\Omega)}\|w^H\|_{H^1(\Omega)}, \tag{29}$$

$$|B(\mathcal{I}_H u_0, w^H) - B_{0,H}(\mathcal{I}_H u_0, w^H)| \leq CH^\ell \|u_0(t)\|_{W^{\ell+1,p}(\Omega)}\|w^H\|_{H^1(\Omega)}, \tag{30}$$

$$|B(\mathcal{I}_H u_0, w^H) - B_{0,H}(\mathcal{I}_H u_0, w^H)| \leq CH^{\ell+\mu} \|u_0(t)\|_{W^{\ell+1,p}(\Omega)} \tag{31}$$

$$\cdot\Big(\sum_{K\in\mathcal{T}_H} \|w^H\|^2_{H^2(K)}\Big)^{1/2}, \tag{32}$$

where $\mathcal{I}_H : C^0(\overline{\Omega}) \to S_0^\ell(\Omega, \mathcal{T}_H)$ is the usual nodal interpolant.

For linear parabolic problems, we can derive fully discrete convergence results in both space and time. Furthermore we can perform this a priori convergence analysis for various class of time integrators including "explicit stabilized Runge-Kutta method". The strategy is to first derive fully discrete error estimates in space. In a second step, using semigroup techniques, fully-discrete space time error estimates can be obtained. In contrast fully discrete space-time estimates could be obtained at once starting directly from a time-discrete numerical method instead of first considering (9). With such a strategy we need however new error estimates for each new time-integrator while with the former approach we can derive error estimates for classes of time integrators "at once". In this section we follow the finding of [9].

3.1 Fully Discrete a Priori Convergence Rates in Space

The starting point of the analysis is to define an appropriate elliptic projection: for all $t \in (0, T)$, let $\Pi_H u^0(t) \in S_0^\ell(\Omega, \mathcal{T}_H)$ be the solution of the problem

$$B_H(\Pi_H u^0(t), z^H) = B(u^0(t), z^H), \qquad \forall z^H \in S_0^\ell(\Omega, \mathcal{T}_H), \quad t \in (0, T), \tag{33}$$

where $u^0(t)$ is the solution of the homogenized problem (5). Thanks to the ellipticity and continuity of B_H, the above problem is well-posed. Using (33), denoting by

$\mathcal{I}_H u^0$ the standard nodal interpolant of u^0 we get for all $z^H \in S_0^\ell(\Omega, \mathcal{T}_H)$,

$$\begin{aligned} B_H(\Pi_H u^0 - \mathcal{I}_H u^0, z^H) &= B(u^0 - \mathcal{I}_H u^0, z^H) \\ &+ B(\mathcal{I}_H u^0, z^H) - B_{0,H}(\mathcal{I}_H u^0, z^H) \\ &+ B_{0,H}(\mathcal{I}_H u^0, z^H) - B_H(\mathcal{I}_H u^0, z^H). \end{aligned} \tag{34}$$

Assuming enough regularity of the homogenized solution, the first two terms of the above inequality are bounded by $CH^\ell \|u_0(t)\|_{W^{\ell+1,p}} \|z^H\|_{H^1(\Omega)}$ using the continuity of B and standard results for nodal interpolant [22] (first term) and (30) (second term). In view of (26) and (27), the definition (15) for $p = p' = 2$ and the assumption (**Q1**) on the quadrature formula we have for the third term

$$|B_{0,H}(\mathcal{I}_H u^0, z^H) - B_H(\mathcal{I}_H u^0, z^H)| \leqslant r_{HMM}(\nabla \mathcal{I}_H u^0) \|\nabla z^H\|_{L^2(\Omega)}. \tag{35}$$

We note that we can further decompose $r_{HMM}(\nabla \mathcal{I}_H u^0)$ as

$$r_{HMM}(\nabla \mathcal{I}_H u^0) \leqslant \sup_{K \in \mathcal{T}_H, x_{K_j} \in K} \|a^0(x_{K_j}) - a^{0,h}(x_{K_j})\|_F \|\nabla \mathcal{I}_H u^0\|_{L^2(\Omega)},$$

where $\|\cdot\|_F$ denotes the Frobenius norm of a matrix. We first have the following lemma.

Lemma 3 *Let $u^0(t)$ be the solution of* (5) *and $\Pi_H u^0(t)$ be the elliptic projection defined in* (33). *Assume that* ($\mathcal{A}_{0-2}$) *and* (***Q1***) *hold. Assume further that the homogenized tensor satisfies $a^0_{ij} \in \mathcal{C}^0([0,T] \times \bar{K})$ for all $K \in \mathcal{T}_H$ and all $i, j = 1, \ldots, d$. Assume further for $\mu = 0$ or 1 and $\ell > d/p$ that*

$$\begin{aligned} &u_0, \partial_t u_0 \in L^2(0,T; W^{\ell+1,p}(\Omega)), \\ &a^0_{ij}, \partial_t a^0_{ij} \in L^\infty(0,T; W^{\ell+\mu,\infty}(\Omega)), \quad \forall i,j = 1 \ldots d. \end{aligned}$$

Then we have for $k = 0, 1$

$$\|\partial_t^k \left(\Pi_H u^0 - u^0\right)\|_{L^2(0,T;H^1(\Omega))} \leqslant C(H^\ell + r_{HMM}(\nabla \mathcal{I}_H u^0)), \tag{36}$$

$$\|\partial_t^k \left(\Pi_H u^0 - u^0\right)\|_{L^2(0,T;L^2(\Omega))} \leqslant C(H^{\ell+\mu} + r_{HMM}(\nabla \mathcal{I}_H u^0)) \; \mu = 0, 1, \tag{37}$$

where we assume that Ω is convex for the estimates (37) *with $\mu = 1$. The constant C is independent of H, h and δ.*

Proof In view of (34) and the bound of the different terms of the right-hand side of this equality, taking $z^H = \Pi_H u^0 - \mathcal{I}_H u^0$, using the ellipticity of B_H and integrating from 0 to T we obtain $\|\Pi_H u^0 - \mathcal{I}_H u^0\|_{L^2(0,T;H^1(\Omega))} \leqslant C(H^\ell + r_{HMM}(\nabla \mathcal{I}_H u^0)))$. The estimate (36) for $k = 0$ follows by using the triangle inequality and the estimate

$\|u^0 - \mathcal{I}_H u^0\|_{L^2(0,T;H^1(\Omega))} \leqslant CH^\ell$. The estimate (36) for $k = 1$ is obtained by differentiating (34) and following the same arguments.

For the estimate (37) $k = 0$ we use the classical Aubin-Nitsche duality argument and consider for almost every $t \in (0, T)$ the solution $\varphi(t) \in H_0^1(\Omega)$ of the problem

$$B(z, \varphi(t)) = (v(t), z), \quad \forall z \in H_0^1(\Omega). \tag{38}$$

Taking $v(t) = z = \Pi_H u^0 - u^0$ using the elliptic projection (33) yields for all φ^H

$$\begin{aligned}(\Pi_H u^0 - u^0, \Pi_H u^0 - u^0) &= B(\Pi_H u^0 - u^0, \varphi - \varphi^H)\\ &\quad + B(\Pi_H u^0 - \mathcal{I}_H u^0, \varphi^H) - B_H(\Pi_H u^0 - \mathcal{I}_H u^0, \varphi^H)\\ &\quad + B(\mathcal{I}_H u^0, \varphi^H) - B_H(\mathcal{I}_H u^0, \varphi^H).\end{aligned} \tag{39}$$

We take $\varphi^H = \mathcal{I}_H \varphi(t)$ use the continuity of B, (29) and (32) to obtain

$$\begin{aligned}(\Pi_H u^0 - u^0, \Pi_H u^0 - u^0) &\leqslant C(H + r_{HMM}(\nabla \mathcal{I}_H u^0))\\ &\quad \cdot \|\Pi_H u^0(t) - u_0(t)\|_{H^1(\Omega)} \|\varphi(t)\|_{H^2(\Omega)}\\ &\quad + (H^{\ell+\mu} + r_{HMM}(\nabla \mathcal{I}_H u^0))) \|u(t)\|_{H^{\ell+1}(\Omega)} \|\varphi(t)\|_{H^2(\Omega)}.\end{aligned}$$

Using $\|\varphi\|_{L^2(0,T;H^2(\Omega))} \leqslant C\|\Pi_H u^0 - u^0\|_{L^2(0,T;L^2(\Omega))}$ that holds thanks to the regularity $a \in (L^\infty(0, T; W^{1,\infty}(\Omega)))^{d\times d}$ of the tensor and the convexity of the polygonal domain Ω gives (37) for $k = 0$. The estimate (37) for $k = 1$ is obtained by differentiating (39) and following the same arguments. □

Remark 1 Under the assumptions of Lemma 3 the Sobolev embedding $H^1(0, T; X)$ into $C^0([0, T]; X)$ (for a given Banach space) allows to deduce

$$\|\Pi_H u^0 - u^0\|_{C^0(0,T;H^1(\Omega))} \leqslant C(H^\ell + r_{HMM}(\nabla \mathcal{I}_H u^0))), \tag{40}$$

$$\|\Pi_H u^0 - u^0\|_{C^0(0,T;L^2(\Omega))} \leqslant C(H^{\ell+\mu} + r_{HMM}(\nabla \mathcal{I}_H u^0))). \tag{41}$$

We state now fully discrete a priori convergence rate in space for the FE-HMM

Theorem 2 *Let $u^0(t)$ be the solution of* (5) *and u^H the solutions of* (9). *Assume the hypotheses of Lemma 3. Then we have the $L^2(H^1)$ and $C^0(L^2)$ estimates*

$$\|u^0 - u^H\|_{L^2([0,T];H^1(\Omega))} \leqslant C(H^\ell + r_{HMM}(\nabla \mathcal{I}_H u^0) + \|g - u_0^H\|_{L^2(\Omega)}), \tag{42}$$

and if $\mu = 1$

$$\|u^0 - u^H\|_{C^0([0,T];L^2(\Omega))} \leqslant C(H^{\ell+1} + r_{HMM}(\nabla \mathcal{I}_H u^0) + \|g - u_0^H\|_{L^2(\Omega)}). \tag{43}$$

If in addition, the tensor is symmetric, then we have the $C^0(H^1)$ estimate

$$\|u^0 - u^H\|_{C^0([0,T];H^1(\Omega))} \leqslant C(H^\ell + r_{HMM}(\nabla \mathcal{I}_H u^0) + \|g - u_0^H\|_{H^1(\Omega)}). \tag{44}$$

The constants C are independent of $H, r_{HMM}(\nabla \mathcal{I}_H u^0)$.

Proof To simplify the notation, we use $r_{HMM} = r_{HMM}(\nabla \mathcal{I}_H u^0)$ in the proof.

Step 1: Estimation of $\|u^H - \Pi_H u^0\|_{L^2(0,T;H^1(\Omega))} + \|u^H - \Pi_H u^0\|_{C^0([0,T];L^2(\Omega))}$. We set $\xi^H(t) = u^H(t) - \Pi_H u^0(t), t \in [0,T]$. In view of the elliptic projection (33), (5) and (9) we have for all $z^H \in S_0^\ell(\Omega, \mathcal{T}_H)$,

$$(\partial_t \xi^H, z^H) + B_H(\xi^H, z^H) = (\partial_t u^0, z^H) - (\partial_t \Pi_H u^0, z^H). \tag{45}$$

We set $z^H = \xi^H$ integrate this equality from 0 to t using the coercivity of $B_H(\cdot,\cdot)$

$$\|\xi^H(t)\|^2_{L^2(\Omega)} + c_1 \int_0^t \|\xi^H(s)\|^2_{H^1(\Omega)} ds \leqslant \|\xi^H(0)\|^2_{L^2(\Omega)} \tag{46}$$

$$+ c_2 \int_0^t \|\partial_t u^0 - \partial_t \Pi_H u^0\|^2_{L^2(\Omega)} ds. \tag{47}$$

Using the decomposition $\xi(0) = (u^0 - \Pi_H u^0)(0) + (u_0^H - g)$, (41) and (37) yields

$$\|\xi(0)\|_{L^2(\Omega)} \leqslant C(H^{\ell+\mu} + r_{HMM}) + \|u_0^H - g\|_{L^2(\Omega)}. \tag{48}$$

Using (48) and (37) gives the $L^2(H^1)$ estimate and taking the supremum with respect to t gives the $C^0(L^2)$ estimate. We thus obtain

$$\|u^H - \Pi_H u^0\|_{C^0([0,T];L^2(\Omega))} + \|u^H - \Pi_H u^0\|_{L^2(0,T;H^1(\Omega))} \leqslant C(H^{\ell+\mu} + r_{HMM}) + \|u_0^H - g\|_{L^2(\Omega)}. \tag{49}$$

This last estimate together with Lemma 3 and the triangle inequality gives (42) and (43).

Step 2: Estimation of $\|u^H - \Pi_H u^0\|_{C^0([0,T];H^1(\Omega))}$. For $\xi^H(t) = u^H(t) - \Pi_H u^0(t), t \in [0,T]$, we set $z^H = \partial_t \xi^H$ in (45). Using the symmetry of the tensor, and integrating from 0 to t, we obtain for $0 \leqslant t \leqslant T$

$$2\int_0^t \|\partial_t \xi^H(s)\|^2_{L^2(\Omega)} ds + B_H(\xi^H(t), \xi^H(t)) = B_H(\xi^H(0), \xi^H(0)) + 2\int_0^t (\partial_t u^0 - \partial_t \Pi_H u^0, \partial_t \xi^H)\, ds.$$

Similarity to (46) we obtain

$$\int_0^t \|\partial_t \xi^H(s)\|^2_{L^2(\Omega)} ds + c_1\|\xi^H(t)\|^2_{H^1(\Omega)} \leqslant c_2(\|\xi^H(0)\|^2_{H^1(\Omega)} + \int_0^t \|\xi^H(s)\|^2_{H^1(\Omega)} ds)$$
$$+ \int_0^t \|\partial_t u^0(s) - \partial_t \Pi_H u^0(s)\|^2_{L^2(\Omega)} ds. \quad (50)$$

As before we set $\xi^H(0) = (u^0 - \Pi_H u^0)(0) + (u_0^H - g)$ and (40) gives

$$\|\xi^H(0)\|_{H^1(\Omega)} \leqslant C(H^\ell + r_{HMM} + \|u_0^H - g\|_{H^1(\Omega)}). \quad (51)$$

Taking the supremum with respect to t in (50), and using (51), (49), (36), we deduce

$$\|u^H - \Pi_H u^0\|_{C^0([0,T];H^1(\Omega))} \leqslant C(H^\ell + r_{HMM} + \|u_0^H - g\|_{H^1(\Omega)}).$$

This together with (40) concludes the proof of (44). □

The last step to obtain fully discrete estimates in space is to quantify r_{HMM}. Remember the decomposition $r_{HMM} \leqslant r_{mod} + r_{mic}$ [see (16a), (16b)]. We can again rewrite

$$r_{mod}(\nabla \mathcal{I}_H u^0) \leqslant \sup_{K\in\mathcal{T}_H, x_{K_j}\in K} \|a^0(x_{K_j}) - \bar{a}^0(x_{K_j})\|_F \|\nabla \mathcal{I}_H u^0\|_{L^2(\Omega)}, \quad (52)$$

$$r_{mic}(\nabla \mathcal{I}_H u^0) \leqslant \sup_{K\in\mathcal{T}_H, x_{K_j}\in K} \|\bar{a}^0(x_{K_j}) - a^{0,h}(x_{K_j})\|_F \|\nabla \mathcal{I}_H u^0\|_{L^2(\Omega)}, \quad (53)$$

where we recall that $\bar{a}^0(x_{K_j})$ is defined similarly as $a^{0,h}(x_{K_j})$ [see (25), (14)] but based on exact micro functions, i.e., when $\chi^{\xi}_{K_j}$ is solution of (12) in $W(K_{\delta_j})$. These terms have first been quantified for elliptic problems in [2] and [8, 56]. Using the definition of the cell problem (12) it is not hard to show (for linear problem) that for symmetric tensors $a^\varepsilon(x)$ one has

$$|(\bar{a}^0(x_{K_j}) - a^{0,h}(x_{K_j}))_{mn}| \quad (54)$$
$$= \left|\frac{1}{|K_{\delta_j}|}\int_{K_{\delta_j}} a^\varepsilon(x)\left(\nabla\chi^{\mathbf{e}_n}_{K_j}(x) - \nabla\chi^{\mathbf{e}_n,h}_{K_j}(x)\right)\cdot\left(\nabla\chi^{\mathbf{e}_m}_{K_j}(x) - \nabla\chi^{\mathbf{e}_m,h}_{K_j}(x)\right)dx\right|.$$

Next assuming $|\chi^{\mathbf{e}_n}_{K_j}|_{H^{q+1}(K_{\delta_j})} \leqslant C\,\varepsilon^{-q}\sqrt{|K_{\delta_j}|}$, where C is independent of ε, the quadrature points x_{K_j}, and the domain K_{δ_j} one obtains

$$r_{mic}(\nabla \mathcal{I}_H u^0) \leqslant C\left(\frac{h}{\varepsilon}\right)^{2q}, \quad (55)$$

when using the micro finite element space (11). The justification of the above regularity assumption depends on the boundary conditions used for (11). For Dirichlet boundary conditions and for $q = 1$ the regularity assumption can be established using classical H^2 regularity results [42, Chap. 2.6] provided $|a^{\varepsilon}_{mn}|_{W^{1,\infty}(\Omega)} \leqslant C\varepsilon^{-1}$ for $m, n = 1, \ldots, d$. For periodic boundary conditions the above regularity assumption can be established for any given q, provided $a^{\varepsilon} = a(x, x/\varepsilon) = a(x, y)$ is Y-periodic in y, $\delta/\varepsilon \in \mathbb{N}$, and a^{ε} is sufficiently smooth, by following classical regularity results for periodic problems [19].

We finally come to the modelling error: here we need to assume some structure for the oscillatory tensor such as periodicity or random stationarity. For locally periodic problems assuming $a^{\varepsilon} = a(x, x/\varepsilon) = a(x, y) \quad Y = (0, 1)^d$-periodic in y, that the sampling domain size is such that $\delta/\varepsilon \in \mathbb{N}$ and that periodic micro boundary conditions are used, we have $r_{mod} \leqslant C\delta$ [2, 8]. Furthermore, if we assume a tensor $a(x_{K_j}, x/\varepsilon)$ collocated in the slow variable $x = x_{K_j}$ for the micro and the macro problem, one can show that $r_{mod} = 0$. For Dirichlet boundary condition assuming $\delta > \varepsilon$ the bound $r_{mod} \leqslant C(\delta + \frac{\varepsilon}{\delta})$ can be established [56].

We note that for non-symmetric problems, an expression similar to (54) can still be established [10, 28], replacing the second parenthesis in the right-hand side of (54) by $\left(\nabla \bar{\chi}^{\mathbf{e}_m}_{K_j}(x) - \nabla \bar{\chi}^{\mathbf{e}_m,h}_{K_j}(x)\right)$, where $\nabla \bar{\chi}^{\mathbf{e}_m}_{K_j}, \bar{\chi}^{\mathbf{e}_m,h}_{K_j}$ are exact, respectively FE solutions of the adjoint problem of (12). The rest of the discussion is then similar. Finally we mention that by using a perturbed micro-problem, using a zeroth order term, higher order rates have been obtained in [34] for the modeling error.

3.2 *Fully Discrete A Priori Convergence Rates in Space and Time*

In this section we analyse the time-discretization error, when using various classes of time-integrators for the parabolic problems. We will concentrate on strongly $A(\theta)$-stable implicit Runge-Kutta methods and explicit stabilized (Chebyshev) methods.

Consider a basis $\{\phi_j\}_{j=1}^{M}$ of $S_0^{\ell}(\Omega, \mathcal{T}_H)$ and denote by U^H the column vector of the coefficients of $u^H = \sum_{j=1}^{M} U_j(t)\phi_j$ in this basis. This allows to rewrite (9) as an ordinary differential equation

$$\frac{d}{dt}U^H(t) = A_H U^H(t) + G^H(t) = F(t, U^H(t)), \quad U^H(0) = U_0, \tag{56}$$

where $A_H = M^{-1}\hat{A}_H$ and $G^H(t) = M^{-1}P^H$. The matrice $\hat{A}_H$ is defined by the map $\hat{A}_H : S_0^{\ell}(\Omega, \mathcal{T}_H) \to S_0^{\ell}(\Omega, \mathcal{T}_H)$, where $(-\hat{A}_H v^H, w^H) = B_H(v^H, w^H)$, the mass matrix M is given by $M = ((\phi_j, \phi_i))_{i,j=1}^{M}$ and P^H corresponds to the source term. Of course in practical computations we never invert the matrix M, but instead solve a linear system. In some situation we can also use mass lumping techniques that transform M into a matrix that is trivial to invert [53]. As mentioned in the beginning of Sect. 3, the FE-HMM method and the spatial convergence results can be generalised for time-dependent tensors $a^{\varepsilon}(t,x)$ and time-dependent right-hand side $f(t,x)$. In this situation we would have $B_H(v_H, w_H)$, $A_H = A_H(t)$ and $P^H = P^H(t)$ (see [9] for details).

Resolvent and α-Accretive Operator To apply semi-group techniques to estimate the error when applying a Runge-Kutta method to (56), we need bound on the resolvent of $-A_H$. For the type of ODE (56) originating from a spatial discretisation of a parabolic problem, it can be shown that $-A_H$ (see for example [24]) is a so-called α-accretive operator, i.e., there exist $0 \leqslant \alpha \leqslant \pi/2$ and $C > 0$ such that for all $z \notin S_\alpha$, the operator $zI + A_H(t)$ is an isomorphism and

$$\|(zI + A_H(t))^{-1}\|_{L^2(\Omega)\to L^2(\Omega)} \leqslant \frac{1}{d(z, S_\alpha)} \qquad \text{for all } z \notin S_\alpha, \tag{57}$$

where $d(z, S_\alpha)$ is the distance between z and $S_\alpha = \{\rho e^{i\theta}\ ;\ \rho \geqslant 0, |\theta| \leqslant \alpha\}$. We note that the operator A_H can be extended straightforwardly to a complex Hilbert space based on $S_0^\ell(\Omega, \mathcal{T}_H)$ equipped with the complex scalar product $(u, v) = \int_\Omega u(x)\overline{v}(x)dx$ which is an extension of the usual L^2 scalar product. If we denote by γ_1, γ_2 the coercivity and continuity constant of the bilinear form $B_H(\cdot,\cdot)$, it can be shown that $\alpha \leqslant \arccos(\gamma_1/\gamma_2)$. Hence A_H generates an analytic semi-group in S_α (see [41]).

Runge-Kutta Methods For the time discretisation of (56) we consider an s-stage Runge-Kutta method

$$U_{n+1} = U_n + \Delta t \sum_{i=1}^{s} b_i K_{ni}, \quad U_{ni} = U_n + \Delta t \sum_{j=1}^{s} \gamma_{ij} K_{nj}, \tag{58}$$

$$K_{ni} = F(t_n + c_i \Delta t U_{ni}), \quad i = 1 \ldots s. \tag{59}$$

where γ_{ij}, b_j, c_i with $i, j = 1 \ldots s$ are the coefficients of the method (with $\sum_{j=1}^{s} \gamma_{ij} = c_i$) and $t_n = n\Delta t$. We further define

$$\Gamma = (\gamma_{ij})_{i,j=1}^{s}, \quad b = (b_1, \ldots, b_s)^T, \quad c = (c_1, \ldots, c_s)^T = \Gamma \mathbf{1}, \quad \mathbf{1} = (1, \ldots, 1)^T.$$

The method is said to have "order r" if the error after one step between the exact and the numerical solutions (with the same initial condition) satisfies

$$U_1 - U(t_1) = \mathcal{O}\left(\Delta t^{r+1}\right), \ \text{for} \Delta t \to 0,$$

for all sufficiently differentiable systems of differential equations. We recall that the rational function $R(\Delta t \lambda) = R(z) = 1 + z b^T (I - z\Gamma)^{-1} \mathbf{1}$ obtained after one step Δt of a Runge-Kutta method applied to the scalar problem $dy/dt = \lambda y,\ y(0) = 1,\ \lambda \in \mathbb{C}$ is called the stability function of the method.

Strongly $A(\theta)$-Stable Methods We consider a subclass of implicit Runge-Kutta methods which are of order r and whose stage order (the accuracy of the internal stages) is $r-1$. We further recall that a Runge-Kutta method is strongly $A(\theta)$-stable with $0 \leqslant \theta \leqslant \pi/2$ if $I - z\Gamma$ is a nonsingular matrix in the sector $|\arg(-z)| \leqslant \theta$ and the stability function satisfies $|R(z)| < 1$ in $|\arg(-z)| \leqslant \theta$. Notice that all s-stage Radau Runge-Kutta methods satisfy the above assumptions (with $\theta \geqslant \pi/2$). In particular, for $s = 1$, we retrieve the implicit Euler method. We refer to [35, Sect. IV.3, IV.15] for details on the stability concepts mentioned here.

Under the assumptions of Theorem (2) we have the following theorem.

Theorem 3 *Let $u^0(t)$ be the solution of* (5) *and let u_n^H be a strongly $A(\theta)$ stable Runge-Kutta approximation of order r and stage order $r-1$ of* (56) *with time step Δt. Assume the hypotheses of Theorem* (2), (55), *that $r_{MOD} = 0$ and $a^\varepsilon \in C^r([0,T], L^\infty(\Omega)^{d\times d})$, $\|\partial_t^r u^H(0)\|_{L^2(\Omega)} \leqslant C$. Then,*

$$\max_{0\leqslant n\leqslant N} \|u_n^H - u^0(t_n)\|_{L^2(\Omega)} \leqslant C\left(H^{\ell+1} + \left(\frac{h}{\varepsilon}\right)^{2q} + \Delta t^r\right).$$

Assuming in addition $\|u^H(0) - g\|_{H^1(\Omega)} \leqslant C(H^\ell)$ and a^ε is symmetric, then

$$\sum_{n=0}^{N-1} \Delta t_n \|u_n^H - u^0(t_n)\|^2_{H^1(\Omega)} \leqslant C\left(H^\ell + \left(\frac{h}{\varepsilon}\right)^{2q} + \Delta t^r\right)^2.$$

All the above constants C are independent of $H, h, \varepsilon, \Delta t$.

The idea of the proof is to consider the decomposition: $u_n^H - u^0(t_n) = (u_n^H - u^H(t_n)) + (u^H(t_n) - u^0(t_n))$. Then the first term is estimated using semigroup techniques (for time independent operators) + perturbation techniques (following [44]). The second term is estimated using Theorem 2. We note that the analysis for implicit methods covers variable time step methods under some mild assumptions on the sequence of time-steps [9]. Finally we mention that the bound $\|\partial_t^r u^H(0)\|_{L^2(\Omega)} \leqslant C$ can be established provided that we assume an inverse assumption $\frac{H}{H_K} \leqslant C$ for all $K \in \mathcal{T}_H$ and all $\mathcal{T}_H$ for the macroscopic finite element mesh and appropriate regularity of $\partial_t^k u^0$, $k = 1, \ldots, r$. We refer to [9] for a detailed proof of the above theorem.

Chebyshev Methods Chebyshev methods are a subclass of explicit Runge-Kutta methods with extended stability along the negative real axis suitable for parabolic (advection-diffusion) problems. Such methods have been constructed for order up to $r = 4$ [1, 7, 43, 54]. They are based on s-stage stability functions satisfying

$$|R_s(x)| \leqslant 1 \quad \text{for } x \in [-L_s, 0] \tag{60}$$

with $L_s = Cs^2$, where the constant C depends on the order of the method. First order methods are based on

$$R_s(x) = T_s(1 + x/s^2), \tag{61}$$

where $T_s(\cdot)$ denotes the Chebyshev polynomial of degree s and $L_s = 2s^2$. The corresponding Runge-Kutta method can be efficiently implemented by using the three-term recurrence relation of the Chebyshev polynomials [54]. For stiff diffusion problems, such methods are much more efficient than classical explicit methods. Indeed let ρ_H be the spectral radius of the discretized parabolic problem (depending on the macro spatial meshsize H) and let Δt be the stepsize to achieve the desired accuracy. Using a classical explicit method such as the forward Euler method requires a stepsize δt satisfying the CFL constraints $\delta t \leqslant 2/\rho_H$. The number of function evaluations per time-step Δt (taken here a the measure of the numerical work) is therefore $\Delta t/\delta t \geqslant (\Delta t \rho_H)/2$. Using a Chebyshev method (of order one) with stability function (61) we choose the number of stages s of the method to ensure stability $\Delta t \rho_H \leqslant 2s^2$. As for Chebyshev methods, there is one new function evaluation per stage the total number of function evaluations per time-step Δt is given by $s = \sqrt{(\Delta t \rho_H)/2}$.

Chebyshev method are usually used in a slightly modified form obtained by changing the stability function (61) to

$$R_s(z) = \frac{T_s(\omega_0 + \omega_1 z)}{T_s(\omega_0)}, \quad \text{with} \quad \omega_0 = 1 + \frac{\eta}{s^2}, \quad \omega_1 = \frac{T_s(\omega_0)}{T_s'(\omega_0)}, \tag{62}$$

we obtain the “damped form” of the Chebyshev method. For any fixed $\eta > 0$ (called the damping parameter) we obtain a damped stability function satisfying

$$\sup_{z \in [-L_s, -\gamma],\, s \geqslant 1} |R_s(z)| < 1, \quad \text{for all } \gamma > 0. \tag{63}$$

This modification also ensure that a strip around the negative real axis is contained in the stability domain $\mathcal{S} := \{z \in \mathbb{C}; |R_s(z)| \leqslant 1\}$. The growth on the negative real axis for the damped form is reduced but remains quadratic [51],[35, Chap. IV.2]. For the analysis we assume the order of the Chebyshev method is $r \geqslant 1$ for linear problem, precisely,

$$\lim_{z \to 0} \left| \frac{e^z - R_s(z)}{z^{r+1}} \right| < \infty \quad \text{for all } s \geqslant 1. \tag{64}$$

We also assume that the stability functions are bounded in a neighbourhood of zero uniformly with respect to s, precisely, there exist $\delta > 0$ and $C > 0$ such that

$$|R_s(z)| \leqslant C \text{ for all } |z| \leqslant \delta \text{ and all } s. \tag{65}$$

This can be checked for the Chebyshev methods with stability functions (61), (62).

Theorem 4 *Let $u^0(t)$ be the solution of (5) with $f = 0$ and a time-independent symmetric tensor a^ε. Let u_n^H be a Runge-Kutta-Chebyshev approximation of the corresponding discretized problem (56) with timestep Δt. Assume that the method satisfies (64) (order r), (63) (strong stability) and (65). Assume in addition that the stage number s of the Chebyshev method is chosen such that $\rho_H \Delta t \leqslant L_s$ holds. Assume the hypotheses of Theorem 2 with $\mu = 1$, (55) and that $r_{MOD} = 0$. Then,*

$$\max_{0\leqslant n\leqslant N} \|u_n^H - u^0(t_n)\|_{L^2(\Omega)} \leqslant C\left(H^{\ell+1} + \left(\frac{h}{\varepsilon}\right)^{2q} + \Delta t^r\right).$$

where C is independent of $H, h, \varepsilon, \Delta t$.

The ideas of the proof are as follows. Consider again the decomposition $u_n^H - u^0(t_n) = (u_n^H - u^H(t_n)) + (u^H(t_n) - u^0(t_n))$. The second term is estimated as before using Theorem 2.

For the first term we follow ideas developed in [24, 25] for implicit methods, adapted here for stabilized methods. Using the symmetry of A_H, there exists an orthonormal basis where the operator A_H is in diagonal form. Define next $\varphi_{n,s}(z) = e^{nz} - R_s(z)^n$. Then we have

$$\|\varphi_{n,s}(\Delta t A_H)\|_{L^2(\Omega)\to L^2(\Omega)} = \sup_{z\in sp(A_H)} |\varphi_{n,s}(\Delta t z)|,$$

where $sp(A_H)$ denotes the spectrum of A_H. Using (63)–(65) we show that $|\varphi_{n,s}(z)| \leqslant C_1 n^{-r}$ for all $z \in [-\delta, 0]$, where C_1 is independent of n and s. For the case $z \in [-L_s, -\delta]$ we denote by $\rho < 1$ the quantity in the left-hand side of (63). We can then estimate

$$|\varphi_{n,s}(z)| \leqslant e^{-n|z|} + \rho^n \leqslant e^{-n\gamma} + e^{-n(1-\rho)} \leqslant \frac{(r/e)^r(\gamma^{-r} + (1-\rho)^{-r})}{n^r} = C_2 n^{-r},$$

where we used twice the estimate $e^{-x} \leqslant (\frac{r}{ex})^r$ (valid for $x \geqslant 0$). We have thus $|\varphi_{n,s}(z)| \leqslant Cn^{-r}$ for all $z \in [-L_s, 0]$, hence

$$\|u_n^H - u^H(t_n)\|_{L^2(\Omega)} = \|\varphi_{n,s}(\Delta t A_H)u_0^H\|_{L^2(\Omega)} \leqslant Cn^{-r}\|u_0^H\|_{L^2(\Omega)},$$

where C is independent of n, s. By noting that $n \leqslant T/\Delta t$ we get the result.

4 Fully Discrete Space-Time Error Estimates for Nonlinear Monotone Parabolic Problem

In this section we describe convergence and error estimates for the numerical method (9) applied to the general problem (1). We focus here on a simple time integrator, namely the implicit Euler method and take piecewise linear macro and micro FEM. We consider a uniform subdivision of the time interval $(0, T)$ with time step $\Delta t = T/N$ and discrete time $t_n = n\Delta t$ for $0 \leqslant n \leqslant N$ and $N \in \mathbb{N}_{>0}$. The method then reads as follows: for $0 \leqslant n \leqslant N-1$ find $u_{n+1}^H \in S_0^1(\Omega, \mathcal{T}_H)$ such that

$$\int_\Omega \frac{u_{n+1}^H - u_n^H}{\Delta t} w^H dx + B_H(u_{n+1}^H; w^H) = \int_\Omega f\, w^H dx, \qquad \forall\, w^H \in S_0^1(\Omega, \mathcal{T}_H), \tag{66}$$

with the nonlinear macro map B_H given by

$$B_H(v^H; w^H) = \sum_{K \in \mathcal{T}_H} |K|\, \mathcal{A}_K^{0,h}(\nabla v^H(x_K)) \cdot \nabla w^H(x_K), \qquad v^H, w^H \in S_0^1(\Omega, \mathcal{T}_H), \tag{67}$$

where $\mathcal{A}_K^{0,h}$ is given by (13) with micro problems (12) computed in $S^1(K_\delta, \mathcal{T}_h)$. Here we have just one quadrature point and sampling domain K_δ located at the barycenter of each macro element K. We note that we will sometimes use the short-hand notation $\bar{\partial}_t v_n = (v_{n+1} - v_n)/\Delta t$. The proof of the existence and uniqueness of a numerical solution can be establish similarly to the proof of Theorem 1. Further, the numerical solution $\{u_n^H\}_{n=1}^N$ satisfies the bound

$$\max_{1 \leqslant n \leqslant N} \|u_n^H\|_{L^2(\Omega)}^2 + \sum_{n=1}^N \Delta t \|\nabla u_n^H\|_{L^p(\Omega)}^p \leqslant C(1 + \|f\|_{L^{p'}(\Omega)}^{p'} + \|u_0^H\|_{L^2(\Omega)}^2), \tag{68}$$

where C only depends on $p, \beta, \lambda L, , L_0, \kappa_1, \kappa_2$, the measure of Ω and the Poincaré constant C_P on Ω.

4.1 General Estimates in the $W^{1,p}$ Setting

For the scheme (66), (67) in the general nonlinear monotone setting we have the following fully discrete convergence result.

Theorem 5 *Let $u^0 \in E$ be the solution to the homogenized problem* (5) *and u_n^H the HMM solution obtained from* (66) *with initial conditions u_0^H satisfying $\|g - u_0^H\|_{L^2(\Omega)} \to 0$ for $H \to 0$. Assume that $\mathcal{A}^\varepsilon$ satisfies $(\mathcal{A}_{0-2})$. Let $\mathcal{A}^0$ be Hölder continuous in space, i.e., there exists $0 < \tilde{\gamma} \leqslant 1$ such that*

$$\left|\mathcal{A}^0(x_1, \xi) - \mathcal{A}^0(x_2, \xi)\right| \leqslant C|x_1 - x_2|^{\tilde{\gamma}}(1 + (\kappa_1 + |\xi|)^{p-1}), \ \ \forall\, x_1, x_2 \in \Omega, \forall\, \xi \in \mathbb{R}^d. \tag{69}$$

Assume in addition that the coupling is such that $r_{mod} = 0$. Then we have the convergence

$$\lim_{(\Delta t,H)\to 0}\lim_{h\to 0}\left[\max_{1\leqslant n\leqslant N}\left\|u^0(\cdot,t_n)-u_n^H\right\|_{L^2(\Omega)}+\left\|\nabla u^0-\nabla u^H\right\|_{\widetilde{L^p}(L^p(\Omega)),}\right]=0.$$

where

$$\left\|\nabla u^0-\nabla u^H\right\|^p_{\widetilde{L^p}(L^p(\Omega))}=\sum_{n=0}^{N-1}\int_{t_n}^{t_{n+1}}\left\|\nabla u^0(\cdot,s)-\nabla u^H_{n+1}\right\|^p_{L^p(\Omega)}ds.$$

We sketch the proof of this result.

Step 1: Approximation by smooth function. Due to the low regularity of the true solution we can only rely on a weak approximation in time. Indeed, for u^0 we can only use the formulation $\int_{t_n}^{t_{n+1}}\langle\partial_t u^0(\cdot,s),w^H\rangle ds$ instead of $\int_{t_n}^{t_{n+1}}\int_\Omega \partial_t u^0(x,s)w^H(x)dxds$ that only make sense with additional regularity. We therefore consider $\mathcal{U}\in E$ with $\mathcal{U}\in \mathcal{C}^0([0,T],W_0^{1,p}(\Omega))$ and $\partial_t\mathcal{U}\in \mathcal{C}^0([0,T],L^2(\Omega))$. Further, let $\mathcal{U}^H(\cdot,t)\in S_0^1(\Omega,\mathcal{T}_H)$ be an approximation of $\mathcal{U}(\cdot,t)$ for $t\in[0,T]$ and define $\mathcal{U}_n^H=\mathcal{U}^H(\cdot,t_n)$ for $0\leqslant n\leqslant N$. We will then decompose the error as

$$\left\|u^0(\cdot,t_n)-u_n^H\right\|_{L^2(\Omega)}\leqslant\left\|u^0(\cdot,t_n)-\mathcal{U}_n^H\right\|_{L^2(\Omega)}+\left\|\theta_n^H\right\|_{L^2(\Omega)} \tag{70}$$

$$\left\|\nabla u^0-\nabla u^H\right\|_{\widetilde{L^p}(L^p(\Omega)}\leqslant\left\|\nabla u^0-\mathcal{U}^H\right\|_{\widetilde{L^p}(L^p(\Omega))}+(\sum_{n=0}^{N-1}\Delta t\|\nabla\theta^H_{n+1}\|^p_{L^p(\Omega)})^{1/p}, \tag{71}$$

where $\theta_n^H=u_n^H-\mathcal{U}_n^H$.

Step 2: Density argument, weak approximation in time. To bound the first terms in (70), (71) we use that $\mathcal{U},\partial_t\mathcal{U}\in\mathcal{C}^0([0,T],W^{1,p}(\Omega))$ to obtain for $t_n\leqslant s\leqslant t_{n+1}$

$$\|\nabla\mathcal{U}(\cdot,t_{n+1})-\nabla\mathcal{U}(\cdot,s)\|_{L^p(\Omega)}=\left\|\int_s^{t_{n+1}}\partial_t\nabla\mathcal{U}(\cdot,\tau)d\tau\right\|_{L^p(\Omega)} \tag{72}$$

$$\leqslant\Delta t\|\partial_t\nabla\mathcal{U}\|_{\mathcal{C}^0([0,T],L^p(\Omega))}. \tag{73}$$

Now if we take $\mathcal{U}_n^H=\mathcal{I}_H\mathcal{U}(\cdot,t_n)$ the above inequality together with standard interpolation results yields (72) in time we get that for $s\in[t_n,t_{n+1}]$ and $0\leqslant n\leqslant N-1$

$$\left\|\nabla\mathcal{U}(\cdot,s)-\nabla\mathcal{U}^H_{n+1}\right\|_{L^p(\Omega)}\leqslant\quad C(\Delta t+H)\Big(\|\mathcal{U}\|_{\mathcal{C}^0([0,T],W^{2,p^*}(\Omega))} \tag{74}$$

$$+\quad\|\partial_t\nabla\mathcal{U}\|_{\mathcal{C}^0([0,T],L^p(\Omega))}\Big). \tag{75}$$

We also have

$$\max_{1\leqslant n\leqslant N}\left\|u^0(\cdot,t_n)-\mathcal{U}_n^H\right\|_{L^2(\Omega)}\leqslant C_E\left\|u^0-\mathcal{U}\right\|_E+CH\|\mathcal{U}\|_{C^0([0,T],W^{2,p^*}(\Omega))},\tag{76}$$

where we used the embeddings $E\hookrightarrow C^0([0,T],L^2(\Omega))$ (with operator norm C_E), $W^{1,p}(\Omega)\hookrightarrow L^2(\Omega)$ and standard interpolation estimates. We then choose $\mathcal{U}\in C^\infty(\overline{\Omega}\times[0,T])$ such that $\mathcal{U}(\cdot,t)\in C_0^\infty(\Omega)$ for any $t\in[0,T]$ and $\|u^0-\mathcal{U}\|_E<\eta/2$. Then, using (72), (74) we find that for each $\eta>0$ there exists $D(\eta)$ such that for all $\Delta t,H\leqslant D_1(\eta)$ we have

$$\left\|\nabla u^0-\nabla\mathcal{U}^H\right\|_{\widetilde{L^p}(L^p(\Omega))}\leqslant\eta,\quad\max_{1\leqslant n\leqslant N}\left\|u^0(\cdot,t_n)-\mathcal{U}_n^H\right\|_{L^2(\Omega)}\leqslant(C_E+1)\eta.\tag{77}$$

Step 3: Macro discretization error. We next need to estimate $\theta_n^H=u_n^H-\mathcal{U}_n^H$, $0\leqslant n\leqslant N$. Hölder inequality and the monotonicity estimate (22) gives

$$\sum_{n=0}^{N-1}\Delta t\left\|\nabla\theta_{n+1}^H\right\|_{L^p(\Omega)}^p\leqslant\mathcal{R}(u_n^H,\mathcal{U}_n^H)^{\frac{p(\beta-p)}{\beta}}\tag{78}$$

$$\cdot\left(\sum_{n=0}^{N-1}\Delta t(B_H(u_{n+1}^H;\theta_{n+1}^H)-B_H(\mathcal{U}_{n+1}^H;\theta_{n+1}^H))\right)^{\frac{p}{\beta}},\tag{79}$$

where

$$\mathcal{R}(u_n^H,\mathcal{U}_n^H)=\gamma_c^{-p/\beta}\left(C+\left(\sum_{n=0}^{N-1}\Delta t\left\|\nabla u_{n+1}^H\right\|_{L^p(\Omega)}^p\right)^{\frac{1}{p}}+\left(\sum_{n=0}^{N-1}\Delta t\left\|\nabla\mathcal{U}_{n+1}^H\right\|_{L^p(\Omega)}^p\right)^{\frac{1}{p}}\right),\tag{80}$$

with C depending on $C_d,T,|\Omega|$. We observe that

$$\mathcal{R}(u_n^H,\mathcal{U}_n^H)\leqslant C,\tag{81}$$

where C is independent of $\mathcal{U},\eta,\Delta t,H$ (for small enough discretization parameters). Indeed, using (68) and $\|g-u_0^H\|_{L^2(\Omega)}\to0$ for $H\to0$ we can find H_0 such that for all $H\leqslant H_0$ we have $\left(\sum_{n=0}^{N-1}\Delta t\left\|\nabla u_{n+1}^H\right\|_{L^p(\Omega)}^p\right)^{\frac{1}{p}}\leqslant C$ independently of the initial approximation u_0^H. Using (74) we find that $\left(\sum_{n=0}^{N-1}\Delta t\left\|\nabla\mathcal{U}_{n+1}^H\right\|_{L^p(\Omega)}^p\right)^{1/p}\leqslant\left\|u^0\right\|_E+1$ for all $\Delta t,H\leqslant\min\{H_0,D_1(\eta_0)\}$.

We need next to estimate $B^H(u^H_{n+1};\theta^H_{n+1}) - B^H(\mathcal{U}^H_{n+1};\theta^H_{n+1})$. This is done by a decomposition

$$\sum_{n=0}^{N-1} \Delta t(B^H(u^H_{n+1};\theta^H_{n+1}) - B^H(\mathcal{U}^H_{n+1};\theta^H_{n+1})) = \sum_{n=0}^{N-1} \mathcal{B}^{tot}_n - \sum_{n=0}^{N-1} \Delta t \int_\Omega \bar{\partial}_t \theta^H_n \theta^H_{n+1} dx \tag{82}$$

where $\mathcal{B}^{tot}_n$ contains a number of terms that represent the contribution to the error due to the weak approximation in time, the macroscopic numerical discretization, the time discretization, the quadrature error, the micro and the modelling error [6, Sect. 5.1]. We also have in view of

$$\frac{1}{2}\bar{\partial}_t \|\theta^H_n\|^2_{L^2(\Omega)} \leqslant \int_\Omega \bar{\partial}_t \theta^H_n \theta^H_{n+1} dx, \qquad \text{for } 0 \leqslant n \leqslant N-1, \tag{83}$$

that

$$-\sum_{n=0}^{N-1} \Delta t \int_\Omega \bar{\partial}_t \theta^H_n \theta^H_{n+1} dx \leqslant \frac{1}{2}\|\theta^H_0\|^2_{L^2(\Omega)} - \frac{1}{2}\|\theta^H_N\|^2_{L^2(\Omega)}. \tag{84}$$

The initial error θ^H_0 in (84) can be bounded by using interpolant estimates and the embedding $E \hookrightarrow C^0([0,T], L^2(\Omega))$ as

$$\|\theta^H_0\|_{L^2(\Omega)} \leqslant \|g - u^H_0\|_{L^2(\Omega)} + C_E\|u^0 - \mathcal{U}\|_E + CH\|\mathcal{U}\|_{C^0([0,T],W^{2,p^*}(\Omega))}. \tag{85}$$

Next it can be shown, in view of (85) and the properties of $\mathcal{U}$ derived in step 2, that for $\Delta t, H \leqslant D_2(\eta)$, with $D_2(\eta)$ small enough we have (see [6, Sect. 5.2] for details)

$$\left(\sum_{n=0}^{N-1} \mathcal{B}^{tot}_n - \sum_{n=0}^{N-1} \Delta t \int_\Omega \bar{\partial}_t \theta^H_n \theta^H_{n+1} dx\right)^{\frac{1}{\beta}} \leqslant C\eta \tag{86}$$

Step 3: Upscaling error. First as $r_{mod} = 0$ we have $r_{HMM}(\nabla \mathcal{U}^H_{n+1}) = r_{mic}(\nabla \mathcal{U}^H_{n+1})$, where r_{mic} is given by (16a). Let the macro mesh size $H > 0$, the time step size $\Delta t > 0$ and the micro finite element space in (12) be given. Then, assuming that $\mathcal{A}^\varepsilon$ satisfies $(\mathcal{A}_{0-2})$ it can be shown that for any sequence $\{\mathcal{U}^H_n\}_{1\leqslant n\leqslant N} \subset S^1_0(\Omega, \mathcal{T}_H)$ for which $\sum_{n=0}^{N-1} \Delta t\|\nabla \mathcal{U}^H_{n+1}\|^p_{L^p(\Omega)}$ is bounded independently of the micro mesh size h, we have

$$\lim_{h\to 0}\left(\sum_{n=0}^{N-1} \Delta t\, r_{mic}(\nabla \mathcal{U}^H_{n+1})^{p'}\right)^{\frac{1}{p'}} = 0.$$

This result follows from a density argument, classical FE interpolation results and the general estimate obtained from $(\mathcal{A}_2)$

$$r_{mic}(\nabla v^H) \leqslant C\left[C_d + \left\|\nabla v^H\right\|_{L^p(\Omega)}\right]^{p-1-\gamma} \times \left(\sum_{K\in\mathcal{T}_H} \frac{|K|}{|K_\delta|} \inf_{z^h\in S^1(K_\delta,\mathcal{T}_h)} \left\|\nabla\bar{\chi}_K^{\nabla v^H(x_K)} - \nabla z^h\right\|^p_{L^p(K_\delta)}\right)^{\frac{\gamma}{p}},$$

for any $v^H \in S_0^1(\Omega,\mathcal{T}_H)$, where $\bar{\chi}_K^{\xi}$ solves (12) in $W(K_{\delta_j})$ and C is independent of H, h, δ and ε.

Step 4: Assembling the pieces: convergence in $L^p(W^{1,p})$ and $\mathcal{C}^0(L^2)$ norm. In view of (79), (81), (86) if we set $0 < D_3(\eta) \leqslant \min\{D_1(\eta_0), H_0, D_2(\eta)\}$ then for $\Delta t, H \leqslant D_3(\eta)$ we have

$$\lim_{h\to 0}\left(\sum_{n=0}^{N-1} \Delta t \left\|\nabla\theta^H_{n+1}\right\|^p_{L^p(\Omega)}\right)^{\frac{1}{p}} \leqslant C\eta, \tag{87}$$

where C is independent of $\mathcal{U}, \eta, H, \Delta t, \delta$ and h. Combining this inequality with (71) and the density estimates of step 2 yields the convergence in the $L^p(W^{1,p})$ norm.

Next to derive a bound in the $\mathcal{C}^0(L^2)$, we first observe that (83) together with the monotonicity estimate (22) yield

$$\frac{1}{2}\bar{\partial}_t\left\|\theta_n^H\right\|^2_{L^2(\Omega)} \leqslant \int_\Omega \bar{\partial}_t\theta_n^H\theta_{n+1}^H dx + B_H(u^H_{n+1};\theta^H_{n+1}) - B_H(\mathcal{U}^H_{n+1};\theta^H_{n+1}), \tag{88}$$

Summing this inequality for $n = 0,\ldots,K-1$, taking the maximum over K, using (83) and the monotonicity of B_H from Lemma 2 we get

$$\frac{1}{2}\left\|\theta_K^H\right\|^2_{L^2(\Omega)} - \frac{1}{2}\left\|\theta_0^H\right\|^2_{L^2(\Omega)} \leqslant \sum_{n=0}^{K-1}\mathcal{B}_n^{tot},$$

where $\mathcal{B}_n^{tot}$ is defined in (82). Using then (85) and an estimate similar to (86) we find that

$$\lim_{h\to 0}\left(\max_{1\leqslant n\leqslant N}\left\|\theta_n^H\right\|_{L^2(\Omega)}\right) \leqslant C\eta, \tag{89}$$

for all $\Delta t, H$ small enough, where C is independent of $\mathcal{U}, \eta, H, \Delta t, \delta$ and h. Hence together with (77) this shows the $\mathcal{C}^0(L^2)$ estimate of Theorem 5.

4.2 Convergence for Strongly Monotone and Lipschitz Nonlinear Maps

Optimal convergence rates can be derived for $p = 2$ and $\alpha = 1, \beta = 2$ in $(\mathcal{A}_{1-2})$, i.e., when the nonlinear map $\mathcal{A}^{\varepsilon}(x, \xi)$ is Lipschitz continuous with respect to its first argument and strongly monotone. In this case we can derive optimal macroscopic, microscopic and temporal error estimates without any structural assumptions such as local periodicity or random stationarity of $\mathcal{A}^{\varepsilon}$. Explicit bounds of the modelling error are however derived only for locally periodic data $\mathcal{A}^{\varepsilon}$.

Theorem 6 *For the case $p = 2$ assume that $\mathcal{A}^{\varepsilon}$ satisfies $(\mathcal{A}_{0-2})$ with $\alpha = 1, \beta = 2$. Let u^0 be the solution to the homogenized problem* (5) *and u_n^H the numerical solution obtained from* (66) *with initial condition u_0^H. Provided in addition that*

$$u^0, \partial_t u^0 \in \mathcal{C}^0([0,T], H^2(\Omega)), \quad \partial_t^2 u^0 \in \mathcal{C}^0([0,T], L^2(\Omega)), \tag{90a}$$

$$\mathcal{A}^0(\cdot,\xi) \in W^{1,\infty}(\Omega;\mathbb{R}^d) \text{ with } \left\|\mathcal{A}^0(\cdot,\xi)\right\|_{W^{1,\infty}(\Omega;\mathbb{R}^d)} \leqslant C(L_0 + |\xi|), \quad \forall\, \xi \in \mathbb{R}^d, \tag{90b}$$

then, the following discrete $\mathcal{C}^0(L^2)$ and $L^2(H^1)$ error estimate holds

$$\max_{1\leqslant n\leqslant N} \left\|u^0(\cdot,t_n) - u_n^H\right\|_{L^2(\Omega)} + \left(\sum_{n=1}^{N} \Delta t \left\|\nabla u^0(\cdot,t_n) - \nabla u_n^H\right\|_{L^2(\Omega)}^2\right)^{1/2} \tag{91}$$
$$\leqslant C\left[\Delta t + H + \max_{1\leqslant n\leqslant N} r_{HMM}(\nabla \mathcal{I}_H u^0(\cdot,t_n)) + \left\|g - u_0^H\right\|_{L^2(\Omega)}\right],$$

where $\mathcal{I}_H u^0$ denotes the nodal interpolant of u^0 and C is independent of $\Delta t, H$ and r_{HMM}.

Remark 2 Under additional regularity assumptions, assuming elliptic regularity and quasi-uniform meshes, one can derive the following improved (discrete) $\mathcal{C}^0(L^2)$ error estimate

$$\max_{1\leqslant n\leqslant N} \left\|u^0(\cdot,t_n) - u_n^H\right\|_{L^2(\Omega)} \leqslant C\left[\Delta t + H^2 + \max_{1\leqslant n\leqslant N} r_{HMM}(\nabla \tilde{u}^{H,0}(\cdot,t_n)) + \left\|g - u_0^H\right\|_{L^2(\Omega)}\right],$$

where $\tilde{u}^{H,0}$ is given by an elliptic projection and C is independent of $\Delta t, H$ and r_{HMM} (see [6, Theorem 4.4]).

We sketch the proof of Theorem 6.

Owing to regularity assumptions for u^0, we can use a strong formulation in time

$$\int_\Omega \partial_t u^0(x,t)\, w\, dx + B^0(u^0(\cdot,t); w) = \int_\Omega f\, w\, dx, \qquad \forall\, w \in W_0^{1,2}(\Omega), \forall\, t \in (0,T].$$

Hence the argument density used in Sect. 4.1 is not needed here. We can then directly define $\mathcal{U}_n^H = \mathcal{I}_H u^0(\cdot, t_n)$ and with $\theta_n^H = \mathcal{U}_n^H - u_n^H$ we obtain instead of (82) the following error propagation formula

$$\int_\Omega \bar{\partial}_t \theta_n^H w^H \, dx + \left[B_H(u_{n+1}^H; w^H) - B_H(\mathcal{U}_{n+1}^H; w^H)\right]$$

$$= \int_\Omega \left[\partial_t u^0(x, t_{n+1}) - \bar{\partial}_t u^0(x, t_n)\right] w^H dx \, ds \tag{92a}$$

$$+ \int_\Omega \left[\partial_t u^0(x, t_n) - \bar{\partial}_t \mathcal{U}_n^H\right] w^H dx \tag{92b}$$

$$+ \left[B^0(u^0(\cdot, t_{n+1}); w^H) - B^0(\mathcal{U}_{n+1}^H; w^H)\right] \tag{92c}$$

$$+ \left[B^0(\mathcal{U}_{n+1}^H; w^H) - \hat{B}^0(\mathcal{U}_{n+1}^H; w^H)\right] \tag{92d}$$

$$+ \left[\hat{B}^0(\mathcal{U}_{n+1}^H; w^H) - B_H(\mathcal{U}_{n+1}^H; w^H)\right]. \tag{92e}$$

In the above formula the term (92a) is due to the time discretization error, the terms (92b) and (92c) account for the finite element error at the discrete time levels t_n. The influence of the quadrature formula is captured by (92d). Finally the components (92a)–(92d) are independent of the temporal and macro spatial error, while last term (92e) is only due to the upscaling strategy and averaging techniques used to define and compute numerically the upscaled tensor. All these terms can be estimated quantitatively [6]. If we set $w^H = \theta_{n+1}^H$ use the inequality (83) we obtain

$$\begin{aligned} \frac{1}{2}\bar{\partial}_t \left\| \theta_n^H \right\|_{L^2(\Omega)}^2 + \lambda \left\| \nabla \theta_{n+1}^H \right\|_{L^2(\Omega)}^2 & \\ \leqslant C\Delta t \left\| \partial_t^2 u^0 \right\|_{C^0([0,T],L^2(\Omega))} \left\| \theta_{n+1}^H \right\|_{L^2(\Omega)} & \\ + CH \left\| u^0 \right\|_{C^0([0,T],H^2(\Omega))} \left\| \theta_{n+1}^H \right\|_{L^2(\Omega)} & \\ + CH \left\| u^0 \right\|_{C^0([0,T],H^2(\Omega))} \left\| \nabla \theta_{n+1}^H \right\|_{L^2(\Omega)} & \\ + r_{HMM}(\nabla \mathcal{U}_{n+1}^H) \left\| \nabla \theta_{n+1}^H \right\|_{L^2(\Omega)}. & \end{aligned} \tag{93}$$

Multiplying the above inequality by Δt and summing first from $n = 0, \ldots, K-1 \leqslant N-1$ and taking the maximum over K yields

$$\begin{aligned} \max_{1 \leqslant n \leqslant N} \left\| \theta_N^H \right\|_{L^2(\Omega)}^2 + \lambda \sum_{n=1}^N \Delta t \left\| \nabla \theta_n^H \right\|_{L^2(\Omega)}^2 & \\ \leqslant \left\| \theta_0^H \right\|_{L^2(\Omega)}^2 + C(\Delta t + H + \max_{1 \leqslant n \leqslant N} r_{HMM}(\nabla \mathcal{U}_n^H))^2. & \end{aligned} \tag{94}$$

The classical estimates for nodal interpolant $\|\mathcal{I}_H z - z\|_{H^1(\Omega)} \leqslant CH\|z\|_{H^2(\Omega)}$ for $u^0(\cdot, t_n) - \mathcal{U}_n^H$ together with the regularity of (90a) and the triangle inequality gives finally the estimate of Theorem 6.

Fully Discrete Space-Time Result Recall that $r_{HMM}(\cdot) \leqslant r_{mod}(\cdot) + r_{mic}(\cdot)$ (see (16a), (16b)). Following the results for linear problems (with additional technicalities due to the nonlinear micro-problems) we can estimate both $r_{mod}(\cdot)$ and $r_{mic}(\cdot)$. First under the assumptions of Theorem 6 and assuming that the exact solution of Problem (12) satisfies $\bar{\chi}_K^{\xi} \in H^2(K_\delta)$ and $\left|\bar{\chi}_K^{\xi}\right|_{H^2(K_\delta)} \leqslant C\varepsilon^{-1}(L_0 + |\xi|)\sqrt{|K_\delta|}$ we have the following error estimate for the micro error

$$r_{mic} \leqslant C\frac{h}{\varepsilon},$$

where C is independent of $\Delta t, H, h, \varepsilon, \delta$. By defining a appropriate linear adjoint auxiliary problem derived from (12) and assuming $W^{1,\infty}(K_\delta)$ regularity of the solutions of these (linear) problems one can get the optimal micro error

$$r_{mic} \leqslant C\left(\frac{h}{\varepsilon}\right)^2, \tag{95}$$

with the same rate as for linear problem [2].

For the modelling error we need structural assumptions and assume that $\mathcal{A}^{\varepsilon}(x,\xi) = \mathcal{A}(x, x/\varepsilon, \xi)$ where $\mathcal{A}(x,y,\xi)$ is Y-periodic in y, i.e., $\mathcal{A}^{\varepsilon}$ is locally periodic. Then, for any $v^H \in S_0^1(\Omega, \mathcal{T}_H)$, the modelling error $r_{mod}(\nabla v^H)$ defined in (16b) is bounded by

$$r_{mod}(\nabla v^H) \leqslant \begin{cases} 0, & \text{if } W(K_\delta) = W^1_{per}(K_\delta), \delta/\varepsilon \in \mathbb{N} \text{ and} \\ & \mathcal{A}^{\varepsilon} = \mathcal{A}(x_K, x/\varepsilon, \xi) \text{ collocated at } x_K, \\ C^1_{mod}\,\delta, & \text{if } W(K_\delta) = W^1_{per}(K_\delta), \delta/\varepsilon \in \mathbb{N}, \\ C^2_{mod}(\delta + \sqrt{\varepsilon/\delta}), & \text{if } W(K_\delta) = H^1_0(K_\delta), \delta > \varepsilon, \end{cases} \tag{96}$$

with C^1_{mod} and C^2_{mod} given by

$$C^1_{mod} = C(L_0 + \|\nabla v^H\|_{L^2(\Omega)}), \quad C^2_{mod} = C(C^1_{mod} + \max_{K \in \mathcal{T}_H} \|\bar{\chi}^{\nabla v^H(x_K)}(x_K, \cdot)\|_{W^{1,\infty}(Y)}),$$

where $\chi^{\xi}(x_K, \cdot)$, for $\xi \in \mathbb{R}^d$, $K \in \mathcal{T}_H$, denote the exact solutions to the homogenization cell problems find $\bar{\chi}^{\xi}(x, \cdot) \in W^1_{per}(Y)$ such that

$$\int_Y \mathcal{A}(x, y, \xi + \nabla\bar{\chi}^{\xi}(x,y)) \cdot \nabla z \, dy = 0, \qquad \forall\, z \in W^1_{per}(Y), \tag{97}$$

and C is independent of $\Delta t, H, h, \varepsilon, \delta$ and v^H. We refer to [6] for a detailed proof of these micro and modelling a priori error estimates. We observe that the first two estimates for the modelling error are similar as for linear problem (see Sect. 3.1). A better estimate can however be derived for linear problem for the third case for which it is possible to derive the estimate $(\delta + \varepsilon/\delta)$ (see again Sect. 3.1).

5 A Linearized Method

We consider again nonlinear monotone problems of the type (1) with strongly monotone and Lipschitz continuous maps $\mathcal{A}^\varepsilon(x, \xi)$, i.e., for the case $p = 2$ and $\alpha = 1$, $\beta = 2$ in $(\mathcal{A}_{1-2})$. We further assume that the nonlinear map is of the form $\mathcal{A}^\varepsilon(x, \xi) = a^\varepsilon(x, \xi)\xi$. We first rewrite the method (66) in a slightly different form: find $u^H_{n+1} \in S^1_0(\Omega, \mathcal{T}_H)$ such that

$$\int_\Omega \frac{u^H_{n+1} - u^H_n}{\Delta t} w^H dx + B_H(u^H_{n+1}; w^H) = \int_\Omega f w^H dx, \quad \forall\, w^H \in S^1_0(\Omega, \mathcal{T}_H), \tag{98}$$

with the nonlinear macro map B_H given by

$$B_H(v^H; w^H) = \sum_{K \in \mathcal{T}_H} \frac{|K|}{|K_\delta|} \int_{K_\delta} a^\varepsilon(x, \nabla \hat{v}^h_K) \nabla \hat{v}^h_K dx \cdot \nabla w^H(x_K), \quad v^H, w^H \in S^1_0(\Omega, \mathcal{T}_H), \tag{99}$$

and the micro functions v^h_K are given similarly to (12) by the following problem: find $\hat{v}^h_K$ such that $\hat{v}^h_K - v^H = v^h_K \in S^1(K_\delta, \mathcal{T}_h)$ and

$$\int_{K_\delta} a^\varepsilon(x, \nabla \hat{v}^h_K) \nabla \hat{v}^h_K \cdot \nabla w^h dx = 0, \quad \forall\, w^h \in S^1(K_\delta, \mathcal{T}_h). \tag{100}$$

The equivalence of the above formulation and the one in (66) with micro problems given by (12) is easy to check. Following [12] we propose a linearized scheme. The idea is to decouple the micro-solutions in (99) and to consider

$$B_H(\hat{z}; v^H, w^H) = \sum_{K \in \mathcal{T}_H} \frac{|K|}{|K_\delta|} \int_{K_\delta} a^\varepsilon(x, \nabla z^h_K) \nabla \hat{v}^{h,z^h_K}_K dx \cdot \nabla w^H(x_K), \quad v^H, w^H \in S^1_0(\Omega, \mathcal{T}_H) \tag{101}$$

where for given $\{z^h_K\} \in \prod_{K \in \mathcal{T}_H} S^1(K_\delta, \mathcal{T}_h)$, $\hat{v}^{h,z^h_K}_K$ is such that $\hat{v}^{h,z^h_K}_K - v^H = v^{h,z^h_K}_K \in S^1(K_\delta, \mathcal{T}_h)$ and solution of the *linear micro problem*

$$\int_{K_\delta} a^\varepsilon(x, \nabla z^h_K) \nabla \hat{v}^{h,z^h_K}_K \cdot \nabla w^h dx = 0, \quad \forall\, w^h \in S^1(K_\delta, \mathcal{T}_h). \tag{102}$$

To formalize the numerical method we consider the product of FE spaces

$$\mathcal{S}^{H,h} = S_0^1(\Omega, \mathcal{T}_H) \times \prod_{K \in \mathcal{T}_H} S^1(K_\delta, \mathcal{T}_h), \tag{103}$$

and define $\hat{z} = (z^H, \{z_K^h\}) \in \mathcal{S}^{H,h}$. Next for a given $\hat{u}_1 = (u_1^H, \{u_{1,K}^h\}) \in \mathcal{S}^{H,h}$ we define one step of the method as a map $\mathcal{S}^{H,h} \mapsto \mathcal{S}^{H,h}$ given by $\hat{u}_n = (u_n^H, \{u_{n,K}^h\}) \mapsto \hat{u}_{n+1} = (u_{n+1}^H, \{u_{n+1,K}^h\})$. To compute this map we implement the following two steps:

1. *update the macroscopic state:* find $u_{n+1}^H \in S_0^1(\Omega, \mathcal{T}_H)$, the solution of the linear problem

$$\int_\Omega \frac{1}{\Delta t}(u_{n+1}^H - u_n^H)\, w^H dx + B_H(\hat{u}_n; u_{n+1}^H, w^H) = \int_\Omega f w^H dx, \quad \forall\, w^H \in S_0^1(\Omega, \mathcal{T}_H); \tag{104}$$

2. *update the microscopic states:* for each $K \in \mathcal{T}_H$, compute $\hat{v}_K^{h,u_{n,K}^h}$ such that $\hat{v}_K^{h,u_{n,K}^h} - u_{n+1}^H \in S^1(K_\delta, \mathcal{T}_h)$ and solution of (102) with parameter $u_{n,K}^h$ and update $u_{n+1,K}^h := \hat{v}_K^{h,u_{n,K}^h} - u_{n+1}^H$.

To completely describe the algorithm we still need to discuss the initialization procedure, i.e., how to define $\hat{u}_1 = (u_1^H, \{u_{1,K}^h\}) \in \mathcal{S}^{H,h}$ given the approximation $u_0^H \in S_0^1(\Omega, \mathcal{T}_H)$ of the initial condition $g(x)$ of (5). We suggest to use one step of the nonlinear method (66) to set $\hat{u}_1$. This choice allows to prove optimal convergence rates. In turns out that the trivial initialisation obtained by setting $\hat{u}_0 = (u_0^H, \{0\})$ and using one step of the linearised method to define $\hat{u}_1$ deteriorates the accuracy of the linearised scheme [13]. It is also shown in [13] that the above linearised method is up to ten times faster than the fully nonlinear method (66)–(67).

Well-posedness of the linearized method can be proved assuming that $a^\varepsilon(x, \xi)$ is uniformly elliptic and bounded, i.e.,

$$\lambda_a |\eta|^2 \leqslant a^\varepsilon(x, \xi)\eta \cdot \eta, \ |a^\varepsilon(x, \xi)\eta| \leqslant \Lambda_a |\eta|, \ \forall\, \xi, \eta \in \mathbb{R}^d, \text{ a.e. } x \in \Omega, \varepsilon > 0.$$

It then follows from similar argument as for linear elliptic problem [2] that

$$B_H(\hat{z}; v^H, v^H) \geqslant \lambda_a \left\| \nabla v^H \right\|_{L^2(\Omega)}^2, \quad \left| B_H(\hat{z}; v^H, w^H) \right| \leqslant \frac{\Lambda_a^2}{\lambda_a} \left\| \nabla v^H \right\|_{L^2(\Omega)} \left\| \nabla w^H \right\|_{L^2(\Omega)}.$$

Combining the above estimate for B_H with the existence and uniqueness of the nonlinear initialisation obtained in Sect. 4 allows to prove existence and uniqueness of a solution to (104) and an a priori estimate similar to (68) with a right-hand side simply given by $C(\|f\|_{L^2(\Omega)} + \left\| u_0^H \right\|_{L^2(\Omega)})$.

5.1 A Priori Error Estimates

Fully discrete a priori error estimates of the linearized method can be established following the steps of Sect. 4.2, with nontrivial modifications due to the linearisation procedure. It will be convenient in the sequel to introduce the two following semi-norm on the space $\mathcal{S}^{H,h}$. For $\hat{z} = (z^H, \{z_K^h\}) \in \mathcal{S}^{H,h}$ we therefore define

$$\|\nabla \hat{z}\|_{\mathcal{S}^{H,h}} = \left(\sum_{K\in\mathcal{T}_H} \frac{|K|}{|K_\delta|} \left\|\nabla \hat{z}_K^h\right\|^2_{L^2(K_\delta)} \right)^{1/2}, \qquad \|\nabla \hat{z}\|_{\mathcal{S}^{H,h}_\infty} = \max_{K\in\mathcal{T}_H} \left\|\nabla \hat{z}_K^h\right\|_{L^\infty(K_\delta)},$$

where $\hat{z}_K^h = z_K^h + z^H$ on K_δ. In fact due to the Poincaré (or Poincaré-Wirtinger) inequality, $\|\cdot\|_{\mathcal{S}^{H,h}}$ is a norm. Observe that since $\int_{K_\delta} \nabla z_K^h dx \cdot \nabla z^H(x_K) = 0$ for micro spaces $S^1(K_\delta, \mathcal{T}_h)$ with periodic and Dirichlet boundary conditions we have $\left\|\nabla \hat{z}_K^h\right\|^2_{L^2(K_\delta)} = \left\|\nabla z^H(x_K)\right\|^2_{L^2(K_\delta)} + \left\|\nabla z_K^h\right\|^2_{L^2(K_\delta)}$, which yield for all $\hat{z} = (z^H, \{z_K^h\}) \in \mathcal{S}^{H,h}$ the inequality $\left\|\nabla z^H\right\|_{L^2(\Omega)} \leqslant \|\nabla \hat{z}\|_{\mathcal{S}^{H,h}}$.

Next consider the numerical solution obtained by the *linearized* multiscale method (104) $\hat{u}_n = (u_n^H, \{u_{n,K}^h\}) \in \mathcal{S}^{H,h}$ and set $\hat{u}_{n,K}^h = u_n^H + u_{n,K}^h$ on K_δ. We also define the nodal interpolation associated to the homogenized solution $\mathcal{U}_n^H = \mathcal{I}_H u^0(\cdot, t_n)$ and consider $\hat{\mathcal{U}}_n = (\mathcal{U}_n^H, \{\mathcal{U}_{n,K}^h\}) \in \mathcal{S}^{H,h}$ such that $\hat{\mathcal{U}}_{n,K}^h = \mathcal{U}_{n,K}^h + \mathcal{U}_n^H$ is the solution to the *nonlinear* micro problem (100). Define for $0 \leqslant n \leqslant N$ and $K \in \mathcal{T}_H$

$$\hat{\theta}_n = \hat{u}_n - \hat{\mathcal{U}}_n, \text{ i.e., } \quad \theta_n^H = u_n^H - \mathcal{U}_n^H, \; \hat{\theta}_{n,K}^h = \hat{u}_{n,K}^h - \hat{\mathcal{U}}_{n,K}^h. \tag{105}$$

A formula similar to (92) leads to

$$\begin{aligned} &\frac{1}{2}\bar{\partial}_t \left\|\theta_n^H\right\|^2_{L^2(\Omega)} + \lambda_a \left\|\nabla \hat{\theta}_{n+1}^H\right\|^2_{\mathcal{S}^{H,h}} \\ &\leqslant C(\Delta t + H + r_{HMM}(\nabla \mathcal{U}_{n+1}^H)) \left\|\nabla \hat{\theta}_{n+1}^H\right\|_{L^2(\Omega)} + \left|L_n(\nabla \hat{\theta}_{n+1})\right|, \end{aligned} \tag{106}$$

where the additional term involves a function; $L_n \colon \mathcal{S}^{H,h} \to \mathbb{R}$ defined by

$$L_n(\nabla \hat{w}) = \sum_{K\in\mathcal{T}_H} \frac{|K|}{|K_\delta|} \int_{K_\delta} \left[a^\varepsilon(x, \nabla \hat{\mathcal{U}}_{n,K}^h) - a^\varepsilon(x, \nabla \hat{u}_{n,K}^h)\right] \nabla \hat{\mathcal{U}}_{n,K}^h \cdot \nabla \hat{w}_K^h dx. \tag{107}$$

This term arises from the linearization error and it can be bounded by

$$|L_n(\nabla \hat{w})| \leqslant \mathcal{L}_n \left\|\nabla \hat{\theta}_n\right\|_{\mathcal{S}^{H,h}} \|\nabla \hat{w}\|_{\mathcal{S}^{H,h}}, \tag{108}$$

where $\mathcal{L}_n$ will be discussed below. Hence using Young's inequality we obtain

$$\frac{1}{2}\bar{\partial}_t\|\theta_n^H\|_{L^2(\Omega)}^2 + \lambda_a\left\|\nabla\hat{\theta}_{n+1}^H\right\|_{S^{H,h}}^2$$
$$\leqslant C(\Delta t^2 + H^2 + r_{HMM}(\nabla\mathcal{U}_{n+1}^H)^2) + \frac{\mathcal{L}_n^2}{\lambda_a}\left\|\nabla\hat{\theta}_n\right\|_{S^{H,h}}^2 + \frac{\lambda_a}{2}\left\|\nabla\hat{\theta}_{n+1}\right\|_{S^{H,h}}^2. \tag{109}$$

Recall that we use the fully nonlinear method for the first step. Hence the convergence results of Sect. 4.2 yield

$$\|\theta_1^H\|_{L^2(\Omega)}^2 - \|\theta_0^H\|_{L^2(\Omega)}^2 + \lambda\Delta t\|\nabla\theta_1^H\|_{L^2(\Omega)}^2 \leqslant C\Delta t\big(\Delta t^2 + H^2 + r_{HMM}(\nabla\mathcal{U}_1^H)^2\big), \tag{110}$$

where λ is the monotonicity constant of $\mathcal{A}^\varepsilon$.

Similarly to (94) summing (109) from $n = 1$ to $n = N-1$, adding the term (110) and using the inequality $\|\nabla z^H\|_{L^2(\Omega)} \leqslant \|\nabla\hat{z}\|_{S^{H,h}}$ gives

$$\max_{1\leqslant n\leqslant N}\|\theta_n^H\|_{L^2(\Omega)}^2 + \lambda\Delta t\|\nabla\theta_1^H\|_{L^2(\Omega)}^2 + C_{\mathcal{L}}\Delta t\sum_{n=2}^{N}\|\nabla\theta_n^H\|_{L^2(\Omega)}^2 \tag{111}$$
$$\leqslant C\left(\Delta t^2 + H^2 + \max_{1\leqslant n\leqslant N} r_{HMM}(\nabla\mathcal{U}_n^H)^2\right) + \|\theta_0^H\|_{L^2(\Omega)}^2 + \frac{2}{\lambda_a}\Delta t\mathcal{L}_1^2\left\|\nabla\hat{\theta}_1\right\|_{S^{H,h}}^2.$$

where $C_{\mathcal{L}} = \lambda_a - \frac{2}{\lambda_a}\max_{2\leqslant n\leqslant N-1}\mathcal{L}_n^2$. Recall that $\hat{\theta}_{1,K}^h = \hat{u}_{1,K}^h - \hat{\mathcal{U}}_{1,K}^h$ where $\hat{u}_{K,1}^h$ and $\hat{\mathcal{U}}_{K,1}^h$ are solutions to the nonlinear micro problem (100) constrained by u_1^H and $\mathcal{U}_1^H$, respectively. The difference of two such micro solutions can be estimated by the difference of their respective macro constraints as

$$\left\|\nabla\hat{\theta}_{1,K}^h\right\|_{L^2(K_\delta)} \leqslant \frac{L}{\lambda}\sqrt{|K_\delta|}\left|\nabla u_1^H(x_K) - \nabla\mathcal{U}_1^H(x_K)\right|, \tag{112}$$

hence $\left\|\nabla\hat{\theta}_1\right\|_{S^{H,h}} \leqslant \frac{L}{\lambda}\|\nabla\theta_1^H\|_{L^2(\Omega)}$. Assuming $\mathcal{L}_1$ is bounded and $C_{\mathcal{L}} > 0$ we obtain

$$\max_{1\leqslant n\leqslant N}\|\theta_n^H\|_{L^2(\Omega)}^2 + \Delta t\sum_{n=1}^{N}\left\|\nabla\hat{\theta}_n\right\|_{L^2(\Omega)}^2 \tag{113}$$
$$\leqslant C\left(\Delta t^2 + H^2 + \max_{1\leqslant n\leqslant N} r_{HMM}(\nabla\mathcal{U}_n^H)^2\right) + C\|\theta_0^H\|_{L^2(\Omega)}^2. \tag{114}$$

Finally as $\|\theta_0^H\|_{L^2(\Omega)} \leqslant \|u_0^H - g\|_{L^2(\Omega)} + \|g - \mathcal{U}_0^H\|_{L^2(\Omega)}$ using the bound $\|g - \mathcal{U}_0^H\|_{L^2(\Omega)} \leqslant CH$ gives an estimate similar to (94). In view of (113), classical estimates for nodal interpolants give under the assumptions of Theorem 6, provided $\mathcal{L}_1$ is bounded and $C_{\mathcal{L}} > 0$, the error estimate (91).

We briefly discuss the additional assumptions on $\mathcal{L}_1$ and $C_{\mathcal{L}}$. These assumptions can be derived in two ways [13]. Under some regularity assumptions on the exact solutions of the micro problems (100), assuming that $u^0 \in \mathcal{C}^0([0,T], W^{2,\infty}(\Omega))$ and $\max_{t\in[0,T]} \left|u^0(x,t)\right|_{W^{1,\infty}(\Omega)}$ is small enough (smallness assumption), then there exist H_0, h_0 such that for any $H < H_0, h < h_0$, $C_{\mathcal{L}} > 0$ and $\mathcal{L}_1$ is bounded. Alternatively we can prove the boundedness of $\mathcal{L}_1$ and the positivity of $C_{\mathcal{L}}$ without a smallness assumption on u^0 and without the additional regularity assumption $\mathcal{C}^0([0,T], W^{2,\infty}(\Omega))$ on u^0 if in addition to the assumptions of Theorem 6 we have

$$\max_{\substack{K\in\mathcal{T}_H \\ 1\leq n\leq N-1}} \|e_{n,K}\|_{(L^\infty(K_\delta))^{d\times d}} < \frac{\lambda_a}{2\sqrt{2}}, \tag{115}$$

where the error term $e_{n,K} \in (L^\infty(K_\delta))^{d\times d}$ is given by

$$e_{n,K}(x) = a^\varepsilon(x, \nabla\hat{u}^h_{n,K}) - \int_0^1 a^\varepsilon(x, \nabla\hat{u}^h_{n,K} - \tau\nabla\hat{\theta}^h_{n,K})d\tau, \qquad \text{a.e. } x \in K_\delta. \tag{116}$$

The term (116) represent the linearization error. It has been shown numerically for tensors with various ellipticity constant λ_a that (116) holds if the spatial and temporal discretization parameters are small enough [13]. Optimal (discrete) $\mathcal{C}^0(L^2)$ can also be derived under the same additional assumptions as for the fully nonlinear method. Finally fully discrete results, i.e., quantitative estimates for the component r_{mic} and r_{mod} of r_{HMM} can be obtained similarly as in Sect. 4.2, with similar rates.

6 Conclusion

We have presented a unified framework and analysis for the FE-HMM applied to monotone parabolic problems. We have shown that under the most general assumptions for which homogenization can be established, we can construct an FE-HMM and establish its convergence. Under more restrictive assumptions, e.g. Lipschitz continuous and strongly monotone maps, fully discrete space time a priori error estimates can be derived and in some situation an efficient linearized scheme can be constructed and analyzed. Finally for linear problems we have shown that the FE-HMM can be coupled with classes of Runge-Kutta methods (Radau or Chebyshev methods) and analyzed by combining fully discrete spatial estimates with semi-group techniques in a Hilbert space framework. We have neither discussed implementation issue nor given numerical experiments. This is carefully documented in [6, 9, 13], where the issue of choosing the right coupling of the micro and macro solvers (i.e., the micro boundary conditions) and the size of the sampling domains are discussed. Numerical experiments for non-periodic problems (e.g., log-normal stochastic field) [9] and degenerate problems [13] illustrate the robustness of the numerical homogenization strategy.

Acknowledgement This research is partially supported by the Swiss National Foundation under Grant 200021_150019.

References

1. A. Abdulle, Fourth order Chebyshev methods with recurrence relation. SIAM J. Sci. Comput. **23**, 2041–2054 (2002)
2. A. Abdulle, On a priori error analysis of fully discrete heterogeneous multiscale FEM. Multiscale Model. Simul. **4**, 447–459 (2005)
3. A. Abdulle, The finite element heterogeneous multiscale method: a computational strategy for multiscale PDEs, in *Multiple Scales Problems in Biomathematics, Mechanics, Physics and Numerics*. GAKUTO International Series. Mathematical Sciences and Applications, vol. 31 (Gakkōtosho, Tokyo, 2009), pp. 133–181
4. A. Abdulle, A priori and a posteriori error analysis for numerical homogenization: a unified framework. Ser. Contemp. Appl. Math. CAM **16**, 280–305 (2011)
5. A. Abdulle, M.E. Huber, Error estimates for finite element approximations of nonlinear monotone elliptic problems with application to numerical homogenization. Numer. Methods Partial Differ. Equ. **32**(3), 737–1104
6. A. Abdulle, M.E. Huber, Finite element heterogeneous multiscale method for nonlinear monotone parabolic homogenization problems. ESAIM Math. Model. Numer. Anal. doi: http://dx.doi.org/10.1051/m2an/2016003
7. A. Abdulle, A.A. Medovikov, Second order Chebyshev methods based on orthogonal polynomials. Numer. Math. **90**, 1–18 (2001)
8. A. Abdulle, C. Schwab, Heterogeneous multiscale FEM for diffusion problems on rough surfaces. Multiscale Model. Simul. **3**, 195–220 (2005)
9. A. Abdulle, G. Vilmart, Coupling heterogeneous multiscale FEM with Runge-Kutta methods for parabolic homogenization problems: a fully discrete space-time analysis. Math. Models Methods Appl. Sci. **22**, 1250002/1–40 (2012)
10. A. Abdulle, G. Vilmart, Analysis of the finite element heterogeneous multiscale method for quasilinear elliptic homogenization problems. Math. Comput. **83**, 513–536 (2014)
11. A. Abdulle, W. E, Finite difference heterogeneous multi-scale method for homogenization problems. J. Comput. Phys. **191**, 18–39 (2003)
12. A. Abdulle, Y. Bai, G. Vilmart, An offline-online homogenization strategy to solve quasilinear two-scale problems at the cost of one-scale problems. Int. J. Numer. Methods Eng. **99**, 469–486 (2014)
13. A. Abdulle, M.E. Huber, G. Vilmart, Linearized numerical homogenization method for nonlinear monotone parabolic multiscale problems. Multiscale Model. Simul. **13**, 916–952 (2015)
14. G. Allaire, M. Briane, Multiscale convergence and reiterated homogenisation. Proc. R. Soc. Edinburgh Sect. A **126**, 297–342 (1996)
15. G. Allaire, R. Brizzi, A multiscale finite element method for numerical homogenization. Multiscale Model. Simul. **4**, 790–812 (2005) (electronic)
16. I. Babuška, J.E. Osborn, Generalized finite element methods: their performance and their relation to mixed methods. SIAM J. Numer. Anal. **20**, 510–536 (1983)
17. J.W. Barrett, W.B. Liu, Finite element approximation of the parabolic p-Laplacian. SIAM J. Numer. Anal. **31**, 413–428 (1994)

18. A. Bensoussan, J.-L. Lions, G. Papanicolaou, *Asymptotic Analysis for Periodic Structures* (North-Holland, Amsterdam, 1978)
19. L. Bers, F. John, M. Schechter, Partial differential equations, in *Proceedings of the Summer Seminar*. Lectures in Applied Mathematics, Boulder, CO (1957)
20. M.E. Brewster, G. Beylkin, A multiresolution strategy for numerical homogenization. Appl. Comput. Harmon. Anal. **2**, 327–349 (1995)
21. V. Chiadò Piat, G. Dal Maso, A. Defranceschi, G-convergence of monotone operators. Ann. Inst. H. Poincaré Anal. Non Linéaire **7**, 123–160 (1990)
22. P.G. Ciarlet, *The Finite Element Method for Elliptic Problems*. Studies in Mathematics and Its Applications, vol. 4 (North-Holland, Amsterdam, 1978)
23. P.G. Ciarlet, P.A. Raviart, The combined effect of curved boundaries and numerical integration in isoparametric finite element methods, in *The Mathematical Foundations of the Finite Element Method with Applications to Partial Differential Equations* (Academic, New York, 1972), pp. 409–474
24. M. Crouzeix, *Approximation of Parabolic Equations* (2005). Lecture notes available at http://perso.univ-rennes1.fr/michel.crouzeix/.
25. M. Crouzeix, S. Larsson, S. Piskarëv, V. Thomée, The stability of rational approximations of analytic semigroups. BIT **33**, 74–84 (1993)
26. G. Dal Maso, A. Defranceschi, Correctors for the homogenization of monotone operators. Differ. Integr. Equ. **3**, 1151–1166 (1990)
27. L. Diening, C. Ebmeyer, M. Růžička, Optimal convergence for the implicit space-time discretization of parabolic systems with p-structure. SIAM J. Numer. Anal. **45**, 457–472 (2007) (electronic)
28. R. Du, P. Ming, Heterogeneous multiscale finite element method with novel numerical integration schemes. Commun. Math. Sci. **8**, 863–885 (2010)
29. Y. Efendiev, T.Y. Hou, *Multiscale Finite Element Methods. Theory and Applications*. Surveys and Tutorials in the Applied Mathematical Sciences, vol. 4 (Springer, New York, 2009)
30. Y. Efendiev, A. Pankov, Numerical homogenization of nonlinear random parabolic operators. Multiscale Model. Simul. **2**, 237–268 (2004)
31. B. Engquist, O. Runborg, Wavelet-based numerical homogenization with applications, in *Multiscale and Multiresolution Methods*. Lecture Notes in Computational Science and Engineering, vol. 20 (Springer, Berlin, 2002), pp. 97–148
32. M.G.D. Geers, V.G. Kouznetsova, W.A.M. Brekelmans, Multi-scale computational homogenization: trends and challenges. J. Comput. Appl. Math. **234**, 2175–2182 (2010)
33. A. Gloria, An analytical framework for the numerical homogenization of monotone elliptic operators and quasiconvex energies. Multiscale Model. Simul. **5**, 996–1043 (2006) (electronic)
34. A. Gloria, Reduction of the resonance error. Part 1: approximation of homogenized coefficients. Math. Models Methods Appl. Sci. **21**, 1601–1630 (2011)
35. E. Hairer, G. Wanner, *Solving Ordinary Differential Equations II. Stiff and Differential-Algebraic Problems* (Springer, Berlin/Heidelberg, 1996)
36. P. Henning, M. Ohlberger, Error control and adaptivity for heterogeneous multiscale approximations of nonlinear monotone problems. Discrete Cont. Dyn. Syst. Ser. S **8**, 119–150 (2015)
37. P. Henning, D. Peterseim, Oversampling for the multiscale finite element method. SIAM Multiscale Model. Simul. **11**, 1149–1175 (2013)
38. T.Y. Hou, X.-H. Wu, Z. Cai, Convergence of a multiscale finite element method for elliptic problems with rapidly oscillating coefficients. Math. Comput. **68**, 913–943 (1999)
39. T.J.R. Hughes, G.R. Feijóo, L. Mazzei, J.-B. Quincy, The variational multiscale method—a paradigm for computational mechanics. Comput. Methods Appl. Mech. Eng. **166**, 3–24 (1998)
40. V.V. Jikov, S.M. Kozlov, O.A. Oleinik, *Homogenization of Differential Operators and Integral Functionals* (Springer, Berlin/Heidelberg, 1994)
41. T. Kato, *Perturbation Theory for Linear Operators* (Springer, Berlin, 1976)
42. O.A. Ladyzhenskaya, *The Boundary Value Problems of Mathematical Physics*. Applied Mathematical Sciences, vol. 49 (Springer, New York, 1985)

43. V. Lebedev, How to solve stiff systems of differential equations by explicit methods, in Numerical Methods and Applications, ed. by G.I. Marčuk (CRC Press, Boca Raton, 1994), pp. 45–80
44. C. Lubich, A. Ostermann, Runge-Kutta approximation of quasi-linear parabolic equations. Math. Comput. **64**, 601–627 (1995)
45. A. Målqvist, A. Persson, Multiscale techniques for parabolic equations (2015). arXiv:1504.08140
46. A. Målqvist, D. Peterseim, Localization of elliptic multiscale problems. Math. Comput. **83**, 2583–2603 (2014)
47. P. Ming, P. Zhang, Analysis of the heterogeneous multiscale method for parabolic homogenization problems. Math. Comput. **76**, 153–177 (2007)
48. G. Nguetseng, A general convergence result for a functional related to the theory of homogenization. SIAM J. Math. Anal. **20**, 608–623 (1989)
49. H. Owhadi, L. Zhang, Metric-based upscaling. Commun. Pure Appl. Math. **60**, 675–723 (2007)
50. A. Pankov, *G-Convergence and Homogenization of Nonlinear Partial Differential Operators*. Mathematics and its Applications, vol. 422 (Kluwer Academic Publishers, Dordrecht, 1997)
51. B.P. Sommeijer, J.G. Verwer, A performance evaluation of a class of Runge-Kutta-Chebyshev methods for solving semi-discrete parabolic differential equations. Tech. rep., Stichting Mathematisch Centrum. Numerieke Wiskunde, Amsterdam, 1980. Report NW91/80
52. N. Svanstedt, *G*-convergence of parabolic operators. Nonlinear Anal. **36**, 807–842 (1999)
53. V. Thomée, *Galerkin Finite Element Methods for Parabolic Problems*. Springer Series in Computational Mathematics, vol. 25 (Springer, Berlin, 1997)
54. P. van der Houwen, B.P. Sommeijer, On the internal stage Runge-Kutta methods for large m-values. Z. Angew. Math. Mech. **60**, 479–485 (1980)
55. W. E, B. Engquist, The heterogeneous multiscale methods. Commun. Math. Sci. **1**, 87–132 (2003)
56. W. E, P. Ming, P. Zhang, Analysis of the heterogeneous multiscale method for elliptic homogenization problems. J. Am. Math. Soc. **18**, 121–156 (2005)
57. E. Zeidler, *Nonlinear Functional Analysis and Its Applications. II/B* (Springer, New York, 1990). Nonlinear monotone operators, Translated from the German by the author and Leo F. Boron

Virtual Element Implementation for General Elliptic Equations

Lourenco Beirão da Veiga, Franco Brezzi, Luisa Donatella Marini, and Alessandro Russo

Abstract In the present paper we detail the implementation of the Virtual Element Method for two dimensional elliptic equations in primal and mixed form with variable coefficients.

1 Introduction

The Virtual Element Method (VEM) is a recent generalization of the Finite Element Method that, in addition to other useful features, can easily handle general polygonal and polyhedral meshes. The interest in numerical methods that can use polytopal elements has a long and relevant history. We just recall the review works [3, 4, 14, 21, 22, 26, 27] and the references therein. However, the use of polytopes showed recently a significant growth both in the mathematical and in the engineering literature, with the emergence of a new class of methods where the traditional approach (based on the approximation and/or numerical integration of test and trial functions) was substituted by various alternative strategies based on suitable different formulations. Among these alternative frameworks (all, deep inside, very similar to each other) we could see the (older) Mimetic Finite Differences (see e.g. [9]

L. Beirão da Veiga • A. Russo (✉)
Department of Mathematics and Applications, University of Milano-Bicocca, via Cozzi 57, 20125 Milano, Italy

IMATI-CNR, via Ferrata 1, 27100 Pavia, Italy
e-mail: lourenco.beirao@unimib.it; alessandro.russo@unimib.it

F. Brezzi
Istituto di Matematica Applicata e Tecnologie Informatiche - Pavia, Consiglio Nazionale delle Ricerche, Pavia, Italy
e-mail: brezzi@imati.cnr.it

L.D. Marini
Department of Mathematics, University of Pavia, via Ferrata 1, 27100 Pavia, Italy

IMATI-CNR, via Ferrata 1, 27100 Pavia, Italy
e-mail: marini@imati.cnr.it

G.R. Barrenechea et al. (eds.), *Building Bridges: Connections and Challenges in Modern Approaches to Numerical Partial Differential Equations*, Lecture Notes in Computational Science and Engineering 114,
DOI 10.1007/978-3-319-41640-3_2

and the references therein), the Hybridizable Discontinuous Galerkin (see e.g. [18] and the references therein) the Gradient Schemes (see e.g. [20] and the references therein) the Weak Galerkin Methods (see e.g. [29] and the references therein), and the Hybrid High Order methods (see e.g. [19] and the references therein), together with the main object of the present paper: the Virtual Element Method.

The subject of polygonal and polyhedral mesh generation is a very active area of research on its own. Here we only refer to [28] for a simple and reliable MATLAB polygonal mesh generator in 2D, and to [24] and the references therein for some insights into the issues of the three-dimensional case.

Very briefly, the key idea of the Virtual Element Method is to adopt also non-polynomial shape functions (that are necessary in order to build conforming discrete spaces on complex polytopal grids) but avoiding their explicit computation, not even in an approximate way. This is achieved by introducing the right set of degrees of freedom and defining computable projection operators on polynomial spaces. In the initial paper [6] the Virtual Element Method was presented for the two dimensional Poisson problem in primal form, while the three dimensional case (still for constant coefficients) was discussed later in [1]. In the more recent papers [12] and [11] the Virtual Element Method was then extended to more general elliptic equations (including variable coefficients with the possible presence of convection and reaction term), respectively in primal and mixed form. At the same time, the method has been applied with success to a wide range of other problems. We just recall [2, 5, 7, 10, 13, 15–17, 23, 25].

The present work can be considered as a natural continuation of [8], where all the coding aspects of the model scheme presented in [6] and [1] where detailed. Here we describe all the tools for the practical implementation of the methods analysed in [12] and [11]. Since the assembly of the global matrix follows the same identical procedure as in the Finite Element case, the focus of this work is on the construction of the local matrices. After a brief description of the discrete spaces and the associated degrees of freedom, we detail step by step the implementation of the projection operators and all the other involved matrices. At the end of each part the reader can find an "algorithm" section where the whole procedure is summarized. Although we believe that the VEM is very elegant and, once some familiarity is acquired, quite easy to implement, we advice the reader to look into the previous work [8] before reading the present one.

The paper is organized as follows. After presenting some minimal notation in Sect. 2, we briefly describe in Sect. 3 the problem under consideration, including its primal and mixed variational formulations. In Sects. 4 and 5 we briefly recall the discrete spaces, the degrees of freedom and the construction of the projection operator of [6]. In Sect. 6 we detail the implementation of the method analysed in [12]; a useful summary can be found in Sect. 7. Section 8 is devoted to a brief description of the discrete spaces and of the degrees of freedom introduced in [11], while the implementation aspects are described in Sects. 9 and 10. A useful summary can be found in Sect. 11.

In this paper we have studied in details the implementation of the Virtual Element Method in two dimensions only. The extension to the three dimensional case does not present any major difficulties, as long as all the 2D machinery is developed with respect to each face of a general polyhedron. We will soon release a full MATLAB implementation for both the 2D and the 3D case.

2 Basic Notation

In the present section we introduce some minimal notation needed in the rest of the paper.

2.1 Polynomial Spaces

For a given a domain $\mathcal{D} \subset \mathbb{R}^d$ and an integer $k \geqslant 1$, we will denote by $\mathcal{P}_k(\mathcal{D})$ the linear space of polynomials of degree less than or equal to k. When $d = 2$, the dimension of $\mathcal{P}_k(\mathcal{D})$ will be denoted by n_k:

$$n_k := \dim \mathcal{P}_k(\mathcal{D}) = \frac{(k+1)(k+2)}{2}.$$

2.2 Polygons

A generic polygon will be denoted by E; the number of vertices will be denoted by N_V and the number of edges by N_e. Of course $N_e = N_V$, but it will be useful to keep separate names. The diameter of the polygon E will be denoted by h_E and its centroid by (x_c, y_c). The outward normal to E will be denoted by $\boldsymbol{n}_E$ or simply by $\boldsymbol{n}$ when no confusion can arise. The normal $\boldsymbol{n}_E$ restricted to ad edge e will be indicated by $\boldsymbol{n}_e$.

2.3 Scaled Monomials

Let $\boldsymbol{\alpha} = (\alpha_x, \alpha_y)$ be a multi-index. We define the *scaled monomial* $m_{\boldsymbol{\alpha}}$ on E by:

$$m_{\boldsymbol{\alpha}}(x, y) := \left(\frac{x - x_c}{h_E}\right)^{\alpha_x} \left(\frac{y - y_c}{h_E}\right)^{\alpha_y}. \tag{1}$$

For k an integer, let

$$\mathcal{M}_k(E) := \{m_{\boldsymbol{\alpha}},\ 0 \leqslant |\boldsymbol{\alpha}| \leqslant k\} \tag{2}$$

where $|\boldsymbol{\alpha}| = \alpha_x + \alpha_y$. With a small abuse of notation we will indicate with α (in contrast with boldface $\boldsymbol{\alpha}$) a linear index running from 1 to n_k. Obviously, $\mathcal{M}_k(E)$ is a basis for $\mathcal{P}_k(E)$.

2.4 Functional Spaces

The scalar product in $L^2(\mathcal{D})$ will be denoted by $(\cdot,\cdot)_{0,\mathcal{D}}$ or simply by $(\cdot,\cdot)$ when the domain is clear from the context.

3 The Elliptic Problem

Let $\Omega \subset \mathbb{R}^2$ be a bounded convex polygonal domain with boundary Γ, let κ and γ be smooth functions $\Omega \to \mathbb{R}$ with $\kappa(\boldsymbol{x}) \geqslant \kappa_0 > 0$ for all $\boldsymbol{x} \in \Omega$, and let $\boldsymbol{b}$ be a smooth vector field $\Omega \to \mathbb{R}^2$. We consider the following elliptic problem:

$$\begin{cases} \mathcal{L}p := \operatorname{div}(-\kappa\nabla p + \boldsymbol{b}p) + \gamma\, p = f & \text{in } \Omega \\ \qquad\qquad\qquad\qquad\qquad\quad p = 0 & \text{on } \Gamma. \end{cases} \tag{3}$$

We assume that problem (3) is solvable for any $f \in H^{-1}(\Omega)$, and that the a-priori estimate

$$\|p\|_{1,\Omega} \leqslant C\|f\|_{-1,\Omega} \tag{4}$$

and the regularity estimate

$$\|p\|_{2,\Omega} \leqslant C\|f\|_{0,\Omega} \tag{5}$$

hold with a constant C independent of f. As shown in [12] and [11], these hypotheses are sufficient to prove the convergence of the Virtual Element approximation, both in primal and in mixed form.

3.1 The Primal Variational Formulation

Set:

$$a(p,q) := \int_\Omega \kappa\,\nabla p\cdot\nabla q\,\mathrm{d}\boldsymbol{x}, \quad b(p,q) := -\int_\Omega p\,(\boldsymbol{b}\cdot\nabla q)\,\mathrm{d}\boldsymbol{x},$$
$$c(p,q) := \int_\Omega \gamma\,p\,q\,\mathrm{d}\boldsymbol{x}, \quad (f,q) = \int_\Omega f\,q\,\mathrm{d}\boldsymbol{x},$$

and define

$$B(p,q) := a(p,q) + b(p,q) + c(p,q). \tag{6}$$

The primal variational formulation of problem (3) is then

$$\begin{cases} \text{find } p \in V := H^1_0(\Omega) \quad \text{such that} \\ B(p,q) = (f,q) \quad \text{for all } q \in V. \end{cases} \tag{7}$$

3.2 *The Mixed Variational Formulation*

In order to build the mixed variational formulation of problem (3), we define

$$\nu := \kappa^{-1}, \quad \boldsymbol{\beta} := \kappa^{-1}\boldsymbol{b},$$

and re-write (3) as

$$\boldsymbol{u} = \nu^{-1}(-\nabla p + \boldsymbol{\beta} p), \quad \operatorname{div} \boldsymbol{u} + \gamma p = f \text{ in } \Omega, \quad p = 0 \text{ on } \Gamma. \tag{8}$$

Introducing the spaces

$$\boldsymbol{V} := \boldsymbol{H}(\operatorname{div}; \Omega), \quad \text{and} \quad Q := L^2(\Omega),$$

the mixed variational formulation of problem (3) is:

$$\begin{cases} \text{Find } (\boldsymbol{u}, p) \in \boldsymbol{V} \times Q \text{ such that} \\ (\nu \boldsymbol{u}, \boldsymbol{v}) - (p, \operatorname{div} \boldsymbol{v}) - (\boldsymbol{\beta} \cdot \boldsymbol{v}, p) = 0 \quad \text{for all } \boldsymbol{v} \in \boldsymbol{V}, \\ (\operatorname{div} \boldsymbol{u}, q) + (\gamma p, q) = (f, q) \quad \text{for all } q \in Q. \end{cases} \tag{9}$$

4 Approximation with the Virtual Element Method

The Virtual Element approximation of problems (7) and (9) fits in the classical conforming Galerkin methods: in principle, in both cases we define finite-dimensional subspaces $V_h \subset V$ (for problem (7)) and $\boldsymbol{V}_h \subset \boldsymbol{V}$, $Q_h \subset Q$ (for problem (9)) and we restrict the various bilinear forms to the spaces V_h and $\boldsymbol{V}_h \times Q_h$ respectively. However, given that for the VEM the functions are not explicitly known, we will also have to *approximate* the various bilinear forms.

As usual, the virtual spaces V_h, $\boldsymbol{V}_h$ and Q_h will be defined at the element level, and on the boundary of the elements the degrees of freedom will be chosen in such a way that they will nicely glue together.

Hence, given a polygon E of the decomposition, we will first define the local virtual spaces $V_h(E)$, $\boldsymbol{V}_h(E)$ and $Q_h(E)$ and then we will set

$$V_h = \{p \in V \text{ such that } p_{|E} \in V_h(E)\} \tag{10}$$

$$\boldsymbol{V}_h = \{\boldsymbol{v} \in \boldsymbol{V} \text{ such that } \boldsymbol{v}_{|E} \in \boldsymbol{V}_h(E)\} \tag{11}$$

$$Q_h = \{q \in Q \text{ such that } q_{|E} \in Q_h(E)\}. \tag{12}$$

Also the approximation of the various bilinear forms will be made element by element.

To encourage the reader, we point out that the space Q_h will consist, as usual in finite element methods, of piecewise discontinuous polynomials of degree k.

5 Virtual Element Space for the Primal Formulation

Before defining the local virtual space $V_h(E)$, we need to become familiar with the projection operator Π_k^∇ which will play a major role in the rest of the paper.

The operator Π_k^∇ is the orthogonal projection onto the space of polynomials of degree k with respect to the scalar product $\int_E \nabla p \cdot \nabla q \, \mathrm{d}\boldsymbol{x}$. Given a function $p \in H^1(E)$, the polynomial $\Pi_k^\nabla p$ is defined by the condition

$$\int_E \nabla(\Pi_k^\nabla p - p) \cdot \nabla r_k \, \mathrm{d}\boldsymbol{x} = 0 \quad \text{for all } r_k \in \mathcal{P}_k(E). \tag{13}$$

When r_k is a constant, condition (13) is the identity $0 \equiv 0$ so the polynomial $\Pi_k^\nabla p$ itself is determined up to a constant. This is fixed by imposing an extra condition, for instance,

$$\int_{\partial E} (\Pi_k^\nabla p - p) \, \mathrm{d}s = 0. \tag{14}$$

The following easy lemma will be useful throughout the section:

Lemma 1 *The polynomial $\Pi_k^\nabla p$ depends only on*

- *the value of p on the boundary of E;*
- *the moments of p in E up to order $k-2$.*

Proof By Eqs. (13) and (14) it is clear that the polynomial $\Pi_k^\nabla p$ is completely determined by the integrals

$$\int_E \nabla p \cdot \nabla r_k \, \mathrm{d}\boldsymbol{x} \qquad \text{and} \qquad \int_{\partial E} p \, \mathrm{d}s.$$

The second integral clearly depends only on the value of p on the boundary of E. Concerning the first integral, integrating by parts we have

$$\int_E \nabla p \cdot \nabla r_k \, \mathrm{d}\boldsymbol{x} = -\int_E p \, \Delta r_k \, \mathrm{d}\boldsymbol{x} + \int_{\partial E} p \, \frac{\partial r_k}{\partial n} \, \mathrm{d}s$$

and since $\Delta r_k \in \mathcal{P}_{k-2}(E)$ the proof is completed.

We are now ready to introduce the local virtual space $V_h(E)$. The space $V_h(E)$ consists of functions p_h such that:

- p_h is continuous on E;
- p_h on each edge e is a polynomial of degree k;
- $\Delta p_h \in \mathcal{P}_k(E)$;
- $\int_E p_h \, m_{\boldsymbol{\alpha}} \, \mathrm{d}\boldsymbol{x} = \int_E \Pi_k^{\nabla} p_h \, m_{\boldsymbol{\alpha}} \, \mathrm{d}\boldsymbol{x}$ for $|\boldsymbol{\alpha}| = n_k - 1$ and $|\boldsymbol{\alpha}| = n_k$.

In [1, 8] we have shown the following results:

1. $V_h(E)$ has dimension $N_V + (k-1)N_e + n_{k-2} = kN_V + n_{k-2}$;
2. $\mathcal{P}_k(E) \subset V_h(E)$;
3. for the space $V_h(E)$ we can take the following degrees of freedom:

Boundary degrees of freedom [$N_V + (k-1) \times N_e = k \times N_V$]

- the values of p_h at the N_V vertices of the polygon E;
- for each edge e, the values of p_h at $k-1$ distinct points of e (for instance equispaced points).

Internal degrees of freedom (only for $k > 1$) [n_{k-2}]

- the moments of p_h up to degree $k-2$, i.e. the integrals

$$\frac{1}{|E|} \int_E p_h \, m_{\boldsymbol{\alpha}} \, \mathrm{d}\boldsymbol{x}, \quad |\boldsymbol{\alpha}| \leqslant k-2.$$

We will indicate by $\mathrm{dof}_i(p_h)$ ($i = 1, \ldots, N_{\mathrm{dof}} := \dim V_h(E)$) the degrees of freedom of p_h. We define the *local basis functions* $\phi_i \in V_h(E)$, $i = 1, \ldots, N_{\mathrm{dof}}$, by the property:

$$\mathrm{dof}_i(\phi_j) = \delta_{ij} \quad i,j = 1, \ldots, N_{\mathrm{dof}} \tag{15}$$

so that we have a Lagrange-type decomposition:

$$p_h = \sum_{i=1}^{N_{\mathrm{dof}}} \mathrm{dof}_i(p_h) \, \phi_i. \tag{16}$$

Given a function $p_h \in V_h(E)$, by Lemma 1 the polynomial $\Pi_k^\nabla p_h$ depends only on the value of p_h on the boundary of E and on the moments of p_h in E up to order $k-2$. Hence, the polynomial $\Pi_k^\nabla p_h$ depends only on the degrees of freedom of p_h. In [8] it is shown that also the L^2 projection $\Pi_k^0 p_h$ of a function $p_h \in V_h(E)$ onto $\mathcal{P}_k(E)$ depends only on its degrees of freedom, and all the details to compute and code $\Pi_k^\nabla \phi_i$ and $\Pi_k^0 \phi_i$, for a generic basis function ϕ_i, are given. For the convenience of the reader we report here the various steps. Write

$$\Pi_k^\nabla \phi_i = \sum_{\alpha=1}^{n_k} s_i^\alpha m_\alpha, \quad i = 1, \dots N_{\text{dof}} \tag{17}$$

and define

$$\mathrm{P}_0 \phi_i := \int_{\partial E} \phi_i \, \mathrm{d}s.$$

Then, defining

$$\mathbf{G} = \begin{bmatrix} \mathrm{P}_0 m_1 & \mathrm{P}_0 m_2 & \dots & \mathrm{P}_0 m_{n_k} \\ 0 & (\nabla m_2, \nabla m_2)_{0,E} & \dots & (\nabla m_2, \nabla m_{n_k})_{0,E} \\ \vdots & \vdots & \ddots & \vdots \\ 0 & (\nabla m_{n_k}, \nabla m_2)_{0,E} & \dots & (\nabla m_{n_k}, \nabla m_{n_k})_{0,E} \end{bmatrix}, \tag{18}$$

$$\boldsymbol{b}_i = \begin{bmatrix} \mathrm{P}_0 \phi_i \\ (\nabla m_2, \nabla \phi_i)_{0,E} \\ \vdots \\ (\nabla m_{n_k}, \nabla \phi_i)_{0,E} \end{bmatrix}, \tag{19}$$

for each i, the coefficients s_i^α, $\alpha = 1, \dots, n_k$ are solution of the $n_k \times n_k$ linear system:

$$\mathbf{G} \boldsymbol{s}_i = \boldsymbol{b}_i.$$

Denoting by $\mathbf{B}$ the $n_k \times N_{\text{dof}}$ matrix given by

$$\mathbf{B} := \begin{bmatrix} \boldsymbol{b}_1 \ \boldsymbol{b}_2 \ \dots \ \boldsymbol{b}_{N_{\text{dof}}} \end{bmatrix} = \begin{bmatrix} \mathrm{P}_0 \phi_1 & \dots & \mathrm{P}_0 \phi_{N_{\text{dof}}} \\ (\nabla m_2, \nabla \phi_1)_{0,E} & \dots & (\nabla m_2, \nabla \phi_{N_{\text{dof}}})_{0,E} \\ \vdots & \ddots & \vdots \\ (\nabla m_{n_k}, \nabla \phi_1)_{0,E} & \dots & (\nabla m_{n_k}, \nabla \phi_{N_{\text{dof}}})_{0,E} \end{bmatrix}, \tag{20}$$

the matrix representation $\overset{*}{\boldsymbol{\Pi}}{}_k^{\nabla}$ of the operator Π_k^{∇} acting from $V_h(E)$ to $\mathcal{P}_k(E)$ in the basis $\mathcal{M}_k(E)$ is given by $(\overset{*}{\boldsymbol{\Pi}}{}_k^{\nabla})_{\alpha i} = s_i^{\alpha}$, that is,

$$\overset{*}{\boldsymbol{\Pi}}{}_k^{\nabla} = \mathbf{G}^{-1}\mathbf{B}. \tag{21}$$

We will also need the matrix representation, in the basis (15), of the same operator Π_k^{∇}, this time thought as an operator $V_h(E) \longrightarrow V_h(E)$. Hence, let

$$\Pi_k^{\nabla}\phi_i = \sum_{j=1}^{N_{\text{dof}}} \pi_i^j \phi_j, \quad i = 1, \dots N_{\text{dof}},$$

with

$$\pi_i^j = \text{dof}_j\big(\Pi_k^{\nabla}\phi_i\big).$$

From (17) and (16) we have

$$\Pi_k^{\nabla}\phi_i = \sum_{\alpha=1}^{n_k} s_i^{\alpha} m_{\alpha} = \sum_{\alpha=1}^{n_k} s_i^{\alpha} \sum_{j=1}^{N_{\text{dof}}} \text{dof}_j(m_{\alpha})\,\phi_j$$

so that

$$\pi_i^j = \sum_{\alpha=1}^{n_k} s_i^{\alpha}\, \text{dof}_j(m_{\alpha}). \tag{22}$$

In order to express (22) in matrix form, we define the $N_{\text{dof}} \times n_k$ matrix $\mathbf{D}$ by:

$$\mathbf{D}_{i\alpha} := \text{dof}_i(m_{\alpha}), \quad i = 1, \dots, N_{\text{dof}}, \quad \alpha = 1, \dots, n_k,$$

that is,

$$\mathbf{D} = \begin{bmatrix} \text{dof}_1(m_1) & \text{dof}_1(m_2) & \dots & \text{dof}_1(m_{n_k}) \\ \text{dof}_2(m_1) & \text{dof}_2(m_2) & \dots & \text{dof}_2(m_{n_k}) \\ \vdots & \vdots & \ddots & \vdots \\ \text{dof}_{N_{\text{dof}}}(m_1) & \text{dof}_{N_{\text{dof}}}(m_2) & \dots & \text{dof}_{N_{\text{dof}}}(m_{n_k}) \end{bmatrix}. \tag{23}$$

Equation (22) becomes:

$$\pi_i^j = \sum_{\alpha=1}^{n_k} (\mathbf{G}^{-1}\mathbf{B})_{\alpha i}\mathbf{D}_{j\alpha} = (\mathbf{D}\mathbf{G}^{-1}\mathbf{B})_{ji}.$$

Hence, the *matrix representation* $\mathbf{\Pi}_k^{\nabla}$ of the operator $\Pi_k^{\nabla} : V_h(E) \longrightarrow V_h(E)$ in the basis (15), is given by

$$\mathbf{\Pi}_k^{\nabla} = \mathbf{D}\mathbf{G}^{-1}\mathbf{B} = \mathbf{D}\overset{*}{\mathbf{\Pi}}{}_k^{\nabla}. \tag{24}$$

Remark 1 We point out that, as shown in [8], the matrix $\mathbf{G}$ can be expressed in terms of the matrices $\mathbf{D}$ and $\mathbf{B}$ as

$$\mathbf{G} = \mathbf{B}\mathbf{D}. \tag{25}$$

Always following [8], we can show that also the L^2 projection onto $\mathcal{P}_k(E)$ of a function $p_h \in V_h(E)$ depends only on its degrees of freedom. If we write

$$\Pi_k^0 \phi_i = \sum_{i=1}^{N_{\text{dof}}} t_i^\alpha m_\alpha,$$

and define

$$\mathbf{H} = \begin{bmatrix} (m_1, m_1)_{0,E} & (m_1, m_2)_{0,E} & \dots & (m_1, m_{n_k})_{0,E} \\ (m_2, m_1)_{0,E} & (m_2, m_2)_{0,E} & \dots & (m_2, m_{n_k})_{0,E} \\ \vdots & \vdots & \ddots & \vdots \\ (m_{n_k}, m_1)_{0,E} & (m_{n_k}, m_2)_{0,E} & \dots & (m_{n_k}, m_{n_k})_{0,E} \end{bmatrix}, \tag{26}$$

$$\boldsymbol{c}_i = \begin{bmatrix} (m_1, \phi_i)_{0,E} \\ (m_2, \phi_i)_{0,E} \\ \vdots \\ (m_{n_k}, \phi_i)_{0,E} \end{bmatrix}, \tag{27}$$

then, for each i, the coefficients t_i^α, $\alpha = 1, \dots, n_k$ are solution of the $n_k \times n_k$ linear system:

$$\mathbf{H}\boldsymbol{t}_i = \boldsymbol{c}_i, \tag{28}$$

which descends directly from the definition of the L^2-projection.

We denote by $\mathbf{C}$ the $n_k \times N_{\text{dof}}$ matrix given by

$$\mathbf{C} := \begin{bmatrix} \boldsymbol{c}_1 \ \boldsymbol{c}_2 \ \dots \ \boldsymbol{c}_{N_{\text{dof}}} \end{bmatrix} = \begin{bmatrix} (m_1, \phi_1)_{0,E} & (m_1, \phi_2)_{0,E} & \dots & (m_1, \phi_{N_{\text{dof}}})_{0,E} \\ (m_2, \phi_1)_{0,E} & (m_2, \phi_2)_{0,E} & \dots & (m_2, \phi_{N_{\text{dof}}})_{0,E} \\ \vdots & \vdots & \ddots & \vdots \\ (m_{n_k}, \phi_1)_{0,E} & (m_{n_k}, \phi_2)_{0,E} & \dots & (m_{n_k}, \phi_{N_{\text{dof}}})_{0,E} \end{bmatrix}. \tag{29}$$

The first n_{k-2} lines of the matrix $\mathbf{C}$ can be computed directly from the degrees of freedom, and the resulting matrix is

$$\text{first } n_{k-2} \text{ lines of } \mathbf{C} = |E| \left[\begin{array}{cccccc|cccc} 0 & 0 & \dots & 0 & 0 & & 1 & 0 & \dots & 0 \\ 0 & 0 & \dots & 0 & 0 & & 0 & 1 & \dots & 0 \\ \vdots & \vdots & \ddots & \vdots & \vdots & & \vdots & \vdots & \ddots & \vdots \\ 0 & 0 & \dots & 0 & 0 & & 0 & 0 & \dots & 1 \end{array}\right]$$

where the rightmost block is the identity matrix of size $n_{k-2} \times n_{k-2}$. The last $n_k - n_{k-2}$ lines of the matrix $\mathbf{C}$ correspond to m_α being a monomial of degree $k-1$ or k and we need to resort to the fundamental property

$$\int_E \phi_i \, m_\alpha \, \mathrm{d}\boldsymbol{x} = \int_E \Pi_k^\nabla \phi_i \, m_\alpha \, \mathrm{d}\boldsymbol{x}.$$

Hence in this case we have

$$\mathbf{C}_{\alpha i} = (\mathbf{H}\mathbf{G}^{-1}\mathbf{B})_{\alpha i}, \quad n_{k-2} < \alpha \leqslant n_k.$$

It follows that the matrix representation $\overset{*}{\boldsymbol{\Pi}}{}_k^0$ of the operator Π_k^0 acting from $V_h(E)$ to $\mathcal{P}_k(E)$ in the basis $\mathcal{M}_k(E)$ is given by $(\overset{*}{\boldsymbol{\Pi}}{}_k^0)_{\alpha i} = t_i^\alpha$, that is,

$$\overset{*}{\boldsymbol{\Pi}}{}_k^0 = \mathbf{H}^{-1}\mathbf{C}. \tag{30}$$

Arguing as before, the matrix representation, in the basis (15), of the same operator Π_k^0, this time thought as an operator $V_h(E) \longrightarrow V_h(E)$, is

$$\boldsymbol{\Pi}_k^0 = \mathbf{D}\mathbf{H}^{-1}\mathbf{C} = \mathbf{D}\overset{*}{\boldsymbol{\Pi}}{}_k^0. \tag{31}$$

In a similar fashion we can also compute the matrix representations $\overset{*}{\boldsymbol{\Pi}}{}_{k-1}^0$ and $\boldsymbol{\Pi}_{k-1}^0$ of the L^2 projection onto the space of polynomials of degree $k-1$. To this end, we consider:

- the $n_{k-1} \times n_{k-1}$ matrix $\mathbf{H}'$ obtained by taking the first n_{k-1} rows and the first n_{k-1} columns of the matrix $\mathbf{H}$ defined in (26);
- the $n_{k-1} \times N_{\text{dof}}$ matrix $\mathbf{C}'$ obtained by taking the first n_{k-1} lines of the matrix $\mathbf{C}$ defined in (29);
- the $N_{\text{dof}} \times n_{k-1}$ matrix $\mathbf{D}'$ obtained by taking the first n_{k-1} columns of the matrix $\mathbf{D}$ defined in (23).

Then we have:

$$\overset{*}{\boldsymbol{\Pi}}{}_{k-1}^0 = (\mathbf{H}')^{-1}\mathbf{C}' \quad \text{and} \quad \boldsymbol{\Pi}_{k-1}^0 = \mathbf{D}'\,\overset{*}{\boldsymbol{\Pi}}{}_{k-1}^0.$$

To summarize, given a "virtual" function $p_h \in V_h(E)$, we can compute the polynomials $\Pi_k^\nabla p_h$, $\Pi_k^0 p_h$ and $\Pi_{k-1}^0 p_h$ in terms of its degrees of freedom.

6 VEM Approximation of the Primal Formulation

As shown in [6], the projectors Π_k^∇ and Π_{k-1}^0 allow us to solve the Laplace equation with a reaction term. Indeed, according to [1], if problem (3) reduces to

$$\begin{cases} -\Delta p + \gamma p = f & \text{in } \Omega \\ u = g & \text{on } \partial\Omega \end{cases}$$

then we have

$$a(p,q) := \int_\Omega \nabla p \cdot \nabla q \, \mathrm{d}\boldsymbol{x}, \quad b(p,q) := 0, \quad c(p,q) := \int_\Omega \gamma \, p \, q \, \mathrm{d}\boldsymbol{x}.$$

The local VEM approximation for $a(\cdot,\cdot)$ is

$$a_h^E(p_h, q_h) := \int_E \nabla \Pi_k^\nabla p_h \cdot \nabla \Pi_k^\nabla q_h \, \mathrm{d}\boldsymbol{x} + \mathcal{S}_E\big((I - \Pi_k^\nabla)p_h, (I - \Pi_k^\nabla)q_h\big)$$

where the stability term $\mathcal{S}_E(\cdot,\cdot)$ is the symmetric and positive definite bilinear form which is the identity on the basis function, i.e. $\mathcal{S}_E(\phi_i, \phi_j) = \delta_{ij}$. The local VEM approximation for $c(\cdot,\cdot)$ is

$$c_h^E(p_h, q_h) := \int_E \gamma \, \Pi_{k-1}^0 p_h \, \Pi_{k-1}^0 q_h \, \mathrm{d}\boldsymbol{x}$$

and similarly the load term (f, q_h) is approximated locally by $(f, \Pi_{k-1}^0 q_h)_{0,E}$.

If the diffusion κ is not constant or a first-order term is present, then we cannot simply approximate ∇p_h with $\nabla \Pi_k^\nabla p_h$; as shown in [12], we would loose the optimal convergence rates. Instead, we should approximate

$$\nabla p_h \qquad \text{with} \qquad \Pi_{k-1}^0 \nabla p_h.$$

Note that for $k = 1$ the two approximations of ∇p_h coincide; in fact,

$$\nabla \Pi_1^\nabla p_h = \frac{1}{|E|} \int_E \nabla p_h \, \mathrm{d}\boldsymbol{x} = \Pi_0^0 \nabla p_h.$$

We will see now how to compute $\Pi_{k-1}^0 \nabla p_h$ in terms of the degrees of freedom. To this end, we observe that in order to obtain $\Pi_{k-1}^0 \nabla p_h$, we need to compute

$$\int_E \nabla p_h \cdot \boldsymbol{r}_{k-1} \, \mathrm{d}\boldsymbol{x}$$

where $\boldsymbol{r}_{k-1}$ is any vector whose components are polynomials of degree $k-1$. Integrating by parts, we have

$$\int_E \nabla p_h \cdot \boldsymbol{r}_{k-1}\,\mathrm{d}\boldsymbol{x} = -\int_E p_h \operatorname{div} \boldsymbol{r}_{k-1}\,\mathrm{d}\boldsymbol{x} + \int_{\partial E} p_h\, \boldsymbol{r}_{k-1} \cdot \boldsymbol{n}\,\mathrm{d}s$$

and since $\operatorname{div} \boldsymbol{r}_{k-1} \in \mathcal{P}_{k-2}(E)$, both integrals are directly computable from the degrees of freedom of p_h. In order to find the matrix representations of the operator $\Pi^0_{k-1}\nabla$, we define the $n_{k-1} \times N_{\mathrm{dof}}$ matrix $\overset{*}{\mathbf{\Pi}}{}^{0,x}_{k-1}$ by

$$\Pi^0_{k-1}\phi_{i,x} = \sum_{\alpha=1}^{n_{k-1}} \left(\overset{*}{\mathbf{\Pi}}{}^{0,x}_{k-1}\right)_{\alpha i} m_\alpha. \tag{32}$$

The polynomial $\Pi^0_{k-1}\phi_{i,x}$ is defined by

$$\int_E \Pi^0_{k-1}\phi_{i,x}\, m_\beta\,\mathrm{d}\boldsymbol{x} = \int_E \phi_{i,x}\, m_\beta\,\mathrm{d}\boldsymbol{x}, \quad \beta = 1,\dots,n_{k-1}$$

which becomes the linear system

$$\sum_{\alpha=1}^{n_{k-1}} \left(\overset{*}{\mathbf{\Pi}}{}^{0,x}_{k-1}\right)_{\alpha i} \int_E m_\alpha\, m_\beta\,\mathrm{d}\boldsymbol{x} = \int_E \phi_{i,x}\, m_\beta\,\mathrm{d}\boldsymbol{x}, \quad \beta = 1,\dots,n_{k-1}.$$

The term $\int_E \phi_{i,x}\, m_\beta\,\mathrm{d}\boldsymbol{x}$ can be computed integrating by parts:

$$\int_E \phi_{i,x}\, m_\beta\,\mathrm{d}\boldsymbol{x} = -\int_E \phi_i\, m_{\beta,x}\,\mathrm{d}\boldsymbol{x} + \int_{\partial E} \phi_i\, m_\beta\, \boldsymbol{n}_x. \tag{33}$$

If we define the matrices $\mathbf{E}^x$ and $\mathbf{E}^y$ by

$$\left(\mathbf{E}^x\right)_{i\beta} = \int_E \phi_{i,x}\, m_\beta\,\mathrm{d}\boldsymbol{x}, \quad \left(\mathbf{E}^y\right)_{i\beta} = \int_E \phi_{i,y}\, m_\beta\,\mathrm{d}\boldsymbol{x}, \quad \beta = 1,\dots n_{k-1} \tag{34}$$

then we have:

$$\overset{*}{\mathbf{\Pi}}{}^{0,x}_{k-1} = \hat{\mathbf{H}}^{-1}\mathbf{E}^x, \quad \overset{*}{\mathbf{\Pi}}{}^{0,y}_{k-1} = \hat{\mathbf{H}}^{-1}\mathbf{E}^y$$

where $\hat{\mathbf{H}}$ is the submatrix of $\mathbf{H}$ defined in (26) obtained taking the first n_{k-1} rows and columns of $\mathbf{H}$.

We can now compute the local VEM stiffness matrices for the variable coefficient case.

6.1 Diffusion Term

We have:

$$(\mathbf{K}^a)_{ij} := a_h^E(\phi_j, \phi_i) = \int_E \kappa\, \Pi_{k-1}^0 \nabla\phi_j \cdot \Pi_{k-1}^0 \nabla\phi_i \,\mathrm{d}\boldsymbol{x} + \bar{\kappa}\, \mathcal{S}_E\big((I - \Pi_k^\nabla)\phi_j, (I - \Pi_k^\nabla)\phi_i\big)$$

where $\bar{\kappa}$ is a constant approximation of κ (for instance, the mean value). We compute separately the consistency term and the stability term.

- **consistency term:**

$$(\mathbf{K}_\mathbf{c}^a)_{ij} := \int_E \kappa\, \Pi_{k-1}^0 \nabla\phi_j \cdot \Pi_{k-1}^0 \nabla\phi_i \,\mathrm{d}\boldsymbol{x} = \int_E \kappa \left\{[\Pi_{k-1}^0 \phi_{j,x}][\Pi_{k-1}^0 \phi_{i,x}] + [\Pi_{k-1}^0 \phi_{j,y}][\Pi_{k-1}^0 \phi_{i,y}]\right\} \mathrm{d}\boldsymbol{x}$$

  and

$$\int_E \kappa\, [\Pi_{k-1}^0 \phi_{j,x}][\Pi_{k-1}^0 \phi_{i,x}] \,\mathrm{d}\boldsymbol{x} = \sum_{\alpha,\beta=1}^{n_{k-1}} \big(\overset{*}{\mathbf{\Pi}}{}_{k-1}^{0,x}\big)_{\alpha j} \big(\overset{*}{\mathbf{\Pi}}{}_{k-1}^{0,x}\big)_{\beta i} \int_E \kappa\, m_\alpha\, m_\beta \,\mathrm{d}\boldsymbol{x},$$

$$\int_E \kappa\, [\Pi_{k-1}^0 \phi_{j,y}][\Pi_{k-1}^0 \phi_{i,y}] \,\mathrm{d}\boldsymbol{x} = \sum_{\alpha,\beta=1}^{n_{k-1}} \big(\overset{*}{\mathbf{\Pi}}{}_{k-1}^{0,y}\big)_{\alpha j} \big(\overset{*}{\mathbf{\Pi}}{}_{k-1}^{0,y}\big)_{\beta i} \int_E \kappa\, m_\alpha\, m_\beta \,\mathrm{d}\boldsymbol{x}.$$

  If we define the $n_{k-1} \times n_{k-1}$ matrix $\mathbf{H}^\kappa$ by

$$(\mathbf{H}^\kappa)_{\alpha\beta} := \int_E \kappa\, m_\alpha\, m_\beta \,\mathrm{d}\boldsymbol{x}, \quad 1 \leqslant \alpha, \beta \leqslant n_{k-1},$$

  then we have

$$\mathbf{K}_\mathbf{c}^a = \big(\overset{*}{\mathbf{\Pi}}{}_{k-1}^{0,x}\big)^{\mathrm{T}} \mathbf{H}^\kappa\, \overset{*}{\mathbf{\Pi}}{}_{k-1}^{0,x} + \big(\overset{*}{\mathbf{\Pi}}{}_{k-1}^{0,y}\big)^{\mathrm{T}} \mathbf{H}^\kappa\, \overset{*}{\mathbf{\Pi}}{}_{k-1}^{0,y}$$

  which can be written as

$$\mathbf{K}_\mathbf{c}^a = \left[\big(\overset{*}{\mathbf{\Pi}}{}_{k-1}^{0,x}\big)^{\mathrm{T}} \;\; \big(\overset{*}{\mathbf{\Pi}}{}_{k-1}^{0,y}\big)^{\mathrm{T}}\right] \begin{bmatrix} \mathbf{H}^\kappa & \mathbf{0} \\ \mathbf{0} & \mathbf{H}^\kappa \end{bmatrix} \begin{bmatrix} \overset{*}{\mathbf{\Pi}}{}_{k-1}^{0,x} \\ \overset{*}{\mathbf{\Pi}}{}_{k-1}^{0,y} \end{bmatrix}. \tag{35}$$

- **stability term:**

$$
\begin{aligned}
(\mathbf{K}^a_{\mathsf{s}})_{ij} &:= \bar{\kappa}\, \mathcal{S}_E\big((I-\Pi^\nabla_k)\phi_j, (I-\Pi^\nabla_k)\phi_i\big) \\
&= \bar{\kappa} \sum_{k,\ell=1}^{N_{\text{dof}}} \big(\delta_{jk} - (\boldsymbol{\Pi}^\nabla_k)_{jk}\big)\, \mathcal{S}_E(\phi_k, \phi_\ell)\, \big(\delta_{i\ell} - (\boldsymbol{\Pi}^\nabla_k)_{i\ell}\big) \\
&= \bar{\kappa} \sum_{\ell=1}^{N_{\text{dof}}} \big(\delta_{j\ell} - (\boldsymbol{\Pi}^\nabla_k)_{j\ell}\big)\big(\delta_{i\ell} - (\boldsymbol{\Pi}^\nabla_k)_{i\ell}\big)
\end{aligned}
$$

i.e.

$$
\mathbf{K}^a_{\mathsf{s}} = \bar{\kappa}\, (\mathbf{I} - \boldsymbol{\Pi}^\nabla_k)^{\mathrm{T}} (\mathbf{I} - \boldsymbol{\Pi}^\nabla_k). \tag{36}
$$

If the diffusion κ happens to be a 2×2 symmetric matrix, i.e.

$$
\kappa = \begin{bmatrix} \kappa_{xx} & \kappa_{xy} \\ \kappa_{xy} & \kappa_{yy} \end{bmatrix},
$$

then we can proceed similarly by defining the $n_{k-1} \times n_{k-1}$ matrices $\mathbf{H}^{\kappa_{xx}}$, $\mathbf{H}^{\kappa_{xy}}$ and $\mathbf{H}^{\kappa_{yy}}$ as follows:

$$
(\mathbf{H}^{\kappa_{xx}})_{\alpha\beta} := \int_E \kappa_{xx}\, m_\alpha\, m_\beta \,\mathrm{d}\boldsymbol{x}, \quad (\mathbf{H}^{\kappa_{xy}})_{\alpha\beta} := \int_E \kappa_{xy}\, m_\alpha\, m_\beta \,\mathrm{d}\boldsymbol{x}, \quad \ldots
$$

and the local virtual diffusion consistency matrix $\mathbf{K}^a_{\mathsf{c}}$ can be written as

$$
\mathbf{K}^a_{\mathsf{c}} = \Big[(\overset{*}{\boldsymbol{\Pi}}{}^{0,x}_{k-1})^{\mathrm{T}} \;\; (\overset{*}{\boldsymbol{\Pi}}{}^{0,y}_{k-1})^{\mathrm{T}} \Big] \begin{bmatrix} \mathbf{H}^{\kappa_{xx}} & \mathbf{H}^{\kappa_{xy}} \\ \mathbf{H}^{\kappa_{xy}} & \mathbf{H}^{\kappa_{yy}} \end{bmatrix} \begin{bmatrix} \overset{*}{\boldsymbol{\Pi}}{}^{0,x}_{k-1} \\ \overset{*}{\boldsymbol{\Pi}}{}^{0,y}_{k-1} \end{bmatrix}.
$$

In this case, the stability matrix $\mathbf{K}^a_{\mathsf{s}}$ can still be taken of the form (36), where this time the constant scalar $\bar{\kappa}$ can be defined as the arithmetic mean of the mean values of κ_{xx} and κ_{yy}. Note that here we are not considering the problem of optimizing the stability matrix with respect to the anisotropy of the diffusion matrix κ, but we are only interested in the convergence as h goes to zero.

6.2 Transport Term

The local VEM approximation for the transport term is

$$
b^E_h(p_h, q_h) := -\int_E \Pi^0_{k-1} p_h\, (\boldsymbol{b} \cdot \Pi^0_{k-1} \nabla q_h)\, \mathrm{d}\boldsymbol{x}
$$

and the corresponding local matrix is

$$(\mathsf{K}^{b})_{ij} := b_h^E(\phi_j, \phi_i) = -\int_E \Pi^0_{k-1}\phi_j \,(\boldsymbol{b}\cdot \Pi^0_{k-1}\nabla\phi_i)\;\mathrm{d}\boldsymbol{x}.$$

Define the $n_{k-1} \times n_{k-1}$ matrices H^{b_x} and H^{b_y} by

$$(\mathsf{H}^{b_x})_{\alpha\beta} := \int_E \boldsymbol{b}_x\, m_\alpha\, m_\beta\, \mathrm{d}\boldsymbol{x}, \quad (\mathsf{H}^{b_y})_{\alpha\beta} := \int_E \boldsymbol{b}_y\, m_\alpha\, m_\beta\, \mathrm{d}\boldsymbol{x}.$$

By (32) we have

$$\begin{aligned}\boldsymbol{b}\cdot[\Pi^0_{k-1}\nabla\phi_i] &= \boldsymbol{b}_x[\Pi^0_{k-1}\nabla\phi_{i,x}] + \boldsymbol{b}_y[\Pi^0_{k-1}\nabla\phi_{i,y}]\\ &= \boldsymbol{b}_x\sum_{\beta=1}^{n_{k-1}}(\overset{*}{\mathbf{\Pi}}{}^{\mathbf{0},x}_{\boldsymbol{k-1}})_{\beta i}\, m_\beta + \boldsymbol{b}_y\sum_{\beta=1}^{n_{k-1}}(\overset{*}{\mathbf{\Pi}}{}^{\mathbf{0},y}_{\boldsymbol{k-1}})_{\beta i}\, m_\beta\end{aligned}$$

so that

$$\begin{aligned}
&-\int_E \Pi^0_{k-1}\phi_j\,(\boldsymbol{b}\cdot\Pi^0_{k-1}\nabla\phi_i)\;\mathrm{d}\boldsymbol{x} =\\
&-\int_E\left[\sum_{\alpha=1}^{n_{k-1}}(\mathbf{\Pi}^{\mathbf{0}}_{\boldsymbol{k-1}})_{\alpha j}\, m_\alpha\right]\left[\boldsymbol{b}_x\sum_{\beta=1}^{n_{k-1}}(\overset{*}{\mathbf{\Pi}}{}^{\mathbf{0},x}_{\boldsymbol{k-1}})_{\beta i}\, m_\beta + \boldsymbol{b}_y\sum_{\beta=1}^{n_{k-1}}(\overset{*}{\mathbf{\Pi}}{}^{\mathbf{0},y}_{\boldsymbol{k-1}})_{\beta i}\, m_\beta\right]\mathrm{d}\boldsymbol{x} =\\
&-\int_E\left\{\boldsymbol{b}_x\sum_{\alpha,\beta=1}^{n_{k-1}}(\mathbf{\Pi}^{\mathbf{0}}_{\boldsymbol{k-1}})_{\alpha j}\,(\overset{*}{\mathbf{\Pi}}{}^{\mathbf{0},x}_{\boldsymbol{k-1}})_{\beta i}\, m_\beta m_\alpha + \boldsymbol{b}_y\sum_{\alpha,\beta=1}^{n_{k-1}}(\mathbf{\Pi}^{\mathbf{0}}_{\boldsymbol{k-1}})_{\alpha j}\,(\overset{*}{\mathbf{\Pi}}{}^{\mathbf{0},y}_{\boldsymbol{k-1}})_{\beta i}\, m_\beta m_\alpha\right\}\mathrm{d}\boldsymbol{x} =\\
&-\sum_{\alpha,\beta=1}^{n_{k-1}}(\mathbf{\Pi}^{\mathbf{0}}_{\boldsymbol{k-1}})_{\alpha j}\,(\overset{*}{\mathbf{\Pi}}{}^{\mathbf{0},x}_{\boldsymbol{k-1}})_{\beta i}\int_E \boldsymbol{b}_x m_\beta m_\alpha\,\mathrm{d}\boldsymbol{x} - \sum_{\alpha,\beta=1}^{n_{k-1}}(\mathbf{\Pi}^{\mathbf{0}}_{\boldsymbol{k-1}})_{\alpha j}\,(\overset{*}{\mathbf{\Pi}}{}^{\mathbf{0},y}_{\boldsymbol{k-1}})_{\beta i}\int_E \boldsymbol{b}_y m_\beta m_\alpha\,\mathrm{d}\boldsymbol{x} =\\
&-\sum_{\alpha,\beta=1}^{n_{k-1}}(\mathbf{\Pi}^{\mathbf{0}}_{\boldsymbol{k-1}})_{\alpha j}\,(\overset{*}{\mathbf{\Pi}}{}^{\mathbf{0},x}_{\boldsymbol{k-1}})_{\beta i}\,(\mathsf{H}^{b_x})_{\alpha\beta} - \sum_{\alpha,\beta=1}^{n_{k-1}}(\mathbf{\Pi}^{\mathbf{0}}_{\boldsymbol{k-1}})_{\alpha j}\,(\overset{*}{\mathbf{\Pi}}{}^{\mathbf{0},y}_{\boldsymbol{k-1}})_{\beta i}\,(\mathsf{H}^{b_y})_{\alpha\beta} =\\
&-\left[(\overset{*}{\mathbf{\Pi}}{}^{\mathbf{0},x}_{\boldsymbol{k-1}})^{\mathrm{T}}\mathsf{H}^{b_x}\mathbf{\Pi}^{\mathbf{0}}_{\boldsymbol{k-1}} + (\overset{*}{\mathbf{\Pi}}{}^{\mathbf{0},y}_{\boldsymbol{k-1}})^{\mathrm{T}}\mathsf{H}^{b_y}\mathbf{\Pi}^{\mathbf{0}}_{\boldsymbol{k-1}}\right]_{ij} =\\
&-\left[\left((\overset{*}{\mathbf{\Pi}}{}^{\mathbf{0},x}_{\boldsymbol{k-1}})^{\mathrm{T}}\mathsf{H}^{b_x} + (\overset{*}{\mathbf{\Pi}}{}^{\mathbf{0},y}_{\boldsymbol{k-1}})^{\mathrm{T}}\mathsf{H}^{b_y}\right)\mathbf{\Pi}^{\mathbf{0}}_{\boldsymbol{k-1}}\right]_{ij}.
\end{aligned}$$

Hence the elementary VEM matrix for the transport term is

$$\mathbf{K}^b = -\Big((\overset{*}{\mathbf{\Pi}}{}^{0,x}_{k-1})^{\mathrm{T}}\mathbf{H}^{b_x} + (\overset{*}{\mathbf{\Pi}}{}^{0,y}_{k-1})^{\mathrm{T}}\mathbf{H}^{b_y}\Big)\mathbf{\Pi}^{0}_{k-1}. \tag{37}$$

6.3 Reaction Term

The local VEM approximation for the reaction term is

$$c_h^E(p_h, q_h) := \int_E \gamma\, [\Pi^0_{k-1} p_h]\, [\Pi^0_{k-1} q_h]\, \mathrm{d}\boldsymbol{x}$$

and in matrix form

$$(\mathbf{K}^c)_{ij} := c_h^E(\phi_j, \phi_i) = \int_E \gamma\, [\Pi^0_{k-1}\phi_j]\, [\Pi^0_{k-1}\phi_i]\, \mathrm{d}\boldsymbol{x}.$$

Define the matrix

$$(\mathbf{H}^{\gamma})_{\alpha\beta} := \int_E \gamma\, m_\alpha m_\beta\, \mathrm{d}\boldsymbol{x}$$

and we have immediately

$$(\mathbf{K}^c)_{ij} = \int_E \gamma \Big[\sum_{\alpha=1}^{n_{k-1}} (\mathbf{\Pi}^0_{k-1})_{\alpha j}\, m_\alpha\Big]\Big[\sum_{\beta=1}^{n_{k-1}} (\mathbf{\Pi}^0_{k-1})_{\beta i}\, m_\beta\Big]\, \mathrm{d}\boldsymbol{x} =$$

$$\sum_{\alpha,\beta=1}^{n_{k-1}} (\mathbf{\Pi}^0_{k-1})_{\alpha j} (\mathbf{\Pi}^0_{k-1})_{\beta i} \int_E \gamma\, m_\alpha m_\beta\, \mathrm{d}\boldsymbol{x} = \big[(\mathbf{\Pi}^0_{k-1})^{\mathrm{T}} \mathbf{H}^{\gamma}\, \mathbf{\Pi}^0_{k-1}\big]_{ij}$$

i.e.

$$\mathbf{K}^c = (\mathbf{\Pi}^0_{k-1})^{\mathrm{T}} \mathbf{H}^{\gamma}\, \mathbf{\Pi}^0_{k-1}. \tag{38}$$

7 Algorithm for the Primal Formulation

For the convenience of the reader, we summarize the results of the previous Section in form of an algorithm ready to be implemented.

7.1 Projectors

1. Compute the $n_k \times N_{\text{dof}}$ matrix $\mathbf{B}$ given by

$$\mathbf{B} = \begin{bmatrix} \mathrm{P}_0\phi_1 & \dots & \mathrm{P}_0\phi_{N_{\text{dof}}} \\ (\nabla m_2, \nabla\phi_1)_{0,E} & \dots & (\nabla m_2, \nabla\phi_{N_{\text{dof}}})_{0,E} \\ \vdots & \ddots & \vdots \\ (\nabla m_{n_k}, \nabla\phi_1)_{0,E} & \dots & (\nabla m_{n_k}, \nabla\phi_{N_{\text{dof}}})_{0,E} \end{bmatrix},$$

where the terms of type $(\nabla m_\alpha, \nabla\phi_i)_{0,E}$ can be determined as shown in Lemma 1.
2. Compute the $N_{\text{dof}} \times n_k$ matrix $\mathbf{D}$ defined by:

$$\mathbf{D}_{i\alpha} = \mathrm{dof}_i(m_\alpha), \quad i = 1, \dots, N_{\text{dof}}, \; \alpha = 1, \dots, n_k.$$

3. Set

$$\mathbf{G} = \mathbf{BD}. \tag{39}$$

Note that the $n_k \times n_k$ matrix $\mathbf{G}$ can be computed independently (see (18)), and (39) can be used as a check of the correctness of the code.
4. Set

$$\overset{*}{\mathbf{\Pi}}{}_k^{\nabla} = \mathbf{G}^{-1}\mathbf{B} \quad \text{and} \quad \mathbf{\Pi}_k^{0} = \mathbf{D}\overset{*}{\mathbf{\Pi}}{}_k^{0}.$$

5. Compute the $n_k \times n_k$ matrix $\mathbf{H}$ defined by:

$$\mathbf{H}_{\alpha\beta} = \int_E m_\alpha m_\beta \,\mathrm{d}\boldsymbol{x} \quad \alpha, \beta = 1, \dots, n_k.$$

6. Compute the $n_k \times N_{\text{dof}}$ matrix $\mathbf{C}$ defined by

$$\mathbf{C}_{\alpha i} = \int_E m_\alpha \,\phi_i \,\mathrm{d}\boldsymbol{x}, \quad \alpha = 1, \dots, n_k, \; i = 1, \dots, N_{\text{dof}}.$$

The matrix $\mathbf{C}$ has the following structure:

- first n_{k-2} lines of $\mathbf{C} = |E| \left[\begin{array}{cccccc|cccc} 0 & 0 & \dots & 0 & 0 & & 1 & 0 & \dots & 0 \\ 0 & 0 & \dots & 0 & 0 & & 0 & 1 & \dots & 0 \\ \vdots & \vdots & \ddots & \vdots & \vdots & & \vdots & \vdots & \ddots & \vdots \\ 0 & 0 & \dots & 0 & 0 & & 0 & 0 & \dots & 1 \end{array}\right]$ where the last block is the identity matrix of size $n_{k-2} \times n_{k-2}$;
- last $n_k - n_{k-2}$ lines of $\mathbf{C}$:

$$\mathbf{C}_{\alpha i} = (\mathbf{H}\,\overset{*}{\mathbf{\Pi}}{}_k^{\nabla})_{\alpha i}, \quad n_{k-2} < \alpha \leqslant n_k.$$

7. Set

$$\overset{*}{\Pi}{}^0_k = \mathbf{H}^{-1}\mathbf{C} \quad \text{and} \quad \Pi^0_k = \mathbf{D}\overset{*}{\Pi}{}^0_k.$$

8. Compute the $N_{\text{dof}} \times n_{k-1}$ matrices $\mathbf{E}^x$ and $\mathbf{E}^y$ (see (33) and (34)) by

$$\left(\mathbf{E}^x\right)_{i\beta} = \int_E \phi_{i,x}\, m_\beta \,\mathrm{d}\boldsymbol{x}, \quad \left(\mathbf{E}^y\right)_{i\beta} = \int_E \phi_{i,y}\, m_\beta \,\mathrm{d}\boldsymbol{x}.$$

9. Set

$$\overset{*}{\Pi}{}^{0,x}_{k-1} = \hat{\mathbf{H}}^{-1}\mathbf{E}^x, \quad \overset{*}{\Pi}{}^{0,y}_{k-1} = \hat{\mathbf{H}}^{-1}\mathbf{E}^y$$

where $\hat{\mathbf{H}}$ is the submatrix of $\mathbf{H}$ obtained by taking the first n_{k-1} rows and columns of $\mathbf{H}$.

7.2 Coefficient Matrices

Compute the $n_{k-1} \times n_{k-1}$ matrices

$$(\mathbf{H}^\kappa)_{\alpha\beta} = \int_E \kappa\, m_\alpha\, m_\beta \,\mathrm{d}\boldsymbol{x}, \tag{40}$$

$$(\mathbf{H}^{b_x})_{\alpha\beta} = \int_E \boldsymbol{b}_x\, m_\alpha\, m_\beta \,\mathrm{d}\boldsymbol{x}, \quad (\mathbf{H}^{b_y})_{\alpha\beta} = \int_E \boldsymbol{b}_y\, m_\alpha\, m_\beta \,\mathrm{d}\boldsymbol{x}, \tag{41}$$

$$(\mathbf{H}^\gamma)_{\alpha\beta} = \int_E \gamma\, m_\alpha m_\beta \,\mathrm{d}\boldsymbol{x}. \tag{42}$$

7.3 Local Stiffness Matrices

Finally, set

$$\mathbf{K}^a = \left[(\overset{*}{\Pi}{}^{0,x}_{k-1})^{\mathrm{T}} \;\; (\overset{*}{\Pi}{}^{0,y}_{k-1})^{\mathrm{T}}\right] \begin{bmatrix} \mathbf{H}^\kappa & \mathbf{0} \\ \mathbf{0} & \mathbf{H}^\kappa \end{bmatrix} \begin{bmatrix} \overset{*}{\Pi}{}^{0,x}_{k-1} \\ \overset{*}{\Pi}{}^{0,y}_{k-1} \end{bmatrix} + \bar{\kappa}\,(\mathbf{I} - \Pi^\nabla_k)^{\mathrm{T}}(\mathbf{I} - \Pi^\nabla_k)$$

$$\mathbf{K}^b = -\left((\overset{*}{\Pi}{}^{0,x}_{k-1})^{\mathrm{T}}\mathbf{H}^{b_x} + (\overset{*}{\Pi}{}^{0,y}_{k-1})^{\mathrm{T}}\mathbf{H}^{b_y}\right) \Pi^0_{k-1}$$

$$\mathbf{K}^c = (\Pi^0_{k-1})^{\mathrm{T}}\mathbf{H}^\gamma\, \Pi^0_{k-1}.$$

8 Virtual Element Spaces for the Mixed Formulation

Before defining the virtual space $\boldsymbol{V}_h(E)$, we need to study certain spaces of polynomials which will play a major role in the definition of the degrees of freedom.

We start by defining an easily computable basis $\{\boldsymbol{m}_I\}$ for $[\mathcal{P}_k(E)]^2$. Let I be an index running from 1 to $2\,n_k = \dim[\mathcal{P}_k(E)]^2$. Set:

$$\begin{cases} \boldsymbol{m}_I := \begin{bmatrix} m_I \\ 0 \end{bmatrix} & \text{if } 1 \leqslant I \leqslant n_k \\ \boldsymbol{m}_I := \begin{bmatrix} 0 \\ m_{I-n_k} \end{bmatrix} & \text{if } n_k + 1 \leqslant I \leqslant 2n_k. \end{cases}$$

We introduce the (vector) polynomial spaces

$$\mathcal{G}_k^{\nabla}(E) := \nabla \mathcal{P}_{k+1}(E)$$

and

$$\mathcal{G}_k^{\perp}(E) := L^2\text{-orthogonal complement of } \mathcal{G}_k^{\nabla}(E) \text{ in } [\mathcal{P}_k(E)]^2$$

or, more generally,

$$\mathcal{G}_k^{\oplus}(E) := \text{any complement of } \mathcal{G}_k^{\nabla}(E) \text{ in } [\mathcal{P}_k(E)]^2.$$

An easy computation shows that

$$\dim \mathcal{G}_k^{\nabla}(E) = n_k^{\nabla} := n_k + (k+1) \quad \text{and} \quad \dim \mathcal{G}_k^{\oplus}(E) = n_k^{\oplus} := n_k - (k+1).$$

We construct now a basis for $\mathcal{G}_k^{\nabla}(E)$ and $\mathcal{G}_k^{\oplus}(E)$. It is easy to check that a basis for $\mathcal{G}_k^{\nabla}(E)$ is given by

$$\boldsymbol{g}_{\alpha}^{\nabla,k} := \nabla m_{\alpha+1}, \quad \alpha = 1, \ldots, n_k^{\nabla}.$$

Let now the $n_k^{\nabla} \times 2n_k$ matrix $\mathbf{T}^{\nabla}$ be such that

$$\boldsymbol{g}_{\alpha}^{\nabla,k} = \sum_{I=1}^{2n_k} \mathbf{T}_{\alpha I}^{\nabla} \boldsymbol{m}_I, \quad \alpha = 1, \ldots, n_k^{\nabla}.$$

A way to obtain a basis in $\mathcal{G}_k^{\oplus}(E)$ is to complete the matrix $\mathbf{T}^{\nabla}$ with a $n_k^{\oplus} \times 2n_k$ matrix $\mathbf{T}^{\oplus}$ to form a non-singular $(n_k^{\nabla} + n_k^{\oplus} = 2n_k) \times 2n_k$ square matrix $\mathbf{T} = \begin{bmatrix} \mathbf{T}^{\nabla} \\ \mathbf{T}^{\oplus} \end{bmatrix}$.

A basis for $\mathcal{G}_k^{\oplus}(E)$ is then given by

$$\boldsymbol{g}_\gamma^{\oplus,k} := \sum_{I=1}^{2n_k} \mathbf{T}_{\gamma I}^{\oplus} \boldsymbol{m}_I, \quad \gamma = 1, \ldots, n_k^{\oplus}.$$

An obvious way of constructing the matrix $\mathbf{T}$ is to define the rows of $\mathbf{T}^{\oplus}$ as a basis for the kernel of $\mathbf{T}^{\nabla}$. This can be easily done symbolically in MATLAB:

```
TO = null(TN)'; T = [TN; TO]; go = T*m;
```

where TN $= \mathbf{T}^{\nabla}$ and TO $= \mathbf{T}^{\oplus}$. In the appendix we present the basis so obtained up to $k = 5$.

8.1 The Space $V_h(E)$

We are ready now to define the local VEM space $\boldsymbol{V}_h(E)$ which consists of functions $\boldsymbol{v}_h$ such that:

- $\boldsymbol{v}_h \in \boldsymbol{H}(\mathrm{div}; E) \cap \boldsymbol{H}(\mathrm{rot}; E)$;
- $\boldsymbol{v}_h \cdot \boldsymbol{n}_e$ is a polynomial of degree k on each edge e;
- $\mathrm{div}\, \boldsymbol{v}_h \in \mathcal{P}_k(E)$;
- $\mathrm{rot}\, \boldsymbol{v}_h \in \mathcal{P}_{k-1}(E)$.

In [11] we have shown the following results:

1. the dimension of $\boldsymbol{V}_h(E)$ on a polygon E is

$$N_{\mathrm{dof}} := \dim \boldsymbol{V}_h(E) = N_e \times (k+1) + \dim \mathcal{G}_{k-1}^{\nabla}(E) + \dim \mathcal{G}_k^{\oplus}(E)$$
$$= N_e \times (k+1) + n_{k-1}^{\nabla} + n_k^{\oplus} = N_e \times (k+1) + 2n_k - k - 2$$

2. $[\mathcal{P}_k(E)]^2 \subset \boldsymbol{V}_h(E)$;
3. for the space $\boldsymbol{V}_h(E)$ we can take the following degrees of freedom:

   - **Edge dofs $[N_e \times (k+1)]$**
     Since on each edge $\boldsymbol{v}_h \cdot \boldsymbol{n}_e$ is a polynomial of degree k and no continuity is enforced at the vertices, we need to identify a polynomial of degree k on each edge without using the values at the vertices.
     This can be done in several ways, the most natural being taking the value of $\boldsymbol{v}_h \cdot \boldsymbol{n}_e$ at $k+1$ internal distinct $\{\boldsymbol{x}_\ell^e\}$ points of the edge e, obtained by subdividing e in $k+2$ equal parts:

$$\mathrm{dof}_\ell^e(\boldsymbol{v}_h) := (\boldsymbol{v}_h \cdot \boldsymbol{n}_e)(\boldsymbol{x}_\ell^e), \quad \ell = 1, \ldots, k+1.$$

This choice automatically ensures the continuity of $\boldsymbol{v}_h \cdot \boldsymbol{n}_e$ across two adjacent elements.

- **Internal ∇ dofs $[\boldsymbol{n}_{k-1}^{\nabla} = \boldsymbol{n}_k - 1]$**
  Let α be an index running from 1 to $\dim \mathcal{G}_{k-1}^{\nabla}(E) = n_{k-1}^{\nabla}$. We define:

$$\mathrm{dof}_{\alpha}^{\nabla}(\boldsymbol{v}_h) := \frac{1}{|E|}\int_E \boldsymbol{v}_h \cdot \boldsymbol{g}_{\alpha}^{\nabla,k-1}\,\mathrm{d}\boldsymbol{x}, \qquad \boldsymbol{g}_{\alpha}^{\nabla,k-1} \in \mathcal{G}_{k-1}^{\nabla}(E).$$

- **Internal ⊕ dofs $[\boldsymbol{n}_k^{\oplus} = \boldsymbol{n}_k - (\boldsymbol{k}+1)]$**
  Let γ be an index running from 1 to $\dim \mathcal{G}_k^{\oplus}(E) = n_k^{\oplus}$. We define:

$$\mathrm{dof}_{\gamma}^{\oplus}(\boldsymbol{v}_h) := \frac{1}{|E|}\int_E \boldsymbol{v}_h \cdot \boldsymbol{g}_{\gamma}^{\oplus,k}\,\mathrm{d}\boldsymbol{x}, \qquad \boldsymbol{g}_{\gamma}^{\oplus,k} \in \mathcal{G}_k^{\oplus}(E).$$

Let i be an index running through all dofs. We define $\boldsymbol{\phi}_i \in \boldsymbol{V}_h(E)$ by

$$\mathrm{dof}_j(\boldsymbol{\phi}_i) = \delta_{ij}, \quad j = 1, \ldots, N_{\mathrm{dof}}$$

in such a way that we have again a Lagrange-type identity:

$$\boldsymbol{v}_h = \sum_{i=1}^{N_{\mathrm{dof}}} \mathrm{dof}_i(\boldsymbol{v}_h)\,\boldsymbol{\phi}_i.$$

8.2 *The Space $Q_h(E)$*

As promised, the space $Q_h(E)$ is simply the space $\mathcal{P}_k(E)$ and as basis functions we take the set of scaled monomials $\mathcal{M}_k(E)$ defined in (2).

9 VEM Approximation of the Mixed Formulation

As show in [11], the VEM approximation of problem (9) is

$$\begin{cases} \text{Find } (\boldsymbol{u}_h, p_h) \in \boldsymbol{V}_h \times Q_h \text{ such that} \\ \displaystyle\sum_E \left\{ a_h^E(\boldsymbol{u}_h, \boldsymbol{v}_h) - (p_h, \mathrm{div}\,\boldsymbol{v}_h)_{0,E} - (\boldsymbol{\beta} \cdot \Pi_k^0 \boldsymbol{v}_h, p_h)_{0,E} \right\} = 0 & \text{for all } \boldsymbol{v}_h \in \boldsymbol{V}_h, \\ \displaystyle\sum_E (\mathrm{div}\,\boldsymbol{u}_h, q_h)_{0,E} + (\gamma p_h, q_h)_{0,\Omega} = (f, q_h)_{0,\Omega} & \text{for all } q_h \in Q_h \end{cases}$$

where

$$a_h^E(\boldsymbol{u}_h, \boldsymbol{v}_h) := (\nu\, \Pi_k^0 \boldsymbol{u}_h, \Pi_k^0 \boldsymbol{v}_h)_{0,E} + \mathcal{S}_E\big((I - \Pi_k^0)\boldsymbol{u}_h, (I - \Pi_k^0)\boldsymbol{v}_h\big).$$

The symmetric and positive bilinear form $\mathcal{S}_E(\cdot,\cdot)$, needed for the stability of the method, is defined by requiring

$$\mathcal{S}_E(\boldsymbol{\phi}_i, \boldsymbol{\phi}_j) = \overline{\nu}|E|\,\delta_{ij},$$

with $\overline{\nu}$ = mean value of ν on E, or $\overline{\nu} = \nu(x_c, y_c)$. The corresponding local stiffness matrices are obtained by restricting all integrals to E and by setting $\boldsymbol{u}_h = \boldsymbol{\phi}_j$, $\boldsymbol{v}_h = \boldsymbol{\phi}_i$, $p_h = m_\alpha$, $q_h = m_\beta$.

9.1 Computation of the L^2-projection in $V_h(E)$

Let $\boldsymbol{\phi}_i$ be a basis function for $\boldsymbol{V}_h(E)$. We need to compute $\Pi_k^0 \boldsymbol{\phi}_i \in [\mathcal{P}_k(E)]^2$. We shall write $\Pi_k^0 \boldsymbol{\phi}_i$ in terms of the basis $\{\boldsymbol{g}_I^k\} = \{\boldsymbol{g}_\alpha^{\nabla,k}, \boldsymbol{g}_\gamma^{\oplus,k}\}$ of $[\mathcal{P}_k(E)]^2$:

$$\Pi_k^0 \boldsymbol{\phi}_i = \sum_{\alpha=1}^{n_k^\nabla} s_i^\alpha\, \boldsymbol{g}_\alpha^{\nabla,k} + \sum_{\gamma=1}^{n_k^\oplus} s_i^\gamma\, \boldsymbol{g}_\gamma^{\oplus,k} = \sum_{I=1}^{2n_k} s_i^I\, \boldsymbol{g}_I^k. \tag{43}$$

Multiplying by $\{\boldsymbol{g}_\beta^{\nabla,k}, \boldsymbol{g}_\gamma^{\oplus,k}\}$ and integrating, we get a linear system in the unknowns $\{s_i^\alpha, s_i^\gamma\} = s_i^I$ (note that $\int_E \Pi_k^0 \boldsymbol{\phi}_i \cdot \boldsymbol{p}_k \,\mathrm{d}\boldsymbol{x} = \int_E \boldsymbol{\phi}_i \cdot \boldsymbol{p}_k \,\mathrm{d}\boldsymbol{x}$):

$$\begin{cases} \displaystyle\sum_{\alpha=1}^{n_k^\nabla} s_i^\alpha \int_E \boldsymbol{g}_\alpha^{\nabla,k} \cdot \boldsymbol{g}_\beta^{\nabla,k} \,\mathrm{d}\boldsymbol{x} + \sum_{\gamma=1}^{n_k^\oplus} s_i^\gamma \int_E \boldsymbol{g}_\gamma^{\oplus,k} \cdot \boldsymbol{g}_\beta^{\nabla,k} \,\mathrm{d}\boldsymbol{x} &= \displaystyle\int_E \boldsymbol{\phi}_i \cdot \boldsymbol{g}_\beta^{\nabla,k} \,\mathrm{d}\boldsymbol{x} \\ \displaystyle\sum_{\alpha=1}^{n_k^\nabla} s_i^\alpha \int_E \boldsymbol{g}_\alpha^{\nabla,k} \cdot \boldsymbol{g}_\delta^{\oplus,k} \,\mathrm{d}\boldsymbol{x} + \sum_{\gamma=1}^{n_k^\oplus} s_i^\gamma \int_E \boldsymbol{g}_\gamma^{\oplus,k} \cdot \boldsymbol{g}_\delta^{\oplus,k} \,\mathrm{d}\boldsymbol{x} &= \displaystyle\int_E \boldsymbol{\phi}_i \cdot \boldsymbol{g}_\delta^{\oplus,k} \,\mathrm{d}\boldsymbol{x}. \end{cases}$$

Set

$$\mathsf{G}_{IJ} := \int_E \boldsymbol{g}_I^k \cdot \boldsymbol{g}_J^k \,\mathrm{d}\boldsymbol{x},$$

and define the $2n_k \times N_{\mathrm{dof}}$ matrices

$$[\overset{*}{\boldsymbol{\Pi}}{}_k^0]_{Ii} := s_i^I \quad \text{and} \quad \mathsf{B}_{Ii} := \int_E \boldsymbol{\phi}_i \cdot \boldsymbol{g}_I^k \,\mathrm{d}\boldsymbol{x}. \tag{44}$$

We have

$$\sum_{J=1}^{2n_k} \mathbf{G}_{IJ}\,[\overset{*}{\boldsymbol{\Pi}}{}^{0}_{k}]_{Ji} = \mathbf{B}_{Ii} \quad \text{i.e.} \quad \mathbf{G}\,\overset{*}{\boldsymbol{\Pi}}{}^{0}_{k} = \mathbf{B} \quad \text{so that} \quad \overset{*}{\boldsymbol{\Pi}}{}^{0}_{k} = \mathbf{G}^{-1}\,\mathbf{B}.$$

We split $\mathbf{B}$ as

$$\mathbf{B} = \begin{bmatrix} \mathbf{B}^{\nabla} \\ \mathbf{B}^{\oplus} \end{bmatrix}.$$

We start from $\mathbf{B}^{\nabla}_{\beta i} = \int_E \boldsymbol{\phi}_i \cdot \boldsymbol{g}_{\beta}^{\nabla,k}\,\mathrm{d}\boldsymbol{x}$. Since

$$\boldsymbol{g}_{\beta}^{\nabla,k} = \nabla m_{\beta+1},$$

we have

$$\begin{aligned}\mathbf{B}^{\nabla}_{\beta i} = \int_E \boldsymbol{\phi}_i \cdot \nabla m_{\beta+1}\,\mathrm{d}\boldsymbol{x} &= -\int_E \operatorname{div}\boldsymbol{\phi}_i\, m_{\beta+1}\,\mathrm{d}\boldsymbol{x} + \int_{\partial E} \boldsymbol{\phi}_i \cdot \boldsymbol{n}_E\, m_{\beta+1}\,\mathrm{d}s \\ &=: \mathbf{B}^{\nabla}_1 + \mathbf{B}^{\nabla}_2.\end{aligned}$$

The term $\mathbf{B}^{\nabla}_2$ can be readily computed because $\boldsymbol{\phi}_i \cdot \boldsymbol{n}$ is a known polynomial on the boundary of E. Concerning the term $\mathbf{B}^{\nabla}_1$, we first observe that we can directly compute $\operatorname{div}\boldsymbol{\phi}_i \in \mathcal{P}_k(E)$. In fact, write $\operatorname{div}\boldsymbol{\phi}_i$ as

$$\operatorname{div}\boldsymbol{\phi}_i = \sum_{\sigma=1}^{n_k} d_i^{\sigma}\, m_{\sigma},$$

multiply by m_τ and integrate over E:

$$\sum_{\sigma=1}^{n_k} d_i^{\sigma} \int_E m_\sigma m_\tau\,\mathrm{d}\boldsymbol{x} = \int_E \operatorname{div}\boldsymbol{\phi}_i\, m_\tau\,\mathrm{d}\boldsymbol{x}.$$

Define the $n_k \times n_k$ matrix $\mathbf{H}$ (as already done in (26)) by

$$\mathbf{H}_{\sigma\tau} := \int_E m_\sigma m_\tau\,\mathrm{d}\boldsymbol{x},$$

and the $n_k \times N_{\mathrm{dof}}$ matrices $\mathbf{V}$ and $\mathbf{W}$ as

$$\mathbf{V}_{\sigma i:} = d_i^{\sigma}, \quad \mathbf{W}_{\tau i} := \int_E \operatorname{div}\boldsymbol{\phi}_i\, m_\tau\,\mathrm{d}\boldsymbol{x} \tag{45}$$

so that

$$\mathbf{H}\mathbf{V} = \mathbf{W} \quad \text{and} \quad \mathbf{V} = \mathbf{H}^{-1}\mathbf{W}.$$

Now,

$$\begin{aligned}\mathbf{W}_{\tau i} = \int_E \operatorname{div} \boldsymbol{\phi}_i \, m_\tau \, \mathrm{d}\boldsymbol{x} &= -\int_E \boldsymbol{\phi}_i \cdot \nabla m_\tau \, \mathrm{d}\boldsymbol{x} + \int_{\partial E} \boldsymbol{\phi}_i \cdot \boldsymbol{n}_E \, m_\tau \, \mathrm{d}s \\ &=: [\mathbf{W_1}]_{\tau i} + [\mathbf{W_2}]_{\tau i}.\end{aligned}$$

Observing that

$$\nabla m_\tau = \boldsymbol{g}_{\tau-1}^{\nabla,k-1},$$

we have

$$[\mathbf{W_1}]_{\tau i} = -|E| \operatorname{dof}^{\boldsymbol{g}}_{\tau-1}(\boldsymbol{\phi}_i) = \begin{cases} -|E| & \text{if } i \text{ corresponds to } \tau - 1 \\ 0 & \text{otherwise.} \end{cases} \tag{46}$$

Concerning the term $\mathbf{W_2}$, we observe that it can be immediately computed since $\boldsymbol{\phi}_i \cdot \boldsymbol{n}_E$ is a known polynomial on the boundary. Consider now $\mathbf{B}_1^{\nabla}$:

$$[\mathbf{B}_1^{\nabla}]_{\beta i} = -\int_E \operatorname{div} \boldsymbol{\phi}_i \, m_{\beta+1} \, \mathrm{d}\boldsymbol{x} = -\sum_{\sigma=1}^{n_k} d_i^\sigma \int_E m_\sigma \, m_{\beta+1} \, \mathrm{d}\boldsymbol{x}.$$

Define the $n_k^\nabla \times n_k$ matrix

$$\mathbf{H}^{\#}_{\beta\sigma} := \int_E m_\sigma \, m_{\beta+1} \, \mathrm{d}\boldsymbol{x}.$$

Obviously, most of the entries of the matrix $\mathbf{H}^{\#}$ are also entries of the matrix $\mathbf{H}$ already computed. Then

$$-\int_E \operatorname{div} \boldsymbol{\phi}_i \, m_{\beta+1} \, \mathrm{d}\boldsymbol{x} = -[\mathbf{H}^{\#}\mathbf{V}]_{\beta i} = -[\mathbf{H}^{\#}\mathbf{H}^{-1}\mathbf{W}]_{\beta i}$$

so that

$$\mathbf{B}_1^{\nabla} = -\mathbf{H}^{\#}\mathbf{H}^{-1}(\mathbf{W_1} + \mathbf{W_2}).$$

Concerning the term $\mathbf{B}^{\oplus}$, we simply observe that

$$\mathbf{B}^{\oplus}_{\delta i} = \int_E \boldsymbol{\phi}_i \cdot \boldsymbol{g}_\delta^{\oplus,k} = |E| \operatorname{dof}^{\oplus}_\delta(\boldsymbol{\phi}_i) = \begin{cases} |E| & \text{if } \delta \text{ corresponds to } i \\ 0 & \text{otherwise.} \end{cases}$$

We will also need $\Pi_k^0 \boldsymbol{\phi}_i$ in terms of the basis $\{\boldsymbol{\phi}_i\}$ itself. To this end, we define π_i^j as

$$\Pi_k^0 \boldsymbol{\phi}_i = \sum_{j=1}^{N_{\mathrm{dof}}} \pi_i^j \boldsymbol{\phi}_j \quad \text{or } \pi_i^j := \mathrm{dof}_j(\Pi_k^0 \boldsymbol{\phi}_i) \tag{47}$$

and the $N_{\mathrm{dof}} \times N_{\mathrm{dof}}$ matrix $\boldsymbol{\Pi}_k^0$ as

$$[\boldsymbol{\Pi}_k^0]_{ji} := \pi_i^j.$$

From (43) we have

$$\Pi_k^0 \boldsymbol{\phi}_i = \sum_{I=1}^{2n_k} s_i^I \boldsymbol{g}_I^k = \sum_{I=1}^{2n_k} s_i^I \left[\sum_{j=1}^{N_{\mathrm{dof}}} \mathrm{dof}_j(\boldsymbol{g}_I^k) \boldsymbol{\phi}_j \right] = \sum_{j=1}^{N_{\mathrm{dof}}} \left[\sum_{I=1}^{2n_k} s_i^I \mathrm{dof}_j(\boldsymbol{g}_I^k) \right] \boldsymbol{\phi}_j,$$

and comparing with (47) we obtain

$$\pi_i^j = \sum_{I=1}^{2n_k} s_i^I \mathrm{dof}_j(\boldsymbol{g}_I^k).$$

If we define the $N_{\mathrm{dof}} \times 2n_k$ matrix

$$\mathbf{D}_{jI} := \mathrm{dof}_j(\boldsymbol{g}_I^k)$$

we have:

$$\boldsymbol{\Pi}_k^0 = \mathbf{D} \overset{*}{\boldsymbol{\Pi}}{}_k^0 \quad \text{i.e.} \quad \boldsymbol{\Pi}_k^0 = \mathbf{D}\mathbf{G}^{-1}\mathbf{B}.$$

We observe that

$$\mathbf{G}_{IJ} = \int_E \boldsymbol{g}_I^k \cdot \boldsymbol{g}_J^k \, \mathrm{d}\boldsymbol{x}, \quad \text{and} \quad \boldsymbol{g}_J^k = \sum_{i=1}^{N_{\mathrm{dof}}} \mathrm{dof}_i(\boldsymbol{g}_J^k) \boldsymbol{\phi}_i$$

so that

$$\mathbf{G}_{IJ} = \sum_{i=1}^{N_{\mathrm{dof}}} \mathrm{dof}_i(\boldsymbol{g}_J^k) \int_E \boldsymbol{g}_I^k \cdot \boldsymbol{\phi}_i \, \mathrm{d}\boldsymbol{x} = \sum_{i=1}^{N_{\mathrm{dof}}} \mathbf{D}_{iJ} \mathbf{B}_{Ii} \quad \text{hence} \quad \mathbf{G} = \mathbf{B}\mathbf{D}. \tag{48}$$

We have the following useful identities:

$$\overset{*}{\boldsymbol{\Pi}}{}_k^0 \mathbf{D} = \mathbf{I} \quad \text{since} \quad \overset{*}{\boldsymbol{\Pi}}{}_k^0 \mathbf{D} = \mathbf{G}^{-1}\mathbf{B}\mathbf{D} = \mathbf{G}^{-1}\mathbf{G} = \mathbf{I}$$

and

$$\mathbf{\Pi}_k^0 \mathbf{D} = \mathbf{D} \quad \text{since} \quad \mathbf{\Pi}_k^0 \mathbf{D} = \mathbf{D} \overset{*}{\mathbf{\Pi}}{}_k^0 \mathbf{D} = \mathbf{D}\mathbf{I} = \mathbf{D}.$$

Another way of arguing is that since Π_k^0 is a projection, then $(\mathbf{\Pi}_k^0)^2 = \mathbf{\Pi}_k^0$. Hence

$$(\mathbf{\Pi}_k^0)^2 = \mathbf{D}\mathbf{G}^{-1}\mathbf{B}\mathbf{D}\mathbf{G}^{-1}\mathbf{B} = \mathbf{D}[\mathbf{G}^{-1}\mathbf{B}\mathbf{D}]\mathbf{G}^{-1}\mathbf{B} = \mathbf{\Pi}_k^0 = \mathbf{D}\mathbf{G}^{-1}\mathbf{B}$$

hence $\mathbf{G}^{-1}\mathbf{B}\mathbf{D}$ must be the identity matrix as stated in (48).

Remark 2 It can be shown that the lower part of the matrix $\mathbf{\Pi}_k^0$ corresponding to the internal dofs (last $n_{k-1}^{\nabla} + n_k^{\oplus}$ rows) is the identity matrix. This property can be exploited in the definition of the stability matrix (50) described below (see [11]).

10 Local Matrices

We are now ready to compute the VEM local matrices for the mixed formulation.

10.1 Term $a_h^E(u_h, v_h)$

The corresponding local matrix is given by

$$\begin{aligned} a_h^E(\boldsymbol{\phi}_i, \boldsymbol{\phi}_j) &= (\nu\, \Pi_k^0 \boldsymbol{\phi}_j, \Pi_k^0 \boldsymbol{\phi}_i)_{0,E} + \mathcal{S}_E\big((I - \Pi_k^0)\boldsymbol{\phi}_j, (I - \Pi_k^0)\boldsymbol{\phi}_i\big) \\ &:= (\mathbf{K}_{\mathbf{c}}^a)_{ij} + (\mathbf{K}_{\mathbf{s}}^a)_{ij}. \end{aligned}$$

Using (43), the *consistency* matrix $\mathbf{K}_{\mathbf{c}}^a$ is given by

$$[\mathbf{K}_{\mathbf{c}}^a]_{ij} = \sum_{I=1}^{2n_k} \sum_{J=1}^{2n_k} s_i^I s_j^J \int_E \nu\, \boldsymbol{g}_I^k \cdot \boldsymbol{g}_J^k \,\mathrm{d}\boldsymbol{x}.$$

Defining the $2n_k \times 2n_k$ matrix $\mathbf{G}^\nu$

$$\mathbf{G}_{IJ}^\nu := \int_E \nu\, \boldsymbol{g}_I^k \cdot \boldsymbol{g}_J^k \,\mathrm{d}\boldsymbol{x},$$

and using (44) we obtain:

$$[\mathbf{K}_{\mathbf{c}}^a]_{ij} = \sum_{I=1}^{2n_k} \sum_{J=1}^{2n_k} [\overset{*}{\mathbf{\Pi}}{}_k^0]_{Ii} [\overset{*}{\mathbf{\Pi}}{}_k^0]_{Jj} \mathbf{G}_{IJ}^\nu$$

i.e.

$$\mathbf{K}^a_{\mathbf{c}} = [\overset{*}{\mathbf{\Pi}}{}^0_k]^{\mathrm{T}} \mathbf{G}^{\nu} \overset{*}{\mathbf{\Pi}}{}^0_k. \tag{49}$$

If $\nu(x) \equiv 1$, i.e. we have the Laplace operator, then $\mathbf{G}^{\nu} = \mathbf{G}$ and

$$\mathbf{K}^a_{\mathbf{c}} = [\mathbf{G}^{-1}\mathbf{B}]^{\mathrm{T}}\, \mathbf{G}\, [\mathbf{G}^{-1}\mathbf{B}] = \mathbf{B}^{\mathrm{T}}\mathbf{G}^{-1}\mathbf{B}.$$

The *stability* matrix $\mathbf{K}^a_{\mathbf{s}}$ can be taken as

$$\mathbf{K}^a_{\mathbf{s}} = \bar{\nu}\, |E|\, (\mathbf{I} - \mathbf{\Pi}^0_k)^{\mathrm{T}}\, (\mathbf{I} - \mathbf{\Pi}^0_k) \tag{50}$$

where $\bar{\nu}$ is a constant approximation of ν.

10.2 Term $-(p_h, div\ \boldsymbol{v}_h)_{0,E}$

By (45) we see that the corresponding local matrix is $-\mathbf{W}^{\mathrm{T}}$ which has already been computed.

The local matrix $\mathbf{K}$ corresponding to $\boldsymbol{\beta} = (0, 0)$ and $\gamma = 0$ is then given by:

$$\mathbf{K} = \begin{bmatrix} \mathbf{K}^a_{\mathbf{c}} + \mathbf{K}^a_{\mathbf{s}} & -\mathbf{W}^{\mathrm{T}} \\ \mathbf{W} & 0 \end{bmatrix}$$

10.3 Term $-(\boldsymbol{\beta} \cdot \Pi^0_k \boldsymbol{v}_h, p_h)_{0,E}$

The corresponding local matrix is

$$\mathbf{T}^{\boldsymbol{\beta}}_{j\sigma} := -\int_E [\boldsymbol{\beta} \cdot \Pi^0_k \boldsymbol{\phi}_j]\, m^k_{\sigma}\, \mathrm{d}\boldsymbol{x} = -\sum_{I=1}^{2n_k} [\overset{*}{\mathbf{\Pi}}{}^0_k]_{Ij} \int_E [\boldsymbol{\beta} \cdot \boldsymbol{g}^k_I]\, m^k_{\sigma}\, \mathrm{d}\boldsymbol{x}.$$

Defining the matrix

$$\mathbf{U}_{I\sigma} := \int_E [\boldsymbol{\beta} \cdot \boldsymbol{g}^k_I]\, m^k_{\sigma}\, \mathrm{d}\boldsymbol{x}$$

we have

$$\mathbf{T}^{\boldsymbol{\beta}} = -(\overset{*}{\mathbf{\Pi}}{}^0_k)^{\mathrm{T}}\mathbf{U}.$$

10.4 Term $(\gamma p_h, q_h)_{0,E}$

The corresponding local matrix is $\mathbf{H}^\gamma$ defined in (38).

10.5 Complete Stiffness Matrix

The local stiffness matrix $\mathbf{K}$ for the complete problem is then given by:

$$\mathbf{K} := \begin{bmatrix} \mathbf{K}_\mathbf{c}^a + \mathbf{K}_\mathbf{s}^a & -\mathbf{W}^\mathrm{T} + \mathbf{T}^\beta \\ \mathbf{W} & \mathbf{H}^\gamma \end{bmatrix}.$$

11 Algorithm for the Mixed Formulation

We summarize here the steps needed to compute the VEM local matrix for the mixed approximation. We indicate in square brackets the size of each matrix.

11.1 L^2 Projection

1. Compute

$$\mathbf{G}_{IJ} = \int_E \boldsymbol{g}_I^k \cdot \boldsymbol{g}_J^k \,\mathrm{d}\boldsymbol{x} \qquad [2n_k \times 2n_k]$$

2. Compute the $[n_k \times N_\mathrm{dof}]$ matrix $\mathbf{W}_1$

$$[\mathbf{W}_1]_{\tau i} = -|E|\,\mathrm{dof}^g_{\tau-1}(\boldsymbol{\phi}_i) = \begin{cases} -|E| & \text{if } i \text{ corresponds to } \tau - 1 \\ 0 & \text{otherwise} \end{cases}$$

3. Compute

$$\mathbf{W}_2 \quad \text{(boundary term)} \qquad [n_k \times N_\mathrm{dof}]$$

4. Set

$$\mathbf{W} = \mathbf{W}_1 + \mathbf{W}_2 \qquad [n_k \times N_\mathrm{dof}]$$

5. Compute

$$\mathbf{H}_{\sigma\tau} = \int_E m_\sigma m_\tau \, \mathrm{d}\boldsymbol{x} \qquad [n_k \times n_k]$$

6. Compute

$$\mathbf{H}^{\#}_{\beta\sigma} = \int_E m_\sigma \, m_{\beta+1} \, \mathrm{d}\boldsymbol{x} \qquad [n_k^{\nabla} \times n_k]$$

7. Set

$$\mathbf{B}_1^{\nabla} = -\mathbf{H}^{\#}\mathbf{H}^{-1}\mathbf{W} \qquad [n_k^{\nabla} \times N_{\mathrm{dof}}]$$

8. Compute

$$\mathbf{B}_2^{\nabla} \quad \text{(boundary term)} \qquad [n_k^{\nabla} \times N_{\mathrm{dof}}]$$

9. Set

$$\mathbf{B}^{\nabla} = \mathbf{B}_1^{\nabla} + \mathbf{B}_2^{\nabla} \qquad [n_k^{\nabla} \times N_{\mathrm{dof}}]$$

10. Compute the $[n_k^{\oplus} \times N_{\mathrm{dof}}]$ matrix $\mathbf{B}^{\oplus}$

$$[\mathbf{B}^{\oplus}]_{\delta i} = |E| \, \mathrm{dof}_\delta^{\oplus}(\boldsymbol{\phi}_i) = |E| \, \delta_{\delta i} = \begin{cases} |E| & \text{if } i \text{ corresponds to } \delta \\ 0 & \text{otherwise} \end{cases}$$

11. Set

$$\mathbf{B} = \begin{bmatrix} \mathbf{B}^{\nabla} \\ \mathbf{B}^{\oplus} \end{bmatrix} \qquad [2n_k \times N_{\mathrm{dof}}]$$

12. Set

$$\overset{*}{\mathbf{\Pi}}{}_k^{\mathbf{0}} = \mathbf{G}^{-1}\mathbf{B} \qquad [2n_k \times N_{\mathrm{dof}}]$$

13. Compute

$$\mathbf{D}_{jI} := \mathrm{dof}_j(\boldsymbol{g}_I^k) \qquad [N_{\mathrm{dof}} \times 2n_k]$$

14. Set

$$\mathbf{\Pi}_k^{\mathbf{0}} = \mathbf{D}\overset{*}{\mathbf{\Pi}}{}_k^{\mathbf{0}} \qquad [2N_{\mathrm{dof}} \times 2N_{\mathrm{dof}}]$$

15. Check that

$$\mathsf{G} = \mathsf{B}\mathsf{D}$$

11.2 Coefficient Matrices

1. Compute

$$\mathsf{G}^{\nu}_{IJ} = \int_E \nu\, \boldsymbol{g}^k_I \cdot \boldsymbol{g}^k_J \,\mathrm{d}\boldsymbol{x} \qquad [2n_k \times 2n_k]$$

2. Define

$$\mathsf{U}_{I\sigma} = \int_E [\boldsymbol{\beta} \cdot \boldsymbol{g}^k_I]\, m^k_\sigma \,\mathrm{d}\boldsymbol{x} \qquad [2n_k \times n_k]$$

3. Set

$$\mathsf{T}^{\beta} = -(\overset{*}{\boldsymbol{\Pi}}{}^{\mathbf{0}}_k)^{\mathrm{T}}\mathsf{U}. \qquad [2n_k \times n_k]$$

4. Define

$$(\mathsf{H}^{\gamma})_{\alpha\beta} := \int_E \gamma\, m_\alpha m_\beta \,\mathrm{d}\boldsymbol{x} \qquad [n_k \times n_k]$$

11.3 Local Matrix

Set

$$\mathsf{K}^a_{\mathsf{c}} = [\overset{*}{\boldsymbol{\Pi}}{}^{\mathbf{0}}_k]^{\mathrm{T}}\mathsf{G}^{\nu}\overset{*}{\boldsymbol{\Pi}}{}^{\mathbf{0}}_k \quad \text{and} \quad \mathsf{K}^a_{\mathsf{s}} = \bar{\nu}\,|E|\,(\mathsf{I} - \boldsymbol{\Pi}^{\mathbf{0}}_k)^{\mathrm{T}}\,(\mathsf{I} - \boldsymbol{\Pi}^{\mathbf{0}}_k).$$

The full local matrix is then

$$\mathsf{K} := \begin{bmatrix} \mathsf{K}^a_{\mathsf{c}} + \mathsf{K}^a_{\mathsf{s}} & -\mathsf{W}^{\mathrm{T}} + \mathsf{T}^{\beta} \\ \mathsf{W} & \mathsf{H}^{\gamma} \end{bmatrix}.$$

Appendix

We list here the basis $\boldsymbol{g}_{\alpha}^{\nabla,k}$ and $\boldsymbol{g}_{\gamma}^{\oplus,k}$ obtained with MATLAB for k up to 5. We point out that in order to have the right scaling, the variable x and y must be replaced by $\left(\frac{x-x_c}{h_E}\right)$ and $\left(\frac{x-y_c}{h_E}\right)$ respectively.

$\boldsymbol{g}_{\alpha}^{\nabla,k}$ $\qquad\qquad\qquad$ $\boldsymbol{g}_{\gamma}^{\oplus,k}$

```
k=1   [           1,           0]   [              -y,          x]
      [           0,           1]
      [         2*x,           0]
      [           y,           x]
      [           0,         2*y]
k=2   [      3*x^2,            0]   [        -(x*y)/2,        x^2]
      [      2*x*y,          x^2]   [        -2*y^2,          x*y]
      [        y^2,        2*x*y]
      [          0,        3*y^2]
k=3   [      4*x^3,            0]   [      -(x^2*y)/3,        x^3]
      [    3*x^2*y,          x^3]   [          -x*y^2,      x^2*y]
      [    2*x*y^2,      2*x^2*y]   [          -3*y^3,      x*y^2]
      [        y^3,      3*x*y^2]
      [          0,        4*y^3]
k=4   [      5*x^4,            0]   [      -(x^3*y)/4,        x^4]
      [    4*x^3*y,          x^4]   [ -(2*x^2*y^2)/3,       x^3*y]
      [ 3*x^2*y^2,       2*x^3*y]   [   -(3*x*y^3)/2,     x^2*y^2]
      [   2*x*y^3,     3*x^2*y^2]   [          -4*y^4,      x*y^3]
      [        y^4,      4*x*y^3]
      [          0,        5*y^4]
k=5   [      6*x^5,            0]   [      -(x^4*y)/5,        x^5]
      [    5*x^4*y,          x^5]   [   -(x^3*y^2)/2,       x^4*y]
      [ 4*x^3*y^2,       2*x^4*y]   [      -x^2*y^3,      x^3*y^2]
      [ 3*x^2*y^3,     3*x^3*y^2]   [      -2*x*y^4,      x^2*y^3]
      [   2*x*y^4,     4*x^2*y^3]   [         -5*y^5,       x*y^4]
      [        y^5,      5*x*y^4]
      [          0,        6*y^5]
```

References

1. B. Ahmad, A. Alsaedi, F. Brezzi, L.D. Marini, A. Russo, Equivalent projectors for virtual element methods. Comput. Math. Appl. **66**(3), 376–391 (2013)
2. P.F. Antonietti, L. Beirão da Veiga, D. Mora, M. Verani, A stream virtual element formulation of the Stokes problem on polygonal meshes. SIAM J. Numer. Anal. **52**(1), 386–404 (2014)
3. M. Arroyo, M. Ortiz, Local maximum-entropy approximation schemes: a seamless bridge between finite elements and meshfree methods. Int. J. Numer. Methods Eng. **65**(13), 2167–2202 (2006)
4. I. Babuska, U. Banerjee, J.E. Osborn, Survey of meshless and generalized finite element methods: a unified approach. Acta Numer. **12**, 1–125 (2003)
5. L. Beirão da Veiga, G. Manzini, A virtual element method with arbitrary regularity. IMA J. Numer. Anal. **34**(2), 759–781 (2014)

6. L. Beirão da Veiga, F. Brezzi, A. Cangiani, G. Manzini, L.D. Marini, A. Russo, Basic principles of virtual element methods. Math. Models Methods Appl. Sci. **23**(1), 199–214 (2013)
7. L. Beirão da Veiga, F. Brezzi, L.D. Marini, Virtual elements for linear elasticity problems. SIAM J. Numer. Anal. **51**(2), 794–812 (2013)
8. L. Beirão da Veiga, F. Brezzi, L.D. Marini, A. Russo, The hitchhiker's guide to the virtual element method. Math. Models Methods Appl. Sci. **24**(8), 1541–1573 (2014)
9. L. Beirão da Veiga, K. Lipnikov, G. Manzini, *The Mimetic Finite Difference Method for Elliptic Problems*. MS&A, Modeling, Simulation and Applications, vol. 11 (Springer, Berlin, 2014)
10. L. Beirão da Veiga, F. Brezzi, L.D. Marini, A. Russo, $H(div)$ and $H(curl)$-conforming VEM. Numer. Math. **133**(2), 303–332 (2015)
11. L. Beirão da Veiga, F. Brezzi, L.D. Marini, A. Russo, Mixed virtual element methods for general second order elliptic problems on polygonal meshes. ESAIM: M2AN **50**(3), 727–747 (2016)
12. L. Beirão da Veiga, F. Brezzi, L.D. Marini, A. Russo virtual element methods for general second order elliptic problems on polygonal meshes. Math. Models Methods. Appl. Sci. **26**(4), 729–750 (2016)
13. M.F. Benedetto, S. Berrone, S. Pieraccini, S. Scialò, The virtual element method for discrete fracture network simulations. Comput. Methods Appl. Mech. Eng. **280**, 135–156 (2014)
14. J.E. Bishop, A displacement-based finite element formulation for general polyhedra using harmonic shape functions. Int. J. Numer. Methods Eng. **97**(1), 1–31 (2014)
15. F. Brezzi, L.D. Marini, Virtual element methods for plate bending problems. Comput. Methods Appl. Mech. Eng. **253**, 455–462 (2013)
16. F. Brezzi, R.S. Falk, L.D. Marini, Basic principles of mixed virtual element methods. ESAIM Math. Model. Numer. Anal. **48**(4), 1227–1240 (2014)
17. C. Chinosi, L.D. Marini, Virtual Element Methods for fourth order problems: L^2 Estimates. Comput. Math. Appl. (2016)
18. B. Cockburn, The hybridizable discontinuous Galerkin methods, in *Proceedings of the International Congress of Mathematicians*, vol. IV (Hindustan Book Agency, New Delhi, 2010), pp. 2749–2775
19. D. Di Pietro, A. Ern, Hybrid high-order methods for variable-diffusion problems on general meshes. C.R. Acad. Sci. Paris. Ser. I **353**, 31–34 (2015)
20. J. Droniou, R. Eymard, T. Gallouët, R. Herbin, Gradient schemes: a generic framework for the discretisation of linear, nonlinear and nonlocal elliptic and parabolic equations. Math. Models Methods Appl. Sci. **23**(13), 2395–2432 (2013)
21. M.S. Floater, Generalized barycentric coordinates and applications. Acta Numer. **24**, 215–258 (2015)
22. T.-P. Fries, T. Belytschko, The extended/generalized finite element method: an overview of the method and its applications. Int. J. Numer. Methods Eng. **84**(3), 253–304 (2010)
23. A.L. Gain, C. Talischi, G.H. Paulino, On the virtual element method for three-dimensional linear elasticity problems on arbitrary polyhedral meshes. Comput. Methods Appl. Mech. Eng. **282**, 132–160 (2014)
24. R.V. Garimella, J. Kim, M. Berndt, Polyhedral mesh generation and optimization for non-manifold domains, in *Proceedings of the 22nd International Meshing Roundtable*, ed. by J. Sarrate, M. Staten (Springer, Berlin, 2013)
25. D. Mora, G. Rivera, R. Rodríguez, A virtual element method for the Steklov eigenvalue problem. Math. Models Methods Appl. Sci. **25**(08), 1421–1445 (2015)
26. N. Sukumar, E.A. Malsch, Recent advances in the construction of polygonal finite element interpolants. Arch. Comput. Methods Eng. **13**(1), 129–163 (2006)
27. C. Talischi, G.H. Paulino, A. Pereira, I.F.M. Menezes, Polygonal finite elements for topology optimization: a unifying paradigm. Int. J. Numer. Methods Eng. **82**(6), 671–698 (2010)
28. C. Talischi, G.H. Paulino, A. Pereira, I.F.M. Menezes, PolyMesher: a general-purpose mesh generator for polygonal elements written in Matlab. J. Struct. Multidiscip. Optim. **45**(3), 309–328 (2012)
29. J. Wang, X. Ye, A weak Galerkin mixed finite element method for second order elliptic problems. Math. Comput. **83**(289), 2101–2126 (2014)

On Quasi-Interpolation Operators in Spline Spaces

Annalisa Buffa, Eduardo M. Garau, Carlotta Giannelli, and Giancarlo Sangalli

Abstract We propose the construction of a class of L^2 stable quasi-interpolation operators onto the space of splines on tensor-product meshes, in any space dimension. The estimate we propose is robust with respect to knot repetition and to knot "vicinity" (up to $p+1$ knots), so it applies to the most general scenario in which the B-spline functions are known to be well defined.

1 Introduction

The use of splines as a tool for the numerical discretization of partial differential equations is experiencing a very fast spreading thanks to the advent of isogeometric analysis [8, 13]. Besides the many engineering applications that are object of study within the isogeometric framework, there is also a renewed attention towards the development of theoretical tools that may provide a clear mathematical understanding and solid groundings for isogeometric methods.

A. Buffa (✉)
Istituto di Matematica Applicata e Tecnologie Informatiche "E. Magenes" (CNR), Pavia, Italy
e-mail: annalisa.buffa@imati.cnr.it

E.M. Garau
Instituto de Matemática Aplicada del Litoral (CONICET-UNL) and Facultad de Ingeniería Química (UNL), Santa Fe, Argentina

Istituto di Matematica Applicata e Tecnologie Informatiche "E. Magenes" (CNR), Pavia, Italy

C. Giannelli
Dipartimento di Matematica e Informatica "U. Dini", Università degli Studi di Firenze, Firenze, Italy

G. Sangalli
Dipartimento di Matematica "F. Casorati", Università degli Studi di Pavia, Pavia, Italy

Istituto di Matematica Applicata e Tecnologie Informatiche "E. Magenes" (CNR), Pavia, Italy

G.R. Barrenechea et al. (eds.), *Building Bridges: Connections and Challenges in Modern Approaches to Numerical Partial Differential Equations*,
Lecture Notes in Computational Science and Engineering 114,
DOI 10.1007/978-3-319-41640-3_3

A state-of-the-art review on the existing mathematical results can be found in the review paper [6] published in 2014. Indeed, several results exists today starting from approximation properties, to wellposedness for various classes of spline discretizations and problems showing that splines, and the isogeometric framework, can be suitably used in the numerical analysis for a variety of PDEs (elliptic, saddle points, Hodge laplacian, etc.). In this paper, we focus on a rather fundamental question that is the approximation properties and the techniques to study them in the most general setting of interests. In [6] and in all the literature until now, approximation properties for splines are analysed under the assumption of local quasi-uniform meshes (see Assumption 3.1 below), possibly in presence of knot repetition. These results are surely useful but fail to analyse the approximation error in the most general framework: indeed, the spline basis remain stable when up to $p + 1$ knots are made closer and closer to each-other (up to becoming coincident) while the interpolation operators proposed become unstable when knot spans collapse to zero.

In the present paper, we fill this gap and we put ourselves in the most general situation. Instead of considering one single choice of interpolation operator, we consider an entire class of operators, mostly inspired by Lee et al. [15] and we analyse there approximation properties under the milder Assumption 3.2 (see below).

Quasi-interpolants in spline spaces are usually defined as linear combination of locally supported functions $\beta \in \mathcal{B}$ that form a convex partition of unity, namely

$$P(f) = \sum_{\beta \in \mathcal{B}} \lambda_\beta(f)\beta\,,$$

where the linear functionals $\lambda_\beta(f)$ may be defined in different ways, by e.g. taking into account the evaluation of the function f, or even related integral/derivative information, at certain points or in regions included in (or close to) the support of β, see for example [10, 15, 16, 18]. The use of spline-based quasi-interpolation schemes is an established technique for the design of effective and reliable approximation algorithms.

In this paper we derive an approximation method in terms of local L^2 projection by exploiting the local stability of the univariate B-spline basis and its tensor-product extension. The stability and approximation properties of the corresponding quasi-interpolation operators are presented. The analysis includes the discussion of mild assumptions on the admissible mesh configuration to be considered.

The remaining of the paper is organised as follows. Section 2 provides a brief overview of isogeometric analysis and introduces some basic notation. In Sect. 3 we state the assumptions on the meshes that we consider. We analyse the stability properties of the B-spline basis in Sect. 4 through some estimations for the inverse of the local Bézier extraction operator. The local approximation method is then presented in Sect. 5, while Sect. 6 defines the locally supported dual basis. Finally, the properties of the quasi-interpolation operator are discussed in Sect. 7.

2 Motivation: The Isogeometric Setting

One of the main motivation of our work is to provide mathematical foundation of isogeometric methods and a rigorous understanding of the properties splines have in practice. To this aim, this section is meant to introduce splines and shortly discuss the isogeometric setting where they are meant to be used.

2.1 Univariate and Tensor-Product B-splines

Let $\Xi := \{\xi_j\}_{j=1}^{n+p+1}$ be a p-open knot (ordered) vector such that

$$0 = \xi_1 = \cdots = \xi_{p+1} < \xi_{p+2} \leqslant \cdots \leqslant \xi_n < \xi_{n+1} = \cdots = \xi_{n+p+1} = 1,$$

where the two positive integers p and n denote a given polynomial degree, and the corresponding number of B-splines defined over the considered knot sequence, respectively. We also introduce the vector $Z := \{\zeta_j\}_{j=1}^{\tilde{n}}$ of knots without repetitions, and denote by m_j the multiplicity of the breakpoint ζ_j, such that

$$\Xi = \{\underbrace{\zeta_1, \ldots, \zeta_1}_{m_1 \text{ times}}, \underbrace{\zeta_2, \ldots, \zeta_2}_{m_2 \text{ times}}, \ldots \underbrace{\zeta_{\tilde{n}}, \ldots, \zeta_{\tilde{n}}}_{m_{\tilde{n}} \text{ times}}\},$$

with $\sum_{i=1}^{\tilde{n}} m_i = n + p + 1$. Note that the two extreme knots are repeated $p + 1$ times, i.e., $m_1 = m_{\tilde{n}} = p + 1$. We assume that an internal knot can be repeated at most $p + 1$ times, i.e., $m_j \leqslant p + 1$, for $j = 2, \ldots, \tilde{n} - 1$.

Let $\{\beta_1, \beta_2, \ldots, \beta_n\}$ be the univariate B-spline basis of degree p associated to the knot vector Ξ, see e.g., [9, 19]. Each B-spline is a piecewise polynomial of degree p on the subdivision $\{\zeta_1, \ldots, \zeta_{\tilde{n}}\}$ and it has $p - m_j$ continuous derivatives at the breakpoint ζ_j. We remark that B-splines are non-negative, locally supported, and form a convex partition of unity, namely

$$\beta_j \geqslant 0, \qquad \operatorname{supp} \beta_j = [\xi_j, \xi_{j+p+1}], \qquad \sum_{i=1}^{n} \beta_i(x) = 1 \quad \forall x \in (0, 1).$$

Let $\mathcal{I}$ be the univariate mesh defined by

$$\mathcal{I} := \{[\zeta_j, \zeta_{j+1}] \,|\, j = 1, \ldots, \tilde{n} - 1\}.$$

For each $I = [\zeta_j, \zeta_{j+1}] \in \mathcal{I}$ there exists a unique $k = \sum_{i=1}^{j} m_j$ such that $I = [\xi_k, \xi_{k+1}]$ and $\xi_k \neq \xi_{k+1}$. The union of the supports of the B-splines acting on I

identifies the *support extension* $\tilde{I}$, namely

$$\tilde{I} := [\xi_{k-p}, \xi_{k+p+1}], \tag{1}$$

Moreover, we define

$$\hat{I} := [\xi_{k-p+1}, \xi_{k+p}]. \tag{2}$$

An example of quadratic B-splines constructed from the open knot vector

$$\Xi = \{0, 0, 0, 1/5, 2/5, 3/5, 3/5, 4/5, 1, 1, 1\}$$

is presented in Fig. 1. Notice that, since the knot $\xi_6 = \xi_7 = \zeta_4 = 3/5$ has multiplicity $m_4 = 2$, the fourth, fifth and sixth functions are only continuous at that point.

In order to define a tensor-product d-variate spline space on $\hat{\Omega} := [0, 1]^d \subset \mathbb{R}^d$, let $\mathbf{p} := (p_1, p_2, \ldots, p_d)$ be the set of polynomial degrees with respect to each coordinate direction. For $i = 1, 2, \ldots, d$, let $\Xi_i := \{\xi_j^{(i)}\}_{j=1}^{n_i+p_i+1}$ be a p_i-open knot vector such that

$$0 = \xi_1^{(i)} = \cdots = \xi_{p_i+1}^{(i)} < \xi_{p_i+2}^{(i)} \leqslant \cdots \leqslant \xi_{n_i}^{(i)} < \xi_{n_i+1}^{(i)} = \cdots = \xi_{n_i+p_i+1}^{(i)} = 1,$$

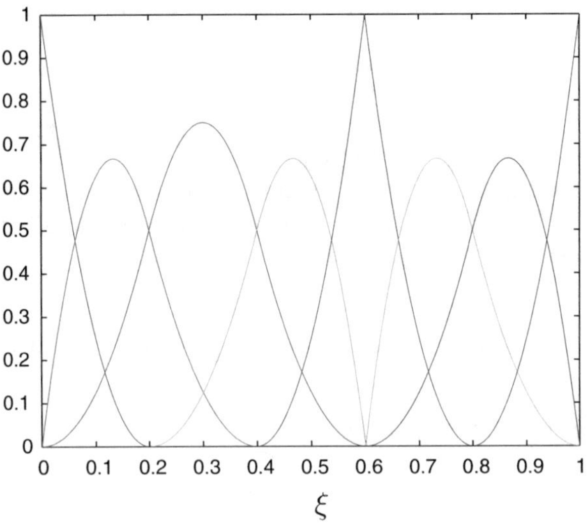

Fig. 1 Quadratic B-splines basis functions constructed from the open knot vector $\Xi = \{0, 0, 0, 1/5, 2/5, 3/5, 3/5, 4/5, 1, 1, 1\}$

where the two extreme knots are repeated $p_i + 1$ times and any internal knot can be repeated at most $p_i + 1$ times. We denote by $\mathbb{V}$ the tensor-product spline space spanned by the B-spline basis $\mathcal{B}$ defined as the tensor-product of the univariate B-spline bases $\mathcal{B}_1, \ldots, \mathcal{B}_d$. Let $\mathcal{Q}$ be tensor-product mesh consisting of the elements $Q = I_1 \times \cdots \times I_d$, where I_i is an element (closed interval) of the i-th univariate mesh, for $i = 1, \ldots, d$.

2.2 The Geometric Map and Isogeometric Refinements

In isogeometric analysis, the physical domain Ω is parametrized by the map $\mathbf{F} : \hat{\Omega} \to \Omega$ given by

$$\mathbf{x} = \mathbf{F}(\hat{\mathbf{x}}), \qquad \hat{\mathbf{x}} \in \hat{\Omega},$$

where $\mathbf{F}$ is a linear combination of the set of B-splines (or their rational extension) defined on an initial, usually coarse, tensor-product grid $\mathcal{Q}_0$. The map $\mathbf{F}$ is assumed to be invertible, with smooth inverse, on each mesh element.

The approximation space on Ω is given by $\operatorname{span}\{\beta \circ \mathbf{F}^{-1}\}_{\beta \in \mathcal{B}}$ as the push-forward of the spline space on $\hat{\Omega}$ and its approximation properties influence the accuracy of the corresponding isogeometric method. Three refinement possibilities are available and are usually indicated as h-refinement (mesh refinement), p-refinement (degree raising) and k-refinement (mesh refinement and degree raising) [1, 2, 4, 11, 13]. The different kinds of refinements are all constructed by applying the standard knot insertion and degree elevation algorithms, see [9, 13]. By exploiting these refinement procedures, refined approximation spaces with various mesh-size, order, and global regularity may be obtained from the initial spline space.

3 Assumptions

The main goal of this article is to build a quasi-interpolant operator for tensor-product spline spaces, assuming that the underlying univariate meshes with respect to the different coordinate directions satisfy one of the following assumptions.

Assumption 3.1 (Local Quasi-Uniformity) *There exists a constant $\theta > 0$ such that*

$$\theta^{-1} \leqslant \frac{\zeta_j - \zeta_{j-1}}{\zeta_{j+1} - \zeta_j} \leqslant \theta, \qquad \forall j = 2, \ldots, \tilde{n} - 1.$$

Assumption 3.2 *There exists a constant $\theta > 0$ such that for every $I \in \mathcal{I}$ and $1 \leqslant j_1, j_2 \leqslant n$,*

$$\theta^{-1} \leqslant \frac{|\operatorname{supp} \beta_{j_1}|}{|\operatorname{supp} \beta_{j_2}|} \leqslant \theta.$$

whenever $\operatorname{supp} \beta_{j_1} \cap \operatorname{supp} \beta_{j_2} \supset I$.

Remark 1 Assumption 3.2 holds if and only if there exists a constant[1] $C_2 > 0$ such that

$$\frac{|\tilde{I}|}{|\operatorname{supp} \beta_j|} \leqslant C_2,$$

for all $I \in \mathcal{I}$ and $1 \leqslant j \leqslant n$ such that $I \subset \operatorname{supp} \beta_j$.

Remark 2 Assumption 3.1 implies Assumption 3.2. On the other hand, Assumption 3.2 allows the shrinking of (at most) $p + 1$ knots and thus, it is weaker than Assumption 3.1. As an example we can consider $p = 2$ and

$$\Xi := \{0, 0, 0, 1/2 - \varepsilon, 1/2 + \varepsilon, 1, 1, 1\},$$

or

$$\Xi := \{0, 0, 0, 1/2 - \varepsilon, 1/2, 1/2 + \varepsilon, 1, 1, 1\},$$

for $0 < \varepsilon < \frac{1}{4}$. In this case, Assumption 3.2 holds but Assumption 3.1 does not, since θ would depend on ε in Assumption 3.1.

4 Some Results in Spline Spaces

In this section, we introduce bounds for the operator performing the change of basis from univariate B-splines restricted to the single knot span to Bernstein polynomials. This operator is commonly known as Bézier extraction operator. We extend such bounds for the tensor-product case and then, we analyse the local stability of the B-spline basis.

[1] This constant depends on the polynomial degree p, since the number of B-spline basis functions acting on a single mesh element is $p + 1$.

4.1 The Inverse of the Local Bézier Extraction Operator

The Bernstein polynomials of degree p on the knot interval $I = [\xi_k, \xi_{k+1}] \in \mathcal{I}$ are defined by

$$B_j^I(x) := \binom{p}{j-1}\left(\frac{x-\xi_k}{\xi_{k+1}-\xi_k}\right)^{j-1}\left(\frac{\xi_{k+1}-x}{\xi_{k+1}-\xi_k}\right)^{p-j+1}, \qquad j = 1, \ldots, p+1.$$

The set $\mathbb{B}_I := \{B_1^I, \ldots, B_{p+1}^I\}$ is a basis for the space $\mathcal{P}_p$ of polynomials of degree at most p over the interval of interest. We also consider the alternative basis $\mathcal{B}_I := \{\beta_1^I, \ldots, \beta_{p+1}^I\}$, consisting of the B-spline basis functions in $\mathcal{B}$ that are nonzero over I. More precisely, we have that

$$\beta_i^I \equiv \beta_{k+i-p-1}, \qquad \text{on } I, \qquad \forall\, i = 1, \ldots, p+1,$$

where $\operatorname{supp} \beta_{k+i-p-1} = [\xi_{k+i-p-1}, \xi_{k+i}]$. Let $D_I = (d_{ij}) \in \mathbb{R}^{(p+1)\times(p+1)}$ be the change of basis matrix such that

$$B_j^I = \sum_{i=1}^{p+1} d_{ji}\beta_i^I, \qquad \text{for } j = 1, \ldots, p+1. \tag{3}$$

For each $j = 1, \ldots, p+1$, the coefficients $\{d_{ji}\}_{i=1}^{p+1}$ can be computed by evaluating the blossom of B_j^I via

$$d_{ji} = B_j^I[\xi_{k+i-p}, \ldots, \xi_{k+i-1}],$$

see, e.g. [17, p. 65]. Thus, we have that (cf. [12, 17])

$$d_{ji} = \frac{1}{(j-1)!(p-j+1)!}\sum_{\sigma\in\Sigma}\left(\prod_{r=1}^{j-1}\frac{\xi_{k+i-\sigma(r)}-\xi_k}{\xi_{k+1}-\xi_k}\prod_{r=j}^{p}\frac{\xi_{k+1}-\xi_{k+i-\sigma(r)}}{\xi_{k+1}-\xi_k}\right),$$

where Σ denotes the set of the permutations in $\{1, \ldots, p\}$. In particular, for $j = 1$ we have that

$$\begin{aligned} d_{11} &= \frac{(\xi_{k+1}-\xi_{k-1})\ldots(\xi_{k+1}-\xi_{k+1-p})}{(\xi_{k+1}-\xi_k)^{p-1}} = \frac{\prod_{r=k+1-p}^{k-1}(\xi_{k+1}-\xi_r)}{(\xi_{k+1}-\xi_k)^{p-1}}, && (i=1) \\ d_{1i} &= 0, && i = 2, \ldots, p+1, \end{aligned} \tag{4}$$

and for $j = p + 1$,

$$d_{p+1,i} = 0, \qquad i = 1, \dots, p,$$

$$d_{p+1,p+1} = \frac{(\xi_{k+2} - \xi_k) \dots (\xi_{k+p} - \xi_k)}{(\xi_{k+1} - \xi_k)^{p-1}} = \frac{\prod_{r=k+2}^{k+p}(\xi_r - \xi_k)}{(\xi_{k+1} - \xi_k)^{p-1}}, \qquad (i = p + 1).$$

Let now consider the case $j = 2, \dots, p$. We may observe that either ξ_k or ξ_{k+1} are in $\{\xi_{k+i-p}, \dots, \xi_{k+i-1}\}$ for i equals to 1 or $p + 1$, respectively. Both ξ_k and ξ_{k+1} belong to this knot interval of interest for all the intermediate cases of $i = 2, \dots, p$. At least one of the two can be then fixed and we may consider the remaining nonzero contributions in the sum over the permutations defining d_{ji}, obtaining

$$\begin{aligned} |d_{ji}| &\leqslant \frac{(p-1)!}{(j-1)!(p-j+1)!} \frac{(\xi_{k+i-1} - \xi_{k+i-p})^{p-1}}{(\xi_{k+1} - \xi_k)^{p-1}} \\ &= \frac{1}{p} \binom{p}{j-1} \frac{(\xi_{k+i-1} - \xi_{k+i-p})^{p-1}}{(\xi_{k+1} - \xi_k)^{p-1}}, \qquad i = 1, \dots, p + 1. \end{aligned}$$

By taking into account that $\frac{1}{p} \sum_{j=2}^{p} \binom{p}{j-1} = \frac{1}{p}(2^p - 2)$, we then obtain

$$\sum_{j=1}^{p+1} |d_{ji}| \leqslant \left(\frac{1}{p}(2^p - 2) + 1 \right) \frac{|\hat{I}|^{p-1}}{|I|^{p-1}}, \qquad i = 1, \dots, p + 1.$$

Thus, we have proved the following result.

Lemma 1 *Let $I \in \mathcal{I}$ and $D_I = (d_{ij}) \in \mathbb{R}^{(p+1)\times(p+1)}$ be the change of basis matrix satisfying* (3). *Then,*

$$\|D_I^T\|_\infty = \max_{i=1,\dots,p+1} \sum_{j=1}^{p+1} |d_{ji}| \leqslant c_p \frac{|\hat{I}|^{p-1}}{|I|^{p-1}}, \tag{5}$$

where $c_p := \frac{1}{p}(2^p - 2) + 1$ and $\hat{I}$ is given by (2).

Whereas Assumption 3.1 allows to bound by above uniformly the right hand side of (5), in the next example we show that this is not the case when only Assumption 3.2 holds.

Example 1 Let $p \geqslant 2$ and let $\varepsilon > 0$. We consider

$$\Xi := \{\underbrace{0, \dots, 0}_{p+1 \text{ times}}, 1/2, 1/2 + \varepsilon, \dots, 1/2 + p\varepsilon, \underbrace{1, \dots, 1}_{p+1 \text{ times}}\}.$$

Note that in this case Assumption 3.2 holds. In particular, we show that it is not possible to bound $\|D_I^T\|_\infty$ uniformly by above, and that, in fact, the behaviour predicted by the right hand side of estimation (5) can be reached. Let $I := [1/2, 1/2 + \varepsilon] = [\xi_{p+2}, \xi_{p+3}]$. We have $\hat{I} = [\xi_3, \xi_{2p+2}] = [0, 1/2 + p\varepsilon]$, and, consequently, $\left(\frac{|\hat{I}|}{|I|}\right)^{p-1} = \mathcal{O}(\varepsilon^{1-p})$ as $\varepsilon \to 0$. According to (4), we then obtain

$$\|D_I^T\|_\infty \geqslant |d_{11}| = \left(\frac{1/2+\varepsilon}{\varepsilon}\right)^{p-1} = \mathcal{O}(\varepsilon^{1-p}), \quad \text{as } \varepsilon \to 0.$$

4.2 Local Stability of the B-spline Basis

Let $Q \in \mathcal{Q}$ and $\mathbf{p} := (p_1, p_2, \dots, p_d)$. We denote by $\mathcal{P}_\mathbf{p}$ the space of tensor-product polynomials with degree at most p_i in the coordinate direction x_i, for $i = 1, 2, \dots, d$. Let $N := \dim \mathcal{P}_\mathbf{p} = \Pi_{i=1}^d (p_i + 1)$. In this section we analyse the local stability of the B-spline basis. More precisely, we study the existence of a constant $C > 0$ (independent of Q) such that

$$\|x\|_\infty \leqslant C \left\| \sum_{j=1}^N x_j \beta_j^Q \right\|_{L^\infty(Q)}, \qquad \forall\, x = (x_1, \dots, x_N) \in \mathbb{R}^N, \tag{6}$$

where $\beta_1^Q, \dots, \beta_N^Q$ are the B-spline basis functions in $\mathcal{B}$ restricted to Q.

Remark 3 (The Inverse of the Local Bézier Extraction Operator) We now generalise the results of Sect. 4.1 to the tensor-product case. Let $Q = I_1 \times \cdots \times I_d \in \mathcal{Q}$ be given. We consider the set $\mathbb{B}_Q := \{B_1^Q, \dots, B_N^Q\}$ of tensor-product Bernstein polynomials on Q, which constitutes a basis for $\mathcal{P}_\mathbf{p}$. On the other hand, we consider the alternative basis $\mathcal{B}_Q := \{\beta_1^Q, \dots, \beta_N^Q\}$, consisting of the B-spline basis functions in $\mathcal{B}$ restricted to Q. Let $D_Q = (d_{ij}) \in \mathbb{R}^{N \times N}$ be the matrix such that

$$B_j^Q = \sum_{i=1}^N d_{ji} \beta_i^Q, \qquad \text{for } j = 1, \dots, N.$$

Notice that D_Q is the matrix for the change of bases and satisfies

$$[f]_{\mathcal{B}_Q} = D_Q^T [f]_{\mathbb{B}_Q}, \qquad \forall f \in \mathcal{P}_\mathbf{p}, \tag{7}$$

where $[f]_{\mathcal{B}_Q}$ and $[f]_{\mathbb{B}_Q}$ denote the vector of coefficients for writing f as a linear combination of the functions in $\mathcal{B}_Q$ and $\mathbb{B}_Q$, respectively.

It is easy to check that $D_Q = D_{I_d} \otimes \cdots \otimes D_{I_1}$ and that

$$\|D_Q^T\|_\infty = \prod_{i=1}^{d} \|D_{I_i}^T\|_\infty,$$

where $D_{I_i}^T$ denotes the corresponding univariate local Bézier extraction operator defined in Sect. 4.1, for $i = 1, \ldots, d$. In view of (5) we have that

$$\|D_Q^T\|_\infty \leqslant C_{\mathbf{p}} \prod_{i=1}^{d} \frac{|\hat{I}_i|^{p_i-1}}{|I_i|^{p_i-1}}, \tag{8}$$

where $C_{\mathbf{p}} := \prod_{i=1}^{d} c_{p_i} = \prod_{i=1}^{d} (\frac{1}{p_i}(2^{p_i} - 2) + 1)$. We remark that (8) generalises Lemma 1 for the tensor-product case. Notice that under Assumption 3.1 (in each coordinate direction), we can bound $\|D_Q^T\|_\infty$ uniformly by a constant which depends only on $\mathbf{p}$ and θ.

Remark 4 (L^∞-local Stability of the Bernstein Basis) Let $Q = [0, 1]^d$. Using the fact that all the norms are equivalent in $\mathcal{P}_{\mathbf{p}}$ we have that there exists a constant $C_{SB} > 0$ depending only on $\mathbf{p}$ such that

$$\|x\|_\infty \leqslant C_{SB} \left\| \sum_{j=1}^{N} x_j B_j^Q \right\|_{L^\infty(Q)}, \tag{9}$$

for all $x = (x_1, \ldots, x_N) \in \mathbb{R}^N$. The same result holds for an arbitrary rectangle $Q \subset \mathbb{R}^d$.

Now, using (7) and (9) we have the following result.

Lemma 2 (L^∞-Local Stability of the B-spline Basis) *Let $Q \in \mathcal{Q}$. Then,*

$$\|x\|_\infty \leqslant C_{SB} \|D_Q^T\|_\infty \left\| \sum_{j=1}^{N} x_j \beta_j^Q \right\|_{L^\infty(Q)},$$

for all $x = (x_1, \ldots, x_N) \in \mathbb{R}^N$.

Under Assumption 3.1, taking into account (8), we can bound $\|D_Q^T\|_\infty$ uniformly by a constant which depends only on $\mathbf{p}$ and θ. In this case, we have that the B-spline basis is L^∞-locally stable (see also [12]). On the other hand, under Assumption 3.2, estimate (6) does not hold, as it is showed in the following example.

Example 2 If we consider again the open-knot vector Ξ of Example 1, we have that

$$B_1^I = d_{11} \beta_1^I,$$

where $d_{11} = \left(\frac{1/2+\varepsilon}{\varepsilon}\right)^{p-1}$. Since $\|B_1^I\|_{L^\infty(I)} = 1$ we obtain

$$\left(\frac{1/2+\varepsilon}{\varepsilon}\right)^{p-1} = \frac{d_{11}}{1} \leqslant \sup_{x\in\mathbb{R}^{p+1}} \frac{\|x\|_\infty}{\left\|\sum_{j=1}^{p+1} x_j\beta_j^I\right\|_{L^\infty(I)}} \longrightarrow \infty,$$

as $\varepsilon \to 0$.

5 Local Approximation Methods

Regarding the stability estimation of Lemma 2, in this section we present a local approximation method.

Lemma 3 (Local L^2-projection) *Let $Q \in \mathcal{Q}$ and let $\Pi_Q : L^1(Q) \to \mathcal{P}_\mathbf{p}$ be the L^2-projection operator defined by*

$$\int_Q (f - \Pi_Q f)g = 0, \qquad \forall\, g \in \mathcal{P}_\mathbf{p}. \tag{10}$$

Then, there exists a constant $C_\Pi > 0$ depending only on **p** *such that*

$$\|\Pi_Q f\|_{L^\infty(Q)} \leqslant C_\Pi |Q|^{-1} \|f\|_{L^1(Q)}, \qquad \forall\, f \in L^1(Q).$$

Proof Let $f \in L^1(Q)$. From the definition of Π_Q it follows that

$$\|\Pi_Q f\|^2_{L^2(Q)} = \int_Q f\Pi_Q f \leqslant \|f\|_{L^1(Q)} \|\Pi_Q f\|_{L^\infty(Q)}.$$

On the other hand, since $\mathcal{P}_\mathbf{p}$ is a finite dimensional space, we have that there exists a constant $C_I > 0$ depending only on **p** such that the following inverse inequality holds:

$$\|g\|_{L^\infty(Q)} \leqslant C_I |Q|^{-\frac{1}{2}} \|g\|_{L^2(Q)}, \qquad \forall\, g \in \mathcal{P}_\mathbf{p}.$$

Therefore,

$$\|\Pi_Q f\|^2_{L^\infty(Q)} \leqslant C_I^2 |Q|^{-1} \|\Pi_Q f\|^2_{L^2(Q)} \leqslant C_I^2 |Q|^{-1} \|f\|_{L^1(Q)} \|\Pi_Q f\|_{L^\infty(Q)},$$

and thus,

$$\|\Pi_Q f\|_{L^\infty(Q)} \leqslant C_I^2 |Q|^{-1} \|f\|_{L^1(Q)}.$$

Remark 5 Notice that

$$\Pi_Q f = \sum_{i=1}^{N} \lambda_i^Q(f)\beta_i^Q, \qquad \forall f \in L^1(Q), \tag{11}$$

where $\lambda^Q(f) := (\lambda_1^Q(f), \ldots, \lambda_N^Q(f))^T$ is the solution of the linear system

$$M_Q \mathbf{x} = F_Q,$$

where

$$M_Q = \left(\int_Q \beta_j^Q \beta_i^Q\right)_{i,j=1,\ldots,N} \in \mathbb{R}^{N\times N}, \quad \text{and} \quad F_Q = \left(\int_Q f\beta_i^Q\right)_{i=1,\ldots,N} \in \mathbb{R}^{N\times 1}.$$

On the other hand, $\{\lambda_i^Q : L^1(Q) \to \mathbb{R} \mid i = 1, \ldots, N\}$ is a dual basis for $\mathcal{B}_Q$ in the sense that

$$\lambda_i^Q(\beta_j^Q) = \delta_{ij}, \qquad i,j = 1, \ldots, N. \tag{12}$$

As a consequence of the L^∞-local stability of the B-spline basis (Lemma 2) we can state the following result.

Theorem 1 *Let $Q \in \mathcal{Q}$ and let $\Pi_Q : L^1(Q) \to \mathcal{P}_{\mathbf{p}}$ be the L^2-projection operator defined by* (10). *Let q be such that $1 \leqslant q \leqslant \infty$. Then,*

$$\|\lambda^Q(f)\|_\infty \leqslant C_\Pi C_{SB} \|D_Q^T\|_\infty |Q|^{-\frac{1}{q}} \|f\|_{L^q(Q)}, \qquad \forall f \in L^q(Q), \tag{13}$$

where $\lambda^Q(f) = (\lambda_1^Q(f), \ldots, \lambda_N^Q(f))^T$ are the coefficients of $\Pi_Q(f)$ with respect to the local B-spline basis $\mathcal{B}_Q$ (cf. (11)*).*

Proof Let q be such that $1 \leqslant q \leqslant \infty$ and $f \in L^q(Q)$. Using Lemmas 2 and 3 we have that

$$\|\lambda^Q(f)\|_\infty \leqslant C_{SB}\|D_Q^T\|_\infty \|\Pi_Q f\|_{L^\infty(Q)} \leqslant C_\Pi C_{SB}\|D_Q^T\|_\infty |Q|^{-1} \|f\|_{L^1(Q)}.$$

Finally, (13) is as consequence of Hölder inequality.

6 Locally Supported Dual Basis for B-splines

The goal of this section is to define a dual basis for the multivariate B-spline basis $\mathcal{B}$, i.e., a set of linear functionals

$$\{\lambda_\beta : L^1(\Omega) \to \mathbb{R} \mid \beta \in \mathcal{B}\},$$

such that $\lambda_{\beta_i}(\beta_j) = \delta_{ij}$, for all $\beta_i, \beta_j \in \mathcal{B}$. More precisely, we are interested in defining such functionals satisfying the following properties:

(i) *Local support:* λ_β is *supported* in $\Lambda_\beta \subset \operatorname{supp}\beta$, i.e.,

$$\forall f \in L^1(\Omega), \quad f_{|\Lambda_\beta} \equiv 0 \qquad \Longrightarrow \qquad \lambda_\beta(f) = 0.$$

(ii) *Dual basis:* For $\beta_i, \beta_j \in \mathcal{B}$, $\lambda_{\beta_i}(\beta_j) = \delta_{ij}$.
(iii) *L^q-Stability:* Let $1 \leqslant q \leqslant \infty$. There exists a constant $C_S > 0$ such that

$$|\lambda_\beta(f)| \leqslant C_S |\operatorname{supp}\beta|^{-\frac{1}{q}} \|f\|_{L^q(\operatorname{supp}\beta)}, \qquad \forall f \in L^q(\Omega),\ \beta \in \mathcal{B}. \tag{14}$$

Remark 6 Condition (iii) will be a key tool for proving the local stability of a quasi-interpolant operator in Sect. 7.

We will use the technique in [15] to define linear functionals $\{\lambda_\beta\}_{\beta\in\mathcal{B}}$ satisfying the desired properties. Roughly speaking, we define the functional λ_β as a convex combination of the local projections onto some $Q \in \mathcal{Q}$ such that $Q \subset \operatorname{supp}\beta$. For $\beta \in \mathcal{B}$, we define

$$\mathcal{Q}_\beta := \{Q \in \mathcal{Q} \,|\, Q \subset \operatorname{supp}\beta\},$$

and for each $Q \in \mathcal{Q}_\beta$, let $\lambda_\beta^Q := \lambda_{i_0}^Q$, where $i_0 = i_0(\beta, Q)$ with $1 \leqslant i_0 \leqslant N$ is such that $\beta_{i_0}^Q \equiv \beta$ on Q. Thus, the functional λ_β is given by

$$\lambda_\beta := \sum_{Q\in\mathcal{Q}_\beta} c_{Q,\beta}\lambda_\beta^Q, \tag{15}$$

where

$$\forall\, Q \in \mathcal{Q}_\beta,\ c_{Q,\beta} \geqslant 0, \qquad \text{and} \qquad \sum_{Q\in\mathcal{Q}_\beta} c_{Q,\beta} = 1. \tag{16}$$

Notice that λ_β is supported in $\Lambda_\beta := \bigcup_{\substack{Q\in\mathcal{Q}_\beta \\ c_{Q,\beta}>0}} Q \subset \operatorname{supp}\beta$, and therefore, condition (i) holds. On the other hand, condition (ii) is a consequence of (12) and (16). In the rest of this section we analyse the validity of (iii).

We propose some possible choices for the coefficients $c_{Q,\beta}$, in order to guarantee the validity of condition (iii) under different assumptions on the underlying univariate meshes.

Case 1: Locally Quasi-Uniform Meshes (Assumption 3.1)

Under the Assumption 3.1 there is no restriction on the choice of the coefficients $c_{Q,\beta}$, i.e., condition (iii) always holds. In fact, taking into account (13), we have that (14) holds with a constant $C_S > 0$ which depends on $\mathbf{p}$ and θ.

Case 2: Non Locally Quasi-Uniform Meshes (Assumption 3.2)

If Assumption 3.1 does not hold, but Assumption 3.2 does, we propose two ways of defining $c_{Q,\beta}$ in order to obtain the validity of condition (iii).

1. Let[2] $c_\beta := \sum_{Q\in\mathcal{Q}_\beta}\prod_{i=1}^{d}\left(\frac{|I_i|}{|\tilde{I}_i|}\right)^{p_i-1+\frac{1}{q}}$ and

$$c_{Q,\beta} := \frac{1}{c_\beta}\prod_{i=1}^{d}\left(\frac{|I_i|}{|\tilde{I}_i|}\right)^{p_i-1+\frac{1}{q}}, \qquad \forall\, Q \in \mathcal{Q}_\beta.$$

Then, using the definition of λ_β given by (15) and the bound for λ_β^Q given by Theorem 1 we have that

$$|\lambda_\beta(f)| \leqslant C_\Pi C_{SB} \sum_{Q\in\mathcal{Q}_\beta} c_{Q,\beta}\|D_Q^T\|_\infty |Q|^{-\frac{1}{q}}\|f\|_{L^q(Q)}.$$

Now, regarding the definition of $c_{Q,\beta}$ and the bound for $\|D_Q^T\|_\infty$ given in (8), we obtain

$$|\lambda_\beta(f)| \leqslant \frac{C_\Pi C_{SB} C_{\mathbf{p}}}{c_\beta}\sum_{Q\in\mathcal{Q}_\beta}|\tilde{Q}|^{-\frac{1}{q}}\|f\|_{L^q(Q)}.$$

Since $|\operatorname{supp}\beta| \leqslant |\tilde{Q}|$, for all $Q \in \mathcal{Q}_\beta$,

$$|\lambda_\beta(f)| \leqslant \frac{C_\Pi C_{SB} C_{\mathbf{p}} (\#\mathcal{Q}_\beta)^{1-\frac{1}{q}}}{c_\beta}|\operatorname{supp}\beta|^{-\frac{1}{q}}\|f\|_{L^q(\operatorname{supp}\beta)}.$$

---

[2]For each Q we consider the representation $Q = I_1 \times \cdots \times I_d$.

Under the Assumption 3.2,[3] we can bound c_β uniformly by below. More precisely, if $p_{\max} := \max_{i=1,\dots,d} p_i$, then

$$\begin{aligned} c_\beta &= \sum_{Q\in\mathcal{Q}_\beta} \prod_{i=1}^{d} \left(\frac{|I_i|}{|\tilde{I}_i|}\right)^{p_i-1+\frac{1}{q}} \geqslant \sum_{Q\in\mathcal{Q}_\beta} \left(\prod_{i=1}^{d} \frac{|I_i|}{|\tilde{I}_i|}\right)^{p_{\max}-1+\frac{1}{q}} \\ &= \sum_{Q\in\mathcal{Q}_\beta} \left(\frac{|Q|}{|\tilde{Q}|}\right)^{p_{\max}-1+\frac{1}{q}} \geqslant (\#\mathcal{Q}_\beta)^{2-p_{\max}-\frac{1}{q}} \left(\sum_{Q\in\mathcal{Q}_\beta} \frac{|Q|}{|\tilde{Q}|}\right)^{p_{\max}-1+\frac{1}{q}} \\ &\geqslant (\#\mathcal{Q}_\beta)^{2-p_{\max}-\frac{1}{q}} \frac{1}{C_2^{d(p_{\max}-1+\frac{1}{q})}} \left(\underbrace{\sum_{Q\in\mathcal{Q}_\beta} \frac{|Q|}{|\operatorname{supp}\beta|}}_{=1}\right)^{p_{\max}-1+\frac{1}{q}}. \end{aligned}$$

Thus, $\frac{1}{c_\beta} \leqslant \frac{C_2^{d(p_{\max}-1+\frac{1}{q})}}{(\#\mathcal{Q}_\beta)^{2-p_{\max}-\frac{1}{q}}}$, and

$$|\lambda_\beta(f)| \leqslant C_\Pi C_{SB} C_{\mathbf{p}} (\#\mathcal{Q}_\beta)^{p_{\max}-1} C_2^{d(p_{\max}-1+\frac{1}{q})} |\operatorname{supp}\beta|^{-\frac{1}{q}} \|f\|_{L^q(\operatorname{supp}\beta)},$$

which in turn implies (14).

2. For each β we associate an element $Q_\beta \in \mathcal{Q}_\beta$ with size equivalent to the size of its support, i.e., such that

$$\frac{|\operatorname{supp}\beta|}{|Q_\beta|} \leqslant C,$$

with a constant C depending on $\mathbf{p}$. For example, we can select $Q_\beta \in \underset{Q\in\mathcal{Q}_\beta}{\arg\max}|Q|$ and in this case $\frac{|\operatorname{supp}\beta|}{|Q_\beta|} \leqslant \#\mathcal{Q}_\beta \leqslant N$.

Under the Assumption 3.2 we have that

$$\|D^T_{Q_\beta}\|_\infty \leqslant C_{\mathbf{p}} \left(\frac{|\tilde{Q}_\beta|}{|Q_\beta|}\right)^{p_{\max}-1} \leqslant C_{\mathbf{p}}(C_2^d C)^{p_{\max}-1}.$$

[3] Without loss of generality, we denote by C_2 the constant in Remark 1 for each coordinate direction, and thus, $\frac{|\tilde{Q}|}{|\operatorname{supp}\beta|} \leqslant C_2^d$, for all $Q \in \mathcal{Q}$ such that $Q \subset \operatorname{supp}\beta$.

In this case, we define $c_{Q,\beta} := \begin{cases} 1, & \text{if } Q = Q_\beta \\ 0, & \text{if } Q \neq Q_\beta \end{cases}$, i.e., $\lambda_\beta = \lambda_\beta^{Q_\beta}$. Finally, (14) follows from Theorem 1.

7 Quasi-Interpolation in Spline Spaces

We consider a dual basis $\{\lambda_\beta\}_{\beta\in\mathcal{B}}$ from (15), satisfying conditions (i)-(ii)-(iii) stated in the previous section, for some q such that $1 \leqslant q \leqslant \infty$. Let $P : L^q(\Omega) \to \mathbb{V} = \operatorname{span}\mathcal{B}$ be given by

$$P(f) := \sum_{\beta\in\mathcal{B}} \lambda_\beta(f)\beta, \qquad \forall f \in L^q(\Omega). \tag{17}$$

The next result states some important properties of P.

Theorem 2 *The following holds:*

(a) *P is a projection on $\mathbb{V}$, i.e., for all $f \in \mathbb{V}$, $P(f) = f$.*
(b) Local stability*: Let $1 \leqslant q \leqslant \infty$. For $Q = I_1 \times \ldots \times I_d \in \mathcal{Q}$, the operator P satisfies*

$$\|Pf\|_{L^q(Q)} \leqslant C_S \|f\|_{L^q(\tilde{Q})}, \qquad \forall f \in L^q(\Omega).$$

where $\tilde{Q} = \tilde{I}_1 \times \ldots \times \tilde{I}_d$ denotes the support extension (see (1)) and $C_S > 0$ is the constant appearing in (14).
(c) Local approximation*: Let $\mathbf{s} := (s_1, \ldots, s_d)$ be such that $0 \leqslant s_i \leqslant p_i + 1$, for $i = 1, \ldots, d$. Then, there exists a constant $C_A > 0$ such that, for $Q = I_1 \times \ldots \times I_d \in \mathcal{Q}$, it holds that*

$$\|f - Pf\|_{L^q(Q)} \leqslant C_A \sum_{i=1}^{d} |\tilde{I}_i|^{s_i} \|D_{x_i}^{s_i} f\|_{L^q(\tilde{Q})}, \qquad \forall f \in W^{q,\mathbf{s}}(\Omega),$$

where $W^{q,\mathbf{s}}(\Omega) := \{f \in L^q(\Omega) : D_{x_i}^{r_i} f \in L^q(\Omega),\ 0 \leqslant r_i \leqslant s_i,\ i = 1, \ldots, d\}$.

Proof (a) It is an immediate consequence of condition (ii).
(b) Let $Q \in \mathcal{Q}$. Then, taking into account the definition of P given by (17), the spline partition-of-unity property and the L^q-stability in (14), we have

$$|P(f)| \leqslant \max_{\substack{\beta\in\mathcal{B} \\ \operatorname{supp}\beta \supset Q}} |\lambda_\beta(f)| \leqslant C_S |Q|^{-\frac{1}{q}} \|f\|_{L^q(\tilde{Q})}, \qquad \text{on } Q.$$

Therefore,

$$\|P(f)\|_{L^q(Q)} \leqslant C_S \|f\|_{L^q(\tilde{Q})}.$$

(c) Let $Q \in \mathcal{Q}$. By the classical polynomial approximation property, there exists $p_{\tilde{Q}} \in \mathcal{P}_{\mathbf{p}}$ such that

$$\|f - p_{\tilde{Q}}\|_{L^q(\tilde{Q})} \leqslant C_T \sum_{i=1}^{d} |\tilde{I}_i|^{s_i} \|D_{x_i}^{s_i} f\|_{L^q(\tilde{Q})}, \tag{18}$$

where the constant $C_T > 0$ only depends on d, $\mathbf{p}$, $\mathbf{s}$ and q. Taking into account the local stability of P given in (b) and (18) we have that

$$\begin{aligned}
\|f - Pf\|_{L^q(Q)} &\leqslant \|f - p_{\tilde{Q}}\|_{L^q(Q)} + \|p_{\tilde{Q}} - Pf\|_{L^q(Q)} \\
&= \|f - p_{\tilde{Q}}\|_{L^q(Q)} + \|P(p_{\tilde{Q}} - f)\|_{L^q(Q)} \\
&\leqslant (1 + C_S)\|f - p_{\tilde{Q}}\|_{L^q(\tilde{Q})} \\
&\leqslant (1 + C_S) C_T \sum_{i=1}^{d} |\tilde{I}_i|^{s_i} \|D_{x_i}^{s_i} f\|_{L^q(\tilde{Q})}.
\end{aligned}$$

Remark 7 The operator defined in [21] and called Bézier projection fits in this framework and consists in a specific choice of coefficients $c_{Q,\beta}$ in (15). Our results of Sect. 6 provide stability for this operator under Assumption 3.1.

8 Conclusions

We have defined a class of quasi-interpolation operators onto spline spaces that enjoy L^2 stability properties and optimal locality properties, under very general assumption on the knot distributions. These operators are proved to deliver optimal approximation properties with respect to h for tensor product spline spaces. It should be noted though that the behaviour of constants with respect to the degree p is not analysed and is likely not optimal.

The class of operators we consider are associated with the construction of a dual basis and for this reason, they can be used in situations that are more general than tensor product B-splines. In particular, following [3] and [5], it is clear that the same construction would provide a dual basis in the case of analysis suitable (or dual compatible) T-splines (see [6] and the references there in). In the same lines, following [20] and [7], our class of operators can be used to provide quasi-interpolation operators for hierarchical splines (see [14, 22]) as well. The analysis presented in this paper can provide a general framework for the study of the local approximation properties for such quasi-interpolants on either T-splines or Hierarchical splines, but such results are beyond the scope of the present contribution.

Acknowledgements Annalisa Buffa and Giancarlo Sangalli were partially supported by the European Research Council through the FP7 ERC Consolidator Grant n.616563 *HIGEOM*, and by the Italian MIUR through the PRIN "Metodologie innovative nella modellistica differenziale numerica". Eduardo M. Garau was partially supported by CONICET through grant PIP 112-2011-0100742, by Universidad Nacional del Litoral through grants CAI+D 500 201101 00029 LI, 501 201101 00476 LI, by Agencia Nacional de Promoción Científica y Tecnológica, through grants PICT-2012-2590 and PICT-2014-2522 (Argentina). Carlotta Giannelli was supported by the project DREAMS (MIUR "Futuro in Ricerca" RBFR13FBI3) and by the Gruppo Nazionale per il Calcolo Scientifico (GNCS) of the Istituto Nazionale di Alta Matematica (INdAM). This support is gratefully acknowledged.

References

1. Y. Bazilevs, L. Beirão da Veiga, J.A. Cottrell, T.J.R. Hughes, G. Sangalli, Isogeometric analysis: approximation, stability and error estimates for h-refined meshes. Math. Models Methods Appl. Sci. **16**, 1031–1090 (2006)
2. L. Beirão da Veiga, A. Buffa, J. Rivas, G. Sangalli, Some estimates for h-p-k-refinement in isogeometric analysis. Numer. Math. **118**, 271–305 (2011)
3. L. Beirão da Veiga, A. Buffa, D. Cho, G. Sangalli, Analysis-suitable T-splines are dual-compatible. Comput. Methods Appl. Mech. Eng. **249/252**, 42–51 (2012)
4. L. Beirão da Veiga, D. Cho, G. Sangalli, Anisotropic NURBS approximation in isogeometric analysis. Comput. Methods Appl. Mech. Eng. **209/212**, 1–11 (2012)
5. L. Beirão da Veiga, A. Buffa, G. Sangalli, R. Vázquez, Analysis-suitable T-splines of arbitrary degree: definition, linear independence and approximation properties. Math. Models Methods Appl. Sci. **23**(2), 1–25 (2013)
6. L. Beirão da Veiga, A. Buffa, G. Sangalli, R.Vázquez, Mathematical analysis of variational isogeometric methods. Acta Numer. **23**, 157–287 (2014)
7. A. Buffa, E.M. Garau, New refinable spaces and local approximation estimates for hierarchical splines. IMA J. Numer Anal. (2016, to appear)
8. J.A. Cottrell, T.J.R. Hughes, Y. Bazilevs, *Isogeometric Analysis: Toward Integration of CAD and FEA* (Wiley, New York, 2009)
9. C. de Boor, *A Practical Guide to Splines*, Revised edition. Applied Mathematical Sciences, vol. 27 (Springer, New York, 2001)
10. C. de Boor, G.J. Fix, Spline approximation by quasi-interpolants. J. Approx. Theory **8**, 19–45 (1973)
11. J. Evans, Y. Bazilevs, I. Babuska, T. Hughes, N-widths, sup-infs, and optimality ratios for the k-version of the isogeometric finite element method. Comput. Methods Appl. Mech. Eng. **198**, 1726–1741 (2009)
12. C. Giannelli, B. Jüttler, H. Speleers, Strongly stable bases for adaptively refined multilevel spline spaces. Adv. Comput. Math. **40**(2), 459–490 (2014)
13. T.J.R. Hughes, J.A. Cottrell, Y. Bazilevs, Isogeometric analysis: CAD, finite elements, NURBS, exact geometry and mesh refinement. Comput. Methods Appl. Mech. Eng. **194**, 4135–4195 (2005)
14. R. Kraft, Adaptive und linear unabhängige multilevel B-Splines und ihre Anwendungen, Ph.D. thesis, Universität Stuttgart, 1998

15. B.-G. Lee, T. Lyche, K. Mørken, Some examples of quasi-interpolants constructed from local spline projectors, in *Mathematical Methods for Curves and Surfaces (Oslo, 2000)*. Innov. Appl. Math. (Vanderbilt University Press, Nashville, TN, 2001), pp. 243–252
16. T. Lyche, L.L. Schumaker, Local spline approximation methods. J. Approx. Theory **15**, 294–325 (1975)
17. H. Prautzsch, W. Boehm, M. Paluszny, *Bézier and B-spline Techniques*. Mathematics and Visualization (Springer, Berlin, 2002), xiv+304 pp. ISBN: 3-540-43761-4
18. P. Sablonnière, Recent progress on univariate and multivariate polynomial and spline quasi-interpolants, in *Trends and Applications in Constructive Approximation*, ed. by M.G. de Brujn, D.H. Mache, J. Szabadoz. ISNM, vol. 151 (Birhäuser Verlag, Basel, 2005), pp. 229–245
19. L.L. Schumaker, *Spline Functions: Basic Theory*. Cambridge Mathematical Library, 3rd edn. (Cambridge University Press, Cambridge, 2007)
20. H. Speleers, C. Manni, Effortless quasi-interpolation in hierarchical spaces. Numer. Math. **132**(1), 55–184 (2016)
21. D.C. Thomas, M.A. Scott, J.A. Evans, K. Tew, E.J. Evans, Bézier projection: a unified approach for local projection and quadrature-free refinement and coarsening of NURBS and T-splines with particular application to isogeometric design and analysis. Comput. Methods Appl. Mech. Eng. **284**, 55–105 (2015)
22. A.-V. Vuong, C. Giannelli, B. Jüttler, B. Simeon, A hierarchical approach to adaptive local refinement in isogeometric analysis. Comput. Methods Appl. Mech. Eng. **200**(49–52), 3554–3567 (2011)

Stabilised Finite Element Methods for Ill-Posed Problems with Conditional Stability

Erik Burman

Abstract In this paper we discuss the adjoint stabilised finite element method introduced in Burman (SIAM J Sci Comput 35(6):A2752–A2780, 2013) and how it may be used for the computation of solutions to problems for which the standard stability theory given by the Lax-Milgram Lemma or the Babuska-Brezzi Theorem fails. We pay particular attention to ill-posed problems that have some conditional stability property and prove (conditional) error estimates in an abstract framework. As a model problem we consider the elliptic Cauchy problem and provide a complete numerical analysis for this case. Some numerical examples are given to illustrate the theory.

1 Introduction

Most methods in numerical analysis are designed making explicit use of the well-posedness [23] of the underlying continuous problem. This is natural as long as the problem at hand indeed is well-posed, but even for well-posed continuous problems the resulting discrete problem may be unstable if the finite element spaces are not well chosen or if the mesh-size is not small enough. This is for instance the case for indefinite problems, such as the Helmholtz problem, or constrained problems such as Stokes' equations. For problems that are ill-posed on the continuous level on the other hand the approach makes less sense and leads to the need of regularization on the continuous level so that the ill-posed problem can be approximated by solving a sequence of well-posed problems. The regularization of the continuous problem can consist for example of Tikhonov regularization [29] or a so-called quasi reversibility method [27]. In both cases the underlying problem is perturbed and the original solution (if it exists) is recovered only in the limit as some regularization parameter goes to zero. The disadvantage of this approach from a numerical analysis perspective is that once the continuous problem has been perturbed to some order, the accuracy

E. Burman (✉)
Department of Mathematics, University College London, London WC1E 6BT, UK
e-mail: e.burman@ucl.ac.uk

G.R. Barrenechea et al. (eds.), *Building Bridges: Connections and Challenges in Modern Approaches to Numerical Partial Differential Equations*,
Lecture Notes in Computational Science and Engineering 114,
DOI 10.1007/978-3-319-41640-3_4

of the computational method must be made to match that of the regularization. The strength of the regularization on the other hand must make the continuous problem stable and damp perturbations induced by errors in measurement data. This leads to a twofold matching problem where the regularization introduces a perturbation of first order, essentially excluding the efficient use of many tools from numerical analysis such as high order methods, adaptivity and stabilisation. The situation is vaguely reminiscent of that in conservation laws where in the beginning low order methods inspired by viscosity solution arguments dominated, to later give way for high resolution techniques, based on flux limiter finite volume schemes or (weakly) consistent stabilised finite element methods such as the Galerkin Least Squares methods (GaLS) or discontinuous Galerkin methods (dG) (see for instance [18] and references therein). These methods allow for high resolution in the smooth zone while introducing sufficient viscous stabilisation in zones with nonlinear phenomena such as shocks or rarefaction waves.

In this paper our aim is to advocate a similar shift towards weakly consistent stabilisation methods for the computation of ill-posed problems. The philosophy behind this is to cast the problem in the form of a constrained optimisation problem, that is first discretized, leading to a possibly unstable discrete problem. The problem is then regularized on the discrete level using techniques known from the theory of stabilised finite element methods. This approach has the following potential advantages some of which will be explored below:

- the optimal scaling of the penalty parameter with respect to the mesh parameter follows from the error analysis;
- for ill-posed problems where a conditional stability estimate holds, error estimates may be derived that are in a certain sense optimal with respect to the discretization parameters;
- discretization errors and perturbation errors may be handled in the same framework;
- a posteriori error estimates may be used to drive adaptivity;
- a range of stabilised finite element methods may be used for the regularization of the discrete problem;
- the theory can be adapted to many different problems.

Stabilised finite element methods represent a general technique for the regularization of the standard Galerkin method in order to improve its stability properties for instance for advection–diffusion problems at high Péclet number or to achieve inf-sup stability for the pressure-velocity coupling in the Stokes' system. To achieve optimal order convergence the stabilisation terms must have some consistency properties, i.e. they decrease at a sufficiently high rate when applied to the exact solution or to any smooth enough function. Such stabilising terms appear to have much in common with Tikhonov regularization in inverse problems, although the connection does not seem to have been made in general. In the recent papers [10, 13] we considered stabilised finite element methods for problems where coercivity fails for the continuous problem and showed that optimal error estimates can be obtained without, or under very weak, conditions on the physical parameters and the mesh parameters, also for problems where the standard Galerkin method may fail.

In the first part of this series [10] we considered the analysis of elliptic problems without coercivity using duality arguments. The second part [13] was consecrated to problems for which coercivity fails, but which satisfy the Babuska-Brezzi Theorem, illustrated by the transport equation. Finally in the note [12] we extended the analysis of [10] to the case of ill-posed problems with some *conditional* stability property.

Our aim in the present essay is to review and unify some of these results and give some further examples of how stabilised methods can be used for the solution of ill-posed problem. To exemplify the theory we will restrict the discussion to the case of scalar second order elliptic problems on the form

$$\mathcal{L}u = f \quad \text{in } \Omega \tag{1}$$

where $\mathcal{L}$ is a linear second order elliptic operator, u is the unknown and f is some known data and Ω is some simply connected, open subset of $\mathbb{R}^d$, $d = 2, 3$. Observe that the operator $\mathcal{L}$ does not necessarily have to be on divergence form, although we will only consider this case here to make the exposition concise (see [30] for an analysis of well-posed elliptic problems on nondivergence form).

The discussion below will also be restricted to finite element spaces that are subsets of $H^1(\Omega)$. For the extension of these results to a nonconforming finite element method we refer to [11].

1.1 Conditional Stability for Ill-Posed Problems

There is a rich literature on conditional stability estimates for ill-posed problems. Such estimates often take the form of three sphere's inequalities or Carleman estimates, we refer the reader to [2] and references therein.

The estimates are conditional, in the sense that they only hold under the condition that the exact solution exists in some Sobolev space V, equipped with scalar product $(\cdot,\cdot)_V$ and associated norm $\|\cdot\|_V := (\cdot,\cdot)_V$. Herein we will only consider the case where $V \equiv H^1(\Omega)$. Then we introduce $V_0 \subset V$ and consider the problem: find $u \in V_0$ such that

$$a(u, w) = l(w), \quad \forall w \in W, \tag{2}$$

Observe that V_0 and W typically are different subsets of $H^1(\Omega)$ and we do not assume that W is a subset or V_0 or vice versa. The operators $a(\cdot,\cdot) : V \times V \to \mathbb{R}$, $l(\cdot) : W \to \mathbb{R}$ denote a bounded bilinear and a bounded linear form respectively. The form $a(\cdot,\cdot)$ is a weak form of $\mathcal{L}u$. We let $\|\cdot\|_C$ denote the norm for which the condition must be satisfied and $\|\cdot\|_S$ denote the norm in which the stability holds.

We then assume that a stability estimate of the following form holds: if for some $x \in V_0$, with $\|x\|_C \leqslant E$ there exist $\varepsilon < 1$ and $r \in W'$ such that

$$\begin{cases} a(x, v) = (r, v)_{\langle W', W\rangle} \ \forall v \in W \\ \|r\|_{W'} \leqslant \varepsilon \end{cases} ; \quad \text{then } \|x\|_S \leqslant \Xi_E(\varepsilon), \tag{3}$$

where $\Xi_E(\cdot) : \mathbb{R}^+ \to \mathbb{R}^+$ is a smooth, positive, function, depending on the problem, $\|\cdot\|_S$ and E, with $\lim_{s\to 0+} \Xi_E(s) = 0$. Depending on the problem different smallness conditions may be required to hold on ε.

The idea is that the stabilised methods we propose may use the estimate (3) *directly* for the derivation of error estimates, without relying on the Lax-Milgram Lemma or the Babuska-Brezzi Theorem. Let us first make two observations valid also for well-posed problems. When the assumptions of the Lax-Milgram's lemma are satisfied (3) holds unconditionally for the energy norm and $\Xi_E(\varepsilon) = C\varepsilon$, for some problem dependent constant C. If for a given problem the adjoint equation $a(v, z) = j(v)$ admits a solution $z \in W$, with $\|z\|_W \leqslant E_j$, for some linear functional $j \in V'$ then

$$|j(x)| = |a(x, z)| = |r(z)| \leqslant E_j \|r\|_{W'} \tag{4}$$

and we see that for this case the condition of the conditional stability applies to the adjoint solution.

Herein we will focus on the case of the elliptic Cauchy problem as presented in [2]. In this problem both Dirichlet and Neumann data are given on a part of the boundary, whereas nothing is known on the complement. We will end this section by detailing the conditional stability (3) of the elliptic Cauchy problem. We give the result here with reduced technical detail and refer to [2] for the exact dependencies of the constants on the physical parameters and the geometry.

1.2 Example: The Elliptic Cauchy Problem

The problem that we are interested in takes the form

$$\begin{cases} -\nabla \cdot (\sigma \nabla u) + cu = f, \text{ in } \Omega \\ u = 0 \text{ on } \Gamma_D \\ \partial_n u = \psi \text{ on } \Gamma_N \end{cases} \tag{5}$$

where $\Omega \subset \mathbb{R}^d$, $d = 2, 3$ is a polyhedral (polygonal) domain with boundary $\partial\Omega$, $\partial_n u := n^T \cdot \nabla u$, (with n the outward pointing normal on $\partial\Omega$), $\sigma \in \mathbb{R}^{d\times d}$ is a symmetric matrix for which $\exists \sigma_0 \in \mathbb{R}$, $\sigma_0 > 0$ such that $y^T \cdot \sigma y > \sigma_0$ for all $y \in \mathbb{R}^d$ and $c \in \mathbb{R}$. By Γ_N, Γ_D we denote polygonal subsets of the boundary $\partial\Omega$, with union $\Gamma_B := \Gamma_D \cup \Gamma_N$ and that overlap on some set of nonzero $(d-1)$-dimensional measure, $\Gamma_S := \Gamma_D \cap \Gamma_N \neq \emptyset$. We denote the complement of the Dirichlet boundary $\Gamma_D' := \partial\Omega \setminus \Gamma_D$, the complement of the Neumann boundary $\Gamma_N' := \partial\Omega \setminus \Gamma_N$ and the complement of their union $\Gamma_B' := \partial\Omega \setminus \Gamma_B$. To exclude the well-posed case, we assume that the $(d-1)$-dimensional measure of Γ_S and Γ_B' is non-zero. The practical interest in (5) stems from engineering problems where the boundary condition, or its data, is unknown on Γ_B', but additional measurements ψ of the fluxes are available on a part of the accessible boundary Γ_S. This results in an ill-posed reconstruction problem, that in practice most likely does not have a solution due to measurement errors in the fluxes [5]. However if the underlying physical process is stable, (in the sense that the problem where boundary data is known is well-posed) we may assume that it allows for a unique solution in the idealized situation of unperturbed data. This is the approach we will take below. To this end we assume that $f \in L^2(\Omega)$, $\psi \in L^2(\Gamma_N)$ and that a unique $u \in H^s(\Omega)$, $s > \frac{3}{2}$ satisfies (5). For the derivation of a weak formulation we introduce the spaces $V_0 := \{v \in H^1(\Omega) : v|_{\Gamma_D} = 0\}$ and $W := \{v \in H^1(\Omega) : v|_{\Gamma_N'} = 0\}$, both equipped with the H^1-norm and with dual spaces denoted by V_0' and W'.

Using these spaces we obtain a weak formulation: find $u \in V_0$ such that

$$a(u, w) = l(w) \quad \forall w \in W, \tag{6}$$

where

$$a(u, w) = \int_\Omega (\sigma \nabla u) \cdot \nabla w + cuw \, dx,$$

and

$$l(w) := \int_\Omega fw \, dx + \int_{\Gamma_N} \psi \, w \, ds.$$

It is known [2, Theorems 1.7 and 1.9 with Remark 1.8] that if there exists a solution $u \in H^1(\Omega)$, to (6), a conditional stability of the form (3) holds provided $0 \leq \varepsilon < 1$ and

$$\begin{aligned} &\|u\|_S := \|u\|_{L^2(\omega)}, \, \omega \subset \Omega : \operatorname{dist}(\omega, \Gamma_B') =: d_{\omega,\Gamma_B'} > 0 \\ &\text{with } \Xi(\varepsilon) = C(E)\varepsilon^\tau, \, C(E) > 0, \, \tau := \tau(d_{\omega,\Gamma_B'}) \in (0,1), \, E = \|u\|_{L^2(\Omega)} \end{aligned} \tag{7}$$

and for

$$\begin{aligned} &\|u\|_S := \|u\|_{L^2(\Omega)} \text{ with } \Xi(\varepsilon) = C_1(E)(|\log(\varepsilon)| + C_2(E))^{-\tau} \\ &\text{with } C_1(E), C_2(E) > 0, \, \tau \in (0,1), \, E = \|u\|_{H^1(\Omega)}. \end{aligned} \tag{8}$$

How to design accurate computational methods that can fully exploit the power of conditional stability estimates for their analysis remains a challenging problem.

Nevertheless the elliptic Cauchy problem is particularly well studied. For pioneering work using logarithmic estimates we refer to [22, 28] and quasi reversibility [26]. For work using regularization and/or energy minimisation see [3, 4, 17, 24, 25]. Recently progress has been made using least squares [19] or quasi reversibility approaches [6–8] inspired by conditional stability estimates [9]. In this paper we draw on our experiences from [11, 12], that appear to be the first works where error estimates for stabilised finite element methods on unstructured meshes have been derived for this type of problem. For simplicity we will only consider the operator $\mathcal{L}u := -\Delta u + cu$, with $c \in \mathbb{R}$ for the discussion below.

2 Discretization of the Ill-Posed Problem

We will here focus on discretizations using finite element spaces, but the ideas in this section are general and may be applied to any finite dimensional space.

We consider the setting of Sect. 1.2. Let $\{\mathcal{T}_h\}_h$ denote a family of quasi uniform, shape regular simplicial triangulations, $\mathcal{T}_h := \{K\}$, of Ω, indexed by the maximum simplex diameter h. The set of faces of the triangulation will be denoted by $\mathcal{F}$ and $\mathcal{F}_I$ denotes the subset of interior faces. The unit normal of a face of the mesh will be denoted n, its orientation is arbitrary but fixed, except on faces in $\partial\Omega$ where the normal is chosen to point outwards from Ω. Now let X_h^k denote the finite element space of continuous, piecewise polynomial functions on $\mathcal{T}_h$,

$$X_h^k := \{v_h \in H^1(\Omega) : v_h|_K \in \mathbb{P}_k(K), \quad \forall K \in \mathcal{T}_h\}.$$

Here $\mathbb{P}_k(K)$ denotes the space of polynomials of degree less than or equal to k on a simplex K. Letting $(\cdot,\cdot)_X$ denote the L^2-scalar product over $X \subset \mathbb{R}^d$ and $\langle\cdot,\cdot\rangle_X$ that over $X \subset \mathbb{R}^{d-1}$, with associated L^2-norms $\|\cdot\|_X$, we define the broken scalar products and the associated norms by,

$$(u_h, v_h)_h := \sum_{K\in\mathcal{T}_h} (u_h, v_h)_K, \quad \|u_h\|_h := (u_h, u_h)_h^{\frac{1}{2}},$$

$$\langle u_h, v_h\rangle_{\mathcal{F}} := \sum_{F\in\mathcal{F}} \langle u_h, v_h\rangle_F, \quad \|u_h\|_{\mathcal{F}} := \langle u_h, u_h\rangle_{\mathcal{F}}^{\frac{1}{2}}.$$

If we consider finite dimensional subspaces $V_h \subset V_0$ and $W_h \subset W$, for instance in the finite element context we may take $V_h := X_h^k \cap V_0$ and $W_h := X_h^k \cap W$, the discrete equivalent of problem (2) (with $g = 0$) reads: find $u_h := \sum_{j=1}^{N_{V_h}} u_j\varphi_j \in V_h$ such that

$$a(u_h, \phi_i) = l(\phi_i), \quad i = 1, \ldots, N_{W_h} \tag{9}$$

where the $\{\varphi_i\}$ and $\{\phi_i\}$ are suitable bases for V_h and W_h respectively and $N_{V_h} := \dim(V_h)$, $N_{W_h} := \dim(W_h)$ This formulation may be written as the linear system

$$AU = L,$$

where A is an $N_{W_h} \times N_{V_h}$ matrix, with coefficients $A_{ij} := a(\varphi_j, \phi_i)$, $U = (u_1, \ldots, u_{N_V})^T$ and $L = (l(\phi_1), \ldots, l(\phi_{N_{W_h}}))^T$. Observe that since we have not assumed $N_{V_h} = N_{W_h}$ this system may not be square, but even if it is, it may have zero eigenvalues. This implies

1. non-uniqueness: there exists $\tilde{U} \in \mathbb{R}^{N_{V_h}} \setminus \{\mathbf{0}\}$ such that $A\tilde{U} = 0$;
2. non-existence: there exists $L \in \mathbb{R}^{N_{W_h}}$ such that $L \notin \mathrm{Im}(A)$.

These two problems actually appear also when discretizing well-posed continuous models. Consider the Stokes' equation for incompressible elasticity, for this problem the well-known challenge is to design a method for which the pressure variable is stable and the velocity field discretely divergence free. Indeed the discrete spaces for pressures and velocities must be well-balanced. Otherwise, there may be spurious pressure modes in the solution, comparable to point 1. above, or if the pressure space is too rich the solution may "lock", implying that only the zero velocity satisfies the divergence free constraint, which is comparable to 2. above. Drawing on the experience of the stabilisation of Stokes' problem this analogy naturally suggests the following approach to the stabilisation of (9).

- Consider (9) of the form $a(u_h, w_h) = l(w_h)$ as the constraint for a minimisation problem;
- minimise some (weakly) consistent stabilisation together with a penalty for the boundary conditions (or other data) under the constraint;
- stabilise the Lagrange multiplier (since discrete inf-sup stability fails in general).

To this end we introduce the Lagrangian functional:

$$\boxed{Ł(u_h, z_h) := \frac{1}{2} s_V(u_h - u, u_h - u) - \frac{1}{2} s_W(z_h, z_h) + a_h(u_h, z_h) - l_h(z_h)} \quad (10)$$

where $s_V(u_h - u, u_h - u)$ and $s_W(z_h, z_h)$ represents a penalty term, imposing measured data through the presence of $s_V(u_h - u, u_h - u)$ and symmetric, weakly consistent stabilisations for the primal and adjoint problems respectively. The forms $a_h(\cdot, \cdot)$ and $l_h(\cdot)$ are discrete realisations of $a(\cdot, \cdot)$ and $l(\cdot)$, that may account for the nonconforming case where $V_h \not\subset V$ and $W_h \not\subset W$.

The discrete method that we propose is given by the Euler-Lagrange equations of (10), find $(u_h, z_h) \in V_h \times W_h$ such that

$$\begin{aligned} a_h(u_h, w_h) - s_W(z_h, w_h) &= l_h(w_h) \\ a_h(v_h, z_h) + s_V(u_h, v_h) &= s_V(u, v_h), \end{aligned} \quad (11)$$

for all $(v_h, w_h) \in V_h \times W_h$. This results in a square linear system regardless of the dimensions of V_h and W_h. Note the appearance of $s_V(u, v_h)$ in the right hand side of the second equation of (11). This means only stabilisations for which $s_V(u, v_h)$ can be expressed using known data may be used. This typically is the case for residual based stabilisations, but also allows for the inclusion of measured data in the computation in a natural fashion. The stabilising terms in (11) are used both to include measurements, boundary conditions and regularization. In order to separate these effects we will sometimes write

$$s_x(\cdot,\cdot) := s_x^D(\cdot,\cdot) + s_x^S(\cdot,\cdot), \quad x = V, W$$

where the s^D contribution is associated with assimilation of data (boundary or measurements) and the s^S contribution is associated with the stabilising terms. For the Cauchy problem $s_V^D(u, v_h)$ depends on ψ and $s_V^S(u, v_h)$ may depend on f as we shall see below.

Observe that the second equation of (11) is a finite element discretization of the dual problem associated to the pde-constraint of (10). Hence, assuming that a unique solution exists for the given data, the solution to approximate is $z = 0$. The discrete function z_h will most likely not be zero, since it is perturbed by the stabilisation operator acting on the solution u_h, which in general does not coincide with the stabilisation acting on u. The precise requirements on the forms will be given in the next section together with the error analysis. We also introduce the following compact form of the formulation (11), find $(u_h, z_h) \in V_h \times W_h$ such that

$$A_h[(u_h, z_h), (v_h, w_h)] = L_h(v_h, w_h) \text{ for all } (v_h, w_h) \in V_h \times W_h, \tag{12}$$

where

$$A_h[(u_h, z_h), (v_h, w_h)] := a_h(u_h, w_h) - s_W(z_h, w_h) + a_h(v_h, z_h) + s_V(u_h, v_h) \tag{13}$$

and

$$L_h(v_h, w_h) := l_h(w_h) + s_V(u, v_h).$$

We will end this section by giving some examples of the construction of the discrete forms. To reduce the amount of generic constants we introduce the notation $a \lesssim b$ for $a \leqslant Cb$ where C denotes a positive constant independent of the mesh-size h.

2.1 Example: Discrete Bilinear Forms and Penalty Terms for the Elliptic Cauchy Problem

For the elliptic Cauchy problem of Sect. 1.2 we define V_h^k and W_h^k to be X_h^k (the superscript will be dropped for general k). Then we use information on the boundary conditions to design a form $a_h(\cdot,\cdot)$ that is both forward and adjoint consistent. A penalty term is also added to enforce the boundary condition.

$$a_h(u_h, v_h) := a(u_h, v_h) - \langle \partial_n u_h, v_h \rangle_{\Gamma_N'} - \langle \partial_n v_h, u_h \rangle_{\Gamma_D} \tag{14}$$

$$s_V^D(u_h, w_h) := \gamma_D \left\langle h^{-1} u_h, w_h \right\rangle_{\Gamma_D} + \gamma_D \langle h \partial_n u_h, \partial_n w_h \rangle_{\Gamma_N}, \tag{15}$$

where $\gamma_D \in \mathbb{R}_+$ denotes a penalty parameter that for simplicity is taken to be the same for all the $s^D(\cdot,\cdot)$ terms, it follows that, if $u = g$ on Γ_D,

$$s_V^D(u, w_h) := \gamma_D \left\langle h^{-1} g, w_h \right\rangle_{\Gamma_D} + \gamma_D \langle h\, \psi, \partial_n w_h \rangle_{\Gamma_N}.$$

The adjoint boundary penalty may then be written

$$s_W^D(z_h, v_h) := \gamma_D \left\langle h^{-1} z_h, v_h \right\rangle_{\Gamma_N'} + \gamma_D \langle h\, \partial_n z_h, \partial_n v_h \rangle_{\Gamma_D'}. \tag{16}$$

We assume that the computational mesh $\mathcal{T}_h$ is such that the boundary subdomains consist of the union of boundary element faces, i.e. the boundaries of Γ_D and Γ_N coincide with element edges. Finally we let $l_h(v_h)$ coincide with $l(v_h)$ for unperturbed data. Observe that there is much more freedom in the choice of the stabilisation for z_h since the exact solution satisfies $z = 0$. We will first discuss the methods so that they are consistent also in the case $z \neq 0$, in order to facilitate the connection to a larger class of control problems. Then we will suggest a stronger stabilisation for z_h.

2.2 Example: Galerkin Least Squares Stabilisation

For the stabilisation term we first consider the classical Galerkin Least Squares stabilisation. Observe that for the finite element spaces considered herein, the GaLS stabilisation in the interior of the elements must be complemented with a jump contribution on the boundary of the element. If C^1-continuous approximation spaces are used this latter contribution may be dropped. First consider the least squares contribution,

$$s_V^S(u_h, v_h) := \gamma_S (h^2 \mathcal{L} u_h, \mathcal{L} v_h)_h + \gamma_S \langle h [\![\partial_n u_h]\!], [\![\partial_n v_h]\!] \rangle_{\mathcal{F}_I}, \ \gamma_S \in \mathbb{R}_+. \tag{17}$$

Here $[\![\partial_n v_h]\!]$ denotes the jump of the normal derivative of v_h over an element face F. It then follows that, considering sufficiently smooth solutions, $u \in H^s(\Omega)$, $s > 3/2$,

$$s_V^S(u, v_h) := \gamma_S (h^2 f, \mathcal{L} v_h)_h.$$

Similarly we define

$$s_W^S(z_h, w_h) = \gamma_S (h^2 \mathcal{L}^* z_h, \mathcal{L}^* w_h)_h + \gamma_S \langle h [\![\partial_n z_h]\!], [\![\partial_n w_h]\!] \rangle_{\mathcal{F}_I}.$$

For symmetric operators $\mathcal{L}$ we see that $s_W^S(\cdot,\cdot) \equiv s_V^S(\cdot,\cdot)$, however in the presence of nonsymmetric terms they must be evaluated separately.

2.3 Example: Continuous Interior Penalty Stabilisation

In this case we may choose the two stabilisations to be the same, $s_W^S(\cdot,\cdot) \equiv s_V^S(\cdot,\cdot)$ and

$$s_V^S(u_h, v_h) = \gamma_S \left\langle h^3 [\![\Delta u_h]\!], [\![\Delta v_h]\!] \right\rangle_{\mathcal{F}_I} + \gamma_S \left\langle h [\![\partial_n u_h]\!], [\![\partial_n v_h]\!] \right\rangle_{\mathcal{F}_I}. \quad (18)$$

2.4 Example: Stronger Adjoint Stabilisation

Observe that since the exact solution satisfies $z = 0$ we can also use the adjoint stabilisation

$$s_W^S(z_h, w_h) = \gamma_S (\nabla z_h, \nabla w_h)_\Omega \quad (19)$$

This simplifies the formulation for non-symmetric problems when the GaLS method is used and reduces the stencil, but the resulting formulation is no longer adjoint consistent and optimal L^2-estimates may no longer be proved in the well-posed case (see [10] for a discussion). In this case the formulation corresponds to a weighted least squares method. This is easily seen by eliminating z_h from the formulation (11).

2.5 Penalty Parameters

Above we have introduced the penalty parameters γ_S and γ_D. The size of these parameters play no essential role for the discussion below. Indeed the convergence orders for unperturbed data are obtained only under the assumption that $\gamma_S, \gamma_D > 0$. Therefore the explicit dependence of the constants in the estimates will not be tracked. Only in some key estimates, relating to stability and preturbed data, will we indicate the dependence on the parameters in terms of $\gamma_{min} := \min(\gamma_S, \gamma_D)$ or $\gamma_{max} := \max(\gamma_S, \gamma_D)$.

3 Hypothesis on Forms and Interpolants

To prepare for the error analysis we here introduce assumptions on the bilinear forms. The key properties that are needed are a discrete stability estimate, that the form $a_h(\cdot,\cdot)$ is continuous on a norm that is controlled by the stabilisation terms and that the finite element residual can be controlled by the stabilisation terms. To simplify the presentation we will introduce the space $H^s(\Omega)$, with $s \in \mathbb{R}_+$ which corresponds to smoother functions than those in V for which $a_h(u, v_h)$ and $s_V(u, v_h)$ always are well defined. This typically allows us to treat the data part s_V^D and the stabilisation part s_V^S together using strong consistency. A more detailed analysis separating the two contributions in s_V and handling the conformity error of a_h for $u \in V$ allows an analysis under weaker regularity assumptions.

Consistency: If $u \in V \cap H^s(\Omega)$ is the solution of (1), then the following Galerkin orthogonality holds

$$a_h(u_h - u, w_h) - s_W(z_h, w_h) = l_h(w_h) - l(w_h), \qquad \text{for all } w_h \in W_h. \tag{20}$$

Stabilisation operators: We consider positive semi-definite, symmetric stabilisation operators, $s_V : V_h \times V_h \mapsto \mathbb{R}$, $s_W : W_h \times W_h \mapsto \mathbb{R}$. We assume that $s_V(u, v_h)$, with u the solution of (2) is explicitly known, it may depend on data from $l(\cdot)$ or measurements of u. Assume that both s_V and s_W define semi-norms on $H^s(\Omega) + V_h$ and $H^s(\Omega) + W_h$ respectively,

$$|v + v_h|_{s_Z} := s_Z(v + v_h, v + v_h)^{\frac{1}{2}}, \forall v \in H^s(\Omega),\ v_h \in Z_h, \text{ with } Z = V, W. \tag{21}$$

Discrete stability: There exists a semi-norm, $|(\cdot,\cdot)|_{\mathcal{L}} : (V_h + H^s(\Omega)) \times (W_h + H^s(\Omega)) \mapsto \mathbb{R}$, such that $|v|_{s_V} + |w|_{s_W} \lesssim |(v, w)|_{\mathcal{L}}$ for $v, w \in (V_h + H^s(\Omega)) \times (W_h + H^s(\Omega))$. The semi-norm $|(\cdot,\cdot)|_{\mathcal{L}}$ satisfies the following stability. There exists $c_s > 0$ independent of h such that for all $(\nu_h, \zeta_h) \in V_h \times W_h$ there holds

$$c_s|(\nu_h, \zeta_h)|_{\mathcal{L}} \leq \sup_{(v_h, w_h) \in V_h \times W_h} \frac{A_h[(\nu_h, \zeta_h), (v_h, w_h)]}{|(v_h, w_h)|_{\mathcal{L}}}. \tag{22}$$

Continuity: There exists interpolation operators $i_V : V \mapsto V_h$ and $i_W : W \mapsto W_h \cap W$ and norms $\|\cdot\|_{*,V}$ and $\|\cdot\|_{*,W}$ defined on $V + V_h$ and W respectively, such that

$$a_h(v - i_V v, w_h) \lesssim \|v - i_V v\|_{*,V} |(0, w_h)|_{\mathcal{L}}, \ \forall v \in V \cap H^s(\Omega),\ w_h \in W_h \tag{23}$$

and for u solution of (2),

$$a(u - u_h, w - i_W w) \lesssim \|w - i_W w\|_{*,W}\ \eta_V(u_h), \ \forall w \in W, \tag{24}$$

where the a posteriori quantity $\eta_V(u_h) : V_h \mapsto \mathbb{R}$ satisfies $\eta_V(u_h) \lesssim |(u - u_h, 0)|_{\mathcal{L}}$ for sufficiently smooth u.

Nonconformity: We assume that the following bounds hold

$$|a_h(u_h, i_W w) - a(u_h, i_W w)| \lesssim \eta_V(u_h) \|w\|_W, \tag{25}$$

and

$$|l_h(i_W w) - l(i_W w)| \leqslant \delta_l(h)\|w\|_W, \tag{26}$$

where $\delta_l : \mathbb{R}^+ \mapsto \mathbb{R}^+$, is some continuous function such that $\lim_{x\to 0^+} \delta_l(x) = \delta_0$, with $\delta_0 = 0$ for unperturbed data.
Also assume that there exists an interpolation operator $r_V : H^1(\Omega) + V_h \mapsto V_0 \cap V_h$ such that

$$\|r_V u_h - u_h\|_S + \|r_V u_h - u_h\|_C + \|r_V u_h - u_h\|_V \lesssim \eta_V(u_h). \tag{27}$$

We assume that r_V has optimal approximation properties in the V-norm and the L^2-norm for functions in $V_0 \cap H^s(\Omega)$.

Approximability: We assume that the interpolants $i_V : V \mapsto V_h$, $i_W : W \mapsto W_h \cap W$ have the following approximation and stability properties. For all $v \in V \cap H^s(\Omega)$ there holds,

$$|(v - i_V v, 0)|_{\mathcal{L}} + \|v - i_V v\|_{*,V} \leqslant C_V(v)h^t, \text{ with } t \geqslant 1. \tag{28}$$

The factor $C_V(v) > 0$ will typically depend on some Sobolev norm of v. For i_W we assume that for some $C_W > 0$ there holds

$$|i_W w|_{\mathcal{L}} + \|w - i_W w\|_{*,W} \leqslant C_W\|w\|_W, \quad \forall w \in W. \tag{29}$$

For smoother functions we assume that i_W has approximation properties similar to (28).

3.1 Satisfaction of the Assumptions for the Methods Discussed

We will now show that the above assumptions are satisfied for the method (14)–(15) associated to the elliptic Cauchy problem of Sect. 1.2. We will assume that $u \in V \cap H^s(\Omega)$ with $s > \frac{3}{2}$. Consider first the bilinear form given by (14). To prove the Galerkin orthogonality an integration by parts shows that

$$\begin{aligned} a_h(u, w_h) = (\mathcal{L}u, w_h) + \langle \partial_n u, w_h \rangle_{\Gamma_N} &= (f, w_h) + \langle \psi, w_h \rangle_{\Gamma_N} \\ &= l(w_h) - l_h(w_h) + a_h(u_h, w_h) - s_W(z_h, w_h). \end{aligned}$$

It is immediate by inspection that the stabilisation operators defined in Sects. 2.2 and 2.3 both define the semi-norm (21). Now define the semi-norm for discrete stability

$$|(u_h, z_h)|_{\mathcal{L}} := \|h\mathcal{L}u_h\|_h + \|h\mathcal{L}^* z_h\|_h + \|h^{\frac{1}{2}} [\![\partial_n u_h]\!]\|_{\mathcal{F}_I} + \|h^{\frac{1}{2}} [\![\partial_n z_h]\!]\|_{\mathcal{F}_I}$$
$$+ \|h^{-\frac{1}{2}} u_h\|_{\Gamma_D} + \|h^{\frac{1}{2}} \partial_n u_h\|_{\Gamma_N} + \|h^{-\frac{1}{2}} z_h\|_{\Gamma_N'} + \|h^{\frac{1}{2}} \partial_n z_h\|_{\Gamma_D'}. \qquad (30)$$

If the adjoint stabilisation (19) is used a term $\|z_h\|_{H^1(\Omega)}$ may be added to the right hand side of (30). Observe that for the GaLS method there holds for $c_s \approx \gamma_{min} > 0$,

$$c_s |(u_h, z_h)|^2_{\mathcal{L}} \leqslant A_h[(u_h, z_h), (u_h, -z_h)]$$

which implies (22). For the CIP-method one may also prove the inf-sup stability (22), we detail the proof in appendix.

For the continuity (23) of the form $a_h(\cdot, \cdot)$ defined by Eq. (14), integrate by parts, from the left factor to the right, with $\phi \in V_h + H^s(\Omega)$ and apply the Cauchy-Schwarz inequality,

$$a_h(\phi, w_h) \leqslant |(\phi, \mathcal{L}^* w_h)_h| + \langle |\phi|, |[\![\partial_n w_h]\!]| \rangle_{\mathcal{F}_I} + |\langle \partial_n \phi, w_h \rangle_{\Gamma_N'}| + |\langle \phi, \partial_n w_h \rangle_{\Gamma_D'}|$$
$$\leqslant \left(\|h^{-1}\phi\|_\Omega + \|h^{-\frac{1}{2}}\phi\|_{\mathcal{F}_I \cup \Gamma_D'} + \|h^{\frac{1}{2}} \partial_n \phi\|_{\Gamma_N'} \right) |(0, w_h)|_{\mathcal{L}}.$$

From this inequality we identify the norm $\|\cdot\|_{*,V}$ to be

$$\|\phi\|_{*,V} := \|h^{-1}\phi\|_\Omega + \|h^{-\frac{1}{2}}\phi\|_{\mathcal{F}_I \cup \Gamma_D'} + \|h^{\frac{1}{2}} \partial_n \phi\|_{\Gamma_N'}.$$

Similarly to prove (24) for the form (14) let $\varphi \in W$ and integrate by parts in $a(u - u_h, \varphi)$, identify the functional $\eta_V(u_h)$ and apply the Cauchy-Schwarz inequality with suitable weights,

$$a(u - u_h, \varphi) = (f, \varphi) + \langle \psi, \varphi \rangle_{\Gamma_N} - a(u_h, \varphi)$$
$$\leqslant |(f - \mathcal{L}u_h, \varphi)_h| + \langle |[\![\partial_n u_h]\!]|, |\varphi| \rangle_{\mathcal{F}_I} + |\langle \psi - \partial_n u_h, \varphi \rangle_{\Gamma_N}|$$
$$\leqslant (\|h^{-1}\varphi\|_\Omega + \|h^{-\frac{1}{2}}\varphi\|_{\mathcal{F}_I \cup \Gamma_N})\, \eta_V(u_h), \qquad (31)$$

where we define

$$\eta_V(u_h) := \|h(f - \mathcal{L}u_h)\|_h + \|h^{\frac{1}{2}} [\![\partial_n u_h]\!]\|_{\mathcal{F}_I} + \|h^{\frac{1}{2}} (\psi - \partial_n u_h)\|_{\Gamma_N} + \|h^{-\frac{1}{2}} u_h\|_{\Gamma_D}$$

with $\eta_V(u_h) = |(u - u_h, 0)|_{\mathcal{L}}$ and we may identify

$$\|\varphi\|_{*,W} := \|h^{-1}\varphi\|_\Omega + \|h^{-\frac{1}{2}}\varphi\|_{\mathcal{F}_I \cup \Gamma_N}.$$

It is important to observe that the continuity (31) holds for the continuous form $a(\phi, \varphi)$, but not for the discrete counterpart $a_h(\phi, \varphi)$, since it is not well defined for $\varphi \in W$.

For the definition of i_V and i_W we may use Scott-Zhang type interpolators, preserving the boundary conditions on V_0 and W, for r_V we use a nodal interpolation operator in the interior such that $r_V u|_{\Gamma_D} = 0$. For $u \in H^s(\Omega)$ with $s > \frac{3}{2}$ the approximation estimate (28) then holds with

$$t := \min(s-1, k) \text{ and } C_V(u) \lesssim \|u\|_{H^{t+1}(\Omega)}. \tag{32}$$

The bound (29) holds by inverse and trace inequalities and the H^1-stability of the Scott-Zhang interpolation operator. It is also known that

$$\|u_h - r_V u_h\|_{H^1(\Omega)} \lesssim \|h^{-\frac{1}{2}} u_h\|_{\Gamma_D} \lesssim \eta_V(u_h).$$

from which (27) follows. The following relation shows (25),

$$|a_h(u_h, i_W w) - a(u_h, i_W w)| = |\underbrace{\langle \partial_n u_h, i_W w\rangle_{\Gamma_N'}}_{=0} + \langle \partial_n i_W w, u_h\rangle_{\Gamma_D}|$$

$$\leqslant \|h^{\frac{1}{2}} \partial_n i_W w\|_{\Gamma_D} \|h^{-\frac{1}{2}} u_h\|_{\Gamma_D} \lesssim \|w\|_{H^1(\Omega)} \eta_V(u_h). \tag{33}$$

Where we used that $i_W w|_{\Gamma_N'} = 0$, since $i_W w \in W$.

4 Error Analysis Using Conditional Stability

We will now derive an error analysis using only the continuous dependence (3). First we prove that assuming smoothness of the exact solution the error converges with the rate h^t in the stabilisation semi-norms defined in Eq. (21), provided that there are no perturbations in data. Then we show that the computational error satisfies a perturbation equation in the form (6), and that the right hand side of the perturbation equation can be upper bounded by the stabilisation semi-norm. Our error bounds are then a consequence of the assumption (3).

Lemma 1 *Let $u \in V_0 \cap H^s(\Omega)$ be the solution of* (2) *and (u_h, z_h) the solution of the formulation* (12). *Assume that* (20), (22), (23) *and* (28) *hold. Then*

$$|(u - u_h, z_h)|_{\mathcal{L}} \lesssim C_V(u)(1 + c_s^{-1}) h^t.$$

Proof Let $\xi_h := u_h - i_V u$. By the triangle inequality

$$|(u - u_h, z_h)|_{\mathcal{L}} \leqslant |(u - i_V u, 0)|_{\mathcal{L}} + |(\xi_h, z_h)|_{\mathcal{L}}$$

and the approximability (28) it is enough to study the error in $|(\xi_h, z_h)|_{\mathcal{L}}$. By the discrete stability (22)

$$c_s|(\xi_h, z_h)|_{\mathcal{L}} \leqslant \sup_{(v_h,w_h)\in V_h\times W_h} \frac{A_h[(\xi_h, z_h), (v_h, w_h)]}{|(v_h, w_h)|_{\mathcal{L}}}.$$

Using Eq. (20) we then have

$$c_s|(\xi_h, z_h)|_{\mathcal{L}} \leqslant \sup_{(v_h,w_h)\in V_h\times W_h} \frac{l_h(w_h) - l(w_h) + a_h(u - i_V u, w_h) + s_V(u - i_V u, v_h)}{|(v_h, w_h)|_{\mathcal{L}}}.$$

Under the assumption of unperturbed data and applying the continuity (23) in the third term of the right hand side and the Cauchy-Schwarz inequality in the last we have

$$a_h(u-i_V u, w_h)+s_V(u-i_V u, v_h) \lesssim (\gamma_{min}^{-\frac{1}{2}}\|u-i_V u\|_{*,V}+\gamma_{max}^{\frac{1}{2}}|(u-i_V u, 0)|_{\mathcal{L}})|(v_h, w_h)|_{\mathcal{L}}$$

and hence

$$c_s|(\xi_h, z_h)|_{\mathcal{L}} \lesssim \gamma_{min}^{-\frac{1}{2}}\|u - i_V u\|_{*,V} + \gamma_{max}^{\frac{1}{2}}|(u - i_V u, 0)|_{\mathcal{L}}.$$

Applying (28) we may deduce

$$c_s|(\xi_h, z_h)|_{\mathcal{L}} \lesssim C_V(u)h^t.$$

□

Theorem 1 *Let $u \in V_0 \cap H^s(\Omega)$ be the solution of* (2) *and (u_h, z_h) the solution of the formulation* (12) *for which* (20)–(29) *hold. Assume that the problem* (2) *has the stability property* (3) *and that u and u_h satisfy the condition for stability. Let c_a define a positive constant depending only on the constants of inequalities* (24), (25), (27) *and* (29) *and define the a posteriori quantity*

$$\eta(u_h, z_h) := \eta_V(u_h) + |z_h|_{s_W}. \tag{34}$$

Then, if $\eta(u_h, z_h) < c_a^{-1}$, there holds

$$\|u - u_h\|_S \lesssim \Xi_E(c_a\eta(u_h, z_h)) + \eta_V(u_h) \tag{35}$$

with Ξ_E independent of h.

For sufficiently smooth u there holds

$$\eta(u_h, z_h) \lesssim C_V(u)(1 + c_s^{-1})h^t. \tag{36}$$

Proof We will first write the error as one V_0-conforming part and one discrete nonconforming part. It then follows that $e := u - u_h = \underbrace{u - r_V u_h}_{=\tilde{e} \in V_0} + \underbrace{r_V u_h - u_h}_{=e_h \in V_h}$.

Observe that

$$\|u - u_h\|_S \leqslant \|u - r_V u_h\|_S + \|u_h - r_V u_h\|_S \leqslant \|\tilde{e}\|_S + \eta_V(u_h).$$

Since both u and u_h satisfy a stability condition it is also satisfied for $\tilde{e}$

$$\|\tilde{e}\|_C \leqslant \|u\|_C + \|u_h\|_C + \|e_h\|_C \lesssim \|u\|_C + \|u_h\|_C + \eta_V(u_h) \lesssim 2E + C_V(u)(1 + c_s^{-1})h^t. \tag{37}$$

Here we used the property that $\eta_V(u_h) \lesssim |(u - u_h, 0)_{\mathcal{L}}| \leqslant C_V(u)(1 + c_s^{-1})h^t$, which follows from Lemma 1. Now observe that

$$a(\tilde{e}, w) = a(e, w) - a(e_h, w) = l(w) - a(u_h, w) - a(e_h, w) \tag{38}$$

and since the right hand side is independent of u we identify $r \in W'$ such that $\forall w \in W$,

$$(r, w)_{\langle W', W \rangle} := l(w) - a(u_h, w) - a(e_h, w). \tag{39}$$

It follows that $\tilde{e}$ satisfies Eq. (6) with right hand side $(r, w)_{\langle W', W \rangle}$. Hence since $\tilde{e}$ satisfies the stability condition estimate (3) holds for $\tilde{e}$. We must then show that $\|r\|_{W'}$ can be made small under mesh refinement. We proceed using an argument similar to that of Strang's lemma and (20) to obtain

$$\begin{aligned}(r, w)_{\langle W', W \rangle} = & \underbrace{a(u - u_h, w - i_W w)}_{T_1} + \underbrace{l(i_W w) - l_h(i_W w)}_{T_2} \\ & + \underbrace{a_h(u_h, i_W w) - a(u_h, i_W w)}_{T_3} - \underbrace{a(e_h, w)}_{T_4} - \underbrace{s_W(z_h, i_W w)}_{T_5}.\end{aligned} \tag{40}$$

We now use the assumptions of Sect. 3 to bound the terms T_1-T_5. First by (24) and (29) there holds

$$T_1 = a(u - u_h, w - i_W w) \lesssim \eta(u_h, 0)\|w - i_W w\|_{*,W} \lesssim \eta(u_h, 0)\|w\|_W.$$

By the assumption of unperturbed data and exact quadrature we have $T_2 = 0$. Using the bound of the conformity error (25) we obtain for T_3

$$T_3 = a_h(u_h, i_W w) - a(u_h, i_W w) \lesssim \eta(u_h, 0)\|w\|_W.$$

For the fourth term we use the continuity of $a(\cdot,\cdot)$, (29) and the properties of r_V to write

$$T_4 = a(e_h, i_W w) \lesssim \|e_h\|_V \|i_W w\|_W \lesssim \eta(u_h, 0)\|w\|_W.$$

Finally we use the Cauchy-Schwarz inequality and the stability of (29) to get the bound

$$T_5 = s_W(z_h, i_W w) \leq |z_h|_{s_W} |i_W w|_{s_W} \lesssim \eta(0, z_h)\|w\|_W.$$

Collecting the above bounds on $T_1, \ldots, T_5$ in a bound for (40) we obtain

$$|(r, w)_{\langle W', W\rangle}| \lesssim \eta(u_h, z_h)\|w\|_W.$$

We conclude that there exists $c_a > 0$ such that $\|r\|_{W'} < c_a \eta(u_h, z_h)$. Applying the conditional stability we obtain the bound

$$\|\tilde{e}\|_S \lesssim \Xi_E(c_a \eta(u_h, z_h))$$

where the constants in Ξ_E are bounded thanks to the assumptions on u and u_h and (37).

The a posteriori estimate (35) follows using the triangle inequality and (27),

$$\|u - u_h\|_S = \|\tilde{e} + e_h\|_S \leq \|\tilde{e}\|_S + \|e_h\|_S \lesssim \Xi_E(\eta(u_h, z_h)) + \eta_V(u_h). \tag{41}$$

The upper bound of (36) is then an immediate consequence of the inequality

$$\eta(u_h, z_h) \leq |(u - u_h, z_h)|_{\mathcal{L}}$$

and Lemma 1. □

Remark 1 Observe that if weak consistency is used for the proof of Lemma 1 and the data and stabilisation parts of the term s_V are treated separately, then we may show that the a posteriori part of Theorem 1 holds assuming only $u \in V$.

4.1 Application of the Theory to the Cauchy Problem

Since the formulation (12) with the forms defined by (14)–(16) and the stabilisations (17), (18) or (19) satisfies the assumptions of Theorem 1 as shown in Sect. 3.1, in principle the error estimates hold for these methods when applied to an elliptic Cauchy problem (5) which admits a unique solution in $V_0 \cap H^s(\Omega)$, $s > \frac{3}{2}$. The order t and the constant $C_V(u)$ of the estimates are given by (32).

However, some important questions are left unanswered related to the a priori bounds on the discrete solution u_h, z_h. Observe that we assumed that the discrete solution u_h satisfies the condition for the stability estimate $\|u_h\|_C \leqslant E$. For the Cauchy problem this means that $\|u_h\|_{H^1(\Omega)} \leqslant E$ uniformly in h. As we shall see below, this bound can be proven only under additional regularity assumptions on u. Nevertheless we can prove sufficient stability on the discrete problem to ensure that the matrix is invertible. We will first show that the $\mathcal{L}$-semi-norm (30) is a norm on $V_h \times W_h$, which immediately implies the existence of a discrete solution through (22).

Lemma 2 *Assume that* $|(\cdot,\cdot)|_{\mathcal{L}}$ *is defined by* (30) *and the penalty operator* (15). *Then* $|(v_h, y_h)|_{\mathcal{L}}$ *is a norm on* $V_h \times W_h$. *Moreover for all* $h > 0$ *and all* $k \geqslant 1$ *there exists* $u_h, z_h \in V_h \times W_h$ *solution to* (12), *with* (14)–(16) *and either* (17) *or* (18) *as primal and adjoint stabilisation or* (19) *for adjoint stabilisation.*

Proof The proof is a consequence of norm equivalence on discrete spaces. We know that $|(v_h, 0)|_{\mathcal{L}}$ is a semi-norm. To show that it is actually norm observe that if $|(v_h, 0)|_{\mathcal{L}} = 0$ then $v_h \in H^2(\Omega)$, $\mathcal{L}v_h|_\Omega = \partial_n u_h|_{\Gamma_N} = u_h|_{\Gamma_D} = 0$. It follows that $v_h \in H^1(\Omega)$ satisfies (6) with zero data. Therefore by (8) $v_h = 0$ and we conclude that $|(v_h, 0)|_{\mathcal{L}}$ is a norm. A similar argument yields the upper bound for y_h. The existence of discrete solution then follows from the inf-sup condition (22). If we assume that $L_h(v_h, w_h) = 0$ we immediately conclude that $|(u_h, z_h)|_{\mathcal{L}} = 0$ by which existence and uniqueness of the discrete solution follows. □

This result also shows that the method has a unique continuation property. This property in general fails for the standard Galerkin method (Christiansen, private communication, 1999).

In the estimate of Theorem 1 above we have assumed that both the exact solution u and the computed approximation u_h satisfy the condition for stability, in particular we need $\|u - r_V u_h\|_C \leqslant E$. Since u is unknown we have no choice but assuming that it satisfies the condition and u_h on the other hand is known so the constant E for u_h or $r_V u_h$ can be checked a posteriori. From a theoretical point of view it is however interesting to ask if the stability of u_h can be deduced from the assumptions on u and the properties of the numerical scheme only. This question in its general form is open. We will here first give a complete answer in the case of piecewise affine approximation of the elliptic Cauchy problem and then make some remarks on the high order case.

Proposition 1 *Assume that* $\|\cdot\|_C$ *is bounded by the* H^1*-norm, that* $u \in H^2(\Omega)$ *is the solution to* (6) *and* $(u_h, z_h) \in V_h \times W_h$, *with* $k = 1$, *is the solution to* (12) *with the bilinear forms defined by* (14)–(15) *and* (18). *Then there holds*

$$\|u_h\|_C \lesssim \|u\|_{H^2(\Omega)}. \tag{42}$$

Proof Observe that by a standard Poincaré inequality followed by a discrete Poincaré inequality for piecewise constant functions [20] we have

$$\|u_h\|_{H^1(\Omega)} \leqslant \|i_V u\|_{H^1(\Omega)} + \|i_V u - u_h\|_{H^1(\Omega)} \lesssim \|u\|_{H^2(\Omega)} + h^{-1}|i_V u - u_h|_{s_V}$$
$$\lesssim \|u\|_{H^2(\Omega)} + h^{-1}|(i_V u - u_h, 0)|_{\mathcal{L}} \lesssim \|u\|_{H^2(\Omega)}.$$

□

A simple way to obtain the conditional stability in the high order case, if the order t is known is to add a term $(h^{2t}\nabla u_h, \nabla v_h)_\Omega$ to $s_V(\cdot,\cdot)$. This term will be weakly consistent to the right order and implies the estimate

$$\|u_h\|_{H^1(\Omega)} \lesssim h^{-t}|u_h|_{s_V} \lesssim \|u\|_{H^{t+1}(\Omega)}.$$

An experimental value for t can be obtained by studying the convergence of $|u_h|_{s_V} + |z_h|_{s_W}$ under mesh refinement. To summarize we present the error estimate that we obtain for the Cauchy problem (5) when piecewise affine approximation is used in the following Corollary to Theorem 1.

Corollary 1 *Let $u \in H^2(\Omega)$ be the solution to the elliptic Cauchy problem* (5) *and $u_h, z_h \in V_h \times W_h$ the solution of* (12), *with* (14)–(16) *and either* (18) *as primal and adjoint stabilisation or* (19) *for adjoint stabilisation. Then the conclusion of Theorem 1 holds with $\|\cdot\|_S := \|\cdot\|_\omega$, with $\omega \subseteq \Omega$, the function Ξ_E and E given by* (7) *or* (8) *and*

$$\eta(u_h, z_h) := \|h(f - \mathcal{L}u_h)\|_h + \|h[\![\partial_n u_h]\!]\|_{\mathcal{F}_I} + \|h^{-\frac{1}{2}}u_h\|_{\Gamma_D} + \|h^{\frac{1}{2}}(\psi - \partial_n u_h)\|_{\Gamma_N} + |z_h|_{s_W}.$$

In particular there holds for h sufficiently small,

$$\|u - u_h\|_\omega \lesssim h^\tau \text{ with } 0 < \tau < 1 \text{ when } dist(\omega, \Gamma_B') > 0 \tag{43}$$

and

$$\|u - u_h\|_\Omega \lesssim (|\log(C_1 h)| + C_2)^{-\tau} \text{ with } 0 < \tau < 1. \tag{44}$$

Proof First observe that it was shown in Sect. 3 that the proposed formulation satisfies the assumptions of Theorem 1. It then only remains to show that the stability condition is uniformly satisfied, but this was shown in Proposition 1. The estimates (43) and (44) are then a consequence of (7), (8), (32) and (36). Observe that by (36) the smallness condition on $\eta(u_h, z_h)$ will be satisfied for h small enough.

□

5 The Effect of Perturbations in Data

We have shown that the proposed stabilised methods can be considered to have a certain optimality with respect to the conditional dependence of the ill-posed problem. In practice however it is important to consider the case of perturbed data. Then it is now longer realistic to assume that an exact solution exists. The above error analysis therefore no longer makes sense. Instead we must include the size of the perturbations, leading to error estimates that measure the relative importance of the discretization error and the error in data. To keep the discussion concise we will present the theory for the Cauchy problem and give full detail only in the case of CIP-stabilisation (the extension to GaLS is straightforward by introducing the perturbations also in the stabilisation s_V^S under additional regularity assumptions.) In the CIP case the perturbations can be included in (12) by assuming that

$$l_h(w) := (f + \delta f, w)_\Omega + \langle \psi + \delta\psi, w\rangle_{\Gamma_N} \tag{45}$$

where δf and $\delta\psi$ denote measurement errors and the unperturbed case still allows for a unique solution. We obtain for (26),

$$|l_h(w_h) - l(w_h)| := |(\delta f, w_h) + \langle \delta\psi, w_h\rangle_{\Gamma_N}| \lesssim \|\delta l\|_{(H^1(\Omega))'} \|w_h\|_{H^1(\Omega)}. \tag{46}$$

Similarly the penalty operator $s_V(u, v_h)$ will be perturbed by a $\delta s(v_h) := \langle h\,\delta\psi, \partial_n v_h\rangle_{\Gamma_N}$, here depending only on $\delta\psi$, but which may depend also on measurement errors in the Dirichlet data. We may then write

$$L_h(w_h, v_h) := l_h(w_h) + s_V^D(u, v_h) + \delta s^D(v_h). \tag{47}$$

Observe that the perturbations must be assumed smooth enough so that the above terms make sense, i.e. in the case of the Cauchy problem, $\delta f \in (H^1(\Omega))'$ and $\delta\psi \in L^2(\Gamma_N)$. It follows that $\delta s^D(v_h) \leq h^{\frac{1}{2}} \|\delta\psi\|_{\Gamma_N} |v_h|_{s_V}$.

A natural question to ask is how the approximate solutions of (12) behaves in the asymptotic limit, in the case where no exact solution exists. In this case we show that a certain norm of the solution must blow up under mesh refinement.

Proposition 2 *Assume that $l_h \in (H^1(\Omega))'$, but no $u \in V_0$ satisfies the equation*

$$a(u, w) = l_h(w), \quad \forall w \in W. \tag{48}$$

Let (u_h, z_h) be the solution of (12) *with the stabilisation chosen to be the CIP-method (Sect. 2.3). Then if $s_V^S \equiv s_W^S$,*

$$\|h^{-\frac{1}{2}} u_h\|_{\Gamma_D} + \|\nabla u_h\|_\Omega + |z_h|_{s_W} \to \infty, \text{ when } h \to 0.$$

If $s_W^S(\cdot,\cdot)$ *is defined by* (19) *then*

$$\|\nabla u_h\|_\Omega \to \infty, \text{ when } h \to 0.$$

Proof Assume that there exists $M \in \mathbb{R}$ such that

$$\|h^{-\frac{1}{2}} u_h\|_{\Gamma_D} + \|\nabla u_h\|_\Omega + |z_h|_{s_W} < M$$

for all $h > 0$. It then follows by weak compactness that we may extract a subsequence $\{u_h\}$ for which $u_h \rightharpoonup v \in V$ as $h \to 0$. We will now show that this function must be a solution of (48), leading to a contradiction. Let $\phi \in C^\infty \cap W$ and consider

$$a(v,\phi) = \lim_{h\to 0} a(u_h,\phi).$$

For the right hand side we observe that

$$a(u_h,\phi) = a_h(u_h,\phi - i_W\phi) + a(u_h,\phi) - a_h(u_h,\phi) \\ + s_W(z_h, i_W\phi) + l_h(i_W\phi - \phi) + l_h(\phi).$$

Now we bound the right hand side term by term. First using an argument similar to that of (31), followed by approximation and trace and inverse inequalities, we have

$$a_h(u_h,\phi - i_W\phi) \lesssim \|\phi - i_W\phi\|_{*,W} |(u_h,0)|_{\mathcal{L}} \leqslant Ch\|\phi\|_{H^2(\Omega)}(\|h^{-\frac{1}{2}}u_h\|_{\Gamma_D} + \|\nabla u_h\|_\Omega). \tag{49}$$

Then using an argument similar to (33) recalling that ϕ is a smooth function we get the bound

$$a(u_h,\phi) - a_h(u_h,\phi) \leqslant h^{\frac{1}{2}}\|\partial_n\phi\|_{\Gamma_D}\|h^{-\frac{1}{2}}u_h\|_{\Gamma_D}. \tag{50}$$

For the adjoint stabilisation, first assume that it is chosen to be the CIP stabilisation and add and subtract ϕ in the right slot to get

$$s_W(z_h, i_W\phi) \leqslant \underbrace{s_W(z_h,\phi)}_{=0} + s_W(z_h, i_W\phi - \phi) \leqslant C|z_h|_{s_W} h\|\phi\|_{H^2(\Omega)}. \tag{51}$$

If the form (19) is used, we first observe that testing (12) with $v_h = u_h, w_h = -z_h$ yields

$$|u_h|_{s_V}^2 + |z_h|_{s_W}^2 = A_h[(u_h,z_h),(u_h,-z_h)] = l_h(z_h) + \langle h\psi, \partial_n u_h\rangle_{\Gamma_N} \\ \leqslant \|l_h\|_{(H^1(\Omega))'}|z_h|_{s_W} + h^{\frac{1}{2}}\|\psi\|_{\Gamma_N}|u_h|_{s_V}.$$

It follows that there exists $M > 0$ such that

$$\|h^{-\frac{1}{2}}u_h\|_{\Gamma_D} + \|h^{\frac{1}{2}}\partial_n u_h\|_{\Gamma_N} + \|h^{-\frac{1}{2}}z_h\|_{\Gamma_N'} + \|z_h\|_{H^1(\Omega)} \leqslant M, \quad \forall h > 0.$$

Assuming also that $\|\nabla u_h\|_\Omega \leqslant M$, we may then extract a subsequence $u_h \rightharpoonup \upsilon \in V$ as $h \to 0$ and $z_h \rightharpoonup \zeta \in W$ as $h \to 0$. Using similar arguments as above we may show that $\exists C > 0$ such that for all $\varphi \in V \cap C^\infty$ there holds

$$a(\varphi, z_h) \leqslant Ch^{\frac{1}{2}}$$

implying that $\zeta = 0$, by (3) and (8). Therefore $s_W(z_h, i_W\phi) \to 0$, for all $\phi \in W$. Observing finally that $l_h(i_W\phi - \phi) \lesssim \|l_h\|_{W'} h \|\phi\|_{H^2(\Omega)}$ we may collect the bounds (49)–(51) to conclude that by density

$$a(\upsilon, \phi) = \lim_{h \to 0} a(u_h, \phi) = l_h(\phi), \quad \forall \phi \in W$$

and hence that υ is a weak solution to (48). This contradicts the assumption that the problem has no solution and we have proved the claim. □

To derive error bounds for the perturbed problem we assume that the W-norm can be bounded by the $\mathcal{L}$-norm, in the following fashion, $\forall w_h \in W_h$

$$\frac{\|w_h\|_W}{|(0, w_h)|_{\mathcal{L}}} \lesssim h^{-\kappa} \tag{52}$$

for some $\kappa \geqslant 0$. We may then prove the following perturbed versions of Lemma 1 and Theorem 1.

Lemma 3 *Assume that the hypothesis of Lemma 1 are satisfied, with $L_h(\cdot)$ defined by* (47) *and* (45). *Also assume that* (52) *holds for some $\kappa \geqslant 0$. Then*

$$|(u - u_h, z_h)|_{\mathcal{L}} \lesssim C_V(u)(1 + c_s^{-1})h^t + c_s^{-1}h^{\frac{1}{2}}\|\delta\psi\|_{\Gamma_N} + c_s^{-1}h^{-\kappa}\|\delta l\|_{(H^1(\Omega))'}.$$

Proof We only show how to modify the proof of Lemma 1 to account for the perturbed data. Observe that the perturbation appears when we apply the Galerkin orthogonality:

$$c_s|(\xi_h, z_h)|_{\mathcal{L}} \leqslant \sup_{\{v_h, w_h\} \in V_h \times W_h} \frac{\delta l(w_h) + a_h(u - i_V u, w_h) + s_V(u - i_V u, v_h) + \delta s^D(v_h)}{|(v_h, w_h)|_{\mathcal{L}}}$$

here $\delta l(w_h) := l_h(w_h) - l(w_h)$. We only need to consider the upper bound of the additional terms related to the perturbations in the following fashion

$$\frac{l_h(w_h) - l(w_h) - \delta s^D(v_h)}{|(v_h, w_h)|_{\mathcal{L}}} \lesssim \|\delta l\|_{(H^1(\Omega))'} \frac{\|w_h\|_{H^1(\Omega)}}{|(v_h, w_h)|_{\mathcal{L}}} + h^{\frac{1}{2}}\|\delta\psi\|_{\Gamma_N} \frac{|v_h|_{s_V}}{|(v_h, w_h)|_{\mathcal{L}}}.$$

The conclusion then follows as in Lemma 1 and by applying the assumption (52) and the fact that the $\mathcal{L}$ semi-norm controls $|v_h|_{s_V}$. □

Remark 2 In two instances we can give the precise value of the power κ. First assume that the adjoint stabilisation is given by Eq. (19) with $|(0, w_h)|_{\mathcal{L}}$ defined by (30) with the added $\|w_h\|_{H^1(\Omega)}$ term. It then follows that (52) holds with $\kappa = 0$. On the other hand if GaLS stabilisation or CIP stabilisation are used also for the adjoint variable and piecewise affine spaces are used for the approximation we know that by a discrete Poincaré inequality [20]

$$\|w_h\|_{H^1(\Omega)} \lesssim h^{-1}|(0, w_h)|_{\mathcal{L}}$$

and therefore $\kappa = 1$ in this case.

Similarly the perturbations will enter the conditional stability estimate and limit the accuracy that can be obtained in the $\|\cdot\|_S$ norm when the result of Theorem 1 is applied.

Theorem 2 *Let u be the solution of* (6) *and (u_h, z_h) the solution of the formulation* (12) *with the right hand side given by* (47). *Assume that the assumptions* (21)–(28) *hold, that the problem* (6) *has the stability property* (3) *and that u satisfies the condition for stability. Let*

$$\eta_\delta(u_h, z_h) := \eta(u_h, z_h) + \|\delta l\|_{W'}$$

with $\eta(u_h, z_h)$ defined by (34). *Then for $\eta_\delta(u_h, z_h)$ small enough, there holds*

$$\|u - u_h\|_S \lesssim \Xi_E(c_{\delta,a}\eta_\delta(u_h, z_h)) + \eta_V(u_h) \tag{53}$$

with Ξ_E dependent on u_h. For sufficiently smooth u there holds

$$\eta_\delta(u_h, z_h) \lesssim C_V(u)(1+c_s^{-1})h^t + c_s^{-1}h^{\frac{1}{2}}\|\delta\psi\|_{\Gamma_N} + (1+c_s^{-1}h^{-\kappa})\|\delta l\|_{(H^1(\Omega))'}. \tag{54}$$

Proof The difference due to the perturbed data appears in the Strang type argument. We only need to study the term T_2 of the Eq. (40) under the assumption (46). Using the H^1-stability of the interpolant i_W we immediately get

$$T_2 = l(i_W w) - l_h(i_W w) \lesssim \|\delta l\|_{W'}\|i_W w\|_W \lesssim \|\delta l\|_{W'}\|w\|_W.$$

It then follows that

$$|(r, w)_{\langle W', W\rangle}| \leqslant (c_a\eta(u_h, z_h) + C_W\|\delta l\|_{W'})\|w\|_W \leqslant c_{\delta,a}\eta_\delta(u_h, z_h)\|w\|_W$$

and assuming that $c_{\delta,a}\eta_\delta(u_h, z_h) < 1$, the a posteriori bound follows by applying the conditional stability (3). For the a priori estimate we apply the result of Lemma 3 and $\|\delta l\|_{W'} \leqslant \|\delta l\|_{(H^1(\Omega))'}$. □

Observe that the function Ξ_E in the error estimate depends on $\|u_h\|_C$ and therefore is not robust. A natural question is how small we can choose h compared to the size of the perturbations before the computational error stagnates or even grows. This leads to a delicate balancing problem since the mesh size must be small so that the residual is small enough, but not too small, since this will make the perturbation terms dominate. Therefore the best we can hope for is a window $0 < h_{min} < h < h_{max}$, within which the estimates (53) and (54) hold. We will explore this below for the approximation of the Cauchy problem using piecewise affine elements.

Corollary 2 *Assume that the hypothesis of Lemma 3 and Theorem 2 are satisfied. Also assume that there exists $h_{min} > 0$ and $C_\delta(u) > 0$ such that*

$$h_{min}^{-\kappa} \|\delta l\|_{(H^1(\Omega))'} + h^{\frac{1}{2}} \|\delta\psi\|_{\Gamma_N} \leqslant C_\delta(u) h \quad for\ h > h_{min} \tag{55}$$

and $h_{max} > 0$ so such that for $h_{min} < h < h_{max}$ there holds $\eta_\delta(u_h, z_h) < c_{\delta,a}^{-1}$. Then for $h_{min} < h < h_{max}$ there exists $\Xi_E(\cdot)$, independent of u_h such that (53) *and* (54) *hold.*

Proof First observe that by Lemma 3 and under the assumption (55) there holds for $h > h_{min}$

$$|(i_V u - u_h, z_h)|_{\mathcal{L}} \lesssim (C_V(u) + C_\delta(u)) c_s^{-1} h$$

It follows by this bound and the discrete Poincaré inequality, that $\|u_h\|_{H^1(\Omega)} \lesssim (C_V(u) + C_\delta(u)) c_s^{-1}$ for $h > h_{min}$. We may conclude that the condition for stability is satisfied for u, u_h and the discrete error $r_V u_h - u_h$. Therefore, since the smallness assumption on $\eta_\delta(u_h, z_h)$ is satisfied for $h < h_{max}$, there exists Ξ_E independent of u_h such that estimates (53) and (54) hold when $h_{min} < h < h_{max}$. □

6 Numerical Examples

Here we will recall some numerical examples from [10] and discuss them in the light of the above analysis. We choose $\Omega = (0, 1) \times (0, 1)$ and limit the study to CIP-stabilisation and the case where the primal and adjoint stabilisations are the same. First we will consider the case of a well-posed but non-coercive convection–diffusion equation, $\mathcal{L} := -\mu\Delta u + \beta\cdot\nabla u$. Then we study the elliptic Cauchy problem with $\mathcal{L}u := -\Delta u$ for unperturbed and perturbed data and finally we revisit the convection-diffusion equation in the framework of the elliptic Cauchy problem and study the effect of the flow characteristics on the stability. All computations were carried out on unstructured meshes. In the convergence plots below the curves have the following characteristics

- piecewise affine approximation: square markers;
- piecewise quadratic approximation: circle markers;

- full line: the stabilisation semi-norm $|u_h|_{S_V} + |z_h|_{S_W}$;
- dashed line: the global L^2-norm;
- dotted line with markers: the local L^2-norm.
- dotted line without markers: reference slopes.

6.1 Convection–Diffusion Problem with Pure Neumann Boundary Conditions

We consider an example given in [16]. The operator is chosen as

$$\mathcal{L}(\cdot) := \nabla \cdot (\mu \nabla(\cdot) + \beta \cdot) \tag{56}$$

with the physical parameters $\mu = 1$,

$$\beta := -100 \begin{pmatrix} x + y \\ y - x \end{pmatrix}$$

(see the left plot of Fig. 1) and the exact solution is given by

$$u(x, y) = 30x(1 - x)y(1 - y). \tag{57}$$

This function satisfies homogeneous Dirichlet boundary conditions and has $\|u\|_\Omega$=1. Note that $\|\beta\|_{L^\infty} = 200$ and $\nabla \cdot \beta = -200$, making the problem strongly noncoercive with a medium high Péclet number. We solve the problem with (non-homogeneous) Neumann-boundary conditions $(\mu \nabla u + \beta u) \cdot n = g$ on $\partial\Omega$. The parameters were set to $\gamma_D = 10$ and $\gamma_S = 0.01$ for piecewise affine approximation and $\gamma_S = 0.001$ for piecewise quadratic approximation. The average value of the approximate solutions has been imposed using a Lagrange multiplier. The right hand side is then chosen as $\mathcal{L}u$ and for the (non-homogeneous) Neumann conditions, a suitable right hand side is introduced to make the boundary penalty term consistent. In the right plot of Fig. 1 we observe optimal convergence rates as predicted by theory (the dual adjoint problem is well-posed, see [10, 15]).

6.2 The Elliptic Cauchy Problem

Here we consider the problem (5) with σ the identity matrix and $c \equiv 0$. We impose the Cauchy data, i.e. both Dirichlet and Neumann data, on boundaries $x = 1$, $0 < y < 1$ and $y = 1$, $0 < x < 1$. We then solve (5) using the method (12) with (14)–(16) and (18) with $k = 1$ and $k = 2$.

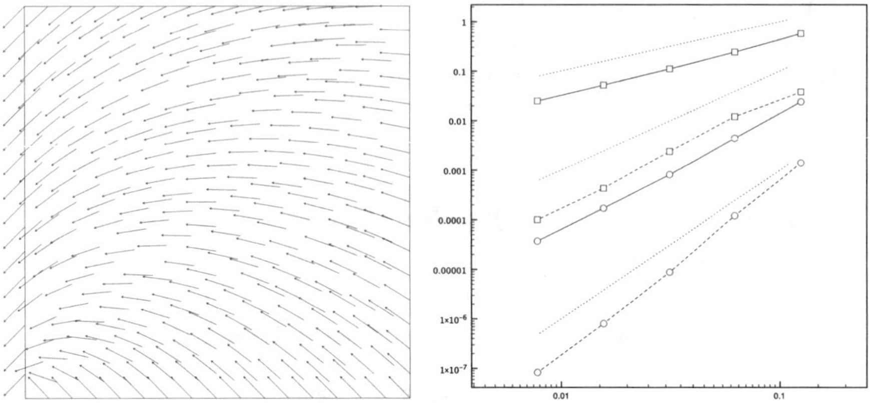

Fig. 1 *Left*: Plot of the velocity vector field. *Right*: Convergence plot, errors against mesh size, *filled lines* $|u_h|_{s_V} + |z_h|_{s_W}$, *dashed lines* L^2-norm error, *dotted lines* reference slopes, from *top* to *bottom* $\mathcal{O}(h)$, $\mathcal{O}(h^2)$, $\mathcal{O}(h^3)$

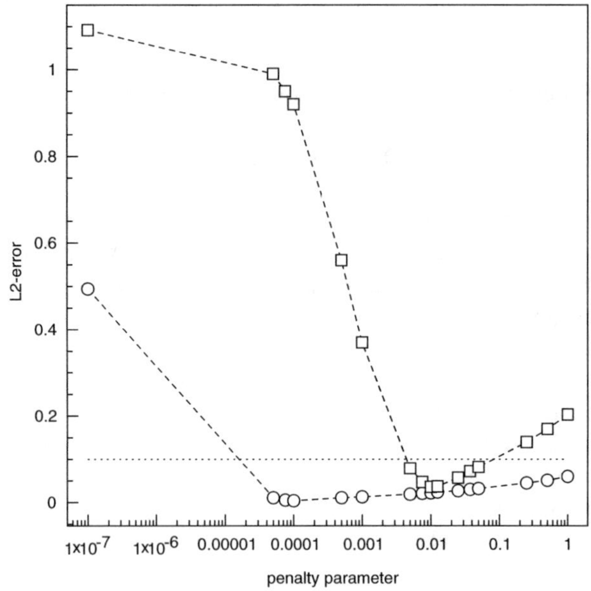

Fig. 2 Study of the global L^2-norm error under variation of the stabilisation parameter, *circles*: affine elements, *squares*: quadratic elements

In Fig. 2, we present a study of the L^2-norm error under variation of the stabilisation parameter. The computations are made on one mesh, with 32 elements per side and the Cauchy problem is solved with $k = 1, 2$ and different values for γ_S with $\gamma_D = 10$ fixed. The level of 10 % relative error is indicated by the horizontal dotted line. Observe that the robustness with respect to stabilisation parameters is

Fig. 3 Contour plots of the interpolated error $i_V - u_h$ (*left plot*) and the error in the dual variable z_h (*right plot*)

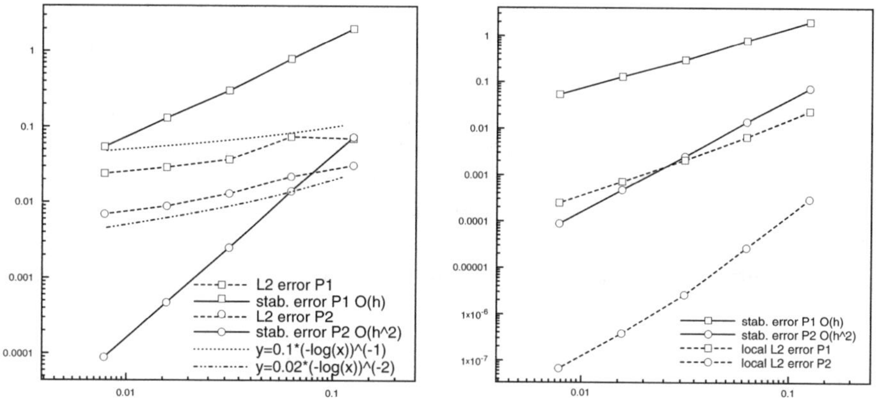

Fig. 4 Convergence under mesh refinement, the same slopes for the stabilization semi-norm are represented in both graphics for reference

much better for second order polynomial approximation. Indeed in that case the 10 % error level is met for all parameter values $\gamma_S \in [2.0E-5, 1]$, whereas in the case of piecewise affine approximation one has to take $\gamma_S \in [0.003, 0.05]$. Similar results for the boundary penalty parameter not reported here showed that the method was even more robust under perturbations of γ_D. In the left plot of Fig. 3 we present the contour plot of the interpolated error $i_V u - u_h$ and in the right, the contour plot of z_h. In both cases the error is concentrated on the boundary where no boundary conditions are imposed for that particular variable.

In Fig. 4 we present the convergence plots for piecewise affine and quadratic approximations. The same stabilisation parameters as in the previous example were used. In both cases we observe the optimal convergence of the stabilisation terms,

$\mathcal{O}(h^k)$, predicted by Lemma 1. For the global L^2-norm of the error we observe experimental convergence of inverse logarithmic type, as predicted by theory. Note that the main effect of increasing the polynomial order is a decrease in the error constant as expected.

For the local L^2-norm error, measured in the subdomain $(0.5, 1)^2$, higher convergence orders, $\mathcal{O}(h^k)$, were obtained in both cases.

6.2.1 The Effect of Perturbations in Data

In this section we will consider some numerical experiments with perturbed data. We consider a perturbation of the form $\delta\psi = \varsigma v_{rand}\psi$ where v_{rand} is a random function defined as a fourth order polynomial on the mesh with random nodal values in $[0, 1]$ and $\varsigma > 0$ gives the relative strength of the perturbation. We consider the same computations as for unperturbed data. In all figures we report the stabilisation semi-norm $|z_h|_{s_W} + |u_h|_{s_V}$ to explore to what extent it can be used as an a posteriori quantity to tune the stabilisation parameter and to detect loss of convergence due to perturbed data.

First we consider the determination of the penalty parameter. First we fix $\gamma_D = 10$. Then, in Fig. 5 we show the results obtained by varying γ_S when the data is perturbed with $\varsigma = 0.01$. We compare the global L^2-error with the stabilisation semi-norm. For the piecewise affine case we observe that the optimal value of the penalty parameter does not change much. It is taken in the interval $[0.01, 0.1]$, which corresponds very well with the minimum of the a posteriori quantity $|z_h|_{s_W} + |u_h|_{s_V}$. For piecewise quadratic approximation there is a stronger difference compared to the unperturbed case. The optimal penalty parameter is now taken in the interval

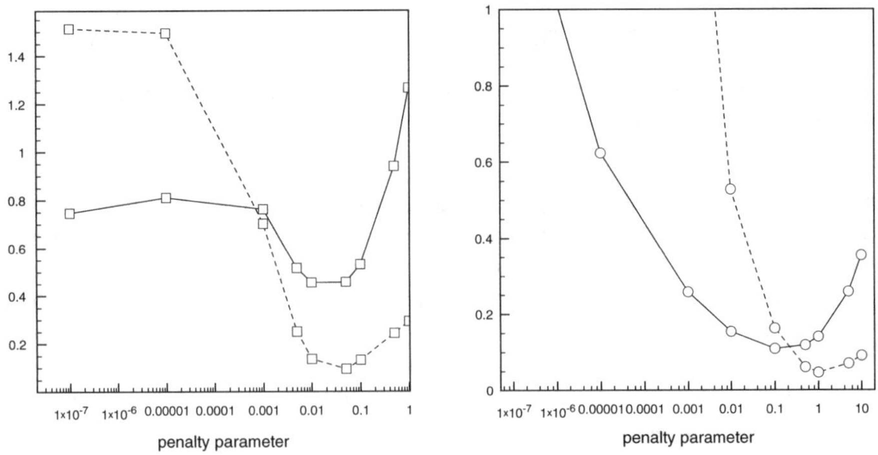

Fig. 5 Variation of the global L^2-error (*dashed line*) and $|z_h|_{s_W} + |u_h|_{s_V}$ (*full line*) against γ. *Left* $k = 1$. *Right* $k = 2$

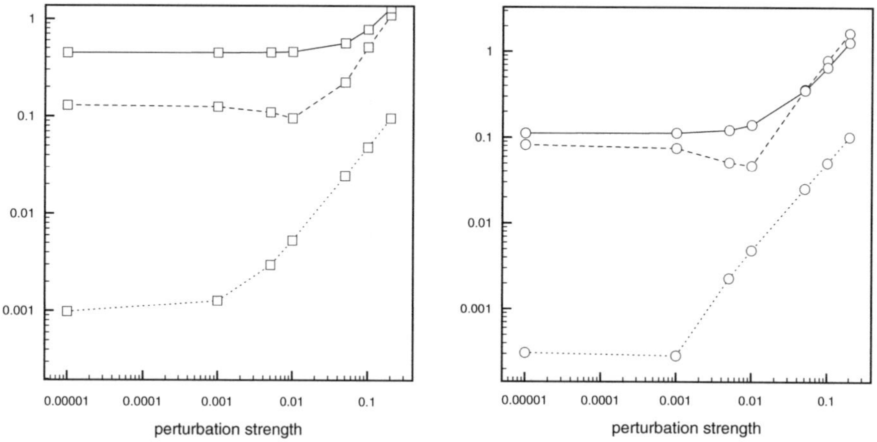

Fig. 6 Variation of the L^2-error (*global dashed line, local dotted line*) and $|z_h|_{sw} + |u_h|_{sv}$ (*full line*) against ς. *Left* $k = 1$. *Right* $k = 2$

[0.5, 5]. The a posteriori quantity takes its minimum value in the interval $[0.1, 0.5]$. From this study we fix the penalty parameter to $\gamma_S = 0.05$ for piecewise affine approximation and to $\gamma_S = 1.0$ in the piecewise quadratic case.

Next we study the sensitivity of the error to variations in the strength of the perturbation, for the chosen penalty parameters. The results are given in Fig. 6. As expected the global L^2-error is minimal for the perturbation $\varsigma = 0.01$. For smaller perturbations it remains approximately constant, but for perturbations larger than 1 % the error growth is linear in ς for all quantities as predicted by theory, assuming the stability condition is satisfied uniformly (see Lemma 3 and Theorem 2.)

Finally we study the convergence under mesh refinement when $\varsigma = 0.01$. The results are presented in Fig. 7. From the theory we expect the reduction of the error to stagnate or even start to grow when $h \lesssim \varsigma$. For the piecewise affine approximation the minimal global L^2-error is 0.065 for $h = 0.015625$ and it follows that the stagnation takes place for $h \approx \varsigma$ in this case. For $k = 2$ the minimal global L^2-error is 0.047 for $h = 0.03125$, that is one refinement level earlier than for the piecewise affine case. In both cases we observe that the convergence of the stabilisation semi-norm degenerates to worse than first order immediately after the critical mesh-size. The dotted lines without markers immediately below the curve representing the a posteriori quantity are reference curves with slopes $O(h^{1.1})$ for affine elements and $O(h^{1.4})$ for quadratic elements $k = 2$. This rate is suboptimal in the latter case, indicating a higher sensibility to perturbations for higher order approximations. It follows that regardless of the smoothness of the (unperturbed) exact solution, high order approximation only pays if perturbations in data are small enough so that they do not dominate before the asymptotic range is reached.

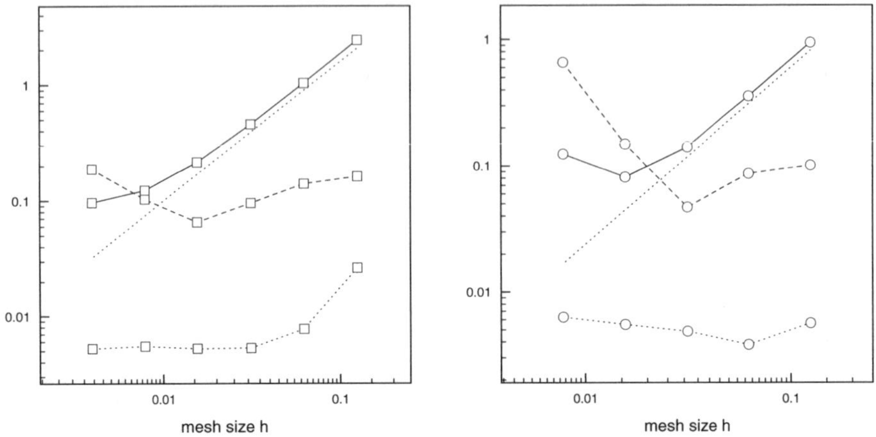

Fig. 7 Variation of the L^2-error (*global dashed line, local dotted line*) and $|z_h|_{sw} + |u_h|_{sv}$ (*full line, with markers*) against h. *Left* $k = 1$, reference $O(h^{1.1})$. *Right* $k = 2$, reference $O(h^{1.4})$

6.3 The Elliptic Cauchy Problem for the Convection–Diffusion Operator

As a last example we consider the Cauchy problem using the noncoercive convection–diffusion operator (56). The stability of the problem depends strongly on where the boundary conditions are imposed in relation to the inflow and outflow boundaries. Strictly speaking this problem is not covered by the theory developed in [2]. Indeed in that work the quantitative unique continuation used the symmetry of the operator. An extension to the convection-diffusion case is likely to be possible, at least in two space dimensions, by combining the results of [1] with those of [2].

To illustrate the dependence of the stability on how boundary data is distributed on inflow and outflow boundaries we propose two configurations. Recalling the left plot of Fig. 1 we observe that the flow enters along the boundaries $y = 0$, $y = 1$ and $x = 1$ and exits on the boundary $x = 0$. Note that the strongest inflow takes place on $y = 0$ and $x = 1$, the flow being close to parallel to the boundary in the right half of the segment $y = 1$. We propose the two different Cauchy problem configurations:

Case 1. We impose Dirichlet and Neumann data on the two inflow boundaries $y = 0$ and $x = 1$.

Case 2. We impose Dirichlet and Neumann data on the two boundaries $x = 0$ and $y = 1$ comprising both inflow and outflow parts.

The gradient penalty operator has been weighted with the Péclet number as suggested in [10], to obtain optimal performance in all regimes. In the first case the main part of the inflow boundary is included in Γ_S whereas in the second case the outflow portion *or* the inflow portion of every streamline are included in the boundary portion Γ_S where data are set. This highlights two different difficulties

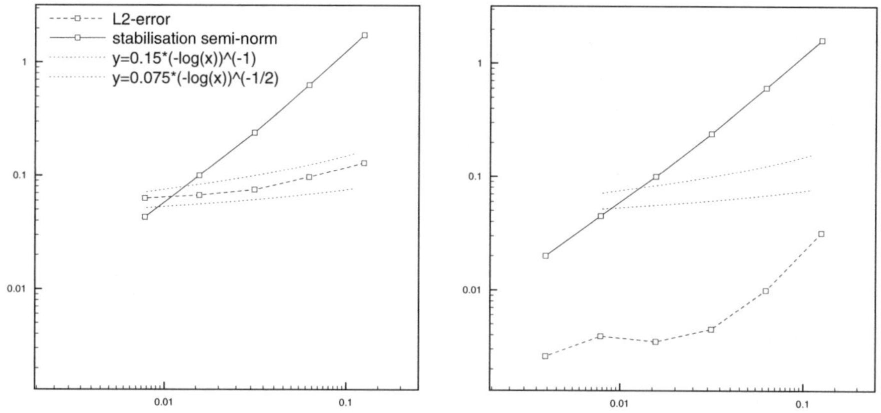

Fig. 8 *Left*: Convergence for Case 1, k=1. *Right*: Convergence for Case 2, k=1

for Cauchy problems for the convection–diffusion operator, in Case 1 the crosswind diffusion must reconstruct missing boundary data whereas in Case 2 we must solve the problem backward along the characteristics, essentially solving a backward heat equation.

In Fig. 8, we report the results on the same sequence of unstructured meshes used in the previous examples for piecewise affine approximations and the two problem configurations. In the left plot of Fig. 8 we see the convergence behaviour for Case 1, when piecewise affine approximation is used. The global L^2-norm error clearly reproduces the inverse logarithmic convergence order predicted by the theory for the symmetric case. In the right plot of Fig. 8 we present the convergence plot for Case 2 (the dotted lines are the same inverse logarithmic reference curves as in the left plot). In this case we see that the convergence initially is approximately linear, similarly as that of the stabilisation term. For finer meshes however the inverse logarithmic error decay is observed, but with a much smaller constant compared to Case 1. In Case 1 the diffusion is important on all scales, since some characteristics have no data neither on inflow or outflow, whereas in Case 2, data is set either on the inflow or the outflow for all characteristics of the flow and the effects of diffusion are therefore much less important, in particular on coarse scales. Indeed the reduced transport problem in the limit of zero diffusivity, is not ill-posed. As the flow is resolved the effect of the diffusion once again dominates and the inverse logarithmic decay reappears.

7 Conclusion

We have proposed a framework using stabilised finite element methods for the approximation of ill-posed problems that have a conditional stability property. The key element is to reformulate the problem as a pde-constrained minimisation problem that is regularized on the discrete level using tools known from the theory of stabilised FEM. Using the conditional stability error estimates are derived that are optimal with respect to the stability of the problem and the approximation properties of the finite element spaces. The effect of perturbations in data may also be accounted for in the framework and leads to limits on the possibility to improve accuracy by mesh refinement. Some numerical examples were presented illustrating different aspects of the theory.

There are several open problems both from theoretical and computational point of view, some of which we will address in future work. Concerning the stabilisation it is not clear if the primal and adjoint stabilisation operators should be chosen to be the same, or not? Does the adjoint consistent choice of stabilisation s_W have any advantages compared to the adjoint stabilisation (19), that gives stronger control of perturbations? Then comes the question of whether or not high order approximation (i.e. polynomials of order higher than one) can be competitive also in the presence of perturbed data? Can the a posteriori error estimate derived in Theorem 1 be used to drive adaptive algorithms? Finally, what is a suitable preconditioner for the linear system? We hope that the present work will help to stimulate discussion on the design of numerical methods for ill-posed problems and provide some new ideas on how to make a bridge between the regularization methods traditionally used and (weakly) consistent stabilised finite element methods.

Appendix

We will here give a proof that the inf-sup stability (22) holds also for the stabilisation (18). We do not track the depedence on γ_D and γ_S.

Proposition 3 *Let $A_h[(\cdot,\cdot),(\cdot,\cdot)]$ be defined by* (13) *with $a_h(\cdot,\cdot)$, $s_W(\cdot,\cdot)$ and $s_V(\cdot,\cdot)$ defined by Eq.* (14)–(16) *and* (18) *(or* (19) *for $s_W(\cdot,\cdot)$). Then the inf-sup condition* (22) *is satisfied for the semi-norm* (30).

Proof We must prove that the L^2-stabilisation of the jump of the Laplacian gives sufficient control for the inf-sup stability of $\mathcal{L}u_h$ evaluated elementwise. It is well known [14] that for the quasi-interpolation operator defined in each node x_i by

$$(I_{os}\Delta u_h)(x_i) := N_i^{-1} \sum_{\{K:x_i\in K\}} \Delta u_h(x_i)|_K,$$

$N_i := \text{card}\{K : x_i \in K\}$ the following discrete interpolation result holds

$$\|h(\Delta u_h - I_{os}\Delta u_h)\|_h \leqslant C_{os} s_V^S(u_h,u_h)^{\frac{1}{2}} \tag{58}$$

as well as following the stabilities obtained using trace inequalities, inverse inequalities and the L^2-stability of I_{os},

$$\|h^{\frac{3}{2}} I_{os} \Delta u_h\|_{\mathcal{F}} + \|h^{\frac{5}{2}} \partial_n I_{os} \Delta u_h\|_{\mathcal{F}} + \|h I_{os} \Delta u_h\|_h + |h^2 I_{os} \Delta u_h|_{sX} \lesssim \|h \Delta u_h\|_h. \tag{59}$$

First observe that by taking $(v_h, w_h) = (u_h, z_h)$ we have

$$|u_h|_{sV}^2 + |z_h|_{sW}^2 = A_h[(u_h, z_h), (u_h, z_h)].$$

Now let $w_h^{\mathcal{L}} = h^2 I_{os} \mathcal{L} u_h = h^2 (I_{os} \Delta u_h + c u_h)$, $v_h^{\mathcal{L}} = h^2 I_{os} \mathcal{L}^* z_h$. Using (59) it is straightforward to show that

$$\|h^{\frac{3}{2}} I_{os} \mathcal{L} u_h\|_{\mathcal{F}} + \|h^{\frac{5}{2}} \partial_n I_{os} \mathcal{L} u_h\|_{\mathcal{F}} + \|h I_{os} \mathcal{L} u_h\|_h + |h^2 I_{os} \mathcal{L} u_h|_{sX} \leqslant \tilde{C}_{os} \|h \mathcal{L} u_h\|_h. \tag{60}$$

Now observe that (for a suitably chosen orientation of the normal on interior faces)

$$\begin{aligned} a_h(u_h, w_h^{\mathcal{L}}) &= \|h \mathcal{L} u_h\|_h^2 + (\mathcal{L} u_h, h^2 (I_{os} \mathcal{L} u_h - \mathcal{L} u_h))_h + \left\langle [\![\partial_n u_h]\!], h^2 I_{os} \mathcal{L} u_h \right\rangle_{\mathcal{F}_I} \\ &\quad + \left\langle \partial_n u_h, h^2 I_{os} \mathcal{L} u_h \right\rangle_{\Gamma_N} + \left\langle \partial_n h^2 I_{os} \mathcal{L} u_h, u_h \right\rangle_{\Gamma_D} \\ &\geqslant \frac{1}{2} \|h \mathcal{L} u_h\|_h^2 - 2 \|h^2 (I_{os} \mathcal{L} u_h - \mathcal{L} u_h)\|_h^2 - 2 \tilde{C}_{os}^{-2} s_V^D(u_h, u_h) \\ &\geqslant \frac{1}{2} \|h \mathcal{L} u_h\|_h^2 - 2 C_{os}^2 s_V^S(u_h, u_h) - 2 \tilde{C}_{os}^{-2} s_V^D(u_h, u_h) \\ &\geqslant \frac{1}{2} \|h \mathcal{L} u_h\|_h^2 - 2 (C_{os}^2 + \tilde{C}_{os}^{-2}) |u_h|_{sV}^2 \end{aligned}$$

and

$$s_W(z_h, w_h^{\mathcal{L}}) \geqslant -\tilde{C}_{os}^{-2} |z_h|_{sW}^2 - \frac{1}{4} \|h \mathcal{L} u_h\|_h^2.$$

Similarly

$$a_h(v_h^{\mathcal{L}}, z_h) \geqslant \frac{1}{2} \|h \mathcal{L}^* z_h\|_h^2 - 2 (C_{os}^2 + \tilde{C}_{os}^{-2}) |z_h|_{sW}^2$$

and

$$s_V(u_h, v_h^{\mathcal{L}}) \geqslant -\tilde{C}_{os}^{-2} |u_h|_{sV}^2 - \frac{1}{4} \|h \mathcal{L}^* z_h\|_h^2.$$

It follows that for some $c_1, c_2 > 0$ there holds

$$|(u_h, z_h)|_{\mathcal{L}}^2 \lesssim A_h[(u_h, z_h), (u_h + c_1 w_h^{\mathcal{L}}, z_h + c_2 v_h^{\mathcal{L}})].$$

We conclude by observing that by inverse inequalities and (60) we have the stability

$$|(u_h + c_1 w_h^{\mathcal{L}}, z_h + c_2 v_h^{\mathcal{L}})|_{\mathcal{L}} \lesssim |(u_h, z_h)|_{\mathcal{L}}.$$

□

References

1. G. Alessandrini, Strong unique continuation for general elliptic equations in 2D. J. Math. Anal. Appl. **386**(2), 669–676 (2012)
2. G. Alessandrini, L. Rondi, E. Rosset, S. Vessella, The stability for the Cauchy problem for elliptic equations. Inverse Prob. **25**(12), 123004, 47 (2009)
3. S. Andrieux, T.N. Baranger, A. Ben Abda, Solving Cauchy problems by minimizing an energy-like functional. Inverse Prob. **22**(1), 115–133 (2006)
4. M. Azaïez, F. Ben Belgacem, H. El Fekih, On Cauchy's problem. II. Completion, regularization and approximation. Inverse Prob. **22**(4), 1307–1336 (2006)
5. F. Ben Belgacem, Why is the Cauchy problem severely ill-posed? Inverse Prob. **23**(2), 823–836 (2007)
6. L. Bourgeois, A mixed formulation of quasi-reversibility to solve the Cauchy problem for Laplace's equation. Inverse Prob. **21**(3), 1087–1104 (2005)
7. L. Bourgeois, Convergence rates for the quasi-reversibility method to solve the Cauchy problem for Laplace's equation. Inverse Prob. **22**(2), 413–430 (2006)
8. L. Bourgeois, J. Dardé, A duality-based method of quasi-reversibility to solve the Cauchy problem in the presence of noisy data. Inverse Prob. **26**(9), 095016, 21 (2010)
9. L. Bourgeois, J. Dardé, About stability and regularization of ill-posed elliptic Cauchy problems: the case of Lipschitz domains. Appl. Anal. **89**(11), 1745–1768 (2010)
10. E. Burman, Stabilized finite element methods for nonsymmetric, noncoercive, and ill-posed problems. Part I: elliptic equations. SIAM J. Sci. Comput. **35**(6), A2752–A2780 (2013)
11. E. Burman, A stabilized nonconforming finite element method for the elliptic Cauchy problem. Math. Comput. ArXiv e-prints (2014). http://dx.doi.org/10.1090/mcom/3092
12. E. Burman, Error estimates for stabilized finite element methods applied to ill-posed problems. C. R. Math. Acad. Sci. Paris **352**(7–8), 655–659 (2014)
13. E. Burman, Stabilized finite element methods for nonsymmetric, noncoercive, and ill-posed problems. Part II: hyperbolic equations. SIAM J. Sci. Comput. **36**(4), A1911–A1936 (2014)
14. E. Burman, M.A. Fernández, P. Hansbo, Continuous interior penalty finite element method for Oseen's equations. SIAM J. Numer. Anal. **44**(3), 1248–1274 (2006)
15. I. Capuzzo-Dolcetta, S. Finzi Vita, Finite element approximation of some indefinite elliptic problems. Calcolo **25**(4), 379–395 (1989/1988)
16. C. Chainais-Hillairet, J. Droniou, Finite-volume schemes for noncoercive elliptic problems with Neumann boundary conditions. IMA J. Numer. Anal. **31**(1), 61–85 (2011)
17. A. Chakib, A. Nachaoui, Convergence analysis for finite element approximation to an inverse Cauchy problem. Inverse Prob. **22**(4), 1191–1206 (2006)
18. B. Cockburn, C. Johnson, C.-W. Shu, E. Tadmor, *Advanced Numerical Approximation of Nonlinear Hyperbolic Equations*. Lecture Notes in Mathematics, vol. 1697 (Springer, Berlin; Centro Internazionale Matematico Estivo (C.I.M.E.), Florence, 1998). Papers from the

C.I.M.E. Summer School held in Cetraro, June 23–28, 1997, ed. by A. Quarteroni, Fondazione C.I.M.E.. [C.I.M.E. Foundation]
19. J. Dardé, A. Hannukainen, N. Hyvönen, An H_{div}-based mixed quasi-reversibility method for solving elliptic Cauchy problems. SIAM J. Numer. Anal. **51**(4), 2123–2148 (2013)
20. S. Brenner, Poincaré-Friedrichs inequalities for piecewise H^1 functions. SIAM J. Numer. Anal. **41**(1), 306–324 (2003)
21. S. Brenner, K. Wang, J. Zhao, Poincaré-Friedrichs inequalities for piecewise H^2 functions. Numer. Funct. Anal. Optim. **25**(5-6), 463–478 (2004)
22. R.S. Falk, P.B. Monk, Logarithmic convexity for discrete harmonic functions and the approximation of the Cauchy problem for Poisson's equation. Math. Comput. **47**(175), 135–149 (1986)
23. J. Hadamard, Sur les problèmes aux derivées partielles et leur signification physique. Bull. Univ. Princeton **13**, 49–52 (1902)
24. W. Han, J. Huang, K. Kazmi, Y. Chen, A numerical method for a Cauchy problem for elliptic partial differential equations. Inverse Prob. **23**(6), 2401–2415 (2007)
25. H. Han, L. Ling, T. Takeuchi, An energy regularization for Cauchy problems of Laplace equation in annulus domain. Commun. Comput. Phys. **9**(4), 878–896 (2011)
26. M.V. Klibanov, F. Santosa, A computational quasi-reversibility method for Cauchy problems for Laplace's equation. SIAM J. Appl. Math. **51**(6), 1653–1675 (1991)
27. R. Lattès, J.-L. Lions, *The Method of Quasi-Reversibility. Applications to Partial Differential Equations*. Modern Analytic and Computational Methods in Science and Mathematics, vol. 18 (Translated from the French edition and edited by Richard Bellman). (American Elsevier Publishing Co., New York, 1969)
28. H.-J. Reinhardt, H. Han, D.N. Hào, Stability and regularization of a discrete approximation to the Cauchy problem for Laplace's equation. SIAM J. Numer. Anal. **36**(3), 890–905 (1999)
29. A.N. Tikhonov, V.Y. Arsenin, in *Solutions of Ill-Posed Problems* (Translated from the Russian, Preface by translation editor Fritz John, Scripta Series in Mathematics). (V. H. Winston and Sons, Washington, DC; Wiley, New York/London, 1977)
30. C. Wang, J. Wang, A primal-dual weak Galerkin finite element method for second order elliptic equations in non-divergence form. ArXiv e-prints (October 2015)

Static Condensation, Hybridization, and the Devising of the HDG Methods

Bernardo Cockburn

Abstract In this paper, we review and refine the main ideas for devising the so-called hybridizable discontinuous Galerkin (HDG) methods; we do that in the framework of steady-state diffusion problems. We begin by revisiting the classic techniques of *static condensation* of continuous finite element methods and that of *hybridization* of mixed methods, and show that they can be reinterpreted as discrete versions of a characterization of the associated exact solution in terms of solutions of Dirichlet boundary-value problems on each element of the mesh which are then patched together by transmission conditions across interelement boundaries. We then define the HDG methods associated to this characterization as those using discontinuous Galerkin (DG) methods to approximate the local Dirichlet boundary-value problems, and using weak impositions of the transmission conditions. We give simple conditions guaranteeing the existence and uniqueness of their approximate solutions, and show that, by their very construction, the HDG methods are amenable to static condensation. We do this assuming that the diffusivity tensor can be inverted; we also briefly discuss the case in which it cannot. We then show how a different characterization of the exact solution, gives rise to a different way of statically condensing an already known HDG method. We devote the rest of the paper to establishing bridges between the HDG methods and other methods (the old DG methods, the mixed methods, the staggered DG method and the so-called Weak Galerkin method) and to describing recent efforts for the construction of HDG methods (one for systematically obtaining superconvergent methods and another, quite different, which gives rise to optimally convergent methods). We end by providing a few bibliographical notes and by briefly describing ongoing work.

Published in LNCSE, vol. XX (2015), pp. 1–45.

B. Cockburn (✉)
School of Mathematics, University of Minnesota, Minneapolis, MN, USA
e-mail: cockburn@math.umn.edu

G.R. Barrenechea et al. (eds.), *Building Bridges: Connections and Challenges in Modern Approaches to Numerical Partial Differential Equations*, Lecture Notes in Computational Science and Engineering 114, DOI 10.1007/978-3-319-41640-3_5

1 Introduction

In this paper, we give a short introduction to the devising of the hybridizable discontinuous Galerkin (HDG) in the framework of the following steady-state diffusion model problem:

$$\mathrm{c}\,\boldsymbol{q} + \nabla u = 0 \qquad \text{in } \Omega \subset \mathbb{R}^d, \tag{1a}$$

$$\nabla \cdot \boldsymbol{q} = f \qquad \text{in } \Omega, \tag{1b}$$

$$u = u_D \qquad \text{on } \partial\Omega. \tag{1c}$$

We assume that the data c ,f and u_D are smooth functions such that the solution itself is smooth. Here c is a matrix-valued function which is symmetric and uniformly positive definite on Ω. We are going to closely follow [33], where the HDG methods were introduced.

Since the HDG methods are discontinuous Galerkin (DG) methods, [25], we begin by defining the DG methods for the above boundary-value problem; we follow [3]. Let us first discretize the domain Ω. We denote a triangulation of the domain Ω by $\Omega_h := \{K\}$ and set $\partial\Omega_h := \{\partial K : \ K \in \Omega_h\}$. The outward unit normal to the element K is denoted by $\boldsymbol{n}$. The set of faces of the element K is denoted by $\mathcal{F}(K)$. An interior face F of the triangulation Ω_h is any set of the form $\partial K^+ \cap \partial K^-$, where $K^\pm$ are elements of Ω_h; we assume that the $(d-1)$-Lebesgue measure of F is not zero. The set of all interior faces is denoted by $\mathcal{F}_h^i$. Similarly, a boundary face F of the triangulation Ω_h is any set of the form $\partial K \cap \partial\Omega$, where K are elements of Ω_h; again, we assume that the $(d-1)$ Lebesgue measure of F is not zero. The set of all boundary faces is denoted by $\mathcal{F}_h^\partial$. The set of interior and boundary faces of the triangulation is denoted by $\mathcal{F}_h$.

The notation associated to the weak formulation of the method is the following. We set

$$(\cdot,\cdot)_{\Omega_h} := \sum_{K\in\Omega_h} (\cdot,\cdot)_K \quad \text{and} \quad \langle\cdot,\cdot\rangle_{\partial\Omega_h} := \sum_{K\in\Omega_h} \langle\cdot,\cdot\rangle_{\partial K},$$

where $(\cdot,\cdot)_K$ denotes the standard $L^2(K)$-inner product, and $\langle\cdot,\cdot\rangle_{\partial K}$ denotes the standard $L^2(\partial K)$-inner product.

We can now introduce the general form of a DG method. The approximate solution $(\boldsymbol{q}_h, u_h)$ given by a DG method is the element of the space $\boldsymbol{V}_h \times W_h$, where

$$\boldsymbol{V}_h := \{\boldsymbol{v} \in \boldsymbol{L}^2(\Omega) : \boldsymbol{v}|_K \in \boldsymbol{V}(K) \ \ \forall K \in \Omega_h\},$$

$$W_h := \{w \in L^2(\Omega) : w|_K \in W(K) \ \forall K \in \Omega_h\}.$$

satisfying the equations

$$(\mathrm{c}\,\boldsymbol{q}_h, \boldsymbol{v})_{\Omega_h} - (u_h, \nabla\cdot\boldsymbol{v})_{\Omega_h} + \langle \hat{u}_h, \boldsymbol{v}\cdot\boldsymbol{n}\rangle_{\partial\Omega_h} = 0,$$
$$-(\boldsymbol{q}_h, \nabla w)_{\Omega_h} + \langle \hat{\boldsymbol{q}}_h\cdot\boldsymbol{n}, w\rangle_{\partial\Omega_h} = (f, w)_{\Omega_h},$$

for all $(\boldsymbol{v}, w) \in \boldsymbol{V}_h \times W_h$, where the numerical traces $\hat{u}_h$ and $\hat{\boldsymbol{q}}_h\cdot\boldsymbol{n}$ are approximations to $u|_{\partial\Omega_h}$ and $\boldsymbol{q}\cdot\boldsymbol{n}|_{\partial\Omega_h}$, respectively. The finite dimensional space $\boldsymbol{V}_h \times W_h$ is chosen so that all the integrals in the above weak formulation are well defined.

It remains to discuss how to choose the numerical traces. To do that, let us begin by introducing some useful notation. The traces of the functions ζ and z defined on $K^\pm \in \Omega_h$ on the boundary $\partial K^\pm$ are denoted by $\zeta^\pm$ and $z^\pm$, respectively. We use the same notation if the functions ζ and z are defined on $\partial\Omega_h$. Thus, we define the jumps of ζ and z across the interior face $F = \partial K^+ \cap \partial K^-$ by

$$[\![\zeta]\!] := \zeta^+\boldsymbol{n}^+ + \zeta^-\boldsymbol{n}^- \quad \text{and} \quad [\![z]\!] := z^+\cdot\boldsymbol{n}^+ + z^-\cdot\boldsymbol{n}^-,$$

respectively, where $\boldsymbol{n}^\pm$ is the outward unit normal to $K^\pm$. On boundary faces F, we simply write

$$[\![\zeta]\!] := \zeta\boldsymbol{n} \quad \text{and} \quad [\![z]\!] := z\cdot\boldsymbol{n},$$

with the obvious notation. We say that the numerical traces are *single-valued* if, on $\mathcal{F}_h^i$, $[\![\hat{u}_h]\!] = \boldsymbol{0}$ and $[\![\hat{\boldsymbol{q}}_h]\!] = 0$.

Slightly extending what was done in [3], the numerical traces $\hat{u}_h$ and (the normal component of) $\hat{\boldsymbol{q}}_h$ are linear mappings $\hat{u}_h : \boldsymbol{H}^1(\Omega_h) \times H^1(\Omega_h) \to L^2(\partial\Omega_h)$ $\hat{\boldsymbol{q}}_h : \boldsymbol{H}^1(\Omega_h) \times H^1(\Omega_h) \to \boldsymbol{L}^2(\partial\Omega_h)$ which approximate the traces of u and (the normal component of) $\boldsymbol{q}$ on $\partial\Omega_h$, respectively. We take these numerical traces to be *consistent*. We say that they are consistent if

$$\hat{u}_h(-\mathrm{a}\nabla v, v) = v|_{\partial\Omega_h}, \qquad \hat{\boldsymbol{q}}_h(-\mathrm{a}\nabla v, v)\cdot\boldsymbol{n} = -(\mathrm{a}\nabla v)\cdot\boldsymbol{n}|_{\partial\Omega_h},$$

whenever $[\![\mathrm{a}\nabla v]\!] = 0$ and $[\![v]\!] = \boldsymbol{0}$ on the interior faces $\mathcal{F}_h^i$. Here $\mathrm{a} := \mathrm{c}^{-1}$. This completes the description of the DG methods.

The HDG methods are the DG methods just described which are amenable to *static condensation*. They are thus efficiently implementable and turn out to be more accurate than its predecessors in many instances. None of them fit in the unifying framework developed in [3], since the numerical trace $\hat{u}_h$ of the HDG methods depends on the approximate flux too. The family of DG methods analyzed in [4] includes some HDG methods.

The paper is organized as follows. In Sect. 2, we show that the classic techniques of *static condensation* of continuous finite element methods and that of *hybridization* of mixed methods, introduced back in 1965 in [55] and [52], respectively, can be *reinterpreted* as discrete versions of a characterization of the associated exact solution expressed in terms of solutions of Dirichlet boundary-value problems on each element of the mesh patched together by transmission conditions across interelement boundaries. In Sect. 3, we use this reinterpretation to *define* the HDG methods associated to this characterization as those using discontinuous Galerkin (DG) methods to approximate the local Dirichlet boundary-value problems, and using weak impositions of the transmission conditions. We show that, by construction, the global problem of these HDG methods *only* involves the approximation to the trace of the scalar variable on the faces of the triangulation. We do this assuming that the diffusivity tensor a is invertible; in Sect. 4, we show that it is trivial to treat the case in which it is not. In Sect. 5, we show that a new characterization of the exact solution, based on the elementwise solution of Neumann boundary-value problems, can be used to produce a different type of static condensation of *already known* HDG methods. In Sect. 6, we establish bridges between the HDG and several other methods and comment on two promising ways of devising new HDG methods. We end by providing a few bibliographical notes and by briefly describing ongoing work.

1.1 Note to the Reader

Engineering and Mathematics Graduate Students interested in numerical methods for partial differential equations should be able to read this paper. An elementary background in finite element methods should be enough since here we focus on the ideas guiding the devising of the methods rather than in their rigorous error analyses.

The material of these notes is strongly related to the one presented at the Durham Symposium entitled "Building bridges: Connections and challenges in modern approaches to numerical partial differential equations" at Durham, U.K., July 8–16, 2014, sponsored by the London Mathematical Society, and EPSRC. I would like to express my gratitude to the organizers, especially to G.R. Barrenechea and E. Georgoulis, for the invitation to talk about HDG methods at that meeting.

These notes have evolved from several short courses the author has given: at the Basque Center of Applied Mathematics, Bilbao, Spain, July 9–17, 2009; at the University of Pavia, May 28–June 1, 2012; at the Department of Mathematics & Statistics of the King Fahad University of Petroleum and Minerals, Dec. 2012; at the International Center for Numerical Methods in Engineering, and Universidad Polytecnica de Catalunya, Barcelona, Spain, July 11–15, 2012; at the US National Conference on Computational Mechanics 12, Raleigh, North Carolina, July 22–25, 2013; and at the Department of Mathematics of the Chinese University of Hong Kong, March 19–21, 2014.

2 Static Condensation and Hybridization

Here we argue that the *static condensation* of the continuous Galerkin method, an implementation technique introduced by R.J. Guyan 1965 in [55], can be reinterpreted as a discrete version of a characterization of the exact solution. We also argue that a similar interpretation can be given to the *static condensation* of a mixed method as proposed by Fraejis de Veubeque also in 1965 [52], who showed that this can be achieved provided the mixed method is *hybridized* first. Although the above-mentioned procedures were carried out in the setting of linear elasticity, we present them for our simpler model problem of steady-state diffusion (1).

We proceed as follows. First, we present a characterization of the exact solution in terms of solutions of local problems patched together by means of transmission and boundary conditions. We then show how the original static condensation of the continuous Galerkin method and that of a mixed method can be thought of as discrete versions of such characterization.

2.1 Static Condensation of the Exact Solution

2.1.1 A Characterization of the Exact Solution

Here, for any given triangulation $\Omega_h := \{K\}$ of Ω, we give a characterization of the exact solution in terms of solutions on each of the elements $K \in \Omega_h$, and a single global problem expressed in terms of transmission and boundary conditions.

Suppose that, for each element $K \in \Omega_h$, we define $(\mathbf{Q}, \mathsf{U})$ as the solution of the local problem

$$\begin{aligned} \mathsf{c}\,\mathbf{Q} + \nabla \mathsf{U} &= 0 && \text{in } K, \\ \nabla \cdot \mathbf{Q} &= f && \text{in } K, \\ \mathsf{U} &= \hat{u} && \text{on } \partial K, \end{aligned}$$

where we want the single-valued function $\hat{u}$ to be such that $(\mathbf{Q}, \mathsf{U}) = (\boldsymbol{q}, u)$ on each element $K \in \Omega_h$. We know that this happens *if and only if* $\hat{u}$ enforces the following transmission and boundary conditions:

$$\begin{aligned} [\![\mathbf{Q}]\!] &= 0 && \text{on } F \in \mathcal{F}_h^i, \\ \hat{u} &= u_D && \text{on } F \in \mathcal{F}_h^\partial. \end{aligned}$$

If we now separate the influence of $\hat{u}$ form that of f, we can easily see that we obtained the following result.

Theorem 1 (Characterization of the Exact Solution) *We have that*

$$(\boldsymbol{q}, u) = (\boldsymbol{Q}, \mathsf{U}) = (\boldsymbol{Q}_{\hat{u}}, \mathsf{U}_{\hat{u}}) + (\boldsymbol{Q}_f, \mathsf{U}_f),$$

where, on the element $K \in \Omega_h$, $(\boldsymbol{Q}_{\hat{u}}, \mathsf{U}_{\hat{u}})$ and $(\boldsymbol{Q}_f, \mathsf{U}_f)$ are the solutions of

$$\begin{aligned} \mathsf{c}\,\boldsymbol{Q}_{\hat{u}} + \nabla \mathsf{U}_{\hat{u}} &= 0 \quad \text{in } K, & \mathsf{c}\,\boldsymbol{Q}_f + \nabla \mathsf{U}_f &= 0 \quad \text{in } K,\\ \nabla \cdot \boldsymbol{Q}_{\hat{u}} &= 0 \quad \text{in } K, & \nabla \cdot \boldsymbol{Q}_f &= f \quad \text{in } K,\\ \mathsf{U}_{\hat{u}} &= \hat{u} \quad \text{on } \partial K, & \mathsf{U}_f &= 0 \quad \text{on } \partial K, \end{aligned}$$

and where $\hat{u}$ is the single-valued function solution of

$$\begin{aligned} -[\![\boldsymbol{Q}_{\hat{u}}]\!] &= [\![\boldsymbol{Q}_f]\!] \quad \text{if } F \in \mathcal{F}_h^i,\\ \hat{u} &= u_D \qquad \text{if } F \in \mathcal{F}_h^\partial. \end{aligned}$$

2.1.2 An Example

Let us illustrate this result with a simple but revealing case. Take $\Omega := (0, 1)$ with $K = (x_{i-1}, x_i)$ for $i = 1, \ldots, N$ where $x_0 = 0$ and $x_N = 1$. For simplicity, we take c to be a constant. We then have that

$$(\boldsymbol{q}, u) = (\mathbf{Q}_{\hat{u}}, \mathsf{U}_{\hat{u}}) + (\mathbf{Q}_f, \mathsf{U}_f),$$

where, for $i = 1, \ldots, N$, the functions $(\mathbf{Q}_{\hat{u}}, \mathsf{U}_{\hat{u}})$ and $(\mathbf{Q}_f, \mathsf{U}_f)$ are the solutions of the local problem

$$\begin{aligned} \mathsf{c}\,\mathbf{Q}_{\hat{u}} + \frac{d}{dx}\mathsf{U}_{\hat{u}} &= 0 \quad \text{in } (x_{i-1}, x_i), & \mathsf{c}\,\mathbf{Q}_f + \frac{d}{dx}\mathsf{U}_f &= 0 \quad \text{in } (x_{i-1}, x_i),\\ \frac{d}{dx}\mathbf{Q}_{\hat{u}} &= 0 \quad \text{in } (x_{i-1}, x_i), & \frac{d}{dx}\mathbf{Q}_f &= f \quad \text{in } (x_{i-1}, x_i),\\ \mathsf{U}_{\hat{u}} &= \hat{u} \quad \text{on } \{x_{i-1}, x_i\}, & \mathsf{U}_f &= 0 \quad \text{on } \{x_{i-1}, x_i\}. \end{aligned}$$

Note that we still do not know the actual values of the function $\hat{u} : \{x_i\}_{i=0}^N \mapsto \mathbb{R}$, but once we obtain them, we can readily get the exact solution $(\boldsymbol{q}, u)$. To find those values, we only have to solve the global problem

$$\begin{aligned} -\mathbf{Q}_{\hat{u}}(x_i^-) + \mathbf{Q}_{\hat{u}}(x_i^+) &= \mathbf{Q}_f(x_i^-) - \mathbf{Q}_f(x_i^+) \quad \text{for } i = 1, \ldots, N-1,\\ \hat{u}(x_i) &= u_D(x_i) \qquad\qquad\qquad \text{for } i = 0, N. \end{aligned}$$

Now, let us solve the local problems and then find the global problem. A simple computation gives that the solutions of the local problems are

$$\mathbf{Q}_{\hat{u}}(x) = -\frac{\hat{u}_i - \hat{u}_{i-1}}{\mathsf{c}\, h_i}, \qquad \mathbf{Q}_f(x) = -\mathsf{c}^{-1} \int_{x_{i-1}}^{x_i} G^i_x(x,s) f(s)\, ds,$$

$$\mathsf{U}_{\hat{u}}(x) = \varphi_i(x)\, \hat{u}_i + \varphi_{i-1}(x)\, \hat{u}_{i-1}, \qquad \mathsf{U}_f(x) = \int_{x_{i-1}}^{x_i} G^i(x,s) f(s)\, ds,$$

where $h_i := x_i - x_{i-1}$ and G^i is the Green's function of the second local problem, namely,

$$G^i(x,s) := \begin{cases} \mathsf{c}\, h_i\, \varphi_i(s)\, \varphi_{i-1}(x) & \text{if } x_{i-1} \leqslant s \leqslant x, \\ \mathsf{c}\, h_i\, \varphi_i(x)\, \varphi_{i-1}(s) & \text{if } x \leqslant s \leqslant x_i. \end{cases}$$

where

$$\varphi_i(s) := \begin{cases} (s - x_{i-1})/h_i & \text{if } x_{i-1} \leqslant s \leqslant x_i, \\ (x_{i+1} - s)/h_{i+1} & \text{if } x_i \leqslant s \leqslant x_{i+1}. \end{cases}$$

As a consequence, the global problem for the values $\{\hat{u}_i\}_{i=0}^N$ is

$$\frac{\hat{u}_i - \hat{u}_{i-1}}{\mathsf{c}\, h_i} - \frac{\hat{u}_{i+1} - \hat{u}_i}{\mathsf{c}\, h_{i+1}} = \int_{x_{i-1}}^{x_{i+1}} \varphi_i(s) f(s)\, ds \quad \text{for } i = 1, \ldots, N-1,$$

$$\hat{u}_j = u_D(x_j) \qquad \text{for } j = 0, N.$$

In other words, the values of the exact solution at the nodes of the triangulation, $\{\hat{u}_i\}_{i=0}^N$, can be obtained by inverting a (symmetric positive definite) tridiagonal matrix of order $N+1$.

2.2 Static Condensation of the Continuous Galerkin Method

Now, we show that a characterization of the continuous Galerkin method similar to that one just obtained for the exact solution can be interpreted as the original static condensation of the method [55].

2.2.1 A Characterization of the Approximate Solution

The continuous Galerkin method provides an approximation to u, u_h, in the space

$$W_h = \{w \in \mathcal{C}^0(\Omega) : w|_K \in W(K)\ \forall K \in \Omega_h\}.$$

It determines it by requiring that it be the only solution in $W_h(u_D)$ of the equation

$$(\mathsf{a}\,\nabla u_h, \nabla w)_\Omega = (f, w)_\Omega \quad \forall w \in W_h(0).$$

where $W_h(g) = \{w \in W_h : w = I_h(g) \text{ on } \partial\Omega\}$, and I_h is a suitably defined interpolation operator.

Now, to obtain our characterization of the approximate solution, we need to split the spaces in a suitable manner. Thus, for each element $K \in \Omega_h$, we define the space associated to the *interior* degrees of freedom,

$$W_0(K) := \{w \in W(K) : w|_{\partial K} = 0\},$$

and the space associated to the degrees of freedom on the *boundary*,

$$W_\partial(K) := \{w \in W(K) : w|_{\partial K} = 0 \implies w|_K = 0\}.$$

Clearly, $W(K) = W_0(K) + W_\partial(K)$ for all $K \in \Omega_h$, and so $W_h = W_{0,h} + W_{\mathcal{F}_h}$ where

$$W_{0,h} := \{w \in W_h : w|_K \in W_0(K)\ \forall K \in \Omega_h\},$$
$$W_{\mathcal{F}_h} := \{w \in W_h : w|_K \in W_\partial(K)\ \forall K \in \Omega_h\}.$$

We also need to introduce the following sets of traces on $\mathcal{F}_h$:

$$M_h := \{w|_{\mathcal{F}_h} : w \in W_h\},$$
$$M_h(g) := \{\mu \in M_h : \mu|_{\partial\Omega} = I_h(g)\}.$$

Note that the trace into $\mathcal{F}_h$ is an isomorphism between $W_{\mathcal{F}_h}$ and M_h.

Suppose that, for each element $K \in \Omega_h$, we define $\mathsf{U} \in W(K)$ as the solution of the local problem

$$\begin{aligned}(\mathsf{a}\,\nabla \mathsf{U}, \nabla w)_K &= (f, w)_K \quad && \forall w \in W_0(K),\\ \mathsf{U} &= \hat{u}_h && \text{on } \partial K,\end{aligned}$$

where we want to chose the function $\hat{u}_h \in M_h$ in such a way that $\mathsf{U} = u_h$ on each element $K \in \Omega_h$. This happens if and only $\hat{u}_h$ is such that

$$\begin{aligned}(\mathsf{a}\,\nabla \mathsf{U}, \nabla w)_\Omega &= (f, w)_\Omega \quad && \forall w \in W_{\mathcal{F}_h},\\ \hat{u}_h &= I_h(u_D) && \text{on } \partial\Omega.\end{aligned}$$

If we separate the influence of $\hat{u}_h$ from that of f in the definition of the local problems, and rework the formulation of the global problem, we get the following result.

Theorem 2 (Characterization of the Continuous Galerkin Method) *The approximation given by the continuous Galerkin method can be written as*

$$u_h = \mathsf{U} = \mathsf{U}_{\hat{u}_h} + \mathsf{U}_f,$$

where, on the element $K \in \Omega_h$, $\mathsf{U}_{\hat{u}_h}$ and U_f are the elements of $W(K)$ that solve the local problems

$$\begin{aligned} (\mathsf{a}\,\nabla \mathsf{U}_{\hat{u}_h}, \nabla w)_K &= 0 \quad \forall w \in W_0(K) \qquad & (\mathsf{a}\,\nabla \mathsf{U}_f, \nabla w)_K &= (f, w)_K \quad \forall w \in W_0(K), \\ \mathsf{U}_{\hat{u}_h} &= \hat{u}_h \quad \text{on } \partial K & \mathsf{U}_f &= 0 \quad \text{on } \partial K, \end{aligned}$$

and $\hat{u}_h$ is the element of $M_h(u_D)$ that solves the global problem

$$(\mathsf{a}\,\nabla \mathsf{U}_{\hat{u}_h}, \nabla \mathsf{U}_\mu)_\Omega = (f, \mathsf{U}_\mu)_\Omega \qquad \forall\, \mu \in M_h(0).$$

Note that, although the static condensation [55] is carried out directly on the stiffness matrix of the method, this result shows how to use (local and global) weak formulations to achieve exactly the same thing.

Proof By the linearity of the problem, we only have to justify the characterization of the function $\hat{u}_h$. Let us start from the fact that $\hat{u}_h$ is the element of $M_h(u_D)$ which solves the global problem

$$(\mathsf{a}\,\nabla \mathsf{U}_{\hat{u}_h}, \nabla w)_\Omega + (\mathsf{a}\,\nabla \mathsf{U}_f, \nabla w)_\Omega = (f, w)_\Omega \qquad \forall\, w \in W_{\mathcal{F}_h}.$$

Now, note that, for any $w \in W_h$, we can define the function w_0 by the equation

$$w = \mathsf{U}_\mu + w_0,$$

where $\mu := w|_{\mathcal{F}_h}$; this readily implies that $w_0 \in W_{0,h}$. If we now insert this expression in the equation and take into consideration the definition of the solution of the local problems, that is, that

$$\begin{aligned} (\mathsf{a}\,\nabla \mathsf{U}_{\hat{u}_h}, \nabla w_0)_\Omega &= 0, \\ (\mathsf{a}\,\nabla \mathsf{U}_f, \nabla \mathsf{U}_\mu)_\Omega &= 0, \\ (\mathsf{a}\,\nabla \mathsf{U}_f, \nabla w_0)_\Omega &= (f, w_0)_\Omega, \end{aligned}$$

we finally get the wanted formulation. This completes the proof. □

2.2.2 The Numerical Trace of the Flux

A quick comparison of the above result with the one for the exact solution, suggests that the global problem for the continuous Galerkin method is a *transmission* condition on a discrete version of the normal component of the flux. This little known fact will allow us to *identify* the numerical trace of the approximate flux for the continuous Galerkin method.

To do this, we first write the global problem in its original form, that is,

$$(\mathrm{a}\,\nabla u_h, \nabla w)_\Omega = (f, w)_\Omega \qquad \forall\ w \in W_{\mathcal{F}_h},$$

and perform a simple integration by parts to get

$$-(\nabla \cdot (\mathrm{a}\,\nabla u_h), w)_{\Omega_h} + \langle (\mathrm{a}\,\nabla u_h) \cdot \boldsymbol{n}, w \rangle_{\partial \Omega_h} = (f, w)_\Omega \qquad \forall\ w \in W_{\mathcal{F}_h},$$

Let us now define, for each element $K \in \Omega_h$, the function $R_h \in W_\partial(K)$ satisfying the equation

$$\langle R_h, w \rangle_{\partial K} = (\nabla \cdot (\mathrm{a}\,\nabla u_h) + f, w)_K \qquad \forall w \in W_\partial(K).$$

Thus, the function R_h is a projection of the *residual* $\nabla \cdot (\mathrm{a}\,\nabla u_h) + f$. With this definition, we get that

$$\langle (-\mathrm{a}\,\nabla u_h) \cdot \boldsymbol{n} + R_h, w \rangle_{\partial \Omega_h} = 0 \qquad \forall\ w \in W_{\mathcal{F}_h},$$

which can be interpreted as a transmission condition forcing the normal component of numerical trace of the flux

$$\hat{\boldsymbol{q}}_h \cdot \boldsymbol{n} := (-\mathrm{a}\,\nabla u_h) \cdot \boldsymbol{n} + R_h \quad \text{on } \partial \Omega_h,$$

to be weakly continuous across interelement boundaries.

2.2.3 Relation with Static Condensation

Let us now show that this characterization is nothing but an application of the well-known technique of *static condensation* [55]. Static condensation was conceived as a way to reducing the size of the stiffness matrix. Indeed, if $[u_h]$ is the vector of degrees of freedom of the approximation u_h, and the matrix equation of the continuous Galerkin method is

$$K\,[u_h] = [f],$$

the static condensation consists in partitioning the vector of degrees of freedom $[u_h]$ into two smaller vectors, namely, the degrees of freedom interior to the elements, $[\mathsf{U}]$, and the degrees of freedom associated to the boundaries of the elements,

$[\hat{u}_h]$, and then *eliminating* $[\mathsf{U}]$ from the equations. Indeed, taking into account this partition, the above equation reads

$$\begin{bmatrix} K_{00} & K_{0\partial} \\ K_{\partial 0} & K_{\partial\partial} \end{bmatrix} \begin{bmatrix} [\mathsf{U}] \\ [\hat{u}_h] \end{bmatrix} = \begin{bmatrix} f_0 \\ f_\partial \end{bmatrix}.$$

By our choice of the degrees of freedom, the matrix K_{00} is easy to invert since it is *block diagonal*, each block being associated to a local problem. We thus get

$$[\mathsf{U}] = -K_{00}^{-1} K_{0\partial} [\hat{u}_h] + K_{00}^{-1} [f_0].$$

We can now eliminate $[\mathsf{U}]$ from the original matrix equation to obtain

$$(-K_{\partial 0}\, K_{00}^{-1}\, K_{0\partial} + K_{\partial\partial})[\hat{u}_h] = -K_{\partial 0}\, K_{00}^{-1} [f_0] + [f_\partial].$$

The matrix in the left-hand side, nowadays called the Schur complement of the matrix K_{00}, is clearly smaller than the original matrix K and is also easier to numerically invert. We have thus shown that our characterization of the approximate solution of the continuous Galerkin method is nothing but another way of carrying out the good, old static condensation. The former expresses in terms of weak formulations what the latter does directly on the matrix equations itself.

2.2.4 An Example

Let us now illustrate this procedure in our simple one-dimensional example. We take

$$W(K) := \mathcal{P}_k(K),$$

where $\mathcal{P}_k(K)$ denotes the space of polynomials of degree at most k defined on the set K. We begin by solving the local problems. If we use the notation $\hat{u}_i = \hat{u}_h(x_i)$ for $i = 0, \ldots, N$, a few manipulations (and the proper choice of the basis functions) allow us to see that the solutions of the local problems are

$$\mathsf{U}_{\hat{u}}(x) = \varphi_i(x)\, \hat{u}_i + \varphi_{i-1}(x)\, \hat{u}_{i-1} \qquad \mathsf{U}_f(x) = \int_{x_{i-1}}^{x_i} G_h^i(x,s) f(s)\, ds,$$

where $h_i := x_i - x_{i-1}$ and G_h^i is the discrete Green's function of the second local problem, namely,

$$G_h^i(x,s) := \frac{h_i}{4\mathsf{a}} \sum_{\ell=1}^{k-1} \frac{1}{2\ell+1} (P_{\ell+1}^i - P_{\ell-1}^i)(x)\, (P_{\ell+1}^i - P_{\ell-1}^i)(s)$$

where $P_n^i(x) := P_n(T^i(x))$, $T^i(\zeta) := (\zeta - (x_i + x_{i-1})/2)/(h_i/2)$ and P_n is the Legendre polynomial of degree n. As a consequence, the global problem for the values $\{\hat{u}_i\}_{i=0}^N$ is

$$\mathsf{a}\,\frac{\hat{u}_i - \hat{u}_{i-1}}{h_i} - \mathsf{a}\,\frac{\hat{u}_{i+1} - \hat{u}_i}{h_{i+1}} = \int_{x_{i-1}}^{x_{i+1}} \varphi_i(s) f(s)\,ds \quad \text{for } i = 1, \ldots, N-1,$$

$$\hat{u}_j = u_D(x_j) \qquad \text{for } j = 0, N.$$

Note that the size of the matrix equation of the global problem is *independent* of the value of the polynomial degree k, a reflection of the effectiveness of the static condensation technique. Note also that the values of the approximate solution at the nodes of the triangulation, $\{\hat{u}_i\}_{i=0}^N$, are actually exact, as expected.

2.3 Static Condensation of Mixed Methods by Hybridization

Next, we show how to extend what was done for the continuous Galerkin method to mixed methods. A particular important point we want to emphasize here is that *hybridization* of a mixed method is what *allows* it to be statically condensed, as first realized in [52].

2.3.1 A Characterization of the Approximate Solution

A mixed method seeks approximations to the flux $\boldsymbol{q} := -\mathsf{a}\nabla u$, $\boldsymbol{q}_h$, and the scalar u, u_h, in the finite dimensional spaces

$$\boldsymbol{\mathcal{V}}_h = \{\boldsymbol{v} \in \boldsymbol{H}(div, \Omega) : \boldsymbol{v}|_K \in \boldsymbol{V}(K) \ \ \forall K \in \Omega_h\}.$$

$$W_h = \{w \in L^2(\Omega) : \quad w|_K \in W(K) \ \forall K \in \Omega_h\},$$

respectively. It determines the function $(\boldsymbol{q}_h, u_h)$ as the only element of $\boldsymbol{\mathcal{V}}_h \times W_h$ satisfying the equations

$$(\mathsf{c}\,\boldsymbol{q}_h, \boldsymbol{v})_\Omega - (u_h, \nabla\cdot\boldsymbol{v})_\Omega = -\langle u_D, \boldsymbol{v}\cdot\boldsymbol{n}\rangle_{\partial\Omega} \quad \forall \boldsymbol{v} \in \boldsymbol{\mathcal{V}}_h,$$

$$(\nabla\cdot\boldsymbol{q}_h, w)_\Omega = (f, w)_\Omega \qquad \forall w \in W_h.$$

For mixed methods, the choice of the finite dimensional space $\boldsymbol{\mathcal{V}}_h \times W_h$ is not simple, but here we assume that *it has* been properly chosen as to define a unique approximate solution.

Now, suppose that, for each element $K \in \Omega_h$, we define $(\mathbf{Q}, \mathsf{U}) \in \boldsymbol{V}(K) \times W(K)$ as the solution of the local problem

$$\begin{aligned}(\mathsf{c}\,\mathbf{Q}, \boldsymbol{v})_K - (\mathsf{U}, \nabla \cdot \boldsymbol{v})_K &= \langle \hat{u}_h, \boldsymbol{v} \cdot \boldsymbol{n} \rangle_{\partial K} \quad \forall \boldsymbol{v} \in \boldsymbol{V}(K),\\ (\nabla \cdot \mathbf{Q}, w)_K &= (f, w)_K \qquad\quad \forall w \in W(K).\end{aligned}$$

This problem is well defined since it is nothing but the application of the mixed method, which we assume to be well defined, to the single element $K \in \Omega_h$. As before, we want to choose the function $\hat{u}_h$ in some finite dimensional space M_h in such a way that $(\mathbf{Q}, \mathsf{U}) = (\boldsymbol{q}_h, u_h)$ on each element $K \in \Omega_h$. For this to hold, we only need to guarantee that

$$\begin{aligned}\mathbf{Q} &\in \boldsymbol{\mathcal{V}}_h,\\ \langle \hat{u}_h, \boldsymbol{v} \cdot \boldsymbol{n} \rangle_{\partial\Omega} &= \langle u_D, \boldsymbol{v} \cdot \boldsymbol{n} \rangle_{\partial\Omega} \quad \forall\ \boldsymbol{v} \in \boldsymbol{\mathcal{V}}_h.\end{aligned}$$

The first property is a transmission condition since it holds if and only if the normal component of $\mathbf{Q}$ is continuous across interelement boundaries. The second condition is nothing but a weak form of the Dirichlet boundary condition.

As for the case of the continuous Galerkin method, the choice of the space M_h has to be made in such a way that the above two conditions do *determine* the numerical trace $\hat{u}_h$. Typically, we take

$$M_h := \{\mu \in L^2(\mathcal{F}_h) :\ \exists\ \boldsymbol{v} \in \boldsymbol{V}_h : \mu = [\![\boldsymbol{v}]\!] \text{ on } \mathcal{F}_h\}.$$

Thus, if we set $M_h(g) := \{\mu \in M_h :\ \langle \mu, \eta \rangle_{\partial\Omega} = \langle g, \eta \rangle_{\partial\Omega}\ \forall \eta \in M_h\}$, the global problem can be expressed as follows:

$$\begin{aligned}\langle \mathbf{Q} \cdot \boldsymbol{n}, \mu \rangle_{\partial\Omega_h} &= 0 \quad \forall\ \mu \in M_h(0),\\ \hat{u}_h &\in M_h(u_D).\end{aligned}$$

Indeed, note that, for any $\mu \in M_h(0)$,

$$\langle \mathbf{Q} \cdot \boldsymbol{n}, \mu \rangle_{\partial\Omega_h} = \langle \mathbf{Q}, \mu \rangle_{\partial\Omega_h \setminus \partial\Omega} = \langle\, [\![\mathbf{Q}]\!], \mu \rangle_{\mathcal{F}_h^i},$$

and if this quantity is zero, we certainly have that $\mathbf{Q} \in \boldsymbol{\mathcal{V}}_h$, as wanted. So, let us assume then that the above global problem for $\hat{u} \in M_h$ is well defined.

So, we have obtained the following result.

Theorem 3 (Characterization of the Mixed Method) *The solution of the mixed method can be written as*

$$(\boldsymbol{q}_h, u_h) = (\boldsymbol{Q}, \mathsf{U}) = (\boldsymbol{Q}_{\hat{u}_h}, \mathsf{U}_{\hat{u}_h}) + (\boldsymbol{Q}_f, \mathsf{U}_f),$$

where, on each element $K \in \Omega_h$, for any $\mu \in L^2(\partial K)$ and $f \in L^2(K)$, the functions $(\boldsymbol{Q}_\mu, \mathsf{U}_\mu)$ and $(\boldsymbol{Q}_f, \mathsf{U}_f)$ are the elements of $\boldsymbol{V}(K) \times W(K)$ which solve the local problems

$$(\mathsf{c}\,\boldsymbol{Q}_\mu, \boldsymbol{v})_K - (\mathsf{U}_\mu, \nabla \cdot \boldsymbol{v})_K = -\langle \mu, \boldsymbol{v} \cdot \boldsymbol{n} \rangle_{\partial K}, \qquad (\mathsf{c}\,\boldsymbol{Q}_f, \boldsymbol{v})_K - (\mathsf{U}_f, \nabla \cdot \boldsymbol{v})_K = 0,$$
$$(\nabla \cdot \boldsymbol{Q}_\mu, w)_K = 0, \qquad (\nabla \cdot \boldsymbol{Q}_f, w)_K = (f, w)_K,$$

for all $(\boldsymbol{v}, w) \in \boldsymbol{V}(K) \times W(K)$, and the function $\hat{u}_h$ is the element of $M_h(u_D)$ which solves the global problem

$$(\mathsf{c}\,\boldsymbol{Q}_{\hat{u}_h}, \boldsymbol{Q}_\mu)_{\Omega_h} = (f, \mathsf{U}_\mu)_{\Omega_h} \quad \forall\, \mu \in M_h(0),$$

Proof We only have to prove that $\hat{u}_h \in M_h(u_D)$ satisfies the equation

$$-\langle \mathbf{Q}_{\hat{u}_h} \cdot \boldsymbol{n}, \mu \rangle_{\partial\Omega_h} = \langle \mathbf{Q}_f \cdot \boldsymbol{n}, \mu \rangle_{\partial\Omega_h} \quad \forall\, \mu \in M_h(0).$$

But, by the definition of the local problems, we have

$$\begin{aligned}
-\langle \mathbf{Q}_{\hat{u}_h} \cdot \boldsymbol{n}, \mu \rangle_{\partial\Omega_h} &= (\mathsf{c}\,\mathbf{Q}_\mu, \mathbf{Q}_{\hat{u}_h})_{\Omega_h},\\
\langle \mathbf{Q}_f \cdot \boldsymbol{n}, \mu \rangle_{\partial\Omega_h} &= -(\mathsf{c}\,\mathbf{Q}_\mu, \mathbf{Q}_f)_{\Omega_h} + (\mathsf{U}_\mu, \nabla \cdot \mathbf{Q}_f)_{\Omega_h}\\
&= -(\mathsf{U}_f, \nabla \cdot \mathbf{Q}_{\hat{u}_h})_{\Omega_h} + (\mathsf{U}_\mu, \nabla \cdot \mathbf{Q}_f)_{\Omega_h}\\
&= (\mathsf{U}_\mu, \nabla \cdot \mathbf{Q}_f)_{\Omega_h}\\
&= (f, \mathsf{U}_\mu)_{\Omega_h},
\end{aligned}$$

and the identity follows. This completes the proof. □

2.3.2 Relation with Static Condensation and Hybridization

Let us now show that what we have done is nothing but the *static condensation* of the *hybridized* version of the mixed method as done by Fraejis de Veubeke in [52]. Suppose that the matrix equation of the mixed method reads

$$\begin{bmatrix} \mathcal{A} & B \\ B^t & 0 \end{bmatrix} \begin{bmatrix} [\boldsymbol{q}_h] \\ [u_h] \end{bmatrix} = \begin{bmatrix} [u_D] \\ [f] \end{bmatrix}.$$

It is not easy to eliminate $[\boldsymbol{q}_h]$ from this equation since the matrix $\mathcal{A}$ is *not* block diagonal because, since $\boldsymbol{q}_h \in \mathcal{V}_h$, its normal component is continuous across inter element boundaries. To overcome this unwanted feature, Fraejis de Veubeque relaxed the continuity condition on $\boldsymbol{q}_h$ and worked with a totally discontinuous approximation $\mathbf{Q}$ instead. Because of this, he had to introduce the *hybrid* unknown $\hat{u}_h$, an approximation to the trace of u on each element; this is why this procedure receives the name of *hybridization* of the mixed method. Finally, in order to guarantee that $\mathbf{Q}$ be identical to the original function $\boldsymbol{q}_h$, he then forced it to have a continuous normal component at the interelement boundaries. This operation resulted the following matrix equation:

$$\begin{bmatrix} A & B & C \\ B^t & 0 & 0 \\ C^t & 0 & 0 \end{bmatrix} \begin{bmatrix} [\mathbf{Q}] \\ [\mathsf{U}] \\ [\hat{u}_h] \end{bmatrix} = \begin{bmatrix} -C_\partial[u_D] \\ [f] \\ 0 \end{bmatrix}.$$

Here, $[\hat{u}_h]$ denotes the digressive freedom of the function $\hat{u}_h$ restricted to the interior faces. On the boundary faces, the relation of $\hat{u}_h$ to u_D is already captured by the right-hand side of the first equation. Note that, since the first two equations define the local problems, we can easily solve them to obtain

$$\begin{bmatrix} [\mathbf{Q}] \\ [\mathsf{U}] \end{bmatrix} = \begin{bmatrix} A & B \\ B^t & 0 \end{bmatrix}^{-1} \begin{bmatrix} -C[\hat{u}_h] - C_\partial[u_D] \\ [f] \end{bmatrix}.$$

The third equation, $C\,[\mathbf{Q}] = 0$ enforces the continuity of the normal component of $\mathbf{Q}$ across inter element boundaries; it is this equation that determines the hybrid unknown in the interior faces, $[\hat{u}_h]$. A few computations show that the resulting matrix equation is of the form

$$H[\,\hat{u}\,] = H_\partial[\,u_D\,] + J\,[\,f\,], \qquad H := C^t E\, C, \qquad E := A^{-1} - A^{-1}B\,(B^tA^{-1}B)^{-1}\,B^tA^{-1},$$

and we see that, as expected, the matrix H is symmetric. Moreover, H is positive definite and E is block-diagonal.

2.3.3 An Example

Next, let us illustrate this procedure in our simple one-dimensional example. We take

$$V(K) \times W(K) := \mathcal{P}_{k+1}(K) \times \mathcal{P}_k(K).$$

We begin by solving the local problems. A little computation gives that the solutions of the local problems are

$$\mathbf{Q}_{\hat{u}}(x) = -\frac{\hat{u}_i - \hat{u}_{i-1}}{\mathsf{c}\, h_i}, \qquad \mathbf{Q}_f(x) = \int_{x_{i-1}}^{x_i} H_h^i(x,s) f(s)\, ds,$$

$$\mathsf{U}_{\hat{u}}(x) = \varphi_i(x)\,\hat{u}_i + \varphi_{i-1}(x)\,\hat{u}_{i-1}, \qquad \mathsf{U}_f(x) = \int_{x_{i-1}}^{x_i} G_h^i(x,s) f(s)\, ds,$$

where $h_i := x_i - x_{i-1}$ and

$$H^i_h(x,s) := \varphi_i(x)\,\varphi_i(s) - \varphi_{i-1}(x)\,\varphi_{i-1}(s) + \frac{1}{2}\sum_{\ell=1}^{k}(P^i_{\ell+1} - P^i_{\ell-1})(x)\,P^i_\ell(s),$$

$$G^i_h(x,s) := \frac{\mathsf{c}\,h_i}{4}\sum_{\ell=1}^{k-1}\frac{1}{2\ell+1}\,(P^i_{\ell+1} - P^i_{\ell-1})(x)\,(P^i_{\ell+1} - P^i_{\ell-1})(s).$$

Let us recall that $P^i_n(x) := P_n(T^i(x))$, $T^i(\zeta) := (\zeta - (x_i + x_{i-1})/2)/(h_i/2)$ and P_n is the Legendre polynomial of degree n. Note that the function G^i_h approximates the Green function G^i whereas $-\mathsf{c}\,H^i_h$ approximates its partial derivative G^i_x. As a consequence, the global problem for the values $\{\hat{u}_i\}_{i=0}^N$ is

$$\frac{\hat{u}_i - \hat{u}_{i-1}}{\mathsf{c}\,h_i} - \frac{\hat{u}_{i+1} - \hat{u}_i}{\mathsf{c}\,h_{i+1}} = \int_{x_{i-1}}^{x_{i+1}} \varphi_i(s) f(s)\,ds \quad \text{for } i = 1,\ldots,N-1,$$

$$\hat{u}_j = u_D(x_j) \quad \text{for } j = 0, N.$$

We thus see that the values of the approximate solution at the nodes of the triangulation, $\{\hat{u}_i\}_{i=0}^N$, are actually exact, as expected.

3 HDG Methods

In this section, we show how to use a discrete version of the characterization of the exact solution obtained in the previous section to devise HDG methods for our model problem (1). The local problems are solved by using a DG method and the transmission conditions by a simple weak formulation. As a consequence, the resulting HDG methods are DG methods whose distinctive feature is that they are amenable to hybridization and hence to static condensation. Let us emphasize that this does not happen by accident, but because they are constructed by using a discrete version of the characterization of the exact solution worked out in the previous section.

After defining the HDG methods, we establish a simple result about the existence and uniqueness of their approximate solution and display some examples. We end by showing several different ways of presenting them which will be useful for relating them to other numerical methods.

We follow closely the work done in 2009 [33] for the original HDG methods, as well as the work done in the 2014 review paper [23] for HDG methods for the Stokes system of incompressible fluid flow.

3.1 Definition

We take the approximate solution of the HDG methods to be the function

$$(\boldsymbol{q}_h, u_h) = (\mathbf{Q}, \mathsf{U}),$$

where, on the element $K \in \Omega_h$, $(\mathbf{Q}, \mathsf{U}) \in \boldsymbol{V}(K) \times W(K)$ is the solution of the local problem

$$\begin{aligned} (\mathrm{c}\,\mathbf{Q}, \boldsymbol{v})_K - (\mathsf{U}, \nabla \cdot \boldsymbol{v})_K + \langle \hat{u}_h, \boldsymbol{v} \cdot \boldsymbol{n} \rangle_{\partial K} &= 0 & \forall \boldsymbol{v} \in \boldsymbol{V}(K), \\ -(\mathbf{Q}, \nabla w)_K + \langle \hat{\mathbf{Q}} \cdot \boldsymbol{n}, w \rangle_{\partial K} &= (f, w)_K & \forall w \in W(K), \end{aligned}$$

where the numerical trace $\hat{\mathbf{Q}}$ has to be suitably chosen. Ideally, the numerical trace of the flux $\hat{\mathbf{Q}}$ should be chosen so that it

1. is consistent,
2. only depends (linearly) on $\mathbf{Q}|_K, \mathsf{U}|_K$ and $\hat{u}_h|_{\partial K}$,
3. renders the local problem solvable.

Our *favorite* choice is

$$\hat{\mathbf{Q}} \cdot \boldsymbol{n} := \mathbf{Q} \cdot \boldsymbol{n} + \tau(\mathsf{U} - \hat{u}_h) \quad \text{on } \partial K,$$

where the function τ is linear. We are also going to require that τ be *symmetric*, that is, that, for all $K \in \Omega_h$,

$$\langle \tau(w), \omega \rangle_{\partial K} = \langle w, \tau(\omega) \rangle_{\partial K} \quad \forall\, w, \omega \in W(K) + M_h(\partial K).$$

Although there are many other choices, we are going to use this one from now on; not only it is very natural but it actually covers *all* the known HDG methods.

To complete the definition of the HDG methods, we take the function $\hat{u}_h$ in the space

$$M_h := \{\mu \in L^2(\mathcal{F}_h) : \ \mu_F \in M(F)\ \forall F \in \mathcal{F}_h\},$$

where $M(F)$ is a suitably chosen finite dimensional space, and require that it be determined as the solution of the following weakly imposed transmission and boundary conditions:

$$\begin{aligned} \langle \mu, [\![\hat{\mathbf{Q}}]\!] \rangle_{\mathcal{F}_h^i} &= \langle \mu, \hat{\mathbf{Q}} \cdot \boldsymbol{n} \rangle_{\partial \Omega_h \setminus \partial \Omega} = 0, \\ \langle \mu, \hat{u}_h \rangle_{\partial \Omega} &= \langle \mu, u_D \rangle_{\partial \Omega}, \end{aligned}$$

for all $\mu \in M_h$. This completes the definition of the HDG methods.

The HDG methods are obtained by choosing different functions τ and different local spaces $V(K)$, $W(K)$ and $M(F)$.

3.2 Existence and Uniqueness

We now provide simple conditions on the local spaces and the function τ ensuring, not only that the local problems are solvable, but that the global problem is also well posed. To do that, we use an *energy* identity we obtain next which will also shed light on the role to the function τ.

Proposition 1 (The Local Energy Identity) *For any element $K \in \Omega_h$, we have*

$$(\mathsf{c}\boldsymbol{Q}, \boldsymbol{Q})_K + \langle(\mathsf{U} - \hat{u}_h), \tau(\mathsf{U} - \hat{u}_h)\rangle_{\partial K} = (f, \mathsf{U})_K - \langle \hat{u}_h, \hat{\boldsymbol{Q}} \cdot \boldsymbol{n}\rangle_{\partial K}.$$

Note that the exact solution satisfies the following energy identity:

$$(\mathsf{c}\boldsymbol{q}, \boldsymbol{q})_K = (f, u)_K - \langle u, \boldsymbol{q} \cdot \boldsymbol{n}\rangle_{\partial K}.$$

Typically, the terms $(\mathsf{c}\boldsymbol{q}, \boldsymbol{q})_K$ and $(\mathsf{c}\,\mathbf{Q}, \mathbf{Q})_K$ are interpreted as the energy stored inside the element K. It is thus reasonable to interpret the term $\langle(\mathsf{U} - \hat{u}_h), \tau(\mathsf{U} - \hat{u}_h)\rangle_{\partial K}$ as an *energy* associated with the jumps $\mathsf{U} - \hat{u}_h$ at the boundary of the element ∂K. Since all energies are nonnegative, we assume that the function τ is such that

$$\langle \tau(w - \mu), w - \mu\rangle_{\partial K} \geqslant 0 \qquad \forall (w, \mu) \in W(K) \times M(\partial K), \tag{2a}$$

where

$$M(\partial K) := \{\mu \in L^2(\partial K) : \ \mu|_F \in M(F), \text{ for any face } F \in \mathcal{F}_h \text{ lying on } \partial K\}. \tag{2b}$$

We now see that the role of τ is to transform the discrepancy between U and $\hat{u}_h$ on ∂K into an energy. Since an increase of energy is typically associated with an enhancement of the stability properties of the numerical method, τ is called the *stabilization* function.

Let us now prove Proposition 1.

Proof If we take $(\boldsymbol{v}, w) := (\mathbf{Q}, \mathsf{U})$ in the equations of the local problems, we get

$$\begin{aligned}(\mathsf{c}\,\mathbf{Q}, \mathbf{Q})_K - (\mathsf{U}, \nabla \cdot \mathbf{Q})_K + \langle \hat{u}_h, \mathbf{Q} \cdot \boldsymbol{n}\rangle_{\partial K} &= 0,\\ -(\mathbf{Q}, \nabla \mathsf{U})_K + \langle \hat{\mathbf{Q}} \cdot \boldsymbol{n}, \mathsf{U}\rangle_{\partial K} &= (f, \mathsf{U})_K,\end{aligned}$$

and adding the two equations, we obtain

$$(\mathsf{c}\,\mathbf{Q}, \mathbf{Q})_K + \langle(\hat{\mathbf{Q}} - \mathbf{Q}) \cdot \boldsymbol{n}, \mathsf{U} - \hat{u}_h\rangle_{\partial K} = (f, \mathsf{U})_K - \langle \hat{\mathbf{Q}} \cdot \boldsymbol{n}, \hat{u}_h\rangle_{\partial K}.$$

The energy identity now follows by simply inserting the definition of the numerical trace $\hat{\mathbf{Q}}$. This completes the proof. $\square$

We are now ready to present our main result. It is a variation of a similar result in [33].

Theorem 4 *Assume that the stabilization function τ satisfies the nonnegativity condition (2). Assume also that, for each element $K \in \Omega_h$, we have that if $(w, \mu) \in W(K) \times M(\partial K)$ is such that*

$$\text{(i)} \quad \langle \tau(w - \mu), w - \mu \rangle_{\partial K} = 0,$$

$$\text{(ii)} \quad (\nabla w, \boldsymbol{v})_K + \langle \mu - w, \boldsymbol{v} \cdot \boldsymbol{n} \rangle_{\partial K} = 0 \ \forall \ \boldsymbol{v} \in \boldsymbol{V}(K),$$

then w is a constant on K and $w = \mu$ on ∂K. Then the approximate solution $(\boldsymbol{q}_h, u_h, \hat{u}_h) \in \boldsymbol{V}_h \times W_h \times M_h$ of the HDG method is well defined.

Note that condition (ii) establishes a relation between the local spaces $\boldsymbol{V}(K)$, $W(K)$ and $M(\partial K)$ and the stabilization function τ *guaranteeing* that the local problems as well as the global problem have a unique solution. Note also that if condition (i) were not necessary to obtain that w is a constant on K and $w = \mu$ on ∂K, we can simply take $\tau \equiv 0$. However, for most cases, without condition (i), the method might simply fail to be well defined. The role of τ, is thus to *prevent* this failure.

Let us now prove Theorem 4.

Proof Since the HDG method defines a finite dimensional square system for the unknowns $(\mathbf{Q}, \mathsf{U}, \hat{u}_h) \in \boldsymbol{V}_h \times W_h \times M_h$, we only have to show that, when we set the data f and u_D to zero, the only solution is the trivial one.

Thus, setting $\mu := \hat{u}_h$ in the transmission condition, and recalling that, by the boundary condition, $\hat{u}_h = 0$ on $\partial\Omega$, we get

$$0 = -\langle \hat{u}_h, \hat{\mathbf{Q}} \cdot \boldsymbol{n} \rangle_{\partial\Omega_h} = (\mathsf{c}\,\mathbf{Q}, \mathbf{Q})_{\Omega_h} + \langle (\mathsf{U} - \hat{u}_h), \tau(\mathsf{U} - \hat{u}_h) \rangle_{\partial\Omega_h},$$

by the energy identity of the previous proposition. By assumption (i), we get that $\mathbf{Q} = 0$ on Ω and that $\langle (\mathsf{U} - \hat{u}_h), \tau(\mathsf{U} - \hat{u}_h) \rangle_{\partial K} = 0$ for any $K \in \Omega_h$. Moreover, the first equation defining the local problem now reads

$$(\nabla \mathsf{U}, \boldsymbol{v})_K + \langle \hat{u}_h - \mathsf{U}, \boldsymbol{v} \cdot \boldsymbol{n} \rangle_{\partial K} = 0 \ \forall \ \boldsymbol{v} \in \boldsymbol{V}(K).$$

By assumption (ii) with $(w, \mu) := (\mathsf{U}, \hat{u}_h)$, we have that, on each element $K \in \Omega_h$, U is a constant on K and that $\mathsf{U} = \hat{u}_h$ on ∂K. As a consequence, U is a constant on Ω and $\mathsf{U} = \hat{u}_h$ on $\mathcal{F}_h$. Since $\hat{u}_h = 0$ on $\partial\Omega$ we finally get that $\mathsf{U} = 0$ on Ω and that $\hat{u}_h = 0$ on $\mathcal{F}_h$. This completes the proof. $\square$

Let us now present an almost direct consequence of the previous result in a case in which the stabilization function τ is very *strong*.

Corollary 1 ([33]) *Assume that the stabilization function* τ *satisfies the nonnegativity condition* (2). *Assume also that, for every element* $K \in \Omega_h$,

(a) $(w, \mu) \in W(K) \times M(\partial K)$: $\langle \tau(w - \mu), w - \mu \rangle_{\partial K} = 0 \implies w = \mu$ *on* ∂K,
(b) $\nabla W(K) \subset \boldsymbol{V}(K)$.

Then the approximate solution $(\boldsymbol{q}_h, u_h, \hat{u}_h) \in \boldsymbol{V}_h \times W_h \times M_h$ *of the HDG method is well defined.*

A remarkable feature of this result is that the method is well defined *completely independently* of the choice of the space M_h. This is a direct consequence of condition (a), which is clearly stronger than condition (i) of Theorem 4 on the stabilization function τ. Thanks to condition (a) , we can replace condition (ii) of Theorem 4 by the simpler condition (b), as we see next.

Proof We only have to show that the assumptions the previous result are satisfied. Since τ is a linear mapping, assumption (a) implies condition (i) of Theorem 4. Now, by assumption (a), if $\langle \tau(w - \mu), w - \mu \rangle_{\partial K} = 0$, we have that $w = \mu$ on ∂K and we get that condition (ii) of Theorem 4 reads

$$(\nabla w, \boldsymbol{v})_K = 0 \ \forall \boldsymbol{v} \in \boldsymbol{V}(K).$$

By assumption (b), we can take $\boldsymbol{v} := \nabla w$ and conclude that w is a constant on K. This implies that the second assumption of Theorem 4 holds. This completes the proof. □

3.3 Characterizations of the HDG Methods

Here, we provide two characterizations of the approximate solution provided by the HDG methods just introduced. We are going to use the set

$$M_h(g) := \{\mu \in M_h : \ \langle \mu, \eta \rangle_{\partial\Omega} := \langle g, \eta \rangle_{\partial\Omega} \forall \eta \in M_h\}.$$

3.3.1 Formulation in Terms of $(\boldsymbol{q}_h, u_h, \hat{u}_h)$

Static Condensation Formulation

The following result reflects the way in which the HDG methods were devised and renders evident the way in which their *implementation* by static condensation can be achieved.

Theorem 5 (First Characterization of HDG Methods) *The approximate solution of the HDG method is given by*

$$(\boldsymbol{q}_h, u_h) = (\boldsymbol{Q}, \mathsf{U}) = (\boldsymbol{Q}_{\hat{u}_h}, \mathsf{U}_{\hat{u}_h}) + (\boldsymbol{Q}_f, \mathsf{U}_f),$$

where, on the element $K \in \Omega_h$, for any $\mu \in L^2(\partial K)$, the function $(\boldsymbol{Q}_\mu, \mathsf{U}_\mu) \in \boldsymbol{V}(K) \times W(K)$ is the solution of the local problem

$$\begin{aligned}
(\mathsf{c}\,\boldsymbol{Q}_\mu, \boldsymbol{v})_K - (\mathsf{U}_\mu, \nabla\cdot\boldsymbol{v})_K + \langle \mu, \boldsymbol{v}\cdot\boldsymbol{n}\rangle_{\partial K} &= 0 \quad \forall \boldsymbol{v} \in \boldsymbol{V}(K),\\
-(\boldsymbol{Q}_\mu, \nabla w)_K + \langle \hat{\boldsymbol{Q}}_\mu\cdot\boldsymbol{n}, w\rangle_{\partial K} &= 0 \quad \forall w \in W(K),\\
\hat{\boldsymbol{Q}}_\mu\cdot\boldsymbol{n} := \boldsymbol{Q}_\mu\cdot\boldsymbol{n} + \tau(\mathsf{U}_\mu - \mu) & \qquad \text{on } \partial K,
\end{aligned}$$

and, for any $f \in L^2(K)$, the function $(\boldsymbol{Q}_f, \mathsf{U}_f) \in \boldsymbol{V}(K) \times W(K)$ is the solution of the local problem

$$\begin{aligned}
(\mathsf{c}\,\boldsymbol{Q}_f, \boldsymbol{v})_K - (\mathsf{U}_f, \nabla\cdot\boldsymbol{v})_K &= 0 && \forall \boldsymbol{v} \in \boldsymbol{V}(K),\\
-(\boldsymbol{Q}_f, \nabla w)_K + \langle \hat{\boldsymbol{Q}}_f\cdot\boldsymbol{n}, w\rangle_{\partial K} &= (f, w)_K && \forall w \in W(K),\\
\hat{\boldsymbol{Q}}_f\cdot\boldsymbol{n} := \boldsymbol{Q}_f\cdot\boldsymbol{n} + \tau(\mathsf{U}_f) & && \text{on } \partial K.
\end{aligned}$$

The function $\hat{u}_h$ is the element of $M_h(u_D)$ such that

$$a_h(\hat{u}_h, \mu) = \ell_h(\mu) \quad \forall\ \mu \in M_h(0),$$

where $a_h(\mu, \lambda) := -\langle \mu, \hat{\boldsymbol{Q}}_\lambda\cdot\boldsymbol{n}\rangle_{\partial\Omega_h}$, and $\ell_h(\mu) := \langle \mu, \hat{\boldsymbol{Q}}_f\cdot\boldsymbol{n}\rangle_{\partial\Omega_h}$. Moreover,

$$a_h(\mu, \lambda) = (\mathsf{c}\,\boldsymbol{Q}_\mu, \boldsymbol{Q}_\lambda)_{\Omega_h} + \langle \mathsf{U}_\mu - \mu, \tau(\mathsf{U}_\lambda - \lambda)\rangle_{\partial\Omega_h}, \quad \ell_h(\mu) = (f, \mathsf{U}_\mu),$$

and $a_h(\cdot,\cdot)$ is symmetric and positive definite on $M_h(0) \times M_h(0)$. Thus, $\hat{u}_h$ minimizes the total energy functional $J_h(\mu) := \frac{1}{2}a_h(\mu, \mu) - \ell_h(\mu)$ over $M_h(u_D)$.

Proof We only need to prove the last two identities and the property of positive definiteness of the bilinear form $a_h(\cdot,\cdot)$.

Let us prove the first identity. If we take $\boldsymbol{v} := \mathbf{Q}_\lambda$ in the first equation defining the first local problem, replace μ by λ in the second equation defining the first local problem and set $w := \mathsf{U}_\mu$, we get

$$\begin{aligned}
(\mathsf{c}\,\mathbf{Q}_\mu, \mathbf{Q}_\lambda)_K - (\mathsf{U}_\mu, \nabla\cdot\mathbf{Q}_\lambda)_K + \langle \mu, \mathbf{Q}_\lambda\cdot\boldsymbol{n}\rangle_{\partial K} &= 0,\\
-(\mathbf{Q}_\lambda, \nabla\mathsf{U}_\mu)_K + \langle \hat{\mathbf{Q}}_\lambda\cdot\boldsymbol{n}, \mathsf{U}_\mu\rangle_{\partial K} &= 0.
\end{aligned}$$

Adding the two equations, we obtain

$$(\mathrm{c}\,\mathbf{Q}_\mu, \mathbf{Q}_\lambda)_K + \langle(\hat{\mathbf{Q}}_\lambda - \mathbf{Q}_\lambda)\cdot \boldsymbol{n}, \mathsf{U}_\mu - \mu\rangle_{\partial K} = -\langle \hat{\mathbf{Q}}_\lambda \cdot \boldsymbol{n}, \mu\rangle_{\partial K}.$$

The first identity follows by inserting the definition of the numerical trace $\hat{\mathbf{Q}}_\lambda$ and adding over the elements $K \in \Omega_h$.

Let us prove the second identity. If we take $\boldsymbol{v} := \mathbf{Q}_f$ in the first equation defining the first local problem and $w := \mathsf{U}_\mu$ in the second equation defining the second local problem, we get

$$(\mathrm{c}\,\mathbf{Q}_\mu, \mathbf{Q}_f)_K - (\mathsf{U}_\mu, \nabla\cdot\mathbf{Q}_f)_K + \langle \mu, \mathbf{Q}_f\cdot\boldsymbol{n}\rangle_{\partial K} = 0,$$

$$-(\mathbf{Q}_f, \nabla \mathsf{U}_\mu)_K + \langle \hat{\mathbf{Q}}_f\cdot\boldsymbol{n}, \mathsf{U}_\mu\rangle_{\partial K} = (f, \mathsf{U}_\mu)_k$$

and if we add the two equations and insert the definition of $\hat{\mathbf{Q}}_f$, we obtain

$$(\mathrm{c}\,\mathbf{Q}_\mu, \mathbf{Q}_f)_K + \langle \tau(\mathsf{U}_f), \mathsf{U}_\mu - \mu\rangle_{\partial K} = (f, \mathsf{U}_\mu)_K - \langle \hat{\mathbf{Q}}_f\cdot\boldsymbol{n}, \mu\rangle_{\partial K}.$$

If we now take $\boldsymbol{v} := \mathbf{Q}_\mu$ in the first equation defining the second local problem and $w := \mathsf{U}_f$ in the second equation defining the first local problem with $\hat{u}_h := \mu$, we get

$$(\mathrm{c}\,\mathbf{Q}_f, \mathbf{Q}_\mu)_K - (\mathsf{U}_f, \nabla\cdot\mathbf{Q}_\mu)_K = 0,$$

$$-(\mathbf{Q}_\mu, \nabla\mathsf{U}_f)_K + \langle \hat{\mathbf{Q}}_\mu\cdot\boldsymbol{n}, \mathsf{U}_f\rangle_{\partial K} = 0,$$

and if we proceed as before, we get

$$(\mathrm{c}\,\mathbf{Q}_f, \mathbf{Q}_\mu)_K + \langle \tau(\mathsf{U}_\mu - \mu), \mathsf{U}_f\rangle_{\partial K} = 0.$$

This implies that

$$-\langle \tau(\mathsf{U}_\mu - \mu), \mathsf{U}_f\rangle_{\partial K} + \langle \tau(\mathsf{U}_f), \mathsf{U}_\mu - \mu\rangle_{\partial K} = (f, \mathsf{U}_\mu)_K - \langle \hat{\mathbf{Q}}_f\cdot\boldsymbol{n}, \mu\rangle_{\partial K},$$

and the result follows by the fact that τ is symmetric.

The fact that $a_h(\cdot,\cdot)$ is symmetric follows from the previous identities and the fact that τ is also symmetric. Finally the fact that it is positive definite on $M_h(0)\times M_h(0)$ follows exactly as in the proof of Theorem 4. This completes the proof. □

Two Compact Formulations

Let us now show how to rewrite the HDG methods in a more *compact* manner. It does not suggest a way to statically condense the methods but it is our *favorite* way of presenting them concisely. It emphasizes the role of the numerical traces of the methods and is suitable for carrying out their analysis. It is the following.

The approximate solution given by the HDG method is the function $(\boldsymbol{q}_h, u_h, \hat{u}_h) \in \boldsymbol{V}_h \times W_h \times M_h(u_D)$ satisfying the equations

$$(\mathrm{c}\,\boldsymbol{q}_h, \boldsymbol{v})_{\Omega_h} - (u_h, \nabla\cdot\boldsymbol{v})_{\Omega_h} + \langle \hat{u}_h, \boldsymbol{v}\cdot\boldsymbol{n}\rangle_{\partial\Omega_h} = 0 \qquad \forall \boldsymbol{v} \in \boldsymbol{V}_h, \tag{3a}$$

$$-(\boldsymbol{q}_h, \nabla w)_{\Omega_h} + \langle \hat{\boldsymbol{q}}_h\cdot\boldsymbol{n}, w\rangle_{\partial\Omega_h} = (f, w)_{\Omega_h} \qquad \forall w \in W_h, \tag{3b}$$

$$\hat{\boldsymbol{q}}_h\cdot\boldsymbol{n} := \boldsymbol{q}_h\cdot\boldsymbol{n} + \tau(u_h - \hat{u}_h) \qquad \text{on } \partial\Omega_h, \tag{3c}$$

$$\langle \mu, \hat{\boldsymbol{q}}_h\cdot\boldsymbol{n}\rangle_{\partial\Omega_h} = 0 \qquad \forall \mu \in M_h(0). \tag{3d}$$

Indeed, note that the first, second and third equations correspond to the definition of the local problems and that the weakly imposed boundary conditions are enforced by requesting that $\hat{u}_h$ be an element of $M_h(u_D)$.

We can also eliminate the numerical trace $\hat{\boldsymbol{q}}_h$ to obtain yet another rewriting of the methods. Once again, it hides the numerical trace of the flux, but emphasizes what we could call the *stabilized mixed method* structure of the methods. The formulation is the following. The approximate solution given by the HDG method is the function $(\boldsymbol{q}_h, u_h, \hat{u}_h) \in \boldsymbol{V}_h \times W_h \times M_h(u_D)$ satisfying the equations

$$\begin{aligned} A_h(\boldsymbol{q}_h, \boldsymbol{v}) + B_h(u_h, \hat{u}_h; \boldsymbol{v}) &= 0 && \forall \boldsymbol{v} \in \boldsymbol{V}_h,\\ -B_h(w, \mu; \boldsymbol{q}_h) + S_h(u_h, \hat{u}_h; w, \mu) &= (f, w)_{\Omega_h} && \forall (w, \mu) \in W_h \times M_h(0), \end{aligned}$$

where

$$\begin{aligned} A_h(\boldsymbol{p}, \boldsymbol{v}) &:= (\mathrm{c}\,\boldsymbol{p}, \boldsymbol{v})_{\Omega_h} && \forall \boldsymbol{p}, \boldsymbol{v} \in \boldsymbol{V}_h,\\ B_h(w, \mu; \boldsymbol{v}) &:= -(w, \nabla\cdot\boldsymbol{v})_{\Omega_h} + \langle \mu, \boldsymbol{v}\cdot\boldsymbol{n}\rangle_{\partial\Omega_h} && \forall (\boldsymbol{v}, w, \mu) \in \boldsymbol{V}_h \times W_h \times M_h,\\ S_h(\omega, \lambda; w, \mu) &:= \langle \tau(\omega - \lambda), w - \mu\rangle_{\partial\Omega_h} && \forall (\omega, \lambda), (w, \mu) \in W_h \times M_h. \end{aligned}$$

Indeed, the first equation follows from the definition of the bilinear forms $A_h(\cdot,\cdot)$ and $B_h(\cdot,\cdot)$. It remains to prove that

$$B_h(w, \mu; \boldsymbol{q}_h) + S_h(u_h, \hat{u}_h; w, \mu) = -(\boldsymbol{q}_h, \nabla w)_{\Omega_h} + \langle \hat{\boldsymbol{q}}_h\cdot\boldsymbol{n}, w\rangle_{\partial\Omega_h} - \langle \hat{\boldsymbol{q}}_h\cdot\boldsymbol{n}, \mu\rangle_{\partial\Omega_h}.$$

But, we have, by the definition of the bilinear forms $B_h(\cdot,\cdot)$ and $S_h(\cdot,\cdot)$, that

$$\begin{aligned} \Theta :=& - B_h(w, \mu; \boldsymbol{q}_h) + S_h(u_h, \hat{u}_h; w, \mu)\\ =& \;(w, \nabla\cdot\boldsymbol{q})_{\Omega_h} - \langle \mu, \boldsymbol{q}\cdot\boldsymbol{n}\rangle_{\partial\Omega_h} + \langle \tau(u_h - \hat{u}_h), w - \mu\rangle_{\partial\Omega_h}\\ =& - (\boldsymbol{q}_h, \nabla w)_{\Omega_h} + \langle \boldsymbol{q}\cdot n + \tau(u_h - \hat{u}_h), w - \mu\rangle_{\partial\Omega_h}, \end{aligned}$$

by integration by parts. The identity we want follows now by using the definition of the numerical trace of the flux.

To end, we note that, thanks to the structure of the methods, it is easy to see that the solution $(\boldsymbol{q}_h, u_h, \hat{u}_h) \in \boldsymbol{V}_h \times W_h \times M_h(u_D)$ minimizes the functional

$$J_h(\boldsymbol{v}, w, \mu) := \frac{1}{2}\{A_h(\boldsymbol{v}, \boldsymbol{v}) + S_h(w, \mu; w, \mu)\} - (f, w)_{\Omega_h} \tag{4a}$$

over all functions $(\boldsymbol{v}, w, \mu)$ in $\boldsymbol{V}_h \times W_h \times M_h(u_D)$ such that

$$A_h(\boldsymbol{v}, \boldsymbol{p}) + B_h(w, \mu; \boldsymbol{p}) = 0 \;\forall \boldsymbol{p} \in \boldsymbol{V}_h. \tag{4b}$$

Note that the last equation can be interpreted as the *elimination* of $\boldsymbol{q}_h$ from the equations. The minimization problem would then be one on the affine space $W_h \times M_h(u_D)$ and would correspond to a problem formulated solely in terms of u_h and $\hat{u}_h$. Next, we explore the static condensation of such reformulation.

3.3.2 Formulation in Terms of $(u_h, \hat{u}_h)$

So, here we *eliminate* the approximate flux $\boldsymbol{q}_h$ from the equations defining the HDG method in order to formulate it solely in terms of $(u_h, \hat{u}_h)$. To achieve this, we simply rewrite $\boldsymbol{q}_h$ as a linear mapping applied to $(u_h, \hat{u}_h)$. This mapping is defined by using the first equation defining the HDG methods. Thus, for any $(w, \mu) \in W_h \times M_h$, we define $\mathbf{Q}_{w,\mu} \in \boldsymbol{V}_h$ as the solution of

$$(\mathsf{c}\,\mathbf{Q}_{w,\mu}, \boldsymbol{v})_{\Omega_h} - (w, \nabla\cdot\boldsymbol{v})_{\Omega_h} + \langle \mu, \boldsymbol{v}\cdot\boldsymbol{n}\rangle_{\partial\Omega_h} = 0 \qquad \forall\, \boldsymbol{v} \in \boldsymbol{V}_h,$$

In this way, we are going to have that $\boldsymbol{q}_h = \mathbf{Q}_{u_h,\hat{u}_h}$. Note that the above equation is nothing but a *rewriting* of Eq. (4b).

Static Condensation Formulation

Using this approach, we obtain the following characterization of the HDG methods. It is useful for their implementation and involves less unknowns than the previous characterization since the unknown for the approximate flux has been eliminated. (Of course, the price to pay for this is that we now we have to work with the mapping $(w, \mu) \mapsto \mathbf{Q}_{w,\mu}$.) This characterization better shows the role of τ as a stabilization function but it hides its relation with the numerical trace of the flux and does not clearly indicate the associated transmission condition.

Theorem 6 (Second Characterization of HDG Methods) *The approximate solution of the HDG method is given by*

$$(\boldsymbol{q}_h, u_h) = (\boldsymbol{Q}, \mathsf{U}) = (\boldsymbol{Q}_{\mathsf{U}_{\hat{u}_h},\hat{u}_h}, \mathsf{U}_{\hat{u}_h}) + (\boldsymbol{Q}_{\mathsf{U}_f,0}, \mathsf{U}_f),$$

where, on the element $K \in \Omega_h$, for any $\mu \in L^2(\partial K)$ and any $f \in L^2(K)$, the functions $\mathsf{U}_\mu, \mathsf{U}_f \in W(K)$ are the solutions of the local problems

$$(\mathsf{c}\,\boldsymbol{Q}_{\mathsf{U}_\mu,\mu}, \boldsymbol{Q}_{w,0})_K + \langle \tau(\mathsf{U}_\mu - \mu), w\rangle_{\partial K} = 0 \qquad \forall w \in W(K),$$
$$(\mathsf{c}\,\boldsymbol{Q}_{\mathsf{U}_f,0}, \boldsymbol{Q}_{w,0})_K + \langle \tau(\mathsf{U}_f), w\rangle_{\partial K} = (f, w)_K \quad \forall w \in W(K),$$

respectively. The function $\hat{u}_h$ is the element of $M_h(u_D)$ such that

$$a_h(\hat{u}_h, \mu) = \ell_h(\mu) \quad \forall\, \mu \in M_h(0),$$

where $a_h(\mu, \lambda) := -\langle \mu, \hat{\boldsymbol{Q}}_{\mathsf{U}_\lambda,\lambda} \cdot \boldsymbol{n}\rangle_{\partial\Omega_h}$ and $\ell_h(\mu) := \langle \mu, \hat{\boldsymbol{Q}}_{\mathsf{U}_f,0} \cdot \boldsymbol{n}\rangle_{\partial\Omega_h}$. Moreover,

$$a_h(\mu, \lambda) = (\mathsf{c}\ \boldsymbol{Q}_{\mathsf{U}_\mu,\mu}, \boldsymbol{Q}_{\mathsf{U}_\lambda,\lambda})_{\partial\Omega_h} + \langle \mathsf{U}_\mu - \mu, \tau(\mathsf{U}_\lambda - \lambda)\rangle_{\partial\Omega_h}, \quad \ell_h(\mu) = (f, \mathsf{U}_\mu),$$

and $a_h(\cdot,\cdot)$ is symmetric and positive definite on $M_h(0) \times M_h(0)$. Thus, $\hat{u}_h$ minimizes the total energy functional $J_h(\mu) := \frac{1}{2} a_h(\mu, \mu) - \ell_h(\mu)$ over $M_h(u_D)$.

Proof This results follows easily from the first characterization of the HDG methods given in Theorem 5. We only have to show that the solutions of the local problems coincide, that is, that $(\mathbf{Q}_{\mathsf{U}_\mu,\mu}, \mathsf{U}_\mu) \in \boldsymbol{V}(K) \times W(K)$ is the solution of

$$(\mathsf{c}\,\mathbf{Q}_{\mathsf{U}_\mu,\mu}, \boldsymbol{v})_K - (\mathsf{U}_\mu, \nabla \cdot \boldsymbol{v})_K + \langle \mu, \boldsymbol{v} \cdot \boldsymbol{n}\rangle_{\partial K} = 0 \quad \forall \boldsymbol{v} \in \boldsymbol{V}(K),$$
$$-(\mathbf{Q}_{\mathsf{U}_\mu,\mu}, \nabla w)_K + \langle \hat{\mathbf{Q}}_{\mathsf{U}_\mu,\mu} \cdot \boldsymbol{n}, w\rangle_{\partial K} = 0 \quad \forall w \in W(K),$$
$$\hat{\mathbf{Q}}_\mu \cdot \boldsymbol{n} := \mathbf{Q}_\mu \cdot \boldsymbol{n} + \tau(\mathsf{U}_\mu - \mu) \qquad \text{on } \partial K,$$

and that $(\mathbf{Q}_{\mathsf{U}_f,0}, \mathsf{U}_f) \in \boldsymbol{V}(K) \times W(K)$ is the solution of

$$(\mathsf{c}\,\mathbf{Q}_{\mathsf{U}_f,0}, \boldsymbol{v})_K - (\mathsf{U}_f, \nabla \cdot \boldsymbol{v})_K = 0 \qquad \forall \boldsymbol{v} \in \boldsymbol{V}(K),$$
$$-(\mathbf{Q}_{\mathsf{U}_f,0}, \nabla w)_K + \langle \hat{\mathbf{Q}}_{\mathsf{U}_f,0} \cdot \boldsymbol{n}, w\rangle_{\partial K} = (f, w)_K \quad \forall w \in W(K),$$
$$\hat{\mathbf{Q}}_f \cdot \boldsymbol{n} := \mathbf{Q}_f \cdot \boldsymbol{n} + \tau(\mathsf{U}_f) \qquad \text{on } \partial K.$$

Since the first equation of these problems is nothing but the definition of the operator $\mathbf{Q}_{w,\mu}$, we only have to show that

$$(\nabla \cdot \mathbf{Q}_{\mathsf{U}_\mu,\mu}, w)_K + \langle \tau(\mathsf{U}_\mu - \mu), w\rangle_{\partial K} = 0 \qquad \forall w \in W(K),$$
$$(\nabla \cdot \mathbf{Q}_{\mathsf{U}_f,0}, w)_K + \langle \tau(\mathsf{U}_f), w\rangle_{\partial K} = (f, w)_K \quad \forall w \in W(K).$$

But, by the definition of $\mathbf{Q}_{w,0}$, we have

$$(\nabla \cdot \mathbf{Q}_{\mathsf{U}_{\mu},\mu}, w)_K = (\mathsf{c}\,\mathbf{Q}_{w,0}, \mathbf{Q}_{\mathsf{U}_{\mu},\mu})_K,$$
$$(\nabla \cdot \mathbf{Q}_{\mathsf{U}_f,0}, w)_K = (\mathsf{c}\,\mathbf{Q}_{w,0}, \mathbf{Q}_{\mathsf{U}_f,0})_K,$$

and the result follows. This completes the proof. □

A Compact Formulation

As we did for the first characterization of the HDG methods, we can rewrite the above result in a *compact* manner as follows. The approximate flux provided by the HDG method is $\boldsymbol{q}_h = \mathbf{Q}_{u_h,\hat{u}_h}$ and $(u_h, \hat{u}_h) \in W_h \times M_h(u_D)$ is the solution of

$$(\mathsf{c}\,\mathbf{Q}_{u_h,\hat{u}_h}, \mathbf{Q}_{w,\mu})_{\Omega_h} + \langle \tau(u_h - \hat{u}_h), w - \mu\rangle_{\partial\Omega_h} = (f, w)_{\Omega_h} \quad \forall (w,\mu) \in W_h \times M_h(0).$$

We immediately see that $(u_h, \hat{u}_h)$ is the only minimum over $W_h \times M_h(0)$ of the *total energy* functional

$$J_h(w,\mu) := \frac{1}{2}\{(\mathsf{c}\,\mathbf{Q}_{w,\mu}, \mathbf{Q}_{w,\mu})_{\Omega_h} + \langle \tau(w-\mu), w-\mu\rangle_{\partial\Omega_h}\} - (f, w)_{\Omega_h}.$$

This minimization problem is *identical* to the minimization (with restrictions) problem (4).

4 HDG Methods Using Only the Tensor $\mathsf{a} := \mathsf{c}^{-1}$

4.1 Motivation

Note that the the first three equations of the weak formulation of the DG methods we have been considering can also be expressed as

$$-(\boldsymbol{g}_h, \boldsymbol{v})_{\Omega_h} - (u_h, \nabla \cdot \boldsymbol{v})_{\Omega_h} + \langle \hat{u}_h, \boldsymbol{v} \cdot \boldsymbol{n}\rangle_{\partial\Omega_h} = 0,$$
$$(\mathsf{c}\,\boldsymbol{q}_h, \boldsymbol{v})_{\Omega_h} = -(\boldsymbol{g}_h, \boldsymbol{v})_{\Omega_h},$$
$$-(\boldsymbol{q}_h, \nabla w)_{\Omega_h} + \langle \hat{\boldsymbol{q}}_h \cdot \boldsymbol{n}, w\rangle_{\partial\Omega_h} = (f, w)_{\Omega_h},$$

where the approximate gradient $\boldsymbol{g}_h$ is taken in $\boldsymbol{V}_h$ If one prefers to work with the tensor $\mathsf{a} := \mathsf{c}^{-1}$, we can simply use the equations

$$-(\boldsymbol{g}_h, \boldsymbol{v})_{\Omega_h} - (u_h, \nabla \cdot \boldsymbol{v})_{\Omega_h} + \langle \hat{u}_h, \boldsymbol{v} \cdot \boldsymbol{n}\rangle_{\partial\Omega_h} = 0,$$
$$(\boldsymbol{q}_h, \boldsymbol{v})_{\Omega_h} = -(\mathsf{a}\,\boldsymbol{g}_h, \boldsymbol{v})_{\Omega_h},$$
$$-(\boldsymbol{q}_h, \nabla w)_{\Omega_h} + \langle \hat{\boldsymbol{q}}_h \cdot \boldsymbol{n}, w\rangle_{\partial\Omega_h} = (f, w)_{\Omega_h},$$

for all $(\boldsymbol{v}, w) \in \boldsymbol{V}_h \times W_h$, where the numerical traces $\hat{u}_h$ and $\hat{\boldsymbol{q}}_h \cdot \boldsymbol{n}$ are approximations to $u|_{\partial\Omega_h}$ and $\boldsymbol{q} \cdot \boldsymbol{n}|_{\partial\Omega_h}$, respectively. The difference between these two DG methods is certainly not abysmal since it consists in picking one of the two ways of relating the approximate gradient $\boldsymbol{g}_h$ to the approximate flux $\boldsymbol{q}_h$, namely,

$$(\mathsf{c}\,\boldsymbol{q}_h, \boldsymbol{v})_{\Omega_h} = -(\boldsymbol{g}_h, \boldsymbol{v})_{\Omega_h} \quad \text{or} \quad (\boldsymbol{q}_h, \boldsymbol{v})_{\Omega_h} = -(\mathsf{a}\,\boldsymbol{g}_h, \boldsymbol{v})_{\Omega_h}.$$

As a consequence, there is a one-to-one correspondence between these two weak formulations, provided both a and c are well defined. Moreover, both formulations coincide whenever a and c are constant on each element $K \in \Omega_h$ which gives rise to super-closeness of their approximations, as noted in [45].

However, if a degenerates and is not invertible at every point, the second formulation might be preferable. This is also what motivated the so-called "extended" form of the mixed methods introduced in [1, 9, 61].

Finally, let us note that in elasticity, $\boldsymbol{g}$ corresponds to the strain, $\boldsymbol{q}$ to the stress, a to the so-called constitutive tensor and c to the so-called compliance tensor. Thus, the HDG methods obtained for linear and nonlinear elasticity, see the HDG methods for elasticity considered in 2008 [85], 2009 [86] and 2014 [53] and in 2015 [59], can be immediately *reduced* to our simpler case; see also the 2006 DG method proposed in [87]. It is well known that to work with the constitutive tensor is usually preferred in the case of nonlinear elasticity. Next, we briefly show how to define and characterize the HDG methods associated with using the tensor $\mathsf{a} := \mathsf{c}^{-1}$.

4.2 *Definition, Existence and Uniqueness*

We take the approximate solution of the HDG methods to be the function

$$(\boldsymbol{q}_h, \boldsymbol{g}_h, u_h) = (\mathbf{Q}, \mathbf{G}, \mathsf{U}),$$

where, on the element $K \in \Omega_h$, $(\mathbf{Q}, \mathbf{G}, \mathsf{U}) \in \boldsymbol{V}(K) \times \boldsymbol{V}(K) \times W(K)$ is the solution of the local problem

$$\begin{aligned}
-(\mathbf{G}, \boldsymbol{v})_K - (\mathsf{U}, \nabla \cdot \boldsymbol{v})_K + \langle \hat{u}_h, \boldsymbol{v} \cdot \boldsymbol{n} \rangle_{\partial K} &= 0 && \forall \boldsymbol{v} \in \boldsymbol{V}(K),\\
(\mathbf{Q}, \boldsymbol{v})_K &= -(\mathsf{a}\,\mathbf{G}, \boldsymbol{v})_K && \forall \boldsymbol{v} \in \boldsymbol{V}(K),\\
-(\mathbf{Q}, \nabla w)_K + \langle \hat{\mathbf{Q}} \cdot \boldsymbol{n}, w \rangle_{\partial K} &= (f, w)_K && \forall w \in W(K),
\end{aligned}$$

where the numerical trace $\hat{\mathbf{Q}}$ is suitably chosen, and $\hat{u}_h \in M_h$ is the solution of the following weakly imposed transmission and boundary conditions:

$$\langle \mu, \hat{\mathbf{Q}} \cdot \boldsymbol{n} \rangle_{\partial\Omega_h \setminus \partial\Omega} = 0,$$
$$\langle \mu, \hat{u}_h \rangle_{\partial\Omega} = \langle \mu, u_D \rangle_{\partial\Omega},$$

for all $\mu \in M_h$. This completes the definition of the HDG methods.

It is not difficult to see that the existence and uniqueness in Theorem 4 and its Corollary 1 do hold *unchanged*.

4.3 Characterizations of the HDG Methods

4.3.1 Formulation in Terms of $(\boldsymbol{q}_h, \boldsymbol{g}_h, u_h, \hat{u}_h)$

Static Condensation Formulation

We have the following result which is analogous to Theorem 5.

Theorem 7 (First Characterization of HDG Methods) *The approximate solution of the HDG method is given by*

$$(\boldsymbol{q}_h, \boldsymbol{g}_h, u_h) = (\boldsymbol{Q}, \boldsymbol{G}, \mathsf{U}) = (\boldsymbol{Q}_{\hat{u}_h}, \boldsymbol{G}_{\hat{u}_h}, \mathsf{U}_{\hat{u}_h}) + (\boldsymbol{Q}_f, \boldsymbol{G}_f, \mathsf{U}_f),$$

where, on the element $K \in \Omega_h$, for any $\mu \in L^2(\partial K)$, the function $(\boldsymbol{Q}_\mu, \boldsymbol{G}_\mu, \mathsf{U}_\mu) \in \boldsymbol{V}(K) \times \boldsymbol{V}(K) \times W(K)$ is the solution of the local problem

$$\begin{aligned}
-(\boldsymbol{G}_\mu, \boldsymbol{v})_K - (\mathsf{U}_\mu, \nabla \cdot \boldsymbol{v})_K + \langle \mu, \boldsymbol{v} \cdot \boldsymbol{n} \rangle_{\partial K} &= 0 && \forall \boldsymbol{v} \in \boldsymbol{V}(K),\\
(\boldsymbol{Q}_\mu, \boldsymbol{v})_K &= -(\mathsf{a}\boldsymbol{G}_\mu, \boldsymbol{v})_K && \forall \boldsymbol{v} \in \boldsymbol{V}(K),\\
-(\boldsymbol{Q}_\mu, \nabla w)_K + \langle \hat{\boldsymbol{Q}}_\mu \cdot \boldsymbol{n}, w \rangle_{\partial K} &= 0 && \forall w \in W(K),\\
\hat{\boldsymbol{Q}}_\mu \cdot \boldsymbol{n} := \boldsymbol{Q}_\mu \cdot \boldsymbol{n} + \tau(\mathsf{U}_\mu - \mu) & && \text{on } \partial K,
\end{aligned}$$

and, for any $f \in L^2(K)$, the function $(\boldsymbol{Q}_f, \boldsymbol{G}_f, \mathsf{U}_f) \in \boldsymbol{V}(K) \times \boldsymbol{V}(K) \times W(K)$ is the solution of the local problem

$$\begin{aligned}
-(\boldsymbol{G}_f, \boldsymbol{v})_K - (\mathsf{U}_f, \nabla \cdot \boldsymbol{v})_K &= 0 && \forall \boldsymbol{v} \in \boldsymbol{V}(K),\\
(\boldsymbol{Q}_f, \boldsymbol{v})_K &= -(\mathsf{a}\boldsymbol{G}_f, \boldsymbol{v})_K && \forall \boldsymbol{v} \in \boldsymbol{V}(K),\\
-(\boldsymbol{Q}_f, \nabla w)_K + \langle \hat{\boldsymbol{Q}}_f \cdot \boldsymbol{n}, w \rangle_{\partial K} &= (f, w)_K && \forall w \in W(K),\\
\hat{\boldsymbol{Q}}_f \cdot \boldsymbol{n} := \boldsymbol{Q}_f \cdot \boldsymbol{n} + \tau(\mathsf{U}_f) & && \text{on } \partial K.
\end{aligned}$$

The function $\hat{u}_h$ is the element of $M_h(u_D)$ such that

$$a_h(\hat{u}_h, \mu) = \ell_h(\mu) \quad \forall\ \mu \in M_h(0),$$

where $a_h(\mu, \lambda) := -\langle \mu, \hat{\boldsymbol{Q}}_\lambda \cdot \boldsymbol{n}\rangle_{\partial\Omega_h}$, and $\ell_h(\mu) := \langle \mu, \hat{\boldsymbol{Q}}_f \cdot \boldsymbol{n}\rangle_{\partial\Omega_h}$. Moreover,

$$a_h(\mu, \lambda) = (\mathsf{a}\boldsymbol{G}_\mu, \boldsymbol{G}_\lambda)_{\partial\Omega_h} + \langle \mathsf{U}_\mu - \mu, \tau(\mathsf{U}_\lambda - \lambda)\rangle_{\partial\Omega_h}, \quad \ell_h(\mu)= (f, \mathsf{U}_\mu),$$

and $a_h(\cdot, \cdot)$ is symmetric and positive definite on $M_h(0) \times M_h(0)$. Thus, $\hat{u}_h$ minimizes the functional $J_h(\mu) := \frac{1}{2}a_h(\mu, \mu) - \ell_h(\mu)$ over $M_h(u_D)$.

Two Compact Formulations

Proceeding as for the first family of HDG methods, we obtain the following two compact formulations. The first emphasized the role of the numerical traces. It reads as follows. The approximate solution given by the HDG method is the function $(\boldsymbol{q}_h, \boldsymbol{g}_h, u_h, \hat{u}_h) \in \boldsymbol{V}_h \times \boldsymbol{V}_h \times W_h \times M_h(u_D)$ satisfying the equations

$$\begin{aligned}
-(\boldsymbol{g}_h, \boldsymbol{v})_{\Omega_h} - (u_h, \nabla \cdot \boldsymbol{v})_{\Omega_h} + \langle \hat{u}_h, \boldsymbol{v} \cdot \boldsymbol{n}\rangle_{\partial\Omega_h} &= 0 && \forall \boldsymbol{v} \in \boldsymbol{V}_h,\\
(\boldsymbol{q}_h, \boldsymbol{v})_{\Omega_h} &= -(\mathsf{a}\boldsymbol{g}_h, \boldsymbol{v})_{\Omega_h} && \forall \boldsymbol{v} \in \boldsymbol{V}_h,\\
-(\boldsymbol{q}_h, \nabla w)_{\Omega_h} + \langle \hat{\boldsymbol{q}}_h \cdot \boldsymbol{n}, w\rangle_{\partial\Omega_h} &= (f, w)_{\Omega_h} && \forall w \in W_h,\\
\hat{\boldsymbol{q}}_h \cdot \boldsymbol{n} := \boldsymbol{q}_h \cdot \boldsymbol{n} + \tau(u_h - \hat{u}_h) & && \text{on } \partial\Omega_h,\\
\langle \mu, \hat{\boldsymbol{q}}_h \cdot \boldsymbol{n}\rangle_{\partial\Omega_h} &= 0 && \forall \mu \in M_h(0).
\end{aligned}$$

The second emphasizes the stabilized mixed structure of the method. It is the following. The approximate solution given by the HDG method is the function $(\boldsymbol{q}_h, u_h, \hat{u}_h) \in \boldsymbol{V}_h \times W_h \times M_h(u_D)$ satisfying the equations

$$\begin{aligned}
A_h(\boldsymbol{g}_h, \boldsymbol{v}) + \mathsf{B}_h(\boldsymbol{q}_h, \boldsymbol{v}) &= 0 && \forall \boldsymbol{v} \in \boldsymbol{V}_h,\\
-\mathsf{B}_h(\boldsymbol{v}, \boldsymbol{g}_h) + B_h(u_h, \hat{u}_h; \boldsymbol{v}) &= 0 && \forall \boldsymbol{v} \in \boldsymbol{V}_h,\\
-B_h(w, \mu; \boldsymbol{q}_h) + S_h(u_h, \hat{u}_h; w, \mu) &= (f, w)_{\Omega_h} && \forall (w, \mu) \in W_h \times M_h(0),
\end{aligned}$$

where

$$\begin{aligned}
A_h(\boldsymbol{p}, \boldsymbol{v}) &:= (\mathsf{a}\boldsymbol{p}, \boldsymbol{v})_{\Omega_h}, && \forall \boldsymbol{p}, \boldsymbol{v} \in \boldsymbol{V}_h,\\
\mathsf{B}_h(\boldsymbol{p}, \boldsymbol{v}) &:= (\boldsymbol{p}, \boldsymbol{v})_{\Omega_h}, && \forall \boldsymbol{p}, \boldsymbol{v} \in \boldsymbol{V}_h,\\
B_h(w, \mu; \boldsymbol{v}) &:= -(w, \nabla \cdot \boldsymbol{v})_{\Omega_h} + \langle \mu, \boldsymbol{v} \cdot \boldsymbol{n}\rangle_{\partial\Omega_h} && \forall (\boldsymbol{v}, w, \mu) \in \boldsymbol{V}_h \times W_h \times M_h,\\
S_h(\omega, \lambda; w, \mu) &:= \langle \tau(\omega - \lambda), w - \mu\rangle_{\partial\Omega_h} && \forall (\omega, \lambda), (w, \mu) \in W_h \times M_h.
\end{aligned}$$

Thanks to the structure of the method, it is easy to see that the solution $(\boldsymbol{g}_h, u_h, \hat{u}_h) \in \boldsymbol{V}_h \times W_h \times M_h(u_D)$ minimizes the functional

$$J_h(\boldsymbol{v}, w, \mu) := \frac{1}{2}\{A_h(\boldsymbol{v}, \boldsymbol{v}) + S_h(w, \mu; w, \mu)\} - (f, w)_{\Omega_h} \tag{5a}$$

over the functions $(\boldsymbol{v}, w, \mu)$ in the space $\boldsymbol{V}_h \times W_h \times M_h(u_D)$ such that there exist $\boldsymbol{q}_h = \boldsymbol{q}_h(\boldsymbol{v}, w, \mu) \in \boldsymbol{V}_h$ such that

$$A_h(\boldsymbol{v},\boldsymbol{p}) + \mathsf{B}_h(\boldsymbol{q}_h,\boldsymbol{p}) = 0 \quad \forall \boldsymbol{p} \in \boldsymbol{V}_h, \tag{5b}$$

$$-\mathsf{B}_h(\boldsymbol{p}, \boldsymbol{v}) + B_h(w, \mu; \boldsymbol{p}) = 0 \quad \forall \boldsymbol{p} \in \boldsymbol{V}_h. \tag{5c}$$

Once again, Note that the last two equations can be interpreted as the *elimination* of $(\boldsymbol{q}_h, \boldsymbol{g}_h)$ from the equations. The minimization problem would then be one on the affine space $W_h \times M_h(u_D)$ and would correspond to a problem formulated solely in terms of u_h and $\hat{u}_h$. Next, we explore such reformulation.

4.3.2 Formulation in Terms of $(u_h, \hat{u}_h)$

We *eliminate* the approximate gradient $\boldsymbol{g}_h$ and the approximate flux $\boldsymbol{q}_h$ from the equations defining the HDG method in order to formulate it solely in terms of $(u_h, \hat{u}_h)$. To achieve that, we simply rewrite $\boldsymbol{g}_h$ and $\boldsymbol{q}_h$ as a linear mappings applied to $(u_h, \hat{u}_h)$. These mappings are defined by using the first equation defining the HDG methods. Thus, for any $(w, \mu) \in W_h \times M_h$, we define $(\mathbf{G}_{w,\mu}, \mathbf{Q}_{w,\mu}) \in \boldsymbol{V}_h \times \boldsymbol{V}_h$ as the solution of

$$-(\mathbf{G}_{w,\mu}, \boldsymbol{v})_{\Omega_h} - (w, \nabla \cdot \boldsymbol{v})_{\Omega_h} + \langle \mu, \boldsymbol{v} \cdot \boldsymbol{n} \rangle_{\partial\Omega_h} = 0 \qquad \forall\, \boldsymbol{v} \in \boldsymbol{V}_h,$$

$$(\mathbf{Q}_{w,\mu}, \boldsymbol{v})_{\Omega_h} = -(\mathsf{a}\, \mathbf{G}_{w,\mu}, \boldsymbol{v})_{\Omega_h} \qquad \forall\, \boldsymbol{v} \in \boldsymbol{V}_h.$$

In this way, we are going to have that $(\boldsymbol{q}_h, \boldsymbol{g}_h) = (\mathbf{Q}_{u_h,\hat{u}_h}, \mathbf{G}_{u_h,\hat{u}_h})$. Note that these two equations are nothing but a rewriting of Eqs. (5b) and (5c).

Static Condensation Formulation

We have the following result.

Theorem 8 (Second Characterization of HDG Methods) *The approximate solution of the HDG method is given by*

$$(\boldsymbol{q}_h, \boldsymbol{g}_h, u_h) = (\boldsymbol{Q}, \boldsymbol{G}, \mathsf{U}) = (\boldsymbol{Q}_{\mathsf{U}_{\hat{u}_h},\hat{u}}, \boldsymbol{G}_{\mathsf{U}_{\hat{u}_h},\hat{u}_h}, \mathsf{U}_{\hat{u}_h}) + (\boldsymbol{Q}_{\mathsf{U}_f,0}, \boldsymbol{G}_{\mathsf{U}_f,0}, \mathsf{U}_f),$$

where, on the element $K \in \Omega_h$*, for any* $\mu \in L^2(\partial K)$ *and* $f \in L^2(K)$*, the functions* $\mathsf{U}_\mu, \mathsf{U}_f \in W(K)$ *are the solutions of the local problems*

$$(\mathsf{a}\, \boldsymbol{G}_{\mathsf{U}_\mu,\mu}, \boldsymbol{G}_{w,0})_K + \langle \tau(\mathsf{U}_\mu - \mu), w\rangle_{\partial K} = 0 \qquad \forall w \in W(K),$$

$$(\mathsf{a}\, \boldsymbol{G}_{\mathsf{U}_f,0}, \boldsymbol{G}_{w,0})_K + \langle \tau(\mathsf{U}_f), w\rangle_{\partial K} = (f, w)_K \quad \forall w \in W(K),$$

respectively. The function $\hat{u}_h$ *is the element of* $M_h(u_D)$ *such that*

$$a_h(\hat{u}_h, \mu) = \ell_h(\mu) \quad \forall\, \mu \in M_h(0),$$

where $a_h(\mu, \lambda) := -\langle \mu, \hat{\boldsymbol{Q}}_{\mathsf{U}_\lambda,\lambda} \cdot \boldsymbol{n}\rangle_{\partial\Omega_h}$*, and* $\ell_h(\mu) := \langle \mu, \hat{\boldsymbol{Q}}_{\mathsf{U}_f,0} \cdot \boldsymbol{n}\rangle_{\partial\Omega_h}$*. Moreover,*

$$a_h(\mu, \lambda) = (\mathsf{a}\boldsymbol{G}_{\mathsf{U}_\mu,\mu}, \boldsymbol{G}_{\mathsf{U}_\lambda,\lambda})_{\partial\Omega_h} + \langle \mathsf{U}_\mu - \mu, \tau(\mathsf{U}_\lambda - \lambda)\rangle_{\partial\Omega_h}, \quad \ell_h(\mu) = (f, \mathsf{U}_\mu),$$

and $a_h(\cdot,\cdot)$ *is symmetric and positive definite on* $M_h(0) \times M_h(0)$*. Thus,* $\hat{u}_h$ *minimizes the functional* $J_h(\mu) := \frac{1}{2} a_h(\mu, \mu) - \ell_h(\mu)$ *over* $M_h(u_D)$*.*

Compact Formulation

Finally, we display the compact form of this formulation of the HDG method. We have that $(\boldsymbol{q}_h, \boldsymbol{g}_h) = (\mathbf{Q}_{u_h,\hat{u}_h}, \mathbf{G}_{u_h,\hat{u}_h})$ where $(u_h, \hat{u}_h) \in W_h \times M_h(u_D)$ is the solution of

$$(\mathsf{a}\, \mathbf{G}_{u_h,\hat{u}_h}, \mathbf{G}_{w,\mu})_{\Omega_h} + \langle \tau(u_h - \hat{u}_h), w - \mu\rangle_{\partial\Omega_h} = (f, w)_{\Omega_h}\ \forall (w, \mu) \in W_h \times M_h(0). \tag{6}$$

In other words, $(u_h, \hat{u}_h)$ is the only minimum over $W_h \times M_h(0)$ of the functional

$$J_h(w, \mu) := \frac{1}{2}\{(\mathsf{a}\, \mathbf{G}_{w,\mu}, \mathbf{G}_{u,\mu})_{\Omega_h} + \langle \tau(w - \mu), w - \mu\rangle_{\partial\Omega_h}\} - (f, w)_{\Omega_h}.$$

This is exactly the minimization problem (5).

5 Using Neumann Instead of Dirichlet Boundary Conditions

In the previous two sections, we have shown how a characterization of the exact solution can be used to *generate* HDG methods. Here we show how a different characterization of the exact solution can be used to produce a different *static condensation*, that is, a different way of *implementing*, an already known HDG method.

We proceed as follows. First, we present a characterization of the exact solution which uses Neumann boundary-value problems *instead* of the Dirichlet boundary-value problems to define the local problems. Then, we consider some HDG methods devised in the previous sections and show how a discrete version of the new characterization of the exact solution is nothing but a new way of implementing them. The resulting form of the HDG method has already been used in the work on multiscale methods in [50]. Recently, two different ways of statically condensing the very same method were proposed in [49].

The idea of using different characterizations of the exact solution to devise HDG methods was introduced back in 2009 in [17] where four different ways were presented to devise HDG methods for the vorticity-velocity-pressure formulation of the Stokes system, as the exact solution could be characterized in terms of four different local problems and transmissions conditions. Just as it happens with the exact solution, the very same HDG method could be obtained by using any of the four ways. In other words, the HDG method could be *hybridized* and then *statically condensed* in each of the above-mentioned four different manners.

5.1 A Second Characterization of the Exact Solution

Let us then show how to use local Neumann boundary-value problems to obtain a characterization of the exact solution.

Suppose that, for every element $K \in \Omega_h$, we define $(\mathbf{Q}, \mathsf{U})$ as the solution of the local problem

$$\begin{aligned} \mathsf{c}\,\mathbf{Q} + \nabla \mathsf{U} &= 0 && \text{in } K,\\ \nabla \cdot \mathbf{Q} &= f + \{\langle \hat{\boldsymbol{q}} \cdot \boldsymbol{n}, 1\rangle_{\partial K} - (f, 1)_K\}/|K| && \text{in } K,\\ \mathbf{Q} \cdot \boldsymbol{n} &= \hat{\boldsymbol{q}} \cdot \boldsymbol{n} && \text{on } \partial K,\\ (\mathsf{U}, 1)_K &= (\overline{u}, 1)_K, \end{aligned}$$

where we want the function $\hat{\boldsymbol{q}}$, which has a single-valued normal component, and the constant $\overline{u}$, to be such $(\boldsymbol{q}, u) = (\mathbf{Q}, \mathsf{U})$ on K. This happens *if and only if* $\hat{\boldsymbol{q}}$ and $\overline{u}$ satisfy the equations

$$\begin{aligned} [\![\mathsf{U}]\!] &= 0 && \text{for } F \in \mathcal{F}_h^i,\\ \langle \hat{\boldsymbol{q}} \cdot \boldsymbol{n}, 1\rangle_{\partial K} &= (f, 1)_K && \text{for } K \in \mathcal{T}_h,\\ \mathsf{U} &= u_D && \text{for } F \in \mathcal{F}_h^{\partial}. \end{aligned}$$

Note that we have to provide the average to U on the element, $\overline{u}$, otherwise the solution U is not uniquely determined. Note also that, we have had to add an additional term to the right-hand side of the second equation in order to ensure that the local problem has a solution for *any* boundary data $\hat{\boldsymbol{q}} \cdot \boldsymbol{n}$. As a consequence, we have to make sure that such term is *zero*. This explains why the global problem consists not only of transmission and boundary conditions, as in the case of Dirichlet boundary-value local problems.

If we now separate the influence of $\hat{\boldsymbol{q}}, \overline{u}$ and f, we readily get the following characterization of the exact solution.

Theorem 9 (Characterization of the Exact Solution) *We have that*

$$(\boldsymbol{q}, u) = (\boldsymbol{Q}, \mathsf{U}) = (\boldsymbol{Q}_{\hat{q}}, \mathsf{U}_{\hat{q}}) + (\boldsymbol{0}, \overline{u}) + (\boldsymbol{Q}_f, \mathsf{U}_f),$$

where $(\boldsymbol{Q}_{\hat{q}}, \mathsf{U}_{\hat{q}})$ *and* $(\boldsymbol{Q}_f, \mathsf{U}_f)$ *are the solution of the local problems*

$$\begin{aligned}
\mathsf{c}\boldsymbol{Q}_{\hat{q}} + \nabla \mathsf{U}_{\hat{q}} &= 0 && \text{in } K, & \mathsf{c}\boldsymbol{Q}_f + \nabla \mathsf{U}_f &= 0 && \text{in } K,\\
\nabla \cdot \boldsymbol{Q}_{\hat{q}} &= \langle \hat{\boldsymbol{q}} \cdot \boldsymbol{n}, 1\rangle_{\partial K}/|K| && \text{in } K, & \nabla \cdot \boldsymbol{Q}_f &= f - (f, 1)_K/|K| && \text{in } K,\\
\boldsymbol{Q}_{\hat{q}} \cdot \boldsymbol{n} &= \hat{\boldsymbol{q}} \cdot \boldsymbol{n} && \text{on } \partial K, & \boldsymbol{Q}_f \cdot \boldsymbol{n} &= 0 && \text{on } \partial K,\\
(\mathsf{U}_{\hat{q}}, 1)_K &= 0, && & (\mathsf{U}_f, 1)_K &= 0. &&
\end{aligned}$$

where the functions $\hat{\boldsymbol{q}} \cdot \boldsymbol{n}$ *and* $\overline{u}$ *are determined as the solution of the equations*

$$\begin{aligned}
-[\![\mathsf{U}_{\hat{q}}]\!] - [\![\overline{u}]\!] &= [\![\mathsf{U}_f]\!] && \text{on } \mathcal{F}_h^i,\\
\langle \hat{\boldsymbol{q}} \cdot \boldsymbol{n}, 1\rangle_{\partial K} &= (f, 1)_K && \text{for } K \in \mathcal{T}_h,\\
\mathsf{U}_{\hat{q}} + \overline{u} &= -\mathsf{U}_f + u_D && \text{on } \mathcal{F}_h^{\partial}.
\end{aligned}$$

5.2 An Example

In the case of our one-dimensional example, this result reads as follows. We have that

$$(\boldsymbol{q}, u) = (\mathbf{Q}_{\hat{q}}, \mathsf{U}_{\hat{q}}) + (\boldsymbol{0}, \overline{u}) + (\mathbf{Q}_f, \mathsf{U}_f),$$

where

$$\begin{aligned}
\mathsf{c}\,\mathbf{Q}_{\hat{q}} + \frac{d}{dx}\mathsf{U}_{\hat{q}} &= 0 && \text{in } (x_{i-1}, x_i), & \mathsf{c}\,\mathbf{Q}_f + \frac{d}{dx}\mathsf{U}_f &= 0 && \text{in } (x_{i-1}, x_i),\\
\frac{d}{dx}\mathbf{Q}_{\hat{q}} &= \frac{1}{h_i}(\hat{q}_i - \hat{q}_{i-1}) && \text{in } (x_{i-1}, x_i), & \frac{d}{dx}\mathbf{Q}_f &= f - \frac{1}{h_i}\int_{x_{i-1}}^{x_i} f && \text{in } (x_{i-1}, x_i),\\
\mathbf{Q}_{\hat{q}} \cdot \boldsymbol{n} &= \hat{\boldsymbol{q}} \cdot \boldsymbol{n} && \text{on } \{x_{i-1}, x_i\}, & \mathbf{Q}_f \cdot \boldsymbol{n} &= 0, && \text{on } \{x_{i-1}, x_i\},\\
\int_{x_{i-1}}^{x_i} \mathsf{U}_{\hat{q}} &= 0, && & \int_{x_{i-1}}^{x_i} \mathsf{U}_f &= 0, &&
\end{aligned}$$

and where the functions $\hat{\boldsymbol{q}}$ and $\overline{u}$ are the solution of

$$\begin{aligned}
\mathsf{U}_{\hat{q}}(x_i^+) - \mathsf{U}_{\hat{q}}(x_i^-) + \overline{u}_{i+1/2} - \overline{u}_{i-1/2} &= -\mathsf{U}_f(x_i^+) + \mathsf{U}_f(x_i^-) && \text{for } i = 1, \dots, N-1,\\
\hat{\boldsymbol{q}}_i - \hat{\boldsymbol{q}}_{i-1} &= \int_{x_{i-1}}^{x_i} f && \text{for } i = 1, \dots, N-1,\\
\mathsf{U}_{\hat{q}}(x_0^+) + \overline{u}_{1/2} &= -\mathsf{U}_f(x_0^+) + u_D(x_0),\\
\mathsf{U}_{\hat{q}}(x_N^-) + \overline{u}_{N-1/2} &= -\mathsf{U}_f(x_0^+) + u_D(x_N).
\end{aligned}$$

Since the solution of the local problems are

$$\mathbf{Q}_{\hat{q}}(x) = \varphi_i(x)\hat{\boldsymbol{q}}_i + \varphi_{i-1}(x)\hat{\boldsymbol{q}}_{i-1}, \qquad \mathbf{Q}_f(x) = -\mathsf{c}^{-1}\int_{x_{i-1}}^{x_i} \mathsf{G}_x^i(x,s) f(s)\, ds,$$

$$\mathsf{U}_{\hat{q}}(x) = \frac{\mathsf{c}\, h_i}{6}\{\psi_i(x)\hat{\boldsymbol{q}}_i - \psi_{i-1}(x)\hat{\boldsymbol{q}}_{i-1}\} \qquad \mathsf{U}_f(x) = \int_{x_{i-1}}^{x_i} \mathsf{G}^i(x,s) f(s)\, ds.$$

where G^i is the Green's function of the second local problem, namely,

$$G^i(x,s) := \begin{cases} \frac{\mathsf{c}\, h_i}{6}\,[1 - 3\varphi_i^2(s) - 3\varphi_{i-1}^2(x)] & \text{if } x_{i-1} \leqslant s \leqslant x,\\ \frac{\mathsf{c}\, h_i}{6}\,[1 - 3\varphi_i^2(x) - 3\varphi_{i-1}^2(s)] & \text{if } x \leqslant s \leqslant x_i. \end{cases}$$

and $\psi_i := 1 - 3\varphi_i^2$, and where the functions $\hat{\boldsymbol{q}}$ and $\overline{u}$ are the solution of

$$\begin{aligned}
\frac{\mathsf{c}\, h_i}{6}(\hat{\boldsymbol{q}}_{i-1} + 2\hat{\boldsymbol{q}}_i) &+ \frac{\mathsf{c}\, h_{i+1}}{6}(2\hat{\boldsymbol{q}}_i + \hat{\boldsymbol{q}}_{i+1})\\
&+ \overline{u}_{i+1/2} - \overline{u}_{i-1/2} = -\mathsf{U}_f(x_i^+) + \mathsf{U}_f(x_i^-) && \text{for } i = 1, \dots, N-1,\\
\hat{\boldsymbol{q}}_i - \hat{\boldsymbol{q}}_{i-1} &= \int_{x_{i-1}}^{x_i} f && \text{for } i = 1, \dots, N-1,\\
\frac{\mathsf{c}\, h_1}{6}(2\hat{\boldsymbol{q}}_0 + \hat{\boldsymbol{q}}_1) + \overline{u}_{1/2} &= -\mathsf{U}_f(x_0^+) + u_D(x_0),\\
\frac{\mathsf{c}\, h_N}{6}(\hat{\boldsymbol{q}}_{N-1} - 2\hat{\boldsymbol{q}}_N) - \overline{u}_{N-1/2} &= \mathsf{U}_f(x_N^-) - u_D(x_N).
\end{aligned}$$

5.3 Another Static Condensation of Known HDG Methods

Let us consider the HDG methods introduced in Sect. 3. Next, we show that those methods can be statically condensed in the way suggested by our new characterization of the exact solution.

5.3.1 Rewriting the Compact Formulation Based on the Numerical Traces

First, we rewrite them in such a way that the numerical trace $\hat{\boldsymbol{q}}_h$, and not $\hat{u}_h$, is an independent unknown. We can do that very easily if we use the compact formulation of those methods based on the numerical traces, (3). It states that the approximate solution given by the HDG method is the function $(\boldsymbol{q}_h, u_h, \hat{u}_h) \in \boldsymbol{V}_h \times W_h \times M_h(u_D)$ satisfying the equations

$$\begin{aligned}
(\mathrm{c}\,\boldsymbol{q}_h, \boldsymbol{v})_{\Omega_h} - (u_h, \nabla\cdot\boldsymbol{v})_{\Omega_h} + \langle \hat{u}_h, \boldsymbol{v}\cdot\boldsymbol{n}\rangle_{\partial\Omega_h} &= 0 && \forall \boldsymbol{v} \in \boldsymbol{V}_h,\\
-(\boldsymbol{q}_h, \nabla w)_{\Omega_h} + \langle \hat{\boldsymbol{q}}_h\cdot\boldsymbol{n}, w\rangle_{\partial\Omega_h} &= (f, w)_{\Omega_h} && \forall w \in W_h,\\
\hat{\boldsymbol{q}}_h\cdot\boldsymbol{n} := \boldsymbol{q}_h\cdot\boldsymbol{n} + \tau(u_h - \hat{u}_h) & && \text{on } \partial\Omega_h,\\
\langle \mu, \hat{\boldsymbol{q}}_h\cdot\boldsymbol{n}\rangle_{\partial\Omega_h} &= 0 && \forall \mu \in M_h(0).
\end{aligned}$$

Now, if we take the stabilization function $\tau(\cdot)$ to be the simple multiplication by the scalar function τ, we have that

$$\hat{u}_h = u_h + \tau^{-1}(\boldsymbol{q}_h\cdot\boldsymbol{n} - \hat{\boldsymbol{q}}_h\cdot\boldsymbol{n}) \qquad \text{on } \partial\Omega_h.$$

If the local space $\boldsymbol{V}(K)\times W(K)$ is such that, for each face F of the element K,

$$\begin{aligned}
\boldsymbol{V}(K)\cdot\boldsymbol{n}|_F &\subset M(F),\\
W(K)|_F &\subset M(F),
\end{aligned}$$

and take τ to be constant on each face of the triangulation, we have that $\hat{\boldsymbol{q}}_h$ belongs to the space

$$\boldsymbol{N}_h := \{\boldsymbol{v} \in \boldsymbol{L}^2(\mathcal{F}_h) :\ \boldsymbol{v}\cdot\boldsymbol{n}|_{\partial K} \in M(\partial K),\ [\![\boldsymbol{v}]\!] = 0 \text{ on } \mathcal{F}_h^i\}.$$

We can thus rewrite the HDG method as follows. The approximate solution given by the HDG method is the function $(\boldsymbol{q}_h, u_h, \hat{\boldsymbol{q}}_h) \in \boldsymbol{V}_h \times W_h \times \boldsymbol{N}_h$ satisfying the equations

$$\begin{aligned}
(\mathrm{c}\,\boldsymbol{q}_h, \boldsymbol{v})_{\Omega_h} - (u_h, \nabla\cdot\boldsymbol{v})_{\Omega_h} + \langle \hat{u}_h, \boldsymbol{v}\cdot\boldsymbol{n}\rangle_{\partial\Omega_h} &= 0 && \forall \boldsymbol{v} \in \boldsymbol{V}_h,\\
-(\boldsymbol{q}_h, \nabla w)_{\Omega_h} + \langle \hat{\boldsymbol{q}}_h\cdot\boldsymbol{n}, w\rangle_{\partial\Omega_h} &= (f, w)_{\Omega_h} && \forall w \in W_h,\\
\hat{u}_h = u_h + \tau^{-1}(\boldsymbol{q}_h\cdot\boldsymbol{n} - \hat{\boldsymbol{q}}_h\cdot\boldsymbol{n}) & && \text{on } \partial\Omega_h,\\
\langle \hat{u}_h, \boldsymbol{v}\cdot\boldsymbol{n}\rangle_{\partial\Omega_h} &= \langle u_D, \boldsymbol{v}\cdot\boldsymbol{n}\rangle_{\partial\Omega} && \forall \boldsymbol{v} \in \boldsymbol{N}_h.
\end{aligned}$$

Note that the last equation enforces both the single-valuedness of $\hat{u}_h$ as well as the Dirichlet boundary conditions of the model problem (1).

5.3.2 The New Static Condensation

So, suppose that, for every element $K \in \Omega_h$, we define $(\mathbf{Q}, \mathsf{U}) \in \boldsymbol{V}(K) \times W(K)$ to be the solution of the local problem

$$\begin{aligned}
(\mathsf{c}\,\mathbf{Q}, \boldsymbol{v})_K - (\mathsf{U}, \nabla\cdot\boldsymbol{v})_K + \langle\hat{\mathsf{U}}, \boldsymbol{v}\cdot\boldsymbol{n}\rangle_{\partial K} &= 0 && \forall \boldsymbol{v}\in \boldsymbol{V}(K),\\
-(\mathbf{Q}, \nabla w)_K + \langle\hat{\boldsymbol{q}}_h\cdot\boldsymbol{n}, w-\overline{w}\rangle_{\partial K} &= (f, w-\overline{w})_{\partial K} && \forall w\in W(K),\\
\hat{\mathsf{U}} &:= \mathsf{U} + \tau^{-1}(\mathbf{Q}-\hat{\boldsymbol{q}}_h)\cdot\boldsymbol{n} && \text{on } \partial K,\\
(\mathsf{U}, 1)_K &= (\overline{u}_h, 1)_K,
\end{aligned}$$

where $\overline{w}|_K := (w,1)_K/|K|$, and where we want to take $\hat{\boldsymbol{q}}_h \in \boldsymbol{N}_h$ and the piecewise constant function $\overline{u}_h$ such that $(\boldsymbol{q}_h, u_h) = (\mathbf{Q}, \mathsf{U})$. Clearly, this happens if we have that $(\hat{\boldsymbol{q}}_h, \overline{u}_h)$ is the solution of the global problem

$$\begin{aligned}
\langle\boldsymbol{v}\cdot\boldsymbol{n}, \hat{\mathsf{U}}\rangle_{\partial\Omega_h} &= \langle\boldsymbol{v}\cdot\boldsymbol{n}, u_D\rangle_{\partial\Omega} && \forall \boldsymbol{v}\in\boldsymbol{N}_h,\\
\langle\hat{\boldsymbol{q}}_h\cdot\boldsymbol{n}, 1\rangle_{\partial K} &= (f, 1)_{\partial K} && \forall\, K\in\Omega_h.
\end{aligned}$$

Separating the influence of $\hat{\boldsymbol{q}}_h$ from that of $\overline{u}_h$ and f, we obtain the following, new static condensation of the HDG method. In what follows, $\overline{W}_h$ denotes the space of real-valued functions which are constant on each element $K \in \Omega_h$.

Theorem 10 (New Static Condensation of HDG Methods) *The approximate solution of the HDG method is*

$$(\boldsymbol{q}_h, u_h) = (\boldsymbol{Q}, \mathsf{U}) = (\boldsymbol{Q}_{\hat{\boldsymbol{q}}_h}, \mathsf{U}_{\hat{\boldsymbol{q}}_h}) + (\mathbf{0}, \overline{u}_h) + (\boldsymbol{Q}_f, \mathsf{U}_f),$$

where, for each element $K \in \Omega_h$, for any $\boldsymbol{\eta} \in \boldsymbol{L}^2(\partial K)$, the function $(\boldsymbol{Q}_{\boldsymbol{\eta}}, \mathsf{U}_{\boldsymbol{\eta}}) \in \boldsymbol{V}(K) \times W(K)$ is the solution of the local problem

$$\begin{aligned}
(\mathsf{c}\,\boldsymbol{Q}_{\boldsymbol{\eta}}, \boldsymbol{v})_K - (\mathsf{U}_{\boldsymbol{\eta}}, \nabla\cdot\boldsymbol{v})_K + \langle\hat{\mathsf{U}}_{\boldsymbol{\eta}}, \boldsymbol{v}\cdot\boldsymbol{n}\rangle_{\partial K} &= 0 && \forall \boldsymbol{v}\in\boldsymbol{V}(K),\\
-(\boldsymbol{Q}_{\boldsymbol{\eta}}, \nabla w)_K + \langle\boldsymbol{\eta}\cdot\boldsymbol{n}, w-\overline{w}\rangle_{\partial K} &= 0 && \forall w\in W(K),\\
\hat{\mathsf{U}}_{\boldsymbol{\eta}} &:= \mathsf{U}_{\boldsymbol{\eta}} + \tau^{-1}(\boldsymbol{Q}_{\boldsymbol{\eta}} - \boldsymbol{\eta})\cdot\boldsymbol{n} && \text{on } \partial K,\\
(\mathsf{U}_{\boldsymbol{\eta}}, 1)_K &= 0,
\end{aligned}$$

and, for any $f \in L^2(K)$, the function $(\boldsymbol{Q}_f, \mathsf{U}_f) \in \boldsymbol{V}(K) \times W(K)$ is the solution of the local problem

$$
\begin{aligned}
(\mathsf{c}\boldsymbol{Q}_f, \boldsymbol{v})_K - (\mathsf{U}_f, \nabla\cdot\boldsymbol{v})_K + \langle \hat{\mathsf{U}}_f, \boldsymbol{v}\cdot\boldsymbol{n}\rangle_{\partial K} &= 0 && \forall \boldsymbol{v} \in \boldsymbol{V}(K),\\
-(\boldsymbol{Q}_f, \nabla w)_K &= (f, w - \overline{w})_{\partial K} && \forall w \in W(K),\\
\hat{\mathsf{U}}_f := \mathsf{U}_f + \tau^{-1}\boldsymbol{Q}_f\cdot\boldsymbol{n} & && \text{on } \partial K,\\
(\mathsf{U}_f, 1)_K = 0, & &&
\end{aligned}
$$

and where $(\hat{\boldsymbol{q}}_h, \overline{u}_h) \in \boldsymbol{N}_h \times \overline{W}_h$ is the solution of the global problem

$$
\begin{aligned}
a_h(\hat{\boldsymbol{q}}_h, \boldsymbol{\nu}) + b_h(\overline{u}_h, \boldsymbol{\nu}) &= \ell_h(\boldsymbol{\nu}) - \langle u_D, \boldsymbol{\nu}\cdot\boldsymbol{n}\rangle_{\partial\Omega} && \forall \boldsymbol{\nu} \in \boldsymbol{N}_h,\\
b_h(\overline{\omega}, \hat{\boldsymbol{q}}_h) &= (f, \overline{\omega})_{\Omega_h} && \forall \overline{\omega} \in \overline{W}_h,
\end{aligned}
$$

where

$$
a_h(\boldsymbol{\eta}, \boldsymbol{\nu}) := -\langle \boldsymbol{\nu}\cdot\boldsymbol{n}, \hat{\mathsf{U}}_{\boldsymbol{\eta}}\rangle_{\partial\Omega_h}, \quad b_h(\overline{\omega}, \boldsymbol{\nu}) := -\langle \boldsymbol{\nu}\cdot\boldsymbol{n}, \overline{\omega}\rangle_{\partial\Omega_h}, \quad \ell_h(\boldsymbol{\nu}) := \langle \boldsymbol{\nu}\cdot\boldsymbol{n}, \hat{\mathsf{U}}_f\rangle_{\partial\Omega_h}.
$$

Moreover,

$$
a_h(\boldsymbol{\eta}, \boldsymbol{\nu}) = (\mathsf{c}\boldsymbol{Q}_{\boldsymbol{\eta}}, \boldsymbol{Q}_{\boldsymbol{\nu}})_{\partial\Omega_h} + \langle (\boldsymbol{Q}_{\boldsymbol{\eta}} - \boldsymbol{\eta})\cdot\boldsymbol{n}, \tau^{-1}(\boldsymbol{Q}_{\boldsymbol{\nu}} - \boldsymbol{\nu})\cdot\boldsymbol{n}\rangle_{\partial\Omega_h}, \quad \ell_h(\boldsymbol{\nu}) = (f, \mathsf{U}_{\boldsymbol{\nu}})_{\Omega_h},
$$

and $\hat{\boldsymbol{q}}_h$ minimizes the complementary energy functional

$$
\mathcal{J}_h(\boldsymbol{\nu}) := \frac{1}{2} a_h(\boldsymbol{\nu}, \boldsymbol{\nu}) - \ell_h(\boldsymbol{\nu}) + \langle u_D, \boldsymbol{\nu}\cdot\boldsymbol{n}\rangle_{\partial\Omega},
$$

over the functions $\boldsymbol{\nu} \in \boldsymbol{N}_h$ such that $b_h(\overline{\omega}, \boldsymbol{\nu}) = (f, \overline{\omega})_{\Omega_h}$ $\forall \overline{\omega} \in \overline{W}_h$.

The proof of this result goes along the very same lines of the proof of the characterization Theorem 5.

5.3.3 The Stabilized Mixed Compact Formulation

Let us end this section by displaying the compact formulation of the method obtained when we eliminate the numerical trace $\hat{u}_h$. Proceeding as for the first characterization, we can obtain that the approximate solution given by the HDG method is the function $(\boldsymbol{q}_h, u_h, \hat{\boldsymbol{q}}_h) \in \boldsymbol{V}_h \times W_h \times \boldsymbol{N}_h$ satisfying the equations

$$
A_h(\boldsymbol{q}_h, \boldsymbol{v}) + S_h(\boldsymbol{q}_h, \hat{\boldsymbol{q}}_h; \boldsymbol{v}, \boldsymbol{\nu}) + B_h(u_h; \boldsymbol{v}, \boldsymbol{\nu}) = -\langle u_D, \boldsymbol{\nu}\cdot\boldsymbol{n}\rangle_{\partial\Omega_h}, \tag{7a}
$$

$$
-B_h(w; \boldsymbol{q}_h, \hat{\boldsymbol{q}}_h) = (f, w)_{\Omega_h}, \tag{7b}
$$

for all $(\boldsymbol{v}, w, \boldsymbol{\nu}) \in \boldsymbol{V}_h \times W_h \times \boldsymbol{N}_h$, where

$$A_h(\boldsymbol{p}, \boldsymbol{v}) := (\mathrm{c}\boldsymbol{p}, \boldsymbol{v})_{\Omega_h} \qquad \forall \boldsymbol{p}, \boldsymbol{v} \in \boldsymbol{V}_h, \tag{7c}$$

$$B_h(w; \boldsymbol{v}, \boldsymbol{\nu}) := (\nabla w, \boldsymbol{v})_{\Omega_h} - \langle w, \boldsymbol{\nu} \cdot \boldsymbol{n} \rangle_{\partial\Omega_h} \qquad \forall (\boldsymbol{v}, w, \boldsymbol{\nu}) \in \boldsymbol{V}_h \times W_h \times \boldsymbol{N}_h, \tag{7d}$$

$$S_h(\boldsymbol{p}, \boldsymbol{\eta}; \boldsymbol{v}, \boldsymbol{\nu}) := \langle (\boldsymbol{p} - \boldsymbol{\eta}) \cdot \boldsymbol{n}, \tau^{-1}(\boldsymbol{v} - \boldsymbol{\nu}) \cdot \boldsymbol{n} \rangle_{\partial\Omega_h} \; \forall (\boldsymbol{p}, \boldsymbol{\eta}), (\boldsymbol{v}, \boldsymbol{\nu}) \in \boldsymbol{V}_h \times \boldsymbol{N}_h. \tag{7e}$$

As a consequence, the solution $(\boldsymbol{q}_h, \hat{\boldsymbol{q}}_h) \in \boldsymbol{V}_h \times \boldsymbol{N}_h$ minimizes the complementary energy functional

$$\mathcal{J}_h(\boldsymbol{v}, \boldsymbol{\nu}) := \frac{1}{2}\{A_h(\boldsymbol{v}, \boldsymbol{v}) + S_h(\boldsymbol{v}, \boldsymbol{\nu}; \boldsymbol{v}, \boldsymbol{\nu})\} + \langle u_D, \boldsymbol{\nu} \cdot \boldsymbol{n} \rangle_{\partial\Omega_h}$$

over all functions $(\boldsymbol{v}, \mu)$ in $\boldsymbol{V}_h \times M_h(u_D)$ such that $B_h(w; \boldsymbol{v}, \boldsymbol{\nu}) = (f, w) \; \forall w \in W_h$.

6 Building Bridges and Constructing Methods

Here, we briefly discuss the evolution of the HDG methods. We being by showing that (some of the earliest) HDG methods can be seen as a particular case of the DG methods introduced in 1998 [24] and analyzed in 2000 [4]. We then recall the strong relation between the HDG and the mixed methods, already pointed out in 2009 [33], and show how this relation drove (and is still driving) the development of superconvergent HDG methods. The bridge built in 2014 [14] between the HDG and the so-called staggered discontinuous Galerkin (SDG), a DG method introduced in 2009 [13] and apparently unrelated to the HDG methods, can be seen as part of this development. We discuss the stabilization introduced by Lehrenfeld (and Schöberl) in 2010 [62]. We end by showing that the so-called Weak-Galerkin methods proposed in 2014 [89] and in 2015 [65, 66], are nothing but rewritings of the HDG methods.

6.1 *Relating HDG to Old DG Methods*

Here, we consider HDG methods whose numerical method defining the local problems is the so-called local discontinuous Galerkin (LDG) method introduced in [24]. The resulting HDG methods are then called the LDG-H methods. For all of them, the stabilization function τ on any face $F \in \mathcal{F}_h$ is a simple multiplication by a constant which we also denote by τ, that is,

$$\hat{\boldsymbol{q}}_h \cdot \boldsymbol{n} := \boldsymbol{q}_h \cdot \boldsymbol{n} + \tau \cdot (u_h - \hat{u}_h) \quad \text{on } \partial\Omega_h.$$

Examples of local spaces, taken from [33], are shown in the table below.

Method	$V(K)$	$W(K)$	$M(F)$
LDG-H	$\boldsymbol{\mathcal{P}}_{k-1}(K)$	$\mathcal{P}_k(K)$	$\mathcal{P}_k(F)$
LDG-H	$\boldsymbol{\mathcal{P}}_k(K)$	$\mathcal{P}_k(K)$	$\mathcal{P}_k(F)$
LDG-H	$\boldsymbol{\mathcal{P}}_k(K)$	$\mathcal{P}_{k-1}(K)$	$\mathcal{P}_k(F)$

In all these cases, we have that the local spaces $\boldsymbol{V}(K) \times W(K)$ are such that, for each face F of the element K,

$$\begin{aligned} \boldsymbol{V}(K) \cdot \boldsymbol{n}|_F &\subset M(F), \\ W(K)|_F &\subset M(F). \end{aligned}$$

This implies that $[\![\boldsymbol{q}_h]\!] \in M_h$ and the transmission condition becomes $[\![\hat{\boldsymbol{q}}_h]\!] = 0$ on $\mathcal{F}_h^i$. This can only hold if and only if, on $\mathcal{F}_h^i$,

$$\begin{aligned} \hat{u}_h &= \frac{\tau^+ {u_h}^+ + \tau^- {u_h}^-}{\tau^+ + \tau^-} + \frac{1}{\tau^+ + \tau^-} [\![\boldsymbol{q}_h]\!], \\ \hat{\boldsymbol{q}}_h &= \frac{\tau^- {\boldsymbol{q}_h}^+ + \tau^+ {\boldsymbol{q}_h}^-}{\tau^+ + \tau^-} + \frac{\tau^+ \tau^-}{\tau^+ + \tau^-} [\![u_h]\!]. \end{aligned}$$

This implies that the DG methods introduced in [24] and analyzed in [4] that have the above choice of numerical traces can be hybridized and then statically condensed. This is why we call these methods the *hybridizable* DG methods.

Note, that none of these LDG-H methods is an LDG method if we take $\tau^\pm \in (0, \infty)$ since for the method to be an LDG method, we must have that $1/(\tau^+ + \tau^-) = 0$. This shows that none of the DG methods considered in [3] is an LDG-H method with finite values of the stabilization function. In fact, these methods can converge faster than any of the DG methods considered therein. For example, in the case in which $\mathsf{c} = \mathsf{Id}$, $\boldsymbol{V}(K) \times W(K) = \boldsymbol{\mathcal{P}}_k(K) \times \mathcal{P}_k(K)$ and $M(F) = \mathcal{P}_k(F)$ this LDG-H method was analyzed in [4], where is was proven that, for arbitrary shape-regular, polyhedral elements, $\boldsymbol{q}_h$ converges with order $k + 1/2$ and u_h with order $k + 1$, for any $k \geqslant 0$, provided τ is of order one. The convergence is in the $L^2(\Omega)$-norm. On the other hand, other LDG-H methods do have the same order of convergence than those considered in [3]. Indeed, by using the same approach in [4], one can easily show that in the case in which $\boldsymbol{V}(K) \times W(K) = \boldsymbol{\mathcal{P}}_{k-1}(K) \times \mathcal{P}_k(K)$ and $M(F) = \mathcal{P}_k(F)$, $\boldsymbol{q}_h$ converges with order k and u_h with order $k + 1$, for any $k \geqslant 1$, provided τ is of order $1/h$. This result holds for meshes made of general shape-regular, polyhedral meshes.

6.2 Relating HDG to Mixed Methods

As pointed out in [33], if the stabilization function τ is taken to be *identically zero* so that $\hat{\boldsymbol{q}}_h \cdot \boldsymbol{n} = \boldsymbol{q}_h \cdot \boldsymbol{n}$ on $\mathcal{F}_h$, *and* the transmission condition implies that $[\![\hat{\boldsymbol{q}}_h]\!] = 0$ on $\mathcal{F}_h^i$, we recover the so-called (hybridized version of the) mixed methods if the mixed method is used to define the local problems; see also [2]. In the table below, we display the main examples of mixed methods with this property when K is a simplex and we compare it with one of the first HDG methods, the LDG-H method.

Method	$V(K)$	$W(K)$	$M(F)$
RT	$\boldsymbol{\mathcal{P}}_k(K) + \boldsymbol{x}\,\widetilde{\mathcal{P}}_k(K)$	$\mathcal{P}_k(K)$	$\mathcal{P}_k(F)$
LDG-H	$\boldsymbol{\mathcal{P}}_k(K)$	$\mathcal{P}_k(K)$	$\mathcal{P}_k(F)$
BDM	$\boldsymbol{\mathcal{P}}_k(K)$	$\mathcal{P}_{k-1}(K)$	$\mathcal{P}_k(F)$

The strong relation between the mixed method and the HDG methods suggested that the HDG methods might share with the mixed methods some of its convergence properties. This was proven to be true for a special LDG-H method obtained by setting $\tau = 0$ on all the faces of the simplex K except one. This method, called the single face-hybridizable (SFH) method, was introduced and analyzed in [30]. Therein, it was shown that the SFH method is strongly related to the RT and BDM mixed methods. Indeed, the bilinear forms $a_h(\cdot,\cdot)$ of the RT, BDM and SFH methods are the same, and the SFH shares with the RT and BDM the same superconvergence properties.

Next, we briefly describe this superconvergence property. For all of the above methods, the local averages of the error $u - u_h$, converge faster than the errors $u - u_h$ and $\boldsymbol{q} - \boldsymbol{q}_h$. As a consequence, we can define, on the each element K, the new approximation $u_h^\star \in W^*(K) := \mathcal{P}_{k+1}(K)$ as the solution of

$$
\begin{aligned}
(\nabla u_h^\star, \nabla w)_K &= -(\mathrm{c}\,\boldsymbol{q}_h, \nabla w)_K \qquad \text{for all } w \in W^*(K),\\
(u_h^\star, 1)_K &= (u_h, 1)_K,
\end{aligned}
$$

Then $u - u_h^\star$ will converge faster than $u - u_h$. The orders of convergence are displayed in the table below; see [30] for the results on the SFH method and [36] for those on the general LGD-H method. The symbol $\star$ indicates that the non-zero values of the stabilization function τ only need to be uniformly bounded by below.

Method	τ	$\boldsymbol{q}_h$	u_h	$\overline{u}_h$	k
RT	0	$k+1$	$k+1$	$k+2$	$\geqslant 0$
SFH	$\star$	$k+1$	$k+1$	$k+2$	$\geqslant 1$
LDG-H	$\mathcal{O}(1)$	$k+1$	$k+1$	$k+2$	$\geqslant 1$
BDM	0	$k+1$	k	$k+2$	$\geqslant 2$
LDG-H	$\mathcal{O}(1/h)$	k	$k+1$	$k+1$	$\geqslant 1$

6.3 The SDG Method as a Limit of SFH Methods

In [14], it was proved that the staggered discontinuous Galerkin (SDG) method, originally introduced in the framework of wave propagation in [13], can be obtained as the limit when the non-zero values of the stabilization function of a *special* SFH method goes to infinity. The special SFH method is obtained as follows. The mesh consists of triangles or tetrahedra subdivided into three triangles or four tetrahedra. On the faces of the bigger simplexes, the stabilization function is not zero; it is equal to zero on all the remaining faces.

By building this bridge between the SDG and the SFH methods, the SDG can now be implemented by hybridization and can share the superconvergence properties of the SFH method. Similarly, the SFH method now share the (related but different) superconvergence property of the SDG method.

6.4 Constructing Superconvergent HDG Methods

The first superconvergent HDG method was the SFH method. A systematic approach to uncover superconvergent HDG methods was undertaken in [39] where the following sufficient conditions were found. The space $\boldsymbol{V}(K) \times W(K)$ must have a subspace $\tilde{\boldsymbol{V}}(K) \times \tilde{W}(K)$ satisfying inclusions

$$\begin{aligned} \mathcal{P}_0(K) &\subset \nabla W(K) \subset \tilde{\boldsymbol{V}}(K), \\ \mathcal{P}_0(K) &\subset \nabla \cdot \boldsymbol{V}(K) \subset \tilde{W}(K), \\ \boldsymbol{V}(K) \cdot \boldsymbol{n} &+ W(K) \subset M(\partial K). \end{aligned}$$

and whose orthogonal complement satisfies the identity

$$\tilde{\boldsymbol{V}}^{\perp} \cdot \boldsymbol{n} \oplus \tilde{W}^{\perp} = M(\partial K).$$

Let us present examples taken from [39] in the case in which K is a cube; the first corresponds to the choice $M(F) = Q^k(F)$ and the second to the choice $M(F) = \mathcal{P}_k(F)$.

In the first example, the HDG method denoted by $\mathbf{HDG}^{Q}_{[k]}$ and the mixed method denoted by $\mathbf{TNT}_{[k]}$ are new. The 7-dimensional space $\boldsymbol{H}^k_7(K)$ is obtained by adding a basis function to the space $\boldsymbol{H}^k_6(K)$. The precise description of these spaces can be found in [39] or, better, in [18], where commuting diagrams for the **TNT** elements on cubes were obtained for the DeRham complex.

In the second example, the HDG method denoted by $\mathbf{HDG}^{P}_{[k]}$ is new. In the corresponding table, we abuse the notation slightly to keep it simple. Thus, by $\mathcal{P}^{k+1}(K)\backslash\tilde{\mathcal{P}}^{k+1}(y,z)$ we mean the span of $\{x^\alpha y^\beta z^\gamma : \alpha + \beta + \gamma \leqslant k+1, \alpha > 0\}$.

$M(F) = \mathcal{Q}^k(F), k \geqslant 1$		
Method	$\boldsymbol{V}(K)$	$W(K)$
$\mathbf{RT}_{[k]}$	$\mathcal{P}^{k+1,k,k}(K)$	$\mathcal{Q}^k(K)$
	$\times\mathcal{P}^{k,k+1,k}(K)$	
	$\times\mathcal{P}^{k,k,k+1}(K)$	
$\mathbf{TNT}_{[k]}$	$\boldsymbol{\mathcal{Q}}^k(K) \oplus \boldsymbol{H}_7^k(K)$	$\mathcal{Q}^k(K)$
$\mathbf{HDG}^{Q}_{[k]}$	$\boldsymbol{\mathcal{Q}}^k(K) \oplus \boldsymbol{H}_6^k(K)$	$\mathcal{Q}^k(K)$

$M(F) = \mathcal{P}_k(F), k \geqslant 1$		
Method	$\boldsymbol{V}(K)$	$W(K)$
$\mathbf{BDFM}_{[k+1]}$	$\mathcal{P}^{k+1}(K)\backslash\tilde{\mathcal{P}}^{k+1}(y,z)$	$\mathcal{P}_k(K)$
	$\times\mathcal{P}^{k+1}(K)\backslash\tilde{\mathcal{P}}^{k+1}(x,z)$	
	$\times\mathcal{P}^{k+1}(K)\backslash\tilde{\mathcal{P}}^{k+1}(x,y)$	
$\mathbf{HDG}^{P}_{[k]}$	$\boldsymbol{\mathcal{P}}^k(K)$	$\mathcal{P}^k(K)$
c	$\oplus\nabla\times(yz\,\tilde{\mathcal{P}}^k(K),0,0)$	
	$\oplus\nabla\times(0,zx\,\tilde{\mathcal{P}}^k(K),0)$	
$\mathbf{BDM}_{[k]}$	$\boldsymbol{\mathcal{P}}^k(K)$	$\mathcal{P}^{k-1}(K)$
$k \geqslant 2$	$\oplus\nabla\times(0,0,xy\,\tilde{\mathcal{P}}^k(y,z))$	
	$\oplus\nabla\times(0,xz\,\tilde{\mathcal{P}}^k(x,y),0)$	
	$\oplus\nabla\times(yz\,\tilde{\mathcal{P}}^k(x,z),0,0)$	

In [39], many new superconvergence HDG methods were found for simplexes, squares, cubes and prisms. For curved elements, see [40].

6.5 The Lehrenfeld-Schöberl Stabilization Function

Let us recall that the case in which $M(F) := \mathcal{P}_k(K)$ and $\boldsymbol{V}(K)\times W(K) := \boldsymbol{\mathcal{P}}_{k-1}(K)\times\mathcal{P}_k(K)$, and the stabilization function τ is the multiplicative stabilization function, namely,

$$\tau(u_h - \hat{u}_h) := \tau \cdot (u_h - \hat{u}_h),$$

the resulting method is an LDG-H method. Moreover, for arbitrary shape-regular, polyhedral elements, we have that $\boldsymbol{q}_h$ converges with order k and u_h with order $k+1$, for any $k \geqslant 0$, provided τ is of order $1/h$. Since the size of the stiffness matrix of the local problem is proportional to the number of faces of the triangulation times the dimension of the space $M(F)$, a reduction of the space $M(F)$ would result in a smaller global problem. The question is if this is possible to achieve without loosing the convergence properties of the method.

In 2010, Ch. Lehrenfeld (and J. Schöberl) [62, Remark 1.2.4] noted that the answer is affirmative, see also [63], if we modify the above stabilization function by simply projecting u_h into M_h:

$$\tau^{\mathrm{LS}}(u_h - \hat{u}_h) := h^{-1} \cdot (P_M(u_h) - \hat{u}_h).$$

The error analysis of this HDG method was carried out in 2014 by I. Oikawa [76] who proved *optimal* orders of convergence for both $\boldsymbol{q}_h$ and u_h for regular-shaped, general polyhedral elements.

For the sake of fairness in the attribution of this simple but remarkable projection, I would like to emphasize that it was announced in 2009 by J. Schöberl in his plenary talk at the ICOSAHOM in Trondheim, Norway; at the 2010 Finite Element Circus in Minneapolis, USA; and then again at Oberwolfach, Germany, February 10–12, 2012; see [84]. I personally knew about it through Ch. Lehrenfeld, who told me about it during a Ph.D. Course in Pavia, May 28- June 1, 2010. At that time, the error estimates obtained later by I. Oikawa [76] were already known to Ch. Lehrenfeld even though he did not include them in [62].

6.6 Relating HDG with the Weak Galerkin Method

So far, no effort has been made to render clear the relation between the HDG and the so-called Weak Galerkin methods. The first Weak Galerkin method was proposed in 2013 [88] in the framework of convection-diffusion-reaction equations. Therein, it is pointed out that the Weak Galerkin is *identical* to some mixed and HDG methods but only in the purely diffusion case and whenever the diffusivity tensor is a constant. This is not an accurate statement which will be discussed elsewhere since it requires addressing issues related to the convective and reaction terms. Instead, here we restrict ourselves to discussing other versions of the Weak Galerkin method devised specifically for steady-state diffusion in [65, 66, 89]. We show that all these Weak Galerkin methods are rewritings of the HDG methods.

The Weak Galerkin method proposed in 2015 [65] (deposited in the archives in 2012), was described therein as identical to the HDG methods for the Poisson equation. Here we show that it is also identical for the model problem under consideration. Indeed, it is nothing but the compact form of the HDG methods (6) in Sect. 4.3.2 using the multiplication stabilization function $\tau(w - \mu) := h^{-1}.(w - \mu)$ and the tensor $\mathsf{a} := \mathsf{c}^{-1}$. Let us point out that, although the HDG methods were introduced in 2009 [33] (submitted in 2007) by using the formulation with the tensor c, the extension to the formulation with $\mathsf{a} := \mathsf{c}^{-1}$ is straightforward. In fact, as argued in Sect. 4.1, these HDG methods can be obtained by reducing to the model problem under consideration the HDG methods for the more difficult problem of linear and nonlinear elasticity. Specifically, the HDG methods for elasticity were obtained in 2008 in [85] and in 2009 [86] (submitted in 2008). The Weak Galerkin method in [65] is thus a simple rewriting of HDG methods.

The Weak Galerkin method proposed in 2015 [66] (deposited in the archives in 2013) is nothing but the compact form of the HDG methods (6) in Sect. 4.3.2 using the Lehrenfeld-Schöberl stabilization function and the tensor $\mathsf{a} := \mathsf{c}^{-1}$. Thus, the Weak Galerkin method in [66] is also a simple rewriting of HDG methods.

Finally, the Weak Galerkin method proposed in 2014 [89] (submitted in 2012) is nothing but the HDG method (7) in Sect. 5.3.3 corresponding to the Lehrenfeld-Schöberl stabilization function and the tensor c. Although the results of Sect. 5.3 have been obtained when the stabilization function is a simple multiplication, the extension to the Lehrenfeld-Schöberl function are straightforward. Indeed, the numerical trace for the HDG method with the Lehrenfeld-Schöberl stabilization is

$$\hat{\boldsymbol{q}}_h \cdot \boldsymbol{n} = \boldsymbol{q}_h \cdot \boldsymbol{n} + \frac{1}{h}(P_M(u_h) - \hat{u}_h)$$

which implies that

$$\hat{u}_h = P_M(u_h) + h(\boldsymbol{q}_h - \hat{\boldsymbol{q}}_h) \cdot \boldsymbol{n}.$$

All the results of Sect. 5.3 now follow from this simple identity and from the fact that $\boldsymbol{V}(K) \cdot \boldsymbol{n}|_F \subset M(F)$ for each face F of the triangulation. In other words, the Weak Galerkin method in [89] is also a simple rewriting of HDG methods.

Let us end by pointing out that, by the previous argument, the Weak Galerkin in [89] is *identical* to the Weak Galerkin method in [66] when the tensors c and a are piecewise constant.

7 Bibliographical Notes and Ongoing Work

After the introduction of the HDG methods in 2009 [33], we have extended the methods to a variety of partial differential equations and introduced a variation of the methods called the embedded discontinuous Galerkn (EDG) methods. The EDG methods were introduced in 2007 [56] in the framework of linear shells, and then analyzed in 2009 [34] for steady-state diffusion. (The HDG and EDG methods were devised almost at the same time but the publication of the HDG methods [33] took much more time than the publication of the EDG methods [34]).

The HDG methods for diffusion were devised and analyzed in [8, 10, 11, 20, 30, 35, 36, 38–41, 58], multigrid methods for them in [42], a posteriori error estimation for HDG methods in [26–28], and the convergence of adaptive HDG methods in [44]. The implementation of the HDG methods in 2D was considered in [60] and in 3D in [54]. The methods have been extended to convection-diffusion in [32, 68, 69, 81], to the Stokes flow of incompressible fluids in [15, 16, 21–23, 37, 70, 71, 80], to the Oseen equations in [7], to the incompressible Navier-Stokes equations in [72, 82], to the compressible Euler and Navier-Stokes Equations [75, 77], to several problems in continuum mechanics in [5, 6, 22, 53, 56, 59, 67, 85, 86], to wave propagation in [19, 73, 74], to the biharmonic in [31] and to scalar conservation laws in [64].

The current search for more efficient, superconvergent or optimally convergent, HDG methods seems to be going in three main directions: (1) The refinement of the sufficient conditions guaranteeing the superconvergence of the HDG methods through the so-called technique of M-decompositions [43], (2) the exploration of the properties of the Lehrenfeld-Schöberl stabilization function [63, 76, 78, 79], and (3) the exploration of the new, remarkable technique for devising numerical traces for the hybrid high-order (HHO) methods [46–48].

In fact, a bridge between the HHO and HDG methods was recently established in [29]. It would also be interesting to establish bridges with other numerical methods like, for example, the SUSHI methods [51], the elements constructed by Christiansen and Gillette [12], the BEM-based methods proposed in [83, 90], and the methods introduced in [57] for multiscale problems.

Acknowledgements The author would like to thank Prof. Henryk K. Stolarski for providing the earliest reference on static condensation. He would also thank Yanlai Chen, Mauricio Flores, Guosheng Fu, Matthias Maier and an anonymous referee for feedback leading to a better presentation of the material in this paper.

References

1. T. Arbogast, M.F. Wheeler, I. Yotov, Mixed finite elements for elliptic problems with tensor coefficients as cell-centered finite differences. SIAM J. Numer. Anal. **34**(2), 828–852 (1997). MR 1442940 (98g:65105)
2. D.N. Arnold, F. Brezzi, Mixed and nonconforming finite element methods: implementation, postprocessing and error estimates. RAIRO **19**, 7–32 (1985)
3. D.N. Arnold, F. Brezzi, B. Cockburn, L.D. Marini, Unified analysis of discontinuous Galerkin methods for elliptic problems. SIAM J. Numer. Anal. **39**, 1749–1779 (2002)
4. P. Castillo, B. Cockburn, I. Perugia, D. Schötzau, An a priori error analysis of the local discontinuous Galerkin method for elliptic problems. SIAM J. Numer. Anal. **38**, 1676–1706 (2000)
5. F. Celiker, B. Cockburn, K. Shi, Hybridizable discontinuous Galerkin methods for Timoshenko beams. J. Sci. Comput. **44**, 1–37 (2010)
6. F. Celiker, B. Cockburn, K. Shi, A projection-based error analysis of HDG methods for Timoshenko beams. Math. Comput. **81**, 131–151 (2012)
7. A. Cesmelioglu, B. Cockburn, Analysis of HDG methods for the Oseen equations. J. Sci. Comput. **55**(2), 392–431 (2013)
8. B. Chabaud, B. Cockburn, Uniform-in-time superconvergence of HDG methods for the heat equation. Math. Comput. **81**, 107–129 (2012)
9. Z. Chen, BDM mixed methods for a nonlinear elliptic problem. J. Comput. Appl. Math. **53**(2), 207–223 (1994). MR 1306126 (95i:65153)
10. Y. Chen, B. Cockburn, Analysis of variable-degree HDG methods for convection-diffusion equations. Part I: general nonconforming meshes. IMA J. Numer. Anal. **32**, 1267–1293 (2012)
11. Y. Chen, B. Cockburn, Analysis of variable-degree HDG methods for convection-diffusion equations. Part II: semimatching nonconforming meshes. Math. Comput. **83**, 87–111 (2014)
12. S.H. Christiansen, A. Gillette, Constructions of some minimal finite element systems, arXiv:1504.04670v1 (2015)
13. E. Chung, B. Engquist, Optimal discontinuous Galerkin methods for the acoustic wave equation in higher dimensions. SIAM J. Numer. Anal. **47**(5), 3820–3848 (2009)

14. E. Chung, B. Cockburn, G. Fu, The staggered DG method is the limit of a hybridizable DG method. SIAM J. Numer. Anal. **52**, 915–932 (2014)
15. B. Cockburn, J. Cui, An analysis of HDG methods for the vorticity-velocity-pressure formulation of the Stokes problem in three dimensions. Math. Comput. **81**, 1355–1368 (2012)
16. B. Cockburn, J. Cui, Divergence-free HDG methods for the vorticity-velocity formulation of the Stokes problem. J. Sci. Comput. **52**, 256–270 (2012)
17. B. Cockburn, J. Gopalakrishnan, The derivation of hybridizable discontinuous Galerkin methods for Stokes flow. SIAM J. Numer. Anal. **47**, 1092–1125 (2009)
18. B. Cockburn, W. Qiu, Commuting diagrams for the TNT elements on cubes. Math. Comput. **83**, 603–633 (2014)
19. B. Cockburn, V. Quenneville-Bélair, Uniform-in-time superconvergence of HDG methods for the acoustic wave equation. Math. Comput. **83**, 65–85 (2014)
20. B. Cockburn, F.-J. Sayas, The devising of symmetric couplings of boundary element and discontinuous Galerkin methods. IMA J. Numer. Anal. **32**, 765–794 (2012)
21. B. Cockburn, F.-J. Sayas, Divergence-conforming HDG methods for Stokes flow. Math. Comput. **83**, 1571–1598 (2014)
22. B. Cockburn, K. Shi, Conditions for superconvergence of HDG methods for Stokes flow. Math. Comput. **82**, 651–671 (2013)
23. B. Cockburn, K. Shi, Devising HDG methods for Stokes flow: an overview. Comput. Fluids **98**, 221–229 (2014)
24. B. Cockburn, C.-W. Shu, The local discontinuous Galerkin method for time-dependent convection-diffusion systems. SIAM J. Numer. Anal. **35**, 2440–2463 (1998)
25. B. Cockburn, C.-W. Shu, Runge-Kutta discontinuous Galerkin methods for convection-dominated problems. J. Sci. Comput. **16**, 173–261 (2001)
26. B. Cockburn, W. Zhang, A posteriori error estimates for HDG methods. J. Sci. Comput. **51**(3), 582–607 (2012)
27. B. Cockburn, W. Zhang, A posteriori error analysis for hybridizable discontinuous Galerkin methods for second order elliptic problems. SIAM J. Numer. Anal. **51**, 676–693 (2013)
28. B. Cockburn, W. Zhang, An a posteriori error estimate for the variable-degree Raviart-Thomas method. Math. Comput. **83**, 1063–1082 (2014)
29. B. Cockburn, D.A. Di Pietro, A. Ern, Bridging the hybrid high-order and hybridizable discontinuous Galerkin methods. ESAIM Math. Model. Numer. Anal. (to appear)
30. B. Cockburn, B. Dong, J. Guzmán, A superconvergent LDG-hybridizable Galerkin method for second-order elliptic problems. Math. Comput. **77**, 1887–1916 (2008)
31. B. Cockburn, B. Dong, J. Guzmán, A hybridizable and superconvergent discontinuous Galerkin method for biharmonic problems. J. Sci. Comput. **40**, 141–187 (2009)
32. B. Cockburn, B. Dong, J. Guzmán, M. Restelli, R. Sacco, Superconvergent and optimally convergent LDG-hybridizable discontinuous Galerkin methods for convection-diffusion-reaction problems. SIAM J. Sci. Comput. **31**, 3827–3846 (2009)
33. B. Cockburn, J. Gopalakrishnan, R. Lazarov, Unified hybridization of discontinuous Galerkin, mixed and continuous Galerkin methods for second order elliptic problems. SIAM J. Numer. Anal. **47**, 1319–1365 (2009)
34. B. Cockburn, J. Guzmán, S.-C. Soon, H.K. Stolarski, An analysis of the embedded discontinuous Galerkin method for second-order elliptic problems. SIAM J. Numer. Anal. **47**, 2686–2707 (2009)
35. B. Cockburn, J. Guzmán, H. Wang, Superconvergent discontinuous Galerkin methods for second-order elliptic problems. Math. Comput. **78**, 1–24 (2009)
36. B. Cockburn, J. Gopalakrishnan, F.-J. Sayas, A projection-based error analysis of HDG methods. Math. Comput. **79**, 1351–1367 (2010)
37. B. Cockburn, J. Gopalakrishnan, N.C. Nguyen, J. Peraire, F.-J. Sayas, Analysis of an HDG method for Stokes flow. Math. Comput. **80**, 723–760 (2011)
38. B. Cockburn, J. Guzmán, F.-J. Sayas, Coupling of Raviart-Thomas and hybridizable discontinuous Galerkin methods with BEM. SIAM J. Numer. Anal. **50**, 2778–2801 (2012)

39. B. Cockburn, W. Qiu, K. Shi, Conditions for superconvergence of HDG methods for second-order elliptic problems. Math. Comput. **81**, 1327–1353 (2012)
40. B. Cockburn, W. Qiu, K. Shi, Conditions for superconvergence of HDG methods on curvilinear elements for second-order elliptic problems. SIAM J. Numer. Anal. **50**, 1417–1432 (2012)
41. B. Cockburn, F.-J. Sayas, M. Solano, Coupling at a distance HDG and BEM. SIAM J. Sci. Comput. **34**, A28–A47 (2012)
42. B. Cockburn, O. Dubois, J. Gopalakrishnan, S. Tan, Multigrid for an HDG method. IMA J. Numer. Anal. **34**(4), 1386–1425 (2014)
43. B. Cockburn, G. Fu, F.J. Sayas, Superconvergence by M-decompositions: general theory for diffusion problems. Math. Comput. (to appear)
44. B. Cockburn, R. Nochetto, W. Zhang, Contraction property of adaptive hybridizable discontinuous Galerkin methods. Math. Comput. **85**, 1113–1141 (2016)
45. B. Cockburn, G. Fu, M.A. Sanchez-Uribe, C. Xiong, Supercloseness of two formulations of DG, HDG and mixed methods (in preparation)
46. D.A. Di Pietro, A. Ern, A hybrid high-order locking-free method for linear elasticity on general meshes. Comput. Methods Appl. Mech. Eng. **283**, 1–21 (2015)
47. D.A. Di Pietro, A. Ern, Hybrid high-order methods for variable-diffusion problems on general meshes. C. R. Acad. Sci Paris Ser. I **353**, 31–34 (2015)
48. D.A. Di Pietro, A. Ern, S. Lemaire, An arbitrary-order and compact-stencil discretization of diffusion on general meshes based on local reconstruction operators. Comput. Methods Appl. Math. **14**(4), 461–472 (2014)
49. D.A. Di Pietro, A. Ern, S. Lemaire, Arbitrary-order mixed method for heterogeneous anisotropic diffusion on general meshes HAL Id: hal-00918482 (2015)
50. Y. Efendiev, R. Lazarov, K. Shi, A multiscale HDG method for second order elliptic equations. Part I. Polynomial and homogenization-based multiscale spaces. SINUM **53**(1), 342–369 (2015)
51. R. Eymard, T. Gallouët, R. Herbin, Discretization of heterogeneous and anisotropic diffusion problems on general nonconforming meshes. SUSHI: a scheme using stabilization and hybrid interfaces. IMA J. Numer. Anal. **30**(4), 1009–1043 (2010)
52. B.M. Fraejis de Veubeke, Displacement and equilibrium models in the finite element method, in *Stress Analysis*, ed. by O.C. Zienkiewicz, G. Holister (Wiley, New York, 1977), pp. 145–197
53. G. Fu, B. Cockburn, H.K. Stolarski, Analysis of an HDG method for linear elasticity. Int. J. Numer. Methods Eng. **102**, 551–575 (2015)
54. Z. Fu, L.F. Gatica, F.-J. Sayas, Algorithm 494: MATLAB tools for HDG in three dimensions. ACM. Trans. Math. Softw. **41**, 21 pp. (2015). Art. No. 20
55. R.J. Guyan, Reduction of stiffness and mass matrices. J. Am. Inst. Aronaut. Astronaut. **3**, 380 (1965)
56. S. Güzey, B. Cockburn, H.K. Stolarski, The embedded discontinuous Galerkin methods: application to linear shells problems. Int. J. Numer. Methods Eng. **70**, 757–790 (2007)
57. C. Harder, D. Paredes, F. Valentin, On a multi scale hybrid-mixed method for advective-reactive dominated problems with heterogenous coefficients. Multiscale Model. Simul. **13**(2), 491–518 (2015)
58. L.N.T. Huynh, N.C. Nguyen, J. Peraire, B.C. Khoo, A high-order hybridizable discontinuous Galerkin method for elliptic interface problems. Int. J. Numer. Methods Eng. **93**, 183–200 (2013)
59. H. Kabaria, A. Lew, B. Cockburn, A hybridizable discontinuous Galerkin formulation for nonlinear elasticity. Comput. Methods Appl. Mech. Eng. **283**, 303–329 (2015)
60. R.M. Kirby, S.J. Sherwin, B. Cockburn, To HDG or to CG: a comparative study. J. Sci. Comput. **51**(1), 183–212 (2012)
61. J. Koebbe, A computationally efficient modification of mixed finite element methods for flow problems with full transmissivity tensors. Numer. Methods Partial Differ. Equ. **9**(4), 339–355 (1993). MR 1223251 (94c:76054)
62. C. Lehrenfeld, Hybrid discontinuous Galerkin methods for solving incompressible flow problems. Diplomigenieur Rheinisch-Westfalishen Technischen Hochchule Aachen (2010)

63. C. Lehrenfeld, J. Schöberl, High order exactly divergence-free hybrid discontinuous Galerkin methods for unsteady incompressible flows. ASC Report No. 27 (2015)
64. D. Moro, N.C. Nguyen, J. Peraire, A hybridized discontinuous Petrov-Galerkin scheme for scalar conservation laws. Int. J. Numer. Methods Eng. **91**, 950–970 (2012)
65. L. Mu, J. Wang, X. Ye, Weak Galerkin finite element methods on polytopal meshes. Int. J. Numer. Anal. Model. **12**, 31–53 (2015)
66. L. Mu, J. Wang, X. Ye, A weak Galerkin finite element method with polynomial reduction. J. Comput. Appl. Math. **285**, 45–58 (2015)
67. N.C. Nguyen, J. Peraire, Hybridizable discontinuous Galerkin methods for partial differential equations in continuum mechanics. J. Comput. Phys. **231**, 5955–5988 (2012)
68. N.C. Nguyen, J. Peraire, B. Cockburn, An implicit high-order hybridizable discontinuous Galerkin method for linear convection-diffusion equations. J. Comput. Phys. **228**, 3232–3254 (2009)
69. N.C. Nguyen, J. Peraire, B. Cockburn, An implicit high-order hybridizable discontinuous Galerkin method for nonlinear convection-diffusion equations. J. Comput. Phys. **228**, 8841–8855 (2009)
70. N.C. Nguyen, J. Peraire, B. Cockburn, A comparison of HDG methods for Stokes flow. J. Sci. Comput. **45**, 215–237 (2010)
71. N.C. Nguyen, J. Peraire, B. Cockburn, A hybridizable discontinuous Galerkin method for Stokes flow. Comput. Methods Appl. Mech. Eng. **199**, 582–597 (2010)
72. N.C. Nguyen, J. Peraire, B. Cockburn, A hybridizable discontinuous Galerkin method for the incompressible Navier-Stokes equations (AIAA Paper 2010-362), in *Proceedings of the 48th AIAA Aerospace Sciences Meeting and Exhibit*, Orlando, FL, January 2010
73. N.C. Nguyen, J. Peraire, B. Cockburn, High-order implicit hybridizable discontinuous Galerkin methods for acoustics and elastodynamics. J. Comput. Phys. **230**, 3695–3718 (2011)
74. N.C. Nguyen, J. Peraire, B. Cockburn, Hybridizable discontinuous Galerkin methods for the time-harmonic Maxwell's equations. J. Comput. Phys. **230**, 7151–7175 (2011)
75. N.C. Nguyen, J. Peraire, B. Cockburn, An implicit high-order hybridizable discontinuous Galerkin method for the incompressible Navier-Stokes equations. J. Comput. Phys. **230**, 1147–1170 (2011)
76. I. Oikawa, A hybridized discontinuous Galerkin method with reduced stabilization. J. Sci. Comput. (published online on December 2014)
77. J. Peraire, N.C. Nguyen, B. Cockburn, A hybridizable discontinuous Galerkin method for the compressible Euler and Navier-Stokes equations (AIAA Paper 2010-363), in *Proceedings of the 48th AIAA Aerospace Sciences Meeting and Exhibit*, Orlando, FL, January 2010
78. W. Qiu, K. Shi, An HDG method for convection-diffusion equations. J. Sci. Comput. **66**, 346–357 (2016)
79. W. Qiu, K. Shi, An HDG method for linear elasticity with strongly symmetric stresses (submitted)
80. S. Rhebergen, B. Cockburn, A space-time hybridizable discontinuous Galerkin method for incompressible flows on deforming domains. J. Comput. Phys. **231**, 4185–4204 (2012)
81. S. Rhebergen, B. Cockburn, Space-time hybridizable discontinuous Galerkin method for the advection-diffusion equation on moving and deforming meshes, in *The Courant-Friedrichs-Lewy (CFL) Condition* (Birkhäuser/Springer, New York, 2013), pp. 45–63
82. S. Rhebergen, B. Cockburn, J.J.W. van der Vegt, A space-time discontinuous Galerkin method for the incompressible Navier-Stokes equations. J. Comput. Phys. **233**, 339–358 (2013)
83. S. Rjasanow, S. Weißer, Higher order BEM-based FEM on polygonal meshes. SIAM J. Numer. Anal. **50**, 2357–2378 (2012)
84. J. Schöberl, private communication, 2015. The 2012 pdf file of J. Schöberl's Oberwolfach talk can be found in http://www.asc.tuwien.ac.at/~schoeberl/wiki/index.php/Talks
85. S.-C. Soon, Hybridizable discontinuous Galerkin methods for solid mechanics. Ph.D. thesis, University of Minnesota, Minneapolis, 2008
86. S.-C. Soon, B. Cockburn, H.K. Stolarski, A hybridizable discontinuous Galerkin method for linear elasticity. Int. J. Numer. Methods Eng. **80**(8), 1058–1092 (2009)

87. A. Ten Eyck, A. Lew, Discontinuous Galerkin methods for non-linear elasticity. Int. J. Numer. Methods Eng. **67**, 1204–1243 (2006)
88. J. Wang, X. Ye, A weak Galerkin finite element method for second order elliptic problems. J. Comput. Appl. Math. **241**, 103–115 (2013)
89. J. Wang, X. Ye, A weak Galerkin mixed finite element method for second order elliptic problems. Math. Comput. **289**, 2101–2126 (2014)
90. S. Weißer, Arbitrary order Trefftz-like basis functions on polygonal meshes and realization in BEM-based FEM. Comput. Math. Appl. **67**, 1390–1406 (2014)

Robust DPG Methods for Transient Convection-Diffusion

Truman Ellis, Jesse Chan, and Leszek Demkowicz

Abstract We introduce two robust DPG methods for transient convection-diffusion problems. Once a variational formulation is selected, the choice of test norm critically influences the quality of a particular DPG method. It is desirable that a test norm produce convergence of the solution in a norm equivalent to L^2 while producing optimal test functions that can be accurately computed and maintaining good conditioning of the optimal test function solve on highly adaptive meshes. Two such *robust* norms are introduced and proven to guarantee close to L^2 convergence of the primary solution variable. Numerical experiments demonstrate robust convergence of the two methods.

1 Introduction

The discontinuous Petrov-Galerkin finite element method presents an attractive new framework for developing robust numerical methods for computational mechanics. DPG contains the promise of being an automated scientific computing technology—it provides stability for any variational formulation, optimal convergence rates in a user-defined norm, virtually no pre-asymptotic stability issues on coarse meshes, and a measure of the error residual which can be used to robustly drive adaptivity. The method also delivers Hermitian positive definite stiffness matrices for any problem, weak enforcement of boundary conditions, and several other attractive properties [9, 11, 12]. For the most recent review of DPG, see [13]. The process of developing robust DPG methods for steady convection-diffusion was explored

T. Ellis • L. Demkowicz (✉)
Institute for Computational Engineering and Sciences, University of Texas at Austin, 201 East 24th St, Stop C0200, Austin, TX 78712, USA
e-mail: truman@ices.utexas.edu; leszek@ices.utexas.edu

J. Chan
Department of Mathematics, Virginia Tech, 460 McBryde Hall, Virginia Tech 225 Stanger Street, Blacksburg, VA 24061-0123, USA
e-mail: jlchan@math.vt.edu

G.R. Barrenechea et al. (eds.), *Building Bridges: Connections and Challenges in Modern Approaches to Numerical Partial Differential Equations*, Lecture Notes in Computational Science and Engineering 114,
DOI 10.1007/978-3-319-41640-3_6

in [7, 14]. In the sense, the main challenge is to come up with a correct test norm. The residual is measured in the dual test norm, and the DPG method minimizes the residual. The residual can be interpreted as a special *energy norm*. In other words, the DPG method delivers an orthogonal projection in the energy norm. The task is especially challenging for singular perturbation problems. Given a trial norm, we strive to determine a quasi-optimal test norm such that the corresponding energy norm is robustly equivalent to the trial norm of choice. An additional difficulty comes from the fact that the optimal test functions should be easily approximated with a simple enrichment strategy. For convection dominated diffusion, this means that the test functions should not develop boundary layers. The task of determining the quasi optimal test norm (we call it a *robust test norm*) leads then to a stability analysis for the adjoint equation which is the subject of this paper. For a more general discussion on the subject, see [13]. We start with an abstract derivation of the DPG framework then define the concept of a robust test norm and specialize to transient convection-diffusion. Two new robust norms are derived and numerical verifications of the theory are presented.

It is worth mentioning connections to other modern stabilized finite element methods. DPG can be thought of as a generalization of least-squares finite element methods, and in fact simplifies to this case when the L^2 topology is chosen for the test space. Connections to multi-scale methods have been studied in [8] and [6]. In the particular case of ultra-weak variational formulation and scaled adjoint graph norm used for the test norm, the DPG method delivers optimal test functions of Barrett and Morton (see [5]). For each trial basis function, these test functions are defined as solutions of the global adjoint equation with the basis function as the forcing term. It can be shown that the Petrov-Galerkin method with the resulting test space delivers L^2-projection. DPG approximates this test space (though not each individual optimal test function) using an enriched (i.e. multiscale) mortar least squares formulation. In a similar sense, DPG can also be interpreted as a Variational Multiscale Method which approximates the fine-scale contribution by enforcing the orthogonality of the fine scales under a specific inner product. The connection to HDG is less clear; DPG shares much with the more classical mortar method, but its connection to HDG has primarily been explored in [22]. Similarities include the use of trace unknowns which live on the mesh skeleton and the use of static condensation to eliminate the interior degrees of freedom, but the means of stabilization for the two methods are very different.

1.1 Space-Time Finite Elements

Most finite element simulations of transient phenomena use a semi-discrete formulation: the PDE is first discretized in space using finite elements and then the leftover system of ordinary differential equations in time is usually solved by a finite difference method. But it is possible to treat time as just another dimension to be discretized with finite elements. Some of the earliest proponents of this approach were Kaczkowski [20], Argyris and Scharpf [2], Fried [17], and Oden [25]. These techniques were built on the underlying concept of Hamilton's principle.

van der Vegt and van der Ven [30] have advanced a space-time discontinuous Galerkin method for 3D inviscid compressible moving boundary problems. Klaij et al. [21] then extended the method to compressible Navier-Stokes while Rhebergen et al. [27] developed the method for incompressible Navier-Stokes. Rhebergen and Cockburn [26] also developed a space-time HDG method for incompressible Navier-Stokes. Tezduyar and Behr [29] develop a deforming-spatial-domain/space-time procedure coupled with Galerkin/least-squares to handle incompressible Navier-Stokes flows with moving boundaries and later Aliabadi and Tezduyar [1] apply the procedure to compressible flows. Hughes and Stewart [19] develop a general space-time multiscale framework for deriving stabilized methods for transient phenomena.

It is possible to use the semi-discrete approach to solving transient problems with DPG, but it doesn't appear to be a natural fit with the adaptive nature of DPG. The Courant-Friedrichs-Lewy (CFL) condition is not binding with implicit time integration schemes, but it can be a guiding principle for temporal accuracy. If we are interested in temporally accurate solutions, we are limited by the fact that our smallest mesh elements (which may be orders of magnitude smaller than the largest elements) are constrained to proceed at a much smaller time step than the mesh as a whole. We can either restrict the whole mesh to the smallest time step, or we can attempt some sort of local time stepping. A space-time DPG formulation presents an attractive choice as we will be able to preserve our natural adaptivity from the steady problems while extending it in time. Thus we achieve an adaptive solution technique for transient problems in a unified framework. This paper expands previous work developing robust DPG methods for steady convection-diffusion to the space-time form.

2 Overview of DPG

2.1 A Generalized Minimum Residual Method

We begin with a well posed variational problem: find $u \in U$ such that

$$b(u, v) = l(v) \quad \forall v \in V,$$

where $b(u, v)$ is a bilinear (sesquilinear) form on $U \times V$ and $l \in V'$. Introducing operator $B : U \to V'$ (V' is the dual space to V) defined by $b(u, v) = \langle Bu, v\rangle_{V' \times V}$, we can reformulate the equation in operator form:

$$Bu = l \in V'.$$

We wish to find the element u_h of a finite dimensional subspace which minimizes the residual $Bu - l$ in V':

$$u_h = \arg\min_{w_h \in U_h} \frac{1}{2} \|Bu - l\|_{V'}^2 .$$

This mathematical framework is very natural, but it is not yet practical as the V' norm is not especially amenable to computations. With the assumption that we are working with Hilbert spaces, we can use the Riesz representation theorem to find a complementary object in V rather than V'. Let $R_V : V \ni v \to (v, \cdot) \in V'$ be the Riesz map, which is an isometry. Then the inverse Riesz map lets us represent our residual in V:

$$u_h = \arg\min_{w_h \in U_h} \frac{1}{2} \left\| R_V^{-1}(Bu - l) \right\|_V^2 .$$

Since this is a convex minimization problem, the solution is given by the critical points where the Gâteaux derivative is zero in all directions $\delta u \in U_h$:

$$\left(R_V^{-1}(Bu_h - l), R_V^{-1} B\delta u \right)_V = 0, \quad \forall \delta u \in U .$$

By definition of the Riesz map this is equivalent to the duality pairing

$$\left\langle Bu_h - l, R_V^{-1} B\delta u_h \right\rangle = 0 \quad \forall \delta u_h \in U_h .$$

We can define an optimal test function $v_{\delta u_h} := R_V^{-1} B\delta u_h$ for each trial function δu_h. This allows us to revert back to our original bilinear form with a finite dimensional set of trial and test functions:

$$b(u_h, v_{\delta u_h}) = l(v_{\delta u_h}).$$

Note that $v_{\delta u_h} \in V$ comes from the auxiliary problem

$$(v_{\delta u_h}, \delta v)_V = \langle R_V v_{\delta u_h}, \delta v \rangle = \langle B\delta u_h, \delta v \rangle = b(\delta u_h, \delta v) \quad \forall \delta v \in V.$$

We might refer to this as an *optimal Petrov-Galerkin* method. We arrive at the same method by realizing the supremum in the inf-sup condition (see [12]), motivating the *optimal* nomenclature. As a minimum residual method, optimal Petrov-Galerkin methods produce Hermitian, positive-definite stiffness matrices since

$$b(u_h, v_{\delta \boldsymbol{u}_h}) = (v_{u_h}, v_{\delta \boldsymbol{u}_h})_V = \overline{(v_{\delta \boldsymbol{u}_h}, v_{u_h})} = \overline{b(\delta u_h, v_{u_h})} .$$

The energy norm of the error is directly related to the residual:

$$\|u_h - u\|_E = \|B(u_h - u)\|_{V'} = \|Bu_h - l\|_{V'} = \left\|R_V^{-1}(Bu_h - l)\right\|_V ,$$

where we designate $R_V^{-1}(Bu_h - l)$ the *error representation function*. This has proven to be a very robust a-posteriori error estimator for driving adaptivity [15].

2.2 Transient Convection-Diffusion

2.2.1 Problem Description

In order to better illustrate choice of the U and V spaces, we introduce the transient convection-diffusion problem. Consider spatial domain Ω and corresponding space-time domain $Q = \Omega \times [0, T]$ with boundary $\Gamma = \Gamma_- \cup \Gamma_+ \cup \Gamma_0 \cup \Gamma_T$ where Γ_- is the inflow boundary ($\boldsymbol{\beta} \cdot \boldsymbol{n}_x < 0$, where $\boldsymbol{\beta}$ is the convection vector and $\boldsymbol{n}_x$ is the outward spatial normal), Γ_+ is the outflow boundary ($\boldsymbol{\beta} \cdot \boldsymbol{n}_x \geq 0$), Γ_0 is the initial time boundary, and Γ_T is the final time boundary. Let $\Gamma_h := \bigcup \partial K$ denote the entire mesh skeleton, where ∂K denotes the boundary of element K. Γ_{h_x} denotes any parts of the skeleton with a nonzero spatial normal and Γ_{h_t} have a nonzero temporal normal.

The transient convection-diffusion equation is

$$\frac{\partial u}{\partial t} + \nabla \cdot (\boldsymbol{\beta} u) - \epsilon \Delta u = f ,$$

where u is the quantity of interest, often interpreted to be a concentration of some quantity, ϵ is the diffusion coefficient, and f is the source term.

We apply flux boundary conditions on the inflow and trace boundary conditions on the outflow

$$\begin{aligned} \operatorname{tr}(\boldsymbol{\beta} \cdot u - \epsilon \nabla u) \cdot \boldsymbol{n}_x &= t_- \quad \text{on} \quad \Gamma_- \\ \operatorname{tr}(u) &= u_+ \quad \text{on} \quad \Gamma_+ \\ \operatorname{tr}(u) &= u_0 \quad \text{on} \quad \Gamma_0 . \end{aligned}$$

We note that Dirichlet boundary conditions also induce Dirichlet boundary conditions for the adjoint problem. Since the direction of convection is reversed for the adjoint convection-diffusion problem, this results in boundary layer adjoint solutions, which must be controlled using special weighted norms [14, 28]. However, since the convection-diffusion operator is not self-adjoint, the Cauchy inflow boundary condition induces a Neumann boundary condition for the adjoint problem. As a result, the adjoint solution does not contain boundary layers, simplifying the construction of a robust DPG method.

2.2.2 Relevant Sobolev Spaces

We begin by defining operators $\nabla_{xt} u := \begin{pmatrix} \nabla u \\ \frac{\partial u}{\partial t} \end{pmatrix}$ and $\nabla_{xt} \cdot \boldsymbol{u} := \nabla \cdot \boldsymbol{u}_x + \frac{\partial u_t}{\partial t}$, where $\boldsymbol{u} = (\boldsymbol{u}_x, u_t)$. We will need the following Sobolev spaces defined on our space-time domain:

$$
\begin{aligned}
H^1(Q) &= \left\{u \in L^2(Q) \; : \; \nabla u \in \mathbf{L}^2(Q)\right\} \\
H^1_{xt}(Q) &= \left\{u \in L^2(Q) \; : \; \nabla_{xt} u \in \mathbf{L}^2(Q)\right\} \\
\boldsymbol{H}(\text{div}, Q) &= \left\{\boldsymbol{\sigma} \in \mathbf{L}^2(Q) \; : \; \nabla \cdot \boldsymbol{\sigma} \in L^2(Q)\right\} \\
\boldsymbol{H}(\text{div}_{xt}, Q) &= \left\{\boldsymbol{\sigma} \in \mathbf{L}^2(Q) \; : \; \nabla_{xt} \cdot \boldsymbol{\sigma} \in L^2(Q)\right\}.
\end{aligned}
$$

We will also need the corresponding broken Sobolev spaces:

$$
\begin{aligned}
H^1(Q_h) &= \left\{u \in L^2(Q) \; : \; u|_K \in H^1(K), \; K \in Q_h\right\} &&= \prod_{K \in Q_h} H^1(K) \\
H^1_{xt}(Q_h) &= \left\{u \in L^2(Q) \; : \; u|_K \in H^1_{xt}(K), \; K \in Q_h\right\} &&= \prod_{K \in Q_h} H^1_{xt}(K) \\
\boldsymbol{H}(\text{div}, Q_h) &= \left\{\boldsymbol{\sigma} \in \mathbf{L}^2(Q) \; : \; u|_K \in \boldsymbol{H}(\text{div}, K), \; K \in Q_h\right\} &&= \prod_{K \in Q_h} \boldsymbol{H}(\text{div}, K) \\
\boldsymbol{H}(\text{div}_{xt}, Q_h) &= \left\{\boldsymbol{\sigma} \in \mathbf{L}^2(Q) \; : \; u|_K \in \boldsymbol{H}(\text{div}_{xt}, K), \; K \in Q_h\right\} &&= \prod_{K \in Q_h} \boldsymbol{H}(\text{div}_{xt}, K).
\end{aligned}
$$

Consider the following trace operators:

$$
\begin{aligned}
\text{tr}^K_{\text{grad}} u &= u|_{\partial K_x} && u \in H^1(K), \\
\text{tr}^K_{\text{div}_{xt}} \boldsymbol{\sigma} &= \boldsymbol{\sigma}|_{\partial K_{xt}} \cdot \boldsymbol{n}_{K_{xt}} && \boldsymbol{\sigma} \in \boldsymbol{H}(\text{div}_{xt}, K),
\end{aligned}
$$

where ∂K_x refers to spatial faces of element K, ∂K_{xt} to the full space-time boundary, and $\boldsymbol{n}_{K_{xt}}$ is the unit outward normal on ∂K_{xt}. The operators tr_{grad} and $\text{tr}_{\text{div}_{xt}}$ perform the same operation element by element to produce the linear maps

$$
\begin{aligned}
\text{tr}_{\text{grad}} &: H^1(Q_h) \to \prod_{K \in Q_h} H^{1/2}(\partial K_x), \\
\text{tr}_{\text{div}_{xt}} &: \boldsymbol{H}(\text{div}_{xt}, Q_h) \to \prod_{K \in Q_h} H^{-1/2}(\partial K_{xt}).
\end{aligned}
$$

Finally, we define spaces of interface functions. In order that our functions be single valued, we use the following definitions:

$$
\begin{aligned}
H^{1/2}(\Gamma_{h_x}) &= \mathrm{tr}_{\mathrm{grad}} H^1(Q)\,, \\
H_{xt}^{-1/2}(\Gamma_h) &= \mathrm{tr}_{\mathrm{div}_{xt}} H(\mathrm{div}_{xt}, Q)\,.
\end{aligned}
$$

For more details on broken and trace Sobolev spaces, see [3].

2.2.3 Variational Formulations

There are many possible manipulations that could be performed before arriving at a variational formulation. We begin by reformulating the problem in terms of the first order system:

$$
\begin{aligned}
\frac{1}{\epsilon}\boldsymbol{\sigma} - \nabla u &= 0 \\
\nabla_{xt} \cdot \begin{pmatrix} \boldsymbol{\beta} u - \boldsymbol{\sigma} \\ u \end{pmatrix} &= f\,.
\end{aligned}
\tag{1}
$$

Multiplying (1) by test functions $\boldsymbol{\tau} \in \mathbf{L}^2(Q)$ and $v \in L^2(Q)$, we obtain the following "trivial" variational formulation equivalent to the strong form:

$$
\begin{aligned}
& u \in H^1_{xt}(Q) && u = u_+ \quad \text{on } \Gamma_+ \\
& && u = u_0 \quad \text{on } \Gamma_0 \\
& \boldsymbol{\sigma} \in \boldsymbol{H}(\mathrm{div}, Q) && (\boldsymbol{\beta} u - \epsilon \nabla u) \cdot \boldsymbol{n} = t_- \quad \text{on } \Gamma_- \\
& \left(\frac{1}{\epsilon}\boldsymbol{\sigma}, \boldsymbol{\tau}\right) - (\nabla u, \boldsymbol{\tau}) && = 0 \quad \forall \boldsymbol{\tau} \in \mathbf{L}^2(Q) \\
& \left(\nabla_{xt} \cdot \begin{pmatrix} \boldsymbol{\beta} u - \boldsymbol{\sigma} \\ u \end{pmatrix}, v\right) && = f \quad \forall v \in L^2(Q)\,.
\end{aligned}
\tag{2}
$$

We can now choose either to relax (integrate by parts and build in the boundary conditions) or strongly enforce each equation. The steady state case and resulting options are explored and analyzed in further detail in [10] and are termed the trivial formulation (don't relax anything), the classical formulation (relax the second equation), the mixed formulation (relax the first equation), and the ultra-weak formulation (relax both equations). The stability constants for the four formulations are related, but the functional settings and norms of convergence change. Early

DPG work emphasized the ultra-weak formulation since in many ways it was the easiest to analyze, though recently the classical formulation has been under very active consideration. In the interests of simpler analysis, we focus on the ultra-weak formulation in this paper:

$$
\begin{aligned}
&u \in L^2(Q)\,,\ \boldsymbol{\sigma} \in \mathbf{L}^2\,(Q) \\
&\left(\frac{1}{\epsilon}\boldsymbol{\sigma},\boldsymbol{\tau}\right) + (u, \nabla\cdot\boldsymbol{\tau}) \quad = 0 \quad \forall\boldsymbol{\tau} \in \boldsymbol{H}(\mathrm{div}, Q)\ :\ \boldsymbol{\tau}\cdot\boldsymbol{n}_x = 0 \text{ on } \Gamma_- \\
&-\left(\begin{pmatrix}\boldsymbol{\beta}u-\boldsymbol{\sigma}\\ u\end{pmatrix}, \nabla_{xt}v\right) \quad = f \quad \forall v \in H^1_{xt}(Q)\ :\ v = 0 \text{ on } \Gamma_+\cup\Gamma_0\,.
\end{aligned}
\tag{3}
$$

We can remove the conditions on the test functions by introducing trace unknowns

$$
\begin{aligned}
\hat{u} &= \mathrm{tr}(u) && \text{on } \partial Q_x,\\
\hat{t} &= \mathrm{tr}\begin{pmatrix}\boldsymbol{\beta}u-\boldsymbol{\sigma}\\ u\end{pmatrix}\cdot\boldsymbol{n}_{xt} && \text{on } \partial Q_{xt}\,.
\end{aligned}
$$

Our new ultra-weak formulation with conforming test functions is

$$
\begin{aligned}
&u \in L^2(Q)\,,\ \boldsymbol{\sigma} \in \mathbf{L}^2\,(Q) && \hat{u} = u_+ \text{ on } \Gamma_+\\
&\hat{u} \in H^{1/2}(\partial Q_x)\,, && \hat{t} = t_- \text{ on } \Gamma_-\,,\quad \hat{t} = -u_0 \text{ on } \Gamma_0\\
&\hat{t} \in H^{-1/2}_{xt}(\partial Q)\,, && \\
&\left(\frac{1}{\epsilon}\boldsymbol{\sigma},\boldsymbol{\tau}\right) + (u, \nabla\cdot\boldsymbol{\tau}) - \langle\hat{u}, \boldsymbol{\tau}\cdot\boldsymbol{n}_x\rangle \quad = 0 \quad \forall\boldsymbol{\tau} \in \boldsymbol{H}(\mathrm{div}, Q) && \\
&-\left(\begin{pmatrix}\boldsymbol{\beta}u-\boldsymbol{\sigma}\\ u\end{pmatrix}, \nabla_{xt}v\right) + \langle\hat{t}, v\rangle \quad = f \quad \forall v \in H^1_{xt}(Q)\,. &&
\end{aligned}
\tag{4}
$$

2.2.4 Broken Test Functions

One of the key insights that led to the development of the DPG framework was the process of breaking test functions, that is testing with functions from larger broken Sobolev spaces, replacing $H^1_{xt}(Q)$ with $H^1_{xt}(Q_h)$ and $\boldsymbol{H}(\mathrm{div}, Q)$ with $\boldsymbol{H}(\mathrm{div}, Q_h)$. Discretizing such spaces is much simpler than standard spaces which require enforcement of global continuity conditions. The cost of introducing broken spaces is that we have to extend our interface unknowns $\hat{u}$ and $\hat{t}$ to live on the mesh skeleton.

Our ultra-weak formulation with broken test functions looks like

$$
\begin{aligned}
& u \in L^2(Q)\,,\ \boldsymbol{\sigma} \in \mathbf{L}^2(Q) && \hat{u} = u_+ \text{ on } \Gamma_+ \\
& \hat{u} \in H^{1/2}(\Gamma_{h_x})\,, && \hat{t} = t_- \text{ on } \Gamma_-\,, \quad \hat{t} = -u_0 \text{ on } \Gamma_0 \\
& \hat{t} \in H_{xt}^{-1/2}(\Gamma_h)\,, && \\
& \left(\frac{1}{\epsilon}\boldsymbol{\sigma}, \boldsymbol{\tau}\right) + (u, \nabla\cdot\boldsymbol{\tau}) - \langle \hat{u}, \boldsymbol{\tau}\cdot\boldsymbol{n}_x\rangle && = 0 \quad \forall \boldsymbol{\tau} \in \boldsymbol{H}(\mathrm{div}, Q_h) \\
& -\left(\begin{pmatrix} \boldsymbol{\beta} u - \boldsymbol{\sigma} \\ u \end{pmatrix}, \nabla_{xt} v\right) + \langle \hat{t}, v\rangle && = f \quad \forall v \in H^1_{xt}(Q_h)\,.
\end{aligned}
\tag{5}
$$

The main consequence of breaking test functions is that it reduces the cost of solving for optimal test functions from a global solve to an embarrassingly parallel solve element-by-element. Now that we've derived a suitable variational formulation, we are left with the task of selecting a test norm with which to compute our optimal test functions.

3 Robust Test Norms

The final unresolved choice is what norm to apply to the V space. This is one of the most important factors in designing a robust DPG method as the corresponding Riesz operator needs to be inverted to solve for the optimal test functions. If the norm produces unresolved boundary layers in the auxiliary problem, then many of the attractive features of DPG may fall apart. This is the primary emphasis of this paper. The problem of constructing stable test norms for steady convection-diffusion was addressed in [7, 14]. In this paper, we extend that work to transient convection-diffusion in space-time.

We define a robust test norm such that the L^2 norm of the solution is bounded by the energy norm of the solution with a constant independent of ϵ. We can rewrite any ultra-weak formulation with broken test functions as the following bilinear form with group variables:

$$
b\left((u, \hat{u}), v\right) = \left(u, A^* v\right)_{L^2} + \langle \hat{u}, [\![v]\!]\rangle_{\Gamma_h}\,,
$$

where A^* represents the adjoint. In the case of convection-diffusion, $u := \{u, \boldsymbol{\sigma}\}$, $\hat{u} := \{\hat{u}, \hat{t}\}$, $v := \{v, \boldsymbol{\tau}\}$.

Note that for conforming v^* satisfying $A^* v^* = u$:

$$
\begin{aligned}
\|u\|^2_{L^2} = b(u, v^*) &= \frac{b(u, v^*)}{\|v^*\|_V}\|v^*\|_V \\
&\leq \sup_{v^*\neq 0}\frac{|b(u, v^*)|}{\|v^*\|}\|v^*\| = \|u\|_E\,\|v^*\|_V\,.
\end{aligned}
$$

This defines a necessary condition for robustness, namely that

$$\|v^*\|_V \lesssim \|u\|_{L^2} \; . \tag{6}$$

If this condition is satisfied, then we get our final result:

$$\|u\|_{L^2} \lesssim \|u\|_E \; .$$

So far, we've assumed that our finite set of optimal test functions are assembled from an infinite dimensional space. In practice, we have found it to be sufficient to use an "enriched" space of higher polynomial dimension than the trial space [18]. This adds an additional requirement when assembling a robust test norm, namely that our optimal test functions should be adequately representable within this enriched space. We illustrate this point by considering three norms which satisfy the above conditions for 1D steady convection-diffusion. The graph norm is $\left(\|A^*v\|_{L^2}^2 + \|v\|_{L^2}^2\right)^{\frac{1}{2}}$:

$$\|(v,\boldsymbol{\tau})\|^2 = \|\nabla\cdot\boldsymbol{\tau} - \boldsymbol{\beta}\cdot\nabla v\|^2 + \left\|\frac{1}{\epsilon}\boldsymbol{\tau} + \nabla v\right\|^2 + \|v\|^2 + \|\boldsymbol{\tau}\|^2 \; .$$

Remark 1 In the DPG technology, the test norm must be *localizable*, i.e.,

$$\|v\|_V^2 = \sum_K \|v\|_{V(K)}^2,$$

where $\|v\|_{V(K)}$ denotes a test norm (and not just a seminorm) for the element test space. In practice this means the addition of properly scaled L^2-terms. Without those terms, we could not invert the Riesz operator on the element level. Addition of the L^2 terms does not necessarily contradict the robustness of the norm, see the discussion in [13] on bounded below operators. An alternate strategy has been explored in [16] where we enforce element conservation property by securing the presence of a constant function in the element test space. The residual is then minimized only over the orthogonal complement to the constants which eliminates the need for adding the L^2-term to the test norm.

The robust norm was derived in [7]:

$$\begin{aligned}\|(v,\boldsymbol{\tau})\|^2 = \|\boldsymbol{\beta}\cdot\nabla v\|^2 + \epsilon\,\|\nabla v\|^2 + \min\left(\frac{\epsilon}{h^2},1\right)\|v\|^2 \\ + \|\nabla\cdot\boldsymbol{\tau}\|^2 + \min\left(\frac{1}{h^2},\frac{1}{\epsilon}\right)\|\boldsymbol{\tau}\|^2 \; .\end{aligned}$$

The case for the coupled robust norm was made in [4]:

$$\begin{aligned}\|(v,\boldsymbol{\tau})\|^2 &= \|\boldsymbol{\beta}\cdot\nabla v\|^2 + \epsilon\,\|\nabla v\|^2 + \min\left(\frac{\epsilon}{h^2},1\right)\|v\|^2\\ &\quad + \|\nabla\cdot\boldsymbol{\tau} - \boldsymbol{\beta}\cdot\nabla v\|^2 + \min\left(\frac{1}{h^2},\frac{1}{\epsilon}\right)\|\boldsymbol{\tau}\|^2 .\end{aligned}$$

The argument for the coupled norm was that in certain cases we noticed pollution of u from errors in $\boldsymbol{\sigma}$, for example at singularities in $\boldsymbol{\sigma}$, u also exhibited degraded quality with the robust norm. The coupled robust norm seemed to relax this behavior, i.e. errors in u appear more independent of errors in $\boldsymbol{\sigma}$.

The bilinear form and test norm define a mapping from input trial functions to an optimal test function:

$$T = R_V^{-1}B : U \to V .$$

Below, we plot the optimal test functions produced given $\epsilon = 10^{-2}$, a representative trial function $u = x - \frac{1}{2}$, and either the graph norm, the robust norm, or the coupled robust norm. Note that the optimal test functions will be different for any other trial function. In the left column, we see the fully resolved *ideal* optimal test function that DPG theory relies on. On the right, we see the approximated optimal test function using a enriched cubic test space (Figs. 1, 2, and 3).

Mathematically, the graph norm satisfies the necessary condition to be a robust norm, but the ideal optimal test functions contain strong boundary layers which can not be realistically approximated with the provided enriched space. If the approximated optimal test functions can not come sufficiently close to the ideal, then the whole DPG theory falls apart. See [18] for more discussion. This provides an additional condition on a test norm before we can truly call it robust: the ideal test functions must be adequately representable within the provided enriched space. This ultimately comes down to an analysis of the relative magnitudes of individual terms within the test norm, usually attempting to bound reactive or convective terms by diffusive terms. The coupled robust norm satisfies condition (6) and also produces relatively smooth optimal test functions that can be sufficiently approximated with a cubic polynomial space. Niemi et al. attempted to approximate boundary layers in optimal shape functions with Shishkin meshes [23, 24].

3.1 *Application to Transient Convection-Diffusion*

Now we present the analysis leading to two robust norms for transient convection-diffusion. Consider the problem with homogeneous boundary conditions:

$$\frac{1}{\epsilon}\boldsymbol{\sigma} - \nabla u = 0$$

$$\frac{\partial u}{\partial t} + \boldsymbol{\beta}\cdot\nabla u - \nabla\cdot\boldsymbol{\sigma} = f$$

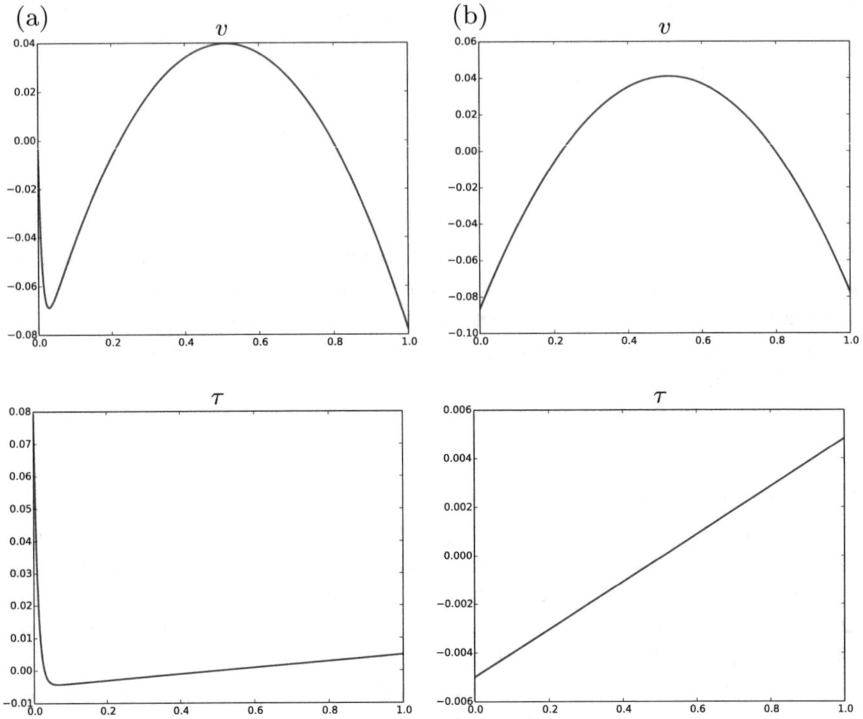

Fig. 1 Graph norm optimal test functions for $u = x - \frac{1}{2}$. (**a**) Ideal. (**b**) Approximated

$$
\begin{aligned}
\beta_n u - \epsilon \frac{\partial u}{\partial n} &= 0 \text{ on } \Gamma_- \\
u &= 0 \text{ on } \Gamma_+ \\
u &= u_0 \text{ on } \Gamma_0 .
\end{aligned}
$$

Let $\tilde{\boldsymbol{\beta}} := \begin{pmatrix} \boldsymbol{\beta} \\ 1 \end{pmatrix}$, then we can rewrite this as

$$
\begin{aligned}
\frac{1}{\epsilon}\boldsymbol{\sigma} - \nabla u &= 0 \\
\tilde{\boldsymbol{\beta}} \cdot \nabla_{xt} u - \nabla \cdot \boldsymbol{\sigma} &= f \\
\beta_n u - \epsilon \frac{\partial u}{\partial n} &= 0 \text{ on } \Gamma_- \\
u &= 0 \text{ on } \Gamma_+ \\
u &= u_0 \text{ on } \Gamma_0 .
\end{aligned}
$$

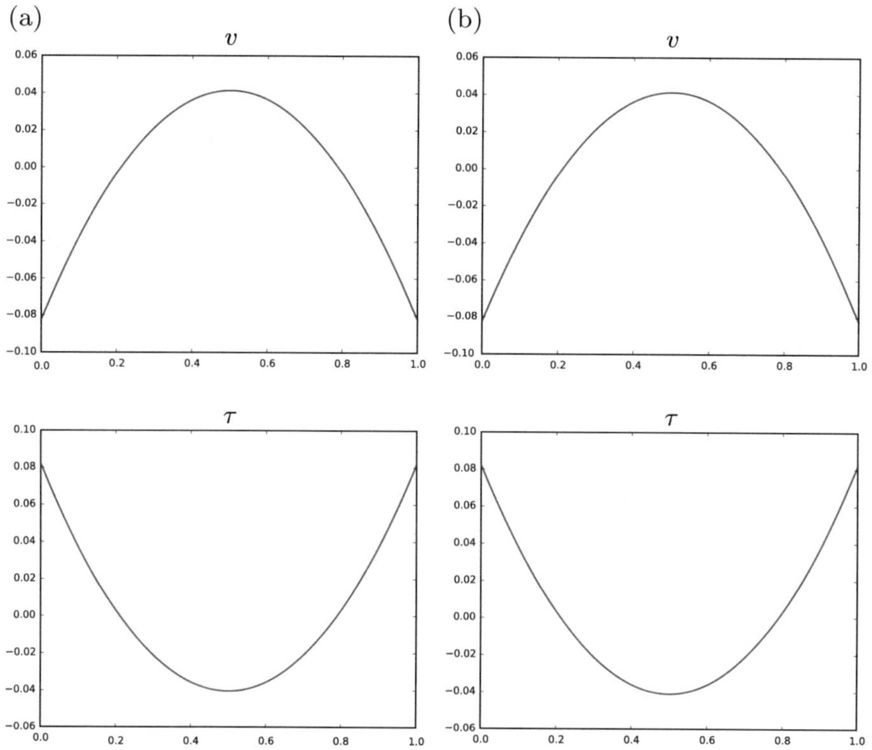

Fig. 2 Robust norm optimal test functions for $u = x - \frac{1}{2}$. (**a**) Ideal. (**b**) Approximated

The adjoint operator A^* is given by

$$A^*(v, \boldsymbol{\tau}) = \left(\frac{1}{\epsilon} \boldsymbol{\tau} + \nabla v, -\tilde{\boldsymbol{\beta}} \cdot \nabla_{xt} v + \nabla \cdot \boldsymbol{\tau} \right).$$

We decompose now the continuous adjoint problem

$$A^*(v, \boldsymbol{\tau}) = (\boldsymbol{f}, g)$$

into two cases a continuous part with forcing term g:

$$\begin{aligned} \frac{1}{\epsilon} \boldsymbol{\tau}_1 + \nabla v_1 &= 0 \\ -\tilde{\boldsymbol{\beta}} \cdot \nabla_{xt} v_1 + \nabla \cdot \boldsymbol{\tau}_1 &= g \\ \boldsymbol{\tau}_1 \cdot \boldsymbol{n}_x &= 0 \text{ on } \Gamma_- \\ v_1 &= 0 \text{ on } \Gamma_+ \\ v_1 &= 0 \text{ on } \Gamma_T \,, \end{aligned}$$

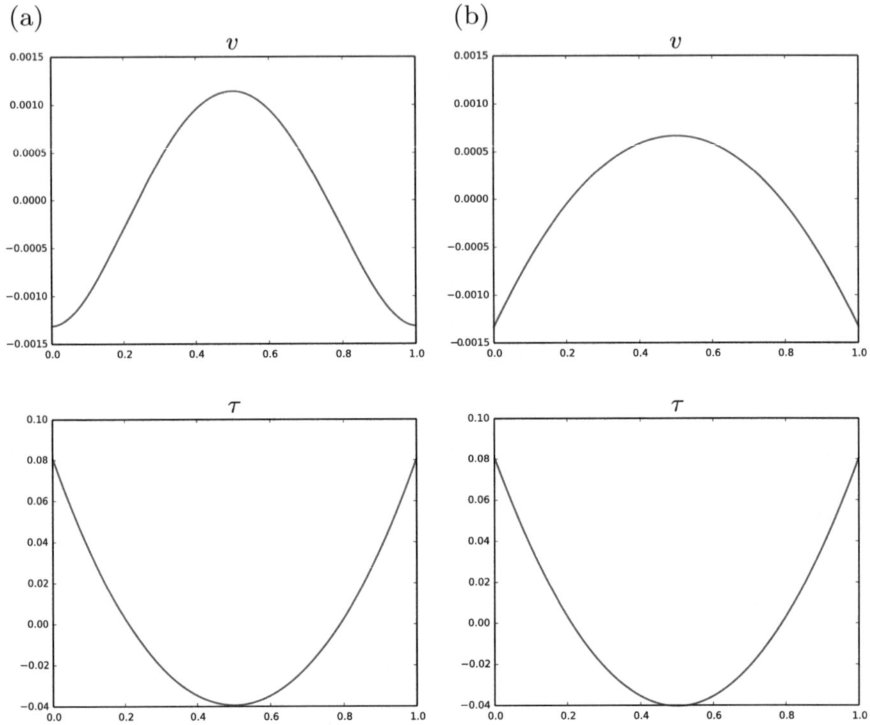

Fig. 3 Coupled robust norm optimal test functions for $u = x - \frac{1}{2}$. (**a**) Ideal. (**b**) Approximated

and a continuous part with forcing $\boldsymbol{f}$:

$$\begin{aligned} \frac{1}{\epsilon}\boldsymbol{\tau}_2 + \nabla v_2 &= \boldsymbol{f} \\ -\tilde{\boldsymbol{\beta}} \cdot \nabla_{xt} v_2 + \nabla \cdot \boldsymbol{\tau}_2 &= 0 \\ \boldsymbol{\tau}_2 \cdot \boldsymbol{n}_x &= 0 \text{ on } \Gamma_- \\ v_2 &= 0 \text{ on } \Gamma_+ \\ v_2 &= 0 \text{ on } \Gamma_T . \end{aligned}$$

(The boundary conditions can be derived by taking the ultra-weak formulation and choosing boundary conditions such that the temporal flux and spatial flux terms $\langle \hat{u}, [\![\tau_n]\!] \rangle_{\Gamma_{out}}$ and $\left\langle \hat{t}_n, [\![v]\!] \right\rangle_{\Gamma_{in}}$ are zero.)

We can then derive that the test norms

$$\|(v,\boldsymbol{\tau})\|_{V,K}^2 := \left\|\tilde{\boldsymbol{\beta}}\cdot\nabla_{xt}v\right\|_K^2 + \epsilon\,\|\nabla v\|_K^2 + \|v\|_K^2 + \|\nabla\cdot\boldsymbol{\tau}\|_K^2 + \frac{1}{\epsilon}\,\|\boldsymbol{\tau}\|_K^2\,, \tag{7}$$

and

$$\|(v,\boldsymbol{\tau})\|_{V,K}^2 := \left\|\tilde{\boldsymbol{\beta}}\cdot\nabla_{xt}v\right\|_K^2 + \epsilon\,\|\nabla v\|_K^2 + \|v\|_K^2 + \left\|\nabla\cdot\boldsymbol{\tau} - \tilde{\boldsymbol{\beta}}\cdot\nabla_{xt}v\right\|_K^2 + \frac{1}{\epsilon}\,\|\boldsymbol{\tau}\|_K^2\,, \tag{8}$$

respectively designated the *robust* test norm and the *coupled robust* test norm, provide the necessary bound $\|v^*\|_V \lesssim \|u\|_{L^2(Q)}$.

In the following lemmas we establish the following bounds:

- Bound on $\|(v_1,\boldsymbol{\tau}_1)\|_V$. Lemma 2 gives $\left\|\tilde{\boldsymbol{\beta}}\cdot\nabla_{xt}v_1\right\| \le \|g\|$. Since $\nabla\cdot\boldsymbol{\tau}_1 = g + \tilde{\boldsymbol{\beta}}\cdot\nabla_{xt}v_1$,

  $$\|\nabla\cdot\boldsymbol{\tau}_1\| \le \|g\| + \left\|\tilde{\boldsymbol{\beta}}\cdot\nabla_{xt}v_1\right\| \le 2\,\|g\|\,.$$

  Or, the fact that $\nabla\cdot\boldsymbol{\tau} - \tilde{\boldsymbol{\beta}}\cdot\nabla_{xt}v_1 = g$ clearly gives

  $$\left\|\nabla\cdot\boldsymbol{\tau} - \tilde{\boldsymbol{\beta}}\cdot\nabla_{xt}v_1\right\| = \|g\|\,.$$

  Lemma 1 gives $\|v_1\|^2 + \epsilon\,\|\nabla v_1\|^2 \le \|g\|^2$. Since $\epsilon^{1/2}\nabla v_1 = -\epsilon^{-1/2}\boldsymbol{\tau}_1$,

  $$\frac{1}{\epsilon}\,\|\boldsymbol{\tau}_1\|^2 \le \|g\|^2\,.$$

  Thus, all $(v_1,\boldsymbol{\tau}_1)$ terms in (7) and (8) are accounted for, guaranteeing at least robust control of u.
- Bound on $\|(v_2,\boldsymbol{\tau}_2)\|_V$. The fact that $\nabla\cdot\boldsymbol{\tau} - \tilde{\boldsymbol{\beta}}\cdot\nabla_{xt}v = 0$ clearly gives

  $$\left\|\nabla\cdot\boldsymbol{\tau} - \tilde{\boldsymbol{\beta}}\cdot\nabla_{xt}v_2\right\| = 0 \le \|\boldsymbol{f}\|\,.$$

  Lemma 1 gives $\|v_2\|^2 + \epsilon\,\|\nabla v_2\|^2 \le \epsilon\,\|\boldsymbol{f}\|^2$. Since $\epsilon^{1/2}\nabla v_2 = \boldsymbol{f} - \epsilon^{-1/2}\boldsymbol{\tau}_2$,

  $$\frac{1}{\epsilon}\,\|\boldsymbol{\tau}_2\|^2 \le (1+\epsilon)\,\|\boldsymbol{f}\|^2\,.$$

We have not been able to develop bounds on $\left\|\tilde{\boldsymbol{\beta}} \cdot \nabla_{xt} v_2\right\|$ and $\|\nabla \cdot \boldsymbol{\tau}\|$ which means that we can not guarantee robust control of $\boldsymbol{\sigma}$ with provided test norms.

We proceed now with the technical estimates.

Lemma 1 *For the duration of this lemma, let* $v := v_1 + v_2$. *Assuming the advection field* $\boldsymbol{\beta}$ *is incompressible, i.e.* $\nabla \cdot \boldsymbol{\beta} = 0$,

$$\|v\|^2 + \epsilon \|\nabla v\|^2 \leq \|g\|^2 + \epsilon \|\boldsymbol{f}\|^2 .$$

Proof Define $w = e^t v$ and note that $\frac{\partial w}{\partial t} = \left(\frac{\partial v}{\partial t} + v\right) e^t$ while all spatial derivatives go through. Multiplying the adjoint by w and integrating over Q gives

$$-\int_Q \tilde{\boldsymbol{\beta}} \cdot \nabla_{xt} v w - \epsilon \Delta v w = \int_Q g w - \epsilon \int_Q \nabla \cdot \boldsymbol{f} w$$

or

$$-\int_Q e^t v \tilde{\boldsymbol{\beta}} \cdot \nabla_{xt} v - \epsilon \int_Q e^t v \Delta v = \int_Q e^t g v - \epsilon \int_Q e^t v \nabla \cdot \boldsymbol{f}.$$

Integrating by parts:

$$\begin{aligned}\int_Q \nabla_{xt} \cdot \left(e^t \tilde{\boldsymbol{\beta}} v\right) v - \int_\Gamma e^t \tilde{\boldsymbol{\beta}} \cdot \boldsymbol{n} v^2 + \epsilon \int_Q e^t \nabla v \cdot \nabla v - \epsilon \int_{\Gamma_x} e^t v \cdot \nabla v \cdot \boldsymbol{n}_x \\ = \int_Q e^t g v + \epsilon \int_Q e^t \nabla v \cdot \boldsymbol{f} - \epsilon \int_{\Gamma_x} e^t v \boldsymbol{f} \cdot \boldsymbol{n}_x.\end{aligned}$$

Note that $\nabla_{xt} \cdot e^t v \tilde{\boldsymbol{\beta}} = e^t (\tilde{\boldsymbol{\beta}} \cdot \nabla_{xt} v + v)$ if $\nabla \cdot \boldsymbol{\beta} = 0$. Moving some terms to the right hand side, we get

$$\begin{aligned}&\int_Q e^t v^2 + \int_Q \epsilon e^t \nabla v \cdot \nabla v \\ &\quad = \int_Q e^t g v + \epsilon \int_Q e^t \nabla v \cdot \boldsymbol{f} - \epsilon \int_{\Gamma_x} e^t v \boldsymbol{f} \cdot \boldsymbol{n}_x \\ &\quad\quad - \int_Q e^t \tilde{\boldsymbol{\beta}} \cdot \nabla_{xt} v v \quad + \int_\Gamma e^t \tilde{\boldsymbol{\beta}} \cdot \boldsymbol{n} v^2 \quad + \epsilon \int_{\Gamma_x} e^t v \cdot \nabla v \cdot \boldsymbol{n}_x.\end{aligned}$$

Note that $1 \leq \|e^t\|_\infty = e^T$. Then

$$\|v\|^2 + \epsilon \|\nabla v\|^2$$

$$\leq e^T \left(\int_Q gv + \epsilon \int_Q \nabla v \cdot \boldsymbol{f} - \epsilon \int_{\Gamma_-} v \underbrace{\boldsymbol{f} \cdot \boldsymbol{n}_x}_{=\tau_n^{\nearrow 0} + \frac{\partial v}{\partial \boldsymbol{n}_x}} - \epsilon \int_{\Gamma_+} \underbrace{v}_{=0} \boldsymbol{f} \cdot \boldsymbol{n}_x \right.$$

$$\left. - \int_Q \tilde{\boldsymbol{\beta}} \cdot \nabla_{xt} vv + \int_\Gamma \tilde{\boldsymbol{\beta}} \cdot \boldsymbol{n} v^2 + \epsilon \int_{\Gamma_-} v \cdot \nabla v \cdot \boldsymbol{n}_x + \epsilon \int_{\Gamma_+} \underbrace{v}_{=0} \frac{\partial v}{\partial \boldsymbol{n}_x} \right)$$

Note: boundary conditions give $\boldsymbol{\tau}_n = 0$ on Γ_- and $v = 0$ on Γ_+

$$= e^T \left(\left(\int_Q gv + \epsilon \int_Q \nabla v \cdot \boldsymbol{f} \cancel{- \epsilon \int_{\Gamma_-} v \frac{\partial v}{\partial \boldsymbol{n}_x} + \epsilon \int_{\Gamma_x} v \frac{\partial v}{\partial \boldsymbol{n}_x}} \right. \right.$$

$$\left. - \frac{1}{2} \int_Q \tilde{\boldsymbol{\beta}} \cdot \nabla_{xt} v^2 + \int_\Gamma \tilde{\boldsymbol{\beta}} \cdot \boldsymbol{n} v^2 \right)$$

Note: $\Gamma_x = \Gamma_- \cup \Gamma_+$ and $v = 0$ on Γ_-

$$= e^T \left(\int_Q gv + \epsilon \int_Q \nabla v \cdot \boldsymbol{f} + \frac{1}{2} \int_Q \cancelto{0}{\nabla_{xt} \cdot \tilde{\boldsymbol{\beta}}} v^2 - \frac{1}{2} \int_\Gamma \tilde{\boldsymbol{\beta}} \cdot \boldsymbol{n} v^2 + \int_\Gamma \tilde{\boldsymbol{\beta}} \cdot \boldsymbol{n} v^2 \right)$$

Note: Integration by parts of $-\frac{1}{2} \int_Q \tilde{\boldsymbol{\beta}} \cdot \nabla_{xt} v^2$ and $\nabla \cdot \boldsymbol{\beta} = 0$

$$= e^T \left(\int_Q gv + \epsilon \int_Q \nabla v \cdot \boldsymbol{f} \right.$$

$$\left. + \frac{1}{2} \left(\int_{\Gamma_0} \underbrace{-v^2}_{\leq 0} + \int_{\Gamma_T} \cancelto{0}{v^2} + \int_{\Gamma_-} \underbrace{\boldsymbol{\beta} \cdot \boldsymbol{n}_x v^2}_{\leq 0} + \int_{\Gamma_+} \boldsymbol{\beta} \cdot \boldsymbol{n}_x \cancelto{0}{v^2} \right) \right)$$

Note: Split boundary term into components, $v = 0$ on Γ_+ and Γ_T

$$\leq e^T \left(\int_Q gv + \epsilon \int_Q \nabla v \cdot \boldsymbol{f} \right)$$

$$\leq e^T \left(\frac{\|g\|^2}{2} + \epsilon \frac{\|\boldsymbol{f}\|^2}{2} + \frac{\|v\|^2}{2} + \epsilon \frac{\|\nabla v\|^2}{2} \right).$$

Note: Young's inequality

Lemma 2 *If* $\left\|\nabla\boldsymbol{\beta} - \frac{1}{2}\nabla\cdot\boldsymbol{\beta}\boldsymbol{I}\right\|_{L^\infty} \leq C_{\boldsymbol{\beta}}$ *we can bound*

$$\left\|\tilde{\boldsymbol{\beta}}\cdot\nabla_{xt}v_1\right\| \lesssim \|g\|\,.$$

Proof Multiply $-\tilde{\boldsymbol{\beta}}\cdot\nabla_{xt}v_1 = g - \nabla\cdot\boldsymbol{\tau}_1$ by $-\tilde{\boldsymbol{\beta}}\cdot\nabla_{xt}v_1$ and integrate over Q to get

$$\left\|\tilde{\boldsymbol{\beta}}\cdot\nabla_{xt}v_1\right\|^2 = -\int_Q g\tilde{\boldsymbol{\beta}}\cdot\nabla_{xt}v_1 + \int_Q \tilde{\boldsymbol{\beta}}\cdot\nabla_{xt}v_1\nabla\cdot\boldsymbol{\tau}_1\,. \tag{9}$$

Note that

$$\frac{1}{\epsilon}\int_Q \tilde{\boldsymbol{\beta}}\cdot\nabla_{xt}v_1\nabla\cdot\boldsymbol{\tau}_1 = -\int_Q \tilde{\boldsymbol{\beta}}\cdot\nabla_{xt}v_1\nabla\cdot\nabla v_1$$

Note: $\boldsymbol{\tau}_1 = \epsilon\nabla v_1$

$$= -\int_{\Gamma_x}\tilde{\boldsymbol{\beta}}\cdot\nabla_{xt}v_1\nabla v_1\cdot\boldsymbol{n}_x + \int_Q \nabla(\tilde{\boldsymbol{\beta}}\cdot\nabla_{xt}v_1)\cdot\nabla v_1$$

Note: Integration by parts

$$= -\int_{\Gamma_x}\tilde{\boldsymbol{\beta}}\cdot\nabla_{xt}v_1\nabla v_1\cdot\boldsymbol{n}_x + \int_Q (\nabla\tilde{\boldsymbol{\beta}}\cdot\nabla_{xt}v_1)\cdot\nabla v_1 + \int_Q \tilde{\boldsymbol{\beta}}\cdot\nabla\nabla_{xt}v_1\cdot\nabla v_1$$

$$= -\int_{\Gamma_x}\tilde{\boldsymbol{\beta}}\cdot\nabla_{xt}v_1\nabla v_1\cdot\boldsymbol{n}_x + \int_Q (\nabla\boldsymbol{\beta}\cdot\nabla v_1)\cdot\nabla v_1 + \frac{1}{2}\int_Q \tilde{\boldsymbol{\beta}}\cdot\nabla_{xt}(\nabla v_1\cdot\nabla v_1)$$

Note: $\nabla\nabla_{xt}v_1\cdot\nabla v_1 = \nabla_{xt}\nabla v_1\cdot\nabla v_1 = \frac{1}{2}\nabla_{xt}(\nabla v_1\cdot\nabla v_1)$

$$= -\int_{\Gamma_x}\tilde{\boldsymbol{\beta}}\cdot\nabla_{xt}v_1\nabla v_1\cdot\boldsymbol{n}_x + \int_Q (\nabla\boldsymbol{\beta}\cdot\nabla v_1)\cdot\nabla v_1 + \frac{1}{2}\int_\Gamma \tilde{\boldsymbol{\beta}}\cdot\boldsymbol{n}(\nabla v_1\cdot\nabla v_1) - \frac{1}{2}\int_Q \nabla_{xt}\cdot\tilde{\boldsymbol{\beta}}(\nabla v_1\cdot\nabla v_1)$$

Note: Integration by parts

$$= -\int_{\Gamma_x}\tilde{\boldsymbol{\beta}}\cdot\nabla_{xt}v_1\nabla v_1\cdot\boldsymbol{n}_x + \int_Q (\nabla\boldsymbol{\beta}\cdot\nabla v_1)\cdot\nabla v_1 + \frac{1}{2}\int_\Gamma \tilde{\boldsymbol{\beta}}\cdot\boldsymbol{n}(\nabla v_1\cdot\nabla v_1) - \frac{1}{2}\int_Q \nabla\cdot\boldsymbol{\beta}(\nabla v_1\cdot\nabla v_1)$$

Note: $\nabla_{xt}\cdot\tilde{\boldsymbol{\beta}} = \nabla\cdot\boldsymbol{\beta}$

$$= -\int_{\Gamma_x} \tilde{\boldsymbol{\beta}} \cdot \nabla_{xt} v_1 \nabla v_1 \cdot \boldsymbol{n}_x + \frac{1}{2}\int_{\Gamma} \tilde{\boldsymbol{\beta}} \cdot \boldsymbol{n}(\nabla v_1 \cdot \nabla v_1)$$

$$+ \int_Q \nabla v_1 (\nabla \boldsymbol{\beta} - \frac{1}{2}\nabla \cdot \boldsymbol{\beta} \boldsymbol{I}) \nabla v_1.$$

Note: $(\nabla \boldsymbol{\beta} \cdot \nabla v_1) \cdot \nabla v_1 - \frac{1}{2}\nabla \cdot \boldsymbol{\beta}(\nabla v_1 \cdot \nabla v_1) = \nabla v_1 (\nabla \boldsymbol{\beta} - \frac{1}{2}\nabla \cdot \boldsymbol{\beta} \boldsymbol{I}) \nabla v_1$

Plugging this into (9), we get

$$\left\| \tilde{\boldsymbol{\beta}} \cdot \nabla_{xt} v_1 \right\|^2 = -\int_Q g\tilde{\boldsymbol{\beta}} \cdot \nabla_{xt} v_1 + \epsilon \int_Q \nabla v_1 (\nabla \boldsymbol{\beta} - \frac{1}{2}\nabla \cdot \boldsymbol{\beta} \boldsymbol{I}) \nabla v_1$$

$$- \epsilon \int_{\Gamma_x} \tilde{\boldsymbol{\beta}} \cdot \nabla_{xt} v_1 \nabla v_1 \cdot \boldsymbol{n}_x + \frac{\epsilon}{2}\int_{\Gamma} \tilde{\boldsymbol{\beta}} \cdot \boldsymbol{n}(\nabla v_1 \cdot \nabla v_1)$$

$$= -\int_Q g\tilde{\boldsymbol{\beta}} \cdot \nabla_{xt} v_1 + \epsilon \int_Q \nabla v_1 (\nabla \boldsymbol{\beta} - \frac{1}{2}\nabla \cdot \boldsymbol{\beta} \boldsymbol{I}) \nabla v_1$$

$$- \epsilon \int_{\Gamma_-} \tilde{\boldsymbol{\beta}} \cdot \nabla_{xt} v_1 \underbrace{\nabla v_1 \cdot \boldsymbol{n}_x}_{=0} - \epsilon \int_{\Gamma_+} \left(\underbrace{\frac{\partial v_1}{\partial t}}_{=0} + \boldsymbol{\beta} \cdot \nabla v_1 \right) \nabla v_1 \cdot \boldsymbol{n}_x$$

Note: $\nabla v_1 \cdot \boldsymbol{n}_x = \tau_{1n} = 0$ on Γ_-, $v_1 = 0$ on Γ_+

$$+ \frac{\epsilon}{2}\int_{\Gamma_-} \underbrace{\boldsymbol{\beta} \cdot \boldsymbol{n}_x}_{<0} (\nabla v_1 \cdot \nabla v_1) + \frac{\epsilon}{2}\int_{\Gamma_+} \boldsymbol{\beta} \cdot \boldsymbol{n}_x (\nabla v_1 \cdot \nabla v_1)$$

$$+ \frac{\epsilon}{2}\int_{\Gamma_0} \underbrace{n_t}_{<0} (\nabla v_1 \cdot \nabla v_1) + \frac{\epsilon}{2}\int_{\Gamma_T} n_t \underbrace{(\nabla v_1 \cdot \nabla v_1)}_{=0}$$

Note: $v_1 = 0$ on Γ_T

$$\leq -\int_Q g\tilde{\boldsymbol{\beta}} \cdot \nabla_{xt} v_1 + \epsilon \int_Q \nabla v_1 (\nabla \boldsymbol{\beta} - \frac{1}{2}\nabla \cdot \boldsymbol{\beta} \boldsymbol{I}) \nabla v_1$$

$$+ \epsilon \int_{\Gamma_+} \left(-\frac{\partial v_1}{\partial \boldsymbol{n}_x} \boldsymbol{\beta} + \frac{1}{2}\boldsymbol{\beta} \cdot \boldsymbol{n}_x \nabla v_1 \right) \cdot \nabla v_1$$

Note: Dropped negative terms from RHS

$$= -\int_Q g\tilde{\boldsymbol{\beta}} \cdot \nabla_{xt} v_1 + \epsilon \int_Q \nabla v_1 (\nabla \boldsymbol{\beta} - \frac{1}{2}\nabla \cdot \boldsymbol{\beta} \boldsymbol{I}) \nabla v_1$$

$$+ \epsilon \int_{\Gamma_+} \left(-\frac{\partial v_1}{\partial \boldsymbol{n}_x} \boldsymbol{\beta} + \frac{1}{2}\boldsymbol{\beta} \cdot \boldsymbol{n}_x \frac{\partial v_1}{\partial \boldsymbol{n}_x} \boldsymbol{n}_x \right) \cdot \frac{\partial v_1}{\partial \boldsymbol{n}_x} \boldsymbol{n}_x$$

Note: $\nabla v_1 \cdot \nabla v_1 = \nabla v_1 \cdot \nabla v_1 \boldsymbol{n}_x \cdot \boldsymbol{n}_x = (\nabla v_1 \cdot \boldsymbol{n}_x \boldsymbol{n}_x) \cdot (\nabla v_1 \cdot \boldsymbol{n}_x \boldsymbol{n}_x)$

$$= -\int_Q g\tilde{\boldsymbol{\beta}} \cdot \nabla_{xt} v_1 + \epsilon \int_Q \nabla v_1 (\nabla \boldsymbol{\beta} - \frac{1}{2} \nabla \cdot \boldsymbol{\beta} \boldsymbol{I}) \nabla v_1$$

$$\underbrace{-\frac{\epsilon}{2} \int_{\Gamma_+} \left(\frac{\partial v_1}{\partial \boldsymbol{n}_x} \right)^2 \boldsymbol{\beta} \cdot \boldsymbol{n}_x}_{<0}$$

$$\leq -\int_Q g\tilde{\boldsymbol{\beta}} \cdot \nabla_{xt} v_1 + \epsilon \int_Q \nabla v_1 (\nabla \boldsymbol{\beta} - \frac{1}{2} \nabla \cdot \boldsymbol{\beta} \boldsymbol{I}) \nabla v_1$$

$$\leq \frac{\|g\|^2}{2} + \frac{\left\| \tilde{\boldsymbol{\beta}} \cdot \nabla_{xt} v_1 \right\|^2}{2} + \epsilon \int_Q \nabla v_1 (\nabla \boldsymbol{\beta} - \frac{1}{2} \nabla \cdot \boldsymbol{\beta} \boldsymbol{I}) \nabla v_1$$

Note: Young's inequality

$$\leq \frac{\|g\|^2}{2} + \frac{\left\| \tilde{\boldsymbol{\beta}} \cdot \nabla_{xt} v_1 \right\|^2}{2} + \epsilon C_{\boldsymbol{\beta}} \|\nabla v_1\|^2$$

Note: Assumption on $\boldsymbol{\beta}$

$$\leq \left(\frac{1}{2} + C_{\boldsymbol{\beta}} \right) \|g\|^2 + \frac{\left\| \tilde{\boldsymbol{\beta}} \cdot \nabla_{xt} v_1 \right\|^2}{2}.$$

In conclusion, with either robust test norm, we can claim the following stability result,

$$\|u - u_h\| \lesssim \|(u, \boldsymbol{\sigma}, \hat{u}, \hat{t}) - (u_h, \boldsymbol{\sigma}_h, \hat{u}_h, \hat{t}_h)\|_E$$
$$= \inf_{(u_h, \boldsymbol{\sigma}_h, \hat{u}_h, \hat{t}_h)} \|(u, \boldsymbol{\sigma}, \hat{u}, \hat{t}) - (u_h, \boldsymbol{\sigma}_h, \hat{u}_h, \hat{t}_h)\|_E .$$

Notice that, contrary to the steady-state case, we have not been able to secure a robust L^2 bound for the stress. The best approximation error in the energy norm can be estimated locally, i.e. element-wise, see [7, 14]. This leads to an ultimate, final h estimate but not necessarily with robust constants. The loss of robustness in the best approximation error estimate is the consequence of rescaling the L^2-terms to avoid boundary layers in the optimal test functions. However, similarly to the steady-state case, with refinements, the mesh-dependent L^2-terms converge to the optimal ones so we hope to regain robustness in the limit. We do not attempt to analyze the best approximation error in this contribution and restrict ourselves to numerical experiments only.

4 Numerical Tests

The norms given in (7) and (8) are robust, but the reaction (zeroth order) terms still dominate the diffusion terms which produces boundary layers in optimal test functions and prohibits their resolution with a simple enrichment strategy. We can mitigate this by introducing mesh-dependent norms:

$$\|(v,\boldsymbol{\tau})\|_{V,K}^2 := \left\|\tilde{\boldsymbol{\beta}}\cdot\nabla_{xt}v\right\|_K^2 + \epsilon\,\|\nabla v\|_K^2 + \min\left(\frac{\epsilon}{h^2},1\right)\|v\|_K^2 + \|\nabla\cdot\boldsymbol{\tau}\|_K^2 + \min\left(\frac{1}{\epsilon},\frac{1}{h^2}\right)\|\boldsymbol{\tau}\|_K^2\,, \tag{10}$$

and

$$\|(v,\boldsymbol{\tau})\|_{V,K}^2 := \left\|\tilde{\boldsymbol{\beta}}\cdot\nabla_{xt}v\right\|_K^2 + \epsilon\,\|\nabla v\|_K^2 + \min\left(\frac{\epsilon}{h^2},1\right)\|v\|_K^2 + \left\|\nabla\cdot\boldsymbol{\tau}-\tilde{\boldsymbol{\beta}}\cdot\nabla_{xt}v\right\|_K^2 + \min\left(\frac{1}{\epsilon},\frac{1}{h^2}\right)\|\boldsymbol{\tau}\|_K^2\,. \tag{11}$$

Note that any version of (7) and (8) with smaller coefficients also satisfies the criteria for robustness. The mesh dependent coefficients were chosen in an attempt to balance the relative size of "reaction" terms like $\|v\|$ which scale like h^d with "diffusive" terms like $\epsilon\,\|\nabla v\|$ which scale like h^{d-2}. This is also the mechanism by which we avoid creating sharp boundary layers in our optimal test functions—by correctly balancing reactive and diffusive terms. In the following numerical experiments, we compute with these mesh dependent norms.

We verify robust convergence of our transient coupled robust norm on an analytical solution (shown in Fig. 4) that decays to a steady state Eriksson-Johnson

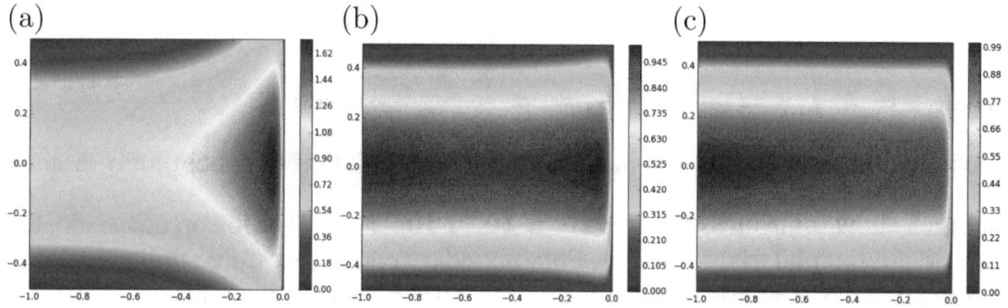

Fig. 4 Transient Eriksson-Johnson solution. (**a**) $t = 0.0$. (**b**) $t = 0.5$. (**c**) $t = 1.0$

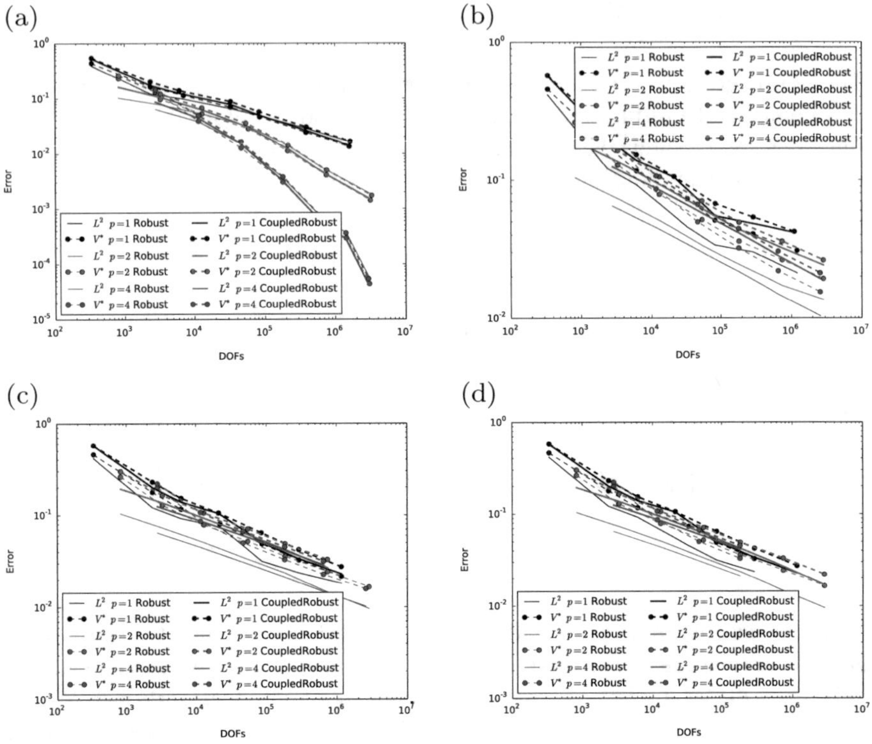

Fig. 5 Convergence to analytical solution. (**a**) $\epsilon = 10^{-2}$. (**b**) $\epsilon = 10^{-4}$. (**c**) $\epsilon = 10^{-6}$. (**d**) $\epsilon = 10^{-8}$

problem:

$$u = \exp(-lt)\left[\exp(\lambda_1 x) - \exp(\lambda_2 x)\right] + \cos(\pi y)\frac{\exp(s_1 x) - \exp(r_1 x)}{\exp(-s_1) - \exp(-r_1)},$$

where $l = 4$, $\lambda_{1,2} = \frac{-1 \pm \sqrt{1-4\epsilon l}}{-2\epsilon}$, $r_1 = \frac{1+\sqrt{1+4\pi^2\epsilon^2}}{2\epsilon}$, and $s_1 = \frac{1-\sqrt{1+4\pi^2\epsilon^2}}{2\epsilon}$. The problem domain is $[-1, 0] \times [-0.5, 0.5]$ and $\boldsymbol{\beta} = \begin{pmatrix} 1 \\ 0 \end{pmatrix}$. Convergence plots presented in Fig. 5 show robustness for $\epsilon = 10^{-2}, 10^{-4}, 10^{-6}, 10^{-8}$ for linear, quadratic, and quartic polynomial trial functions. Flux boundary conditions were applied based on the exact solution at $x = -1$ and $t = 0$ while trace boundary conditions were set at $y = -0.5$, $y = 0.5$, and $x = 0$. An adaptive solve was undertaken using a greedy refinement strategy in which any elements with at least 20 % of the energy error of highest energy error element were refined at each step. See [15] for details on adaptivity within the DPG context.

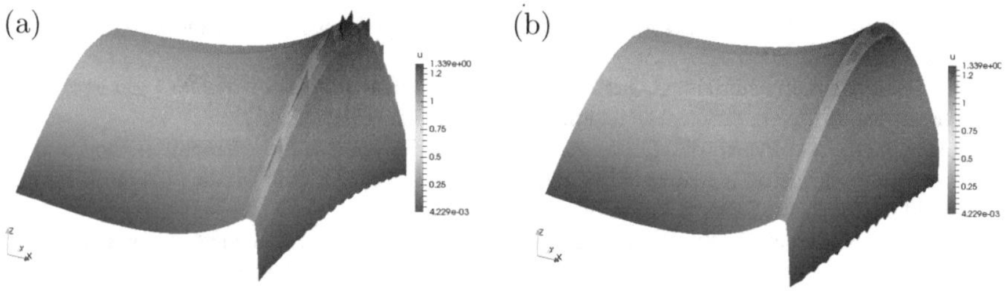

Fig. 6 u at $t = 0.2$ for $\epsilon = 10^{-2}$ and $p = 2$ after four adaptive refinements. (**a**) Robust norm. (**b**) Coupled robust norm

In the plot legends, L^2 indicates $\left(\|u - u_{\text{exact}}\|_L^2 + \|\boldsymbol{\sigma} - \boldsymbol{\sigma}_{\text{exact}}\|_{L^2}\right)^{\frac{1}{2}}$ while V^* indicates the energy error reported by the method. Despite a lack of guaranteed control $\boldsymbol{\sigma}$ by norms (10) and (11), $\|\boldsymbol{\sigma} - \boldsymbol{\sigma}_{\text{exact}}\|_{L^2}$ is included in the L^2 error computation and does appear to be under control in the problems considered here. When plotted in isolation, the L^2 error in $\boldsymbol{\sigma}$ was usually orders of magnitude smaller than $\|u - u_{\text{exact}}\|_{L^2}$.

We provide surface plots of temporal slices of the solution at $t = 0.2$ for the two norms with $\epsilon = 10^{-2}$, and $p = 2$ after four adaptive refinements. The results conform to our previous experience with steady convection-diffusion where the coupled robust norm tends to produce smoother results in regions with sharp gradients (Fig. 6).

5 Conclusions

As expected, convergence of the energy error appears to be a reliable predictor of convergence of the L^2 error. This relation is especially tight for moderate values of ϵ. We've developed two robust test norms for transient convection-diffusion, though neither one guarantees robust control over $\boldsymbol{\sigma}$ as we had with their steady analogs.

References

1. S.K. Aliabadi, T.E. Tezduyar, Space-time finite element computation of compressible flows involving moving boundaries and interfaces. Comput. Methods Appl. Mech. Eng. **107**(1–2), 209–223 (1993)
2. J.H. Argyris, D.W. Scharpf, Finite elements in time and space. Nucl. Eng. Des. **10**(4), 456–464 (1969)
3. C. Carstensen, L.F. Demkowicz, J. Gopalakrishnan, Breaking spaces and forms for the DPG method and applications including Maxwell equations. Technical Report 15-18, ICES (2015)

4. J.L. Chan, A DPG method for convection-diffusion problems. Ph.D. thesis, University of Texas at Austin, 2013
5. J.L. Chan, J. Gopalakrishnan, L.F. Demkowicz, Global properties of DPG test spaces for convection-diffusion problems. Technical Report 13-05, ICES (2013)
6. J.L. Chan, J.A. Evans, W. Qiu, A dual Petrov-Galerkin finite element method for the convection–diffusion equation. Comput. Math. Appl. **68**(11), 1513–1529 (2014). Minimum Residual and Least Squares Finite Element Methods
7. J. Chan, N. Heuer, T. Bui-Thanh, L. Demkowicz, A robust DPG method for convection-dominated diffusion problems II: adjoint boundary conditions and mesh-dependent test norms. Comput. Math. Appl. **67**(4), 771–795 (2014). High-order Finite Element Approximation for Partial Differential Equations
8. A. Cohen, W. Dahmen, G. Welper, Adaptivity and variational stabilization for convection-diffusion equations. ESAIM Math. Model. Numer. Anal. **46**(5), 1247–1273 (2012)
9. W. Dahmen, C. Huang, C. Schwab, G. Welper, Adaptive Petrov-Galerkin methods for first order transport equations. SIAM J. Numer. Anal. **50**(5), 2420–2445 (2012)
10. L.F. Demkowicz, Various variational formulations and closed range theorem. Technical Report 15-03, ICES (2015)
11. L.F. Demkowicz, J. Gopalakrishnan, A class of discontinuous Petrov-Galerkin methods. Part II: optimal test functions. Numer. Methods Partial Differ. Equ. **27**, 70–105 (2011)
12. L.F. Demkowicz, J. Gopalakrishnan, An overview of the DPG method, in *Recent Developments in Discontinuous Galerkin Finite Element Methods for Partial Differential Equations* ed. by X. Feng, O. Karakashian, Y. Xing. IMA Volumes in Mathematics and Its Applications, vol. 157 (Springer, Cham, 2014), pp. 149–180
13. L.F. Demkowicz, J. Gopalakrishnan, Discontinuous Petrov-Galerkin (DPG) method. Technical Report 15-20, ICES (2015)
14. L.F. Demkowicz, N. Heuer, Robust DPG method for convection-dominated diffusion problems. SIAM J. Numer. Anal. **51**(5), 1514–2537 (2013)
15. L.F. Demkowicz, J. Gopalakrishnan, A.H. Niemi, A class of discontinuous Petrov-Galerkin methods. Part III: adaptivity. Appl. Numer. Math. **62**(4), 396–427 (2012)
16. T.E. Ellis, L.F. Demkowicz, J.L. Chan, Locally conservative discontinuous Petrov-Galerkin finite elements for fluid problems. Comput. Math. Appl. **68**(11), 1530–1549 (2014)
17. I. Fried, Finite-element analysis of time-dependent phenomena. AIAA J. **7**(6), 1170–1173 (1969)
18. J. Gopalakrishnan, W. Qiu, An analysis of the practical DPG method. Math. Comput. **83**(286), 537–552 (2014)
19. T.J.R. Hughes, J.R. Stewart, A space-time formulation for multiscale phenomena. J. Comput. Appl. Math. **74**(1–2), 217–229 (1996)
20. Z. Kaczkowski, The method of finite space-time elements in dynamics of structures. J. Tech. Phys. **16**(1), 69–84 (1975)
21. C.M. Klaij, J.J.W. van der Vegt, H. van der Ven, Space-time discontinuous Galerkin method for the compressible Navier-Stokes equations. J. Comput. Phys. **217**(2), 589–611 (2006)
22. D. Moro, N.C. Nguyen, J. Peraire, A hybridized discontinuous Petrov-Galerkin scheme for compressible flows. Master's thesis, Massachusetts Institute of Technology, 2011
23. A.H. Niemi, N.O. Collier, V.M. Calo, Automatically stable discontinuous Petrov-Galerkin methods for stationary transport problems: quasi-optimal test space norm. Comput. Math. Appl. **66**(10), 2096–2113 (2013)
24. A.H. Niemi, N.O. Collier, V.M. Calo, Discontinuous Petrov-Galerkin method based on the optimal test space norm for steady transport problems in one space dimension. J. Comput. Sci. **4**(3), 157–163 (2013)
25. J.T. Oden, A general theory of finite elements. II. Applications. Int. J. Numer. Methods Eng. **1**(3), 247–259 (1969)
26. S. Rhebergen, B. Cockburn, A space-time hybridizable discontinuous Galerkin method for incompressible flows on deforming domains. J. Comput. Phys. **231**(11), 4185–4204 (2012)

27. S. Rhebergen, B. Cockburn, J.J.W. Van Der Vegt, A space-time discontinuous Galerkin method for the incompressible Navier-Stokes equations. J. Comput. Phys. **233**, 339–358 (2013)
28. H.G. Roos, M. Stynes, L. Tobiska, *Robust Numerical Methods for Singularly Perturbed Differential Equations*. Springer Series in Computational Mathematics, vol. 24, 2nd edn. (Springer, Berlin, 2008)
29. T.E. Tezduyar, M. Behr, J. Liou, A new strategy for finite element computations involving moving boundaries and interfaces—the deforming-spatial-domain/space-time procedure: I. The concept and the preliminary numerical tests. Comput. Methods Appl. Mech. Eng. **94**(3), 339–351 (1992)
30. J.J.W. van der Vegt, H. van der Ven, Space-time discontinuous Galerkin finite element method with dynamic grid motion for inviscid compressible flows: I. General formulation. J. Comput. Phys. **182**(2), 546–585 (2002)

A Review of Hybrid High-Order Methods: Formulations, Computational Aspects, Comparison with Other Methods

Daniele A. Di Pietro, Alexandre Ern, and Simon Lemaire

Abstract Hybrid High-Order (HHO) methods are formulated in terms of discrete unknowns attached to mesh faces and cells (hence, the term hybrid), and these unknowns are polynomials of arbitrary order $k \geq 0$ (hence, the term high-order). HHO methods are devised from local reconstruction operators and a local stabilization term. The discrete problem is assembled cellwise, and cell-based unknowns can be eliminated locally by static condensation. HHO methods support general meshes, are locally conservative, and allow for a robust treatment of physical parameters in various situations, e.g., heterogeneous/anisotropic diffusion, quasi-incompressible linear elasticity, and advection-dominated transport. This paper reviews HHO methods for a variable-diffusion model problem with nonhomogeneous, mixed Dirichlet–Neumann boundary conditions, including both primal and mixed formulations. Links with other discretization methods from the literature are discussed.

1 Introduction

Over the last few years, a significant effort has been devoted to devising and analyzing discretization methods for elliptic PDEs on general meshes including nonmatching interfaces and polytopal (polygonal/polyhedral) cells. Such meshes are encountered, e.g., in the context of subsurface flow simulations in saline aquifers and petroleum basins, where polytopal elements and nonmatching interfaces appear to account for eroded layers and fractures. In petroleum reservoir modeling,

D.A. Di Pietro
Institut Montpelliérain Alexander Grothendieck, University of Montpellier, 34095 Montpellier, France
e-mail: daniele.di-pietro@umontpellier.fr

A. Ern (✉) • S. Lemaire
University Paris-Est, CERMICS (ENPC), 6–8 Avenue Blaise Pascal, 77455 Marne-la-Vallée Cedex 2, France
e-mail: alexandre.ern@enpc.fr; simon.lemaire87@gmail.com

G.R. Barrenechea et al. (eds.), *Building Bridges: Connections and Challenges in Modern Approaches to Numerical Partial Differential Equations*, Lecture Notes in Computational Science and Engineering 114,
DOI 10.1007/978-3-319-41640-3_7

polytopal elements can also appear in the near-wellbore regions, where radial meshes are usually employed to account for the (qualitative) features of the solution. A more recent and original application of meshes composed of polytopal elements is adaptive mesh coarsening [2, 7], where a coarse mesh is obtained by element agglomeration from a fine mesh accounting for the geometric details of the domain.

Polytopal discretization methods were first investigated in the framework of lowest-order schemes. In the context of Finite Volume methods, several families of polytopal methods have resulted from the effort to circumvent the superadmissible mesh condition required for the consistency of the classical two-point scheme; cf., in particular, [38, Definition 9.1]. Interestingly, most of these methods possess local conservation properties on the primal mesh and exhibit numerical fluxes without resorting to local reconstructions. We can mention here, e.g., the Mixed and Hybrid Finite Volume (MHFV) schemes of [34, 39] and the Discrete Duality Finite Volume (DDFV) method of [33].

Other families of lowest-order polytopal discretization methods have been obtained by reproducing at the discrete level salient features of the continuous problem. Mimetic Finite Difference (MFD) methods were originally derived by mimicking the Stokes theorem in a discrete setting to formulate discrete counterparts of the usual first-order differential operators combined with constitutive relations and of L^2-products; cf. [16, 17] and also [9] for an overview. Another viewpoint starts from the seminal ideas of Tonti [44] and Bossavit [13] hinging on differential geometry and algebraic topology. Related schemes include the so-called Discrete Geometric Approach (DGA) [22], and more generally, the Compatible Discrete Operator (CDO) framework of [10, 11], cf. also [12], where the building blocks are metric-free discrete differential operators combined with a discrete Hodge operator approximating constitutive relations. Another approach consists in reproducing classical properties of nonconforming and penalized methods on general meshes, as in the Cell-Centered Galerkin (CCG) method [23] and the generalized Crouzeix–Raviart method [31]. The idea is to formulate the method in terms of (possibly incomplete) polynomial spaces so as to re-deploy classical (nonconforming) Finite Element analysis tools.

Recent works have led to unifying frameworks that capture the links among (some of) the above methods. The close relation between MHFV and MFD methods has been investigated in [35], where equivalence at the algebraic level is demonstrated. A unifying viewpoint that encompasses the above and other classical methods has been proposed under the name of Gradient Schemes [36]. Another unifying viewpoint (closely related to Gradient Schemes) is provided by the CDO framework which encompasses vertex-based schemes (such as first-order Lagrange finite elements and nodal MFD) and cell-based schemes (such as MHFV and MFD).

In parallel, high-order polytopal discretization methods have received significant attention over the last few years. Increasing the approximation order can significantly speed up convergence when the solution exhibits sufficient (local) regularity. When this is not the case, the better convergence properties of high-order methods can be recovered using mesh adaption (by local refinement or coarsening). High-order polytopal discretization methods can be obtained by fully nonconforming approaches such as the Discontinuous Galerkin (DG) method; cf. [4] and also [5, 15] for a unified presentation for the Poisson problem, [37] for Friedrichs' systems, [18] for an *hp*-version, and [25] for a comprehensive introduction. An interesting class of DG methods is that of Hybridizable Discontinuous Galerkin (HDG) methods [21] (cf. also [19]). Such methods were originally devised as discrete versions of a characterization of the exact solution in terms of solutions of local problems globally matched through transmission conditions. A similar approach can be found in [45, 46].

Very recent works have developed other viewpoints to achieve high-order polytopal discretizations. A salient example is the Virtual Element (VE) method introduced in [8, 14]. The H^1-conforming VE method takes the steps from the nodal MFD method recast in a Finite Element framework, and can be viewed as a generalization of conforming (Lagrange, Hermite) Finite Element methods. The main idea is to define a local space of basis functions for which only the values of degrees of freedom are known (i.e., no analytical expression is available). Starting from these degrees of freedom, one devises a computable projection onto a polynomial space so as to formulate the local contributions to the discrete problem.

Our focus is here on the Hybrid High-Order (HHO) method introduced in [28, 32]. The term hybrid refers to the fact that the method is originally formulated using discrete unknowns attached to mesh faces and cells. These discrete unknowns are polynomial functions, and the cell-based ones can be eliminated locally by static condensation. The term high-order refers to the fact that the order of the polynomial functions can be an arbitrary integer $k \geq 0$. The main idea of HHO methods consists in locally reconstructing high-order differential operators acting on the face- and cell-based unknowns. The guideline underpinning such reconstructions is an integration by parts formula. These reconstructions are then used to formulate the elementwise contributions to the discrete problem including a high-order stabilization term exhibiting a rich structure coupling locally the face- and cell-based unknowns. Local contributions are conceived so that the only globally coupled unknowns after static condensation are discontinuous polynomials on the mesh skeleton. This is a distinctive feature with respect to the VE method, where H^1-conforming reconstructions are present in the background. A study of the relations between HHO and HDG methods can be found in [20], which also fits into the HHO framework (up to equivalent stabilization) the recent high-order MFD method of [6, 43] (also referred to as nonconforming VE method in subsequent publications). We also mention that HHO methods for polynomial order $k = 0$ are closely related to MHFV (and so to lowest-order MFD); cf., in particular, [32, Sect. 2.5] and [30, Sect. 5.4].

HHO methods offer several assets. Besides supporting general meshes, their construction is dimension-independent, and they are locally conservative [27]. Moreover, they allow for a natural treatment of physical parameters [29], and lead to discretizations that are robust over the entire range of variation of physical parameters in various situations, e.g., heterogeneous/anisotropic diffusion [29], quasi-incompressible linear elasticity [28] and advection-dominated transport [30]. When compared to interior penalty DG methods, HHO methods are also appealing in terms of computational cost. To achieve an order of convergence of $(k+1)$ in the energy norm for a pure diffusion problem in three space dimensions, the globally coupled degrees of freedom for DG grow as $\frac{1}{6}(k+2)(k+3)(k+4)N_{\mathrm{E}}$ with N_{E} the number of mesh elements, whereas for HHO they only grow as $\frac{1}{2}(k+1)(k+2)N_{\mathrm{F}}$ with N_{F} the number of mesh faces (only leading-order terms are considered in the above computations).

The goal of this paper is to provide an up-to-date review of HHO methods, with a particular focus on the various possible formulations and computational aspects. For the sake of simplicity, we focus on a model elliptic problem with possibly heterogeneous/anisotropic diffusion tensor. Most of the results contained herein can be derived from relatively straightforward adaptations of the proofs contained in previous works [1, 20, 26–30, 32]; for the sake of conciseness, we provide bibliographic references for the most technical proofs, while some details are included for those proofs that allow us to highlight the more practical aspects of the method. One novel aspect is that we treat nonhomogeneous mixed Dirichlet–Neumann boundary conditions, while previous work has focused on homogeneous, pure Dirichlet boundary conditions. Another novelty is that we detail the main implementation aspects under the viewpoint of an offline/online decomposition.

The material is organized as follows. Section 2 describes the continuous and discrete settings, including the model problem, the notion of admissible mesh sequence, and the assumptions on the data. Section 3 is devoted to the presentation and analysis of the HHO method in primal form, while Sect. 4 is concerned with the mixed form of the HHO method. Finally, the links between both forms are studied in Sect. 5, while Sect. 6 contains some concluding remarks and perspectives.

2 Continuous and Discrete Settings

This section presents the model problem, the key definitions and notation concerning the mesh, and the assumptions on the data of the model problem.

2.1 Model Problem

Let $\Omega \subset \mathbb{R}^d$, $d \geq 2$, be an open, connected, bounded polytopal domain, with boundary Γ and unit outward normal $\boldsymbol{n}$. We assume that there exists a partition of Γ such that $\Gamma := \Gamma_{\mathrm{d}} \cup \Gamma_{\mathrm{n}}$, with $\Gamma_{\mathrm{d}} \cap \Gamma_{\mathrm{n}} = \varnothing$, and such that the measure of Γ_{d} is nonzero. For any connected subset $X \subset \overline{\Omega}$ with nonzero Lebesgue measure, the inner product and norm of the Lebesgue space $L^2(X)$ are denoted by $(\cdot,\cdot)_X$ and $\|\cdot\|_X$, respectively, with the convention that the index is omitted if $X = \Omega$.

We consider a variable-diffusion model problem with tensor-valued diffusivity $\mathbb{M}$. Throughout the paper, $\mathbb{M}$ is assumed to be symmetric, piecewise Lipschitz on a polytopal partition P_Ω of Ω, and uniformly elliptic, in the sense that, for a.e. $\boldsymbol{x} \in \Omega$,

$$0 < \mu_\flat \le \mathbb{M}(\boldsymbol{x})\boldsymbol{\xi}\cdot\boldsymbol{\xi} \le \mu_\sharp < +\infty, \qquad \forall \boldsymbol{\xi} \in \mathbb{R}^d \text{ such that } |\boldsymbol{\xi}| = 1.$$

The model problem reads: Find $u : \Omega \to \mathbb{R}$ such that

$$\begin{aligned} -\mathrm{div}(\mathbb{M}\nabla u) &= f && \text{in } \Omega, \\ u &= \psi_\partial && \text{on } \Gamma_\mathrm{d}, \\ \mathbb{M}\nabla u\cdot\boldsymbol{n} &= \phi_\partial && \text{on } \Gamma_\mathrm{n}, \end{aligned} \tag{1}$$

where $f \in L^2(\Omega)$, $\psi_\partial = (u_\partial)_{|\Gamma_\mathrm{d}}$ with $u_\partial \in H^1(\Omega)$, and $\phi_\partial \in L^2(\Gamma_\mathrm{n})$ (whenever the measure of Γ_n is nonzero). Henceforth, u is termed the potential. Owing to the nonzero assumption on the measure of Γ_d, we do not consider pure Neumann boundary conditions; the results presented in what follows can be adapted to this case, up to minor modifications. The pure Dirichlet case, corresponding to a $(d-1)$-dimensional zero-measure set Γ_n, is included in the present setting.

2.2 *Admissible Mesh Sequences*

Denoting by $\mathcal{H} \subset \mathbb{R}_*^+$ a countable set of meshsizes having 0 as its unique accumulation point, we consider mesh sequences $(\mathcal{T}_h)_{h\in\mathcal{H}}$ where, for all $h \in \mathcal{H}$, $\mathcal{T}_h = \{T\}$ is a finite collection of nonempty disjoint open polytopes (polygons/polyhedra) T, called *elements* or *cells*, such that $\overline{\Omega} = \bigcup_{T\in\mathcal{T}_h} \overline{T}$ and $h = \max_{T\in\mathcal{T}_h} h_T$ (where h_T stands for the diameter of the element T). Recall that polytopes in $\mathbb{R}^d$ have flat sides.

A hyperplanar closed connected subset F of $\overline{\Omega}$ is called a *face* (for $d > 3$, these geometric objects are also called facets) if it has positive $(d-1)$-dimensional Lebesgue measure and if either (1) there exist $T_1, T_2 \in \mathcal{T}_h$ such that $F = \partial T_1 \cap \partial T_2$ or $F \subset \partial T_1 \cap \partial T_2$ and F is a side of both T_1 and T_2 (and F is termed *interface*), or (2) there exists $T \in \mathcal{T}_h$ such that $F = \partial T \cap \partial\Omega$ or $F \subset \partial T \cap \partial\Omega$ and F is a side of T (and F is termed *boundary face*). Interfaces are collected in the set $\mathcal{F}_h^\mathrm{i}$, boundary faces in $\mathcal{F}_h^\mathrm{b}$, and we let $\mathcal{F}_h := \mathcal{F}_h^\mathrm{i} \cup \mathcal{F}_h^\mathrm{b}$. The diameter of a face $F \in \mathcal{F}_h$ is denoted h_F. For all $T \in \mathcal{T}_h$, $\mathcal{F}_T := \{F \in \mathcal{F}_h \mid F \subset \partial T\}$ denotes the set of faces lying on the boundary of T and, symmetrically, for all $F \in \mathcal{F}_h$, $\mathcal{T}_F := \{T \in \mathcal{T}_h \mid F \subset \partial T\}$ denotes the set gathering the one (if F is a boundary face) or two (if F is an interface) element(s) sharing F. For all $F \in \mathcal{F}_T$, we let $\boldsymbol{n}_{T,F}$ be the unit normal vector to F pointing out of T. Finally, for every interface $F \in \mathcal{F}_h^\mathrm{i}$, an orientation is fixed once and for all by means of a unit normal vector $\boldsymbol{n}_F$.

We adopt the following notion of admissible mesh sequence, cf. [25, Sect. 1.4].

Definition 2.1 (Admissible Mesh Sequence) The mesh sequence $(\mathcal{T}_h)_{h\in\mathcal{H}}$ is *admissible* if, for all $h \in \mathcal{H}$, $\mathcal{T}_h$ admits a matching simplicial submesh $\mathfrak{T}_h$ such that there exists a real number $\gamma > 0$, called *mesh regularity parameter*, independent of h and such that, for all $h \in \mathcal{H}$,

(i) for any simplex $S \in \mathfrak{T}_h$ of diameter h_S and inradius r_S, $\gamma h_S \leq r_S$;
(ii) for all $T \in \mathcal{T}_h$, and all $S \in \mathfrak{T}_T := \{S \in \mathfrak{T}_h \mid S \subseteq T\}$, $\gamma h_T \leq h_S$.

Consequences of Definition 2.1 are that (1) the quantity $\max_{T\in\mathcal{T}_h} \mathrm{card}(\mathcal{F}_T)$ is uniformly bounded with respect to the meshsize, and that (2) mesh faces have a comparable diameter to that of the cells they belong to; cf. [25, Lemmas 1.41 and 1.42]. We add the following notion of compatibility, in order to deal with the partitions associated with the diffusion tensor and with the boundary conditions.

Definition 2.2 (Compatible Mesh Sequence) The mesh sequence $(\mathcal{T}_h)_{h\in\mathcal{H}}$ is *compatible* if, for all $h \in \mathcal{H}$,

(i) $\mathcal{T}_h$ fits the (polytopal) partition P_Ω associated with the diffusion tensor $\mathbb{M}$, meaning that, for all $T \in \mathcal{T}_h$, there is a unique Ω_i in P_Ω containing T;
(ii) $\mathcal{T}_h$ fits the partition $\Gamma = \Gamma_{\mathrm{d}} \cup \Gamma_{\mathrm{n}}$ of the boundary, in the sense that we can define two sets, $\mathcal{F}_h^{\mathrm{d}} := \{F \in \mathcal{F}_h^{\mathrm{b}} \mid F \subseteq \Gamma_{\mathrm{d}}\}$ and $\mathcal{F}_h^{\mathrm{n}} := \{F \in \mathcal{F}_h^{\mathrm{b}} \mid F \subseteq \Gamma_{\mathrm{n}}\}$, such that $\mathcal{F}_h^{\mathrm{d}} \cup \mathcal{F}_h^{\mathrm{n}} = \mathcal{F}_h^{\mathrm{b}}$.

2.3 Broken Polynomial Spaces

For integers $k \geq 0$, $1 \leq l \leq d$, we denote by $\mathbb{P}_l^k$ the vector space spanned by l-variate polynomial functions of total degree $\leq k$ of dimension

$$N_{k,l} := \binom{k+l}{k}. \tag{2}$$

For all $T \in \mathcal{T}_h$, $\mathbb{P}_d^k(T)$ denotes the restriction to T of functions in $\mathbb{P}_d^k$. We also introduce the broken polynomial space

$$\mathbb{P}_d^k(\mathcal{T}_h) := \{v \in L^2(\Omega) \mid v_{|T} \in \mathbb{P}_d^k(T) \text{ for all } T \in \mathcal{T}_h\}.$$

Broken polynomial spaces are special instances of broken Sobolev spaces, for an integer $m \geq 1$:

$$H^m(\mathcal{T}_h) := \{v \in L^2(\Omega) \mid v_{|T} \in H^m(T) \text{ for all } T \in \mathcal{T}_h\}.$$

We use the notation ∇_h to denote the broken gradient operator acting elementwise on functions from broken Sobolev spaces.

We denote by π_h^k the L^2-orthogonal projector onto $\mathbb{P}_d^k(\mathcal{T}_h)$ such that, for all $v \in L^2(\Omega)$ and all $T \in \mathcal{T}_h$, $(\pi_h^k v)_{|T} := \pi_T^k v_{|T}$, where π_T^k is the L^2-orthogonal projector onto $\mathbb{P}_d^k(T)$. Additionally, for all $F \in \mathcal{F}_h$ and all $v \in L^2(F)$, we denote by $\pi_F^k v$ the

L^2-orthogonal projection of v onto $\mathbb{P}^k_{d-1}(F)$, where $\mathbb{P}^k_{d-1}(F)$ is the restriction to F of $\mathbb{P}^k_{d-1} \circ \Xi^{-1}$, with Ξ an affine bijective mapping from $\mathbb{R}^{d-1}$ to the affine hyperplane supporting F.

2.4 *Diffusion Tensor*

We assume, for the sake of simplicity, that $\mathbb{M}$ is piecewise constant on P_Ω, and thus, by Definition 2.2, on $\mathcal{T}_h$ for every $h \in \mathcal{H}$. For $T \in \mathcal{T}_h$, we let $\mathbb{M}_T := \mathbb{M}_{|T}$ (owing to the above assumption, $\mathbb{M}_T$ is a constant matrix), and we denote by $\mu_{\flat,T}$ and $\mu_{\sharp,T}$, respectively, the lowest and largest eigenvalues of $\mathbb{M}_T$. We also introduce the local anisotropy ratio $\rho_T := \mu_{\sharp,T}/\mu_{\flat,T} \geq 1$; the global ratio is defined as $\rho := \max_{T\in\mathcal{T}_h} \rho_T$. Finally, for all $T \in \mathcal{T}_h$ and $F \in \mathcal{F}_T$, we set $\mu_{T,F} := \mathbb{M}_T \boldsymbol{n}_F \cdot \boldsymbol{n}_F > 0$.

In what follows, we often abbreviate as $a \lesssim b$ the inequality $a \leq Cb$, with $C > 0$ independent of the meshsize h and of the diffusion tensor $\mathbb{M}$, but possibly depending on the mesh regularity parameter γ and on the polynomial degree k.

3 The HHO Method in Primal Form

Let $U := H^1(\Omega)$ and $U_0 := \{v \in U \mid v_{|\Gamma_{\rm d}} = 0\}$. The starting point of the HHO method in primal form is the following weak formulation of problem (1): Find $u_0 \in U_0$ such that

$$(\mathbb{M}\nabla u_0, \nabla v) = (f, v) - (\mathbb{M}\nabla u_\partial, \nabla v) + (\phi_\partial, v)_{\Gamma_{\rm n}} \qquad \forall v \in U_0. \tag{3}$$

The solution $u \in U$ is then computed as $u = u_0 + u_\partial$ with u_∂ defined in Sect. 2.1.

3.1 *Discrete Setting*

Let an integer $k \geq 0$ be fixed, and let us consider an admissible and compatible mesh sequence $(\mathcal{T}_h)_{h\in\mathcal{H}}$ in the sense of Definitions 2.1 and 2.2. We further suppose that the assumptions of Sect. 2.4 concerning the diffusion tensor hold.

3.1.1 Discrete Unknowns

We adopt the convention that underlined quantities in roman font (sets, elements from these sets) are hybrid quantities, i.e., quantities featuring both a cell-based and a face-based contribution. We introduce, first locally, then globally, the discrete unknowns associated with the potential.

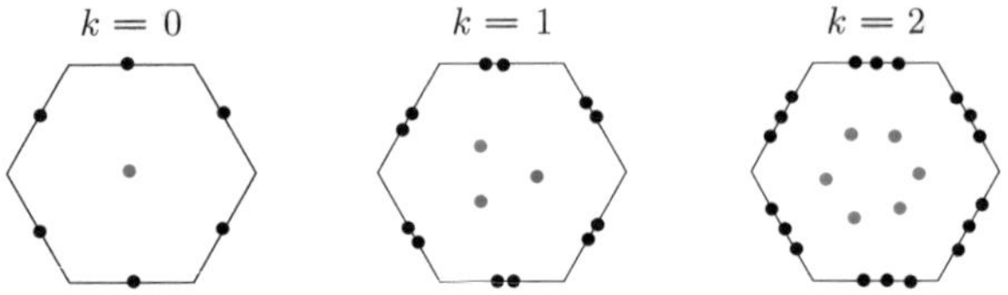

Fig. 1 Degrees of freedom associated with hybrid (cell- and face-based) potential discrete unknowns, $d = 2$, $k \in \{0, 1, 2\}$

Local Definition For $T \in \mathcal{T}_h$, letting

$$U_T^k := \mathbb{P}_d^k(T), \qquad \mathfrak{U}_F^k := \mathbb{P}_{d-1}^k(F) \text{ for all } F \in \mathcal{F}_T, \tag{4}$$

we define the local set of hybrid potential unknowns, cf. Fig. 1, as

$$\underline{U}_T^k := U_T^k \times \mathfrak{U}_{\partial T}^k, \qquad \mathfrak{U}_{\partial T}^k := \bigtimes_{F \in \mathcal{F}_T} \mathfrak{U}_F^k.$$

In the sequel, any element $\underline{v}_T \in \underline{U}_T^k$ is decomposed as $\underline{v}_T := (v_T \in U_T^k, \mathfrak{v}_{\partial T} \in \mathfrak{U}_{\partial T}^k)$, with $\mathfrak{v}_{\partial T} := (\mathfrak{v}_F \in \mathfrak{U}_F^k)_{F \in \mathcal{F}_T}$. We also introduce the local reduction operator $\underline{I}_T^k : H^1(T) \to \underline{U}_T^k$ such that, for all $v \in H^1(T)$, $\underline{I}_T^k v := \left(\pi_T^k v, (\pi_F^k v)_{F \in \mathcal{F}_T}\right)$.

Remark 3.1 (Variant on Cell-Based Unknowns) A variant in the definition of cell-based unknowns is studied in [20], where these unknowns belong to the polynomial space $\mathbb{P}_d^l(T)$ with $l \in \{k-1, k, k+1\}$ (up to some minor adaptations if $k = 0$ and $l = -1$). The choice $l = k - 1$ allows one to establish a link (up to equivalent stabilizations) with the high-order MFD method of [6, 43] (in the case $k = 0$, $l = -1$, one can recover the classical Crouzeix–Raviart element on simplices), while the choice $l = k + 1$ is related to a variant of the HDG method introduced in [42].

Global Definition We define the global set of hybrid potential unknowns as

$$\underline{U}_h^k := U_h^k \times \mathfrak{U}_h^k, \tag{5}$$

with

$$U_h^k := \bigtimes_{T \in \mathcal{T}_h} U_T^k, \qquad \mathfrak{U}_h^k := \bigtimes_{F \in \mathcal{F}_h} \mathfrak{U}_F^k.$$

Observe that $U_h^k = \mathbb{P}_d^k(\mathcal{T}_h)$ and that potential unknowns attached to interfaces are single-valued. Given an element $\underline{v}_h \in \underline{U}_h^k$, we denote v_h and $\mathfrak{v}_h$ its restrictions to U_h^k and $\mathfrak{U}_h^k$, respectively, while, for any $T \in \mathcal{T}_h$, we denote by $\underline{v}_T = (v_T, \mathfrak{v}_{\partial T}) \in \underline{U}_T^k$ its restriction to the element T. To account for (homogeneous) Dirichlet boundary conditions in a strong manner, we introduce the following subspace of $\underline{U}_h^k$:

$$\underline{U}_{h,0}^k := U_h^k \times \mathfrak{U}_{h,0}^k, \qquad \text{with} \qquad \mathfrak{U}_{h,0}^k := \left\{\mathfrak{v}_h \in \mathfrak{U}_h^k \mid \mathfrak{v}_F \equiv 0, \forall F \in \mathcal{F}_h^{\mathrm{d}}\right\}.$$

Finally, we introduce the global reduction operator $\underline{I}_h^k : U \to \underline{U}_h^k$ such that, for all $v \in U$, and for all $T \in \mathcal{T}_h$, $(\underline{I}_h^k v)_{|T} := \underline{I}_T^k v_{|T}$. Single-valuedness at interfaces is ensured by the regularity of functions in U.

3.1.2 Potential Reconstruction Operator

Let $T \in \mathcal{T}_h$. The local potential reconstruction operator $p_T^{k+1} : \underline{U}_T^k \to \mathbb{P}_d^{k+1}(T)$ is defined, for all $\underline{v}_T = (v_T, \mathfrak{v}_{\partial T}) \in \underline{U}_T^k$, as the solution of the well-posed Neumann problem (the usual compatibility condition on the right-hand side is verified)

$$(\mathbb{M}_T \nabla p_T^{k+1} \underline{v}_T, \nabla w)_T = -(v_T, \operatorname{div}(\mathbb{M}_T \nabla w))_T + \sum_{F \in \mathcal{F}_T} (\mathfrak{v}_F, \mathbb{M}_T \nabla w \cdot \boldsymbol{n}_{T,F})_F \quad \forall w \in \mathbb{P}_d^{k+1}(T), \tag{6}$$

which further satisfies $\int_T p_T^{k+1} \underline{v}_T = \int_T v_T$. Computing the operator p_T^{k+1} requires to invert a symmetric positive-definite matrix of size $N_{k+1,d}$, cf. (2), which can be performed effectively via a Cholesky factorization (the cost of such a factorization is roughly $N_{k+1,d}^3/6$ flops). The following result shows that $p_T^{k+1}\underline{I}_T^k$ is the $\mathbb{M}_T$-weighted elliptic projector onto $\mathbb{P}_d^{k+1}(T)$.

Lemma 3.1 (Characterization of $p_T^{k+1}\underline{I}_T^k$ and Polynomial Consistency) *The following holds for all $v \in H^1(T)$:*

$$(\mathbb{M}_T \nabla (v - p_T^{k+1}\underline{I}_T^k v), \nabla w)_T = 0 \qquad \forall w \in \mathbb{P}_d^{k+1}(T). \tag{7}$$

Consequently, for all $v \in \mathbb{P}_d^{k+1}(T)$, we have

$$p_T^{k+1}\underline{I}_T^k v = v. \tag{8}$$

Proof For $v \in H^1(T)$, let us plug $\underline{v}_T := \underline{I}_T^k v = \left(\pi_T^k v, (\pi_F^k v)_{F \in \mathcal{F}_T}\right)$ into (6). Since $\mathbb{M}_T$ is a constant tensor and since $w \in \mathbb{P}_d^{k+1}(T)$, we infer that $\operatorname{div}(\mathbb{M}_T \nabla w) \in \mathbb{P}_d^{k-1}(T) \subset \mathbb{P}_d^k(T)$ and that $\mathbb{M}_T \nabla w_{|F} \cdot \boldsymbol{n}_{T,F} \in \mathbb{P}_{d-1}^k(F)$, which means that, for all $w \in \mathbb{P}_d^{k+1}(T)$,

$$\begin{aligned}(\mathbb{M}_T \nabla p_T^{k+1}\underline{I}_T^k v, \nabla w)_T &= -(v, \operatorname{div}(\mathbb{M}_T \nabla w))_T + \sum_{F \in \mathcal{F}_T} (v, \mathbb{M}_T \nabla w \cdot \boldsymbol{n}_{T,F})_F \\ &= (\mathbb{M}_T \nabla v, \nabla w)_T,\end{aligned}$$

hence concluding the proof of (7). For $v \in \mathbb{P}_d^{k+1}(T)$, we deduce from (7) that $(v - p_T^{k+1}\underline{I}_T^k v) \in \mathbb{P}_d^0(T)$, and we conclude by invoking the relation $\int_T p_T^{k+1}\underline{I}_T^k v = \int_T \pi_T^k v = \int_T v$.

The next result can be found in [29, Lemma 2.1].

Lemma 3.2 (Approximation) *For all $v \in H^{k+2}(T)$, the following holds:*

$$\|v - p_T^{k+1}\underline{\mathrm{I}}_T^k v\|_T + h_T^{1/2}\|v - p_T^{k+1}\underline{\mathrm{I}}_T^k v\|_{\partial T} + h_T\|\nabla(v - p_T^{k+1}\underline{\mathrm{I}}_T^k v)\|_T \\ + h_T^{3/2}\|\nabla(v - p_T^{k+1}\underline{\mathrm{I}}_T^k v)\|_{\partial T} \lesssim \rho_T^{1/2} h_T^{k+2}\|v\|_{H^{k+2}(T)}. \tag{9}$$

In the more general case of a piecewise Lipschitz diffusivity, only approximate polynomial consistency holds, while a factor ρ_T instead of $\rho_T^{1/2}$ appears in the estimate (9) (cf. [29]).

For further use, we define the global potential reconstruction operator

$$p_h^{k+1} : \underline{\mathrm{U}}_h^k \to \mathbb{P}_d^{k+1}(\mathcal{T}_h)$$

such that, for all $\underline{\mathrm{v}}_h \in \underline{\mathrm{U}}_h^k$, and for all $T \in \mathcal{T}_h$, $(p_h^{k+1}\underline{\mathrm{v}}_h)_{|T} := p_T^{k+1}\underline{\mathrm{v}}_T$.

3.1.3 Stabilization

For all $T \in \mathcal{T}_h$, we define the stabilization bilinear form $j_T : \underline{\mathrm{U}}_T^k \times \underline{\mathrm{U}}_T^k \to \mathbb{R}$ such that

$$j_T(\underline{\mathrm{u}}_T, \underline{\mathrm{v}}_T) := \sum_{F \in \mathcal{F}_T} \frac{\mu_{T,F}}{h_F}(\pi_F^k(q_T^{k+1}\underline{\mathrm{u}}_T - \mathrm{u}_F), \pi_F^k(q_T^{k+1}\underline{\mathrm{v}}_T - \mathrm{v}_F))_F, \tag{10}$$

with $q_T^{k+1} : \underline{\mathrm{U}}_T^k \to \mathbb{P}_d^{k+1}(T)$ such that, for all $\underline{\mathrm{w}}_T \in \underline{\mathrm{U}}_T^k$,

$$q_T^{k+1}\underline{\mathrm{w}}_T := w_T + (p_T^{k+1}\underline{\mathrm{w}}_T - \pi_T^k p_T^{k+1}\underline{\mathrm{w}}_T).$$

Notice that j_T is symmetric, positive semi-definite, and polynomially consistent [as a consequence of (8)] in the sense that, for all $v \in \mathbb{P}_d^{k+1}(T)$,

$$j_T(\underline{\mathrm{I}}_T^k v, \underline{\mathrm{w}}_T) = 0 \qquad \forall \underline{\mathrm{w}}_T \in \underline{\mathrm{U}}_T^k. \tag{11}$$

Another important property of j_T is the following approximation property: For all $v \in H^{k+2}(T)$, the following bound holds:

$$j_T(\underline{\mathrm{I}}_T^k v, \underline{\mathrm{I}}_T^k v)^{1/2} \lesssim \mu_{\sharp,T}^{1/2}\rho_T^{1/2} h_T^{k+1}\|v\|_{H^{k+2}(T)}, \tag{12}$$

showing that j_T matches the approximation properties of the gradient of p_T^{k+1}; cf. Lemma 3.2.

3.2 Discrete Problem: Formulation and Key Properties

3.2.1 Formulation

For all $T \in \mathcal{T}_h$, we define the following local bilinear form:

$$a_T : \underline{\mathrm{U}}_T^k \times \underline{\mathrm{U}}_T^k \to \mathbb{R}; \qquad (\underline{\mathrm{u}}_T, \underline{\mathrm{v}}_T) \mapsto (\mathbb{M}_T \nabla p_T^{k+1} \underline{\mathrm{u}}_T, \nabla p_T^{k+1} \underline{\mathrm{v}}_T)_T + j_T(\underline{\mathrm{u}}_T, \underline{\mathrm{v}}_T), \tag{13}$$

with potential reconstruction operator p_T^{k+1} defined by (6) and stabilization bilinear form j_T defined by (10). Introduce now the following global bilinear form obtained by a standard element-by-element assembly procedure:

$$a_h : \underline{\mathrm{U}}_h^k \times \underline{\mathrm{U}}_h^k \to \mathbb{R}; \qquad (\underline{\mathrm{u}}_h, \underline{\mathrm{v}}_h) \mapsto \sum_{T \in \mathcal{T}_h} a_T(\underline{\mathrm{u}}_T, \underline{\mathrm{v}}_T).$$

Then, the (primal) HHO discretization of problem (3) reads: Find $\underline{\mathrm{u}}_{h,0} \in \underline{\mathrm{U}}_{h,0}^k$ such that

$$a_h(\underline{\mathrm{u}}_{h,0}, \underline{\mathrm{v}}_h) = (f, v_h) - a_h(\underline{\mathrm{u}}_{h,\partial}, \underline{\mathrm{v}}_h) + \sum_{F \in \mathcal{F}_h^{\mathrm{n}}} (\phi_\partial, \mathrm{v}_F)_F \qquad \forall \underline{\mathrm{v}}_h \in \underline{\mathrm{U}}_{h,0}^k, \tag{14}$$

where $\underline{\mathrm{u}}_{h,\partial} := \underline{\mathrm{I}}_h^k u_\partial \in \underline{\mathrm{U}}_h^k$ is the reduction of the continuous lifting u_∂ of ψ_∂. The discrete solution $\underline{\mathrm{u}}_h \in \underline{\mathrm{U}}_h^k$ is finally computed as

$$\underline{\mathrm{u}}_h = \underline{\mathrm{u}}_{h,0} + \underline{\mathrm{u}}_{h,\partial}. \tag{15}$$

Remark 3.2 (Discrete Dirichlet Datum) In practical implementation, the continuous lifting u_∂ of the Dirichlet datum is not needed, and one can simply select $\underline{\mathrm{u}}_{h,\partial}$ such that

$$u_{T,\partial} \equiv 0 \quad \forall T \in \mathcal{T}_h, \qquad \mathrm{u}_{F,\partial} = \pi_F^k \psi_\partial \quad \forall F \in \mathcal{F}_h^{\mathrm{d}}, \qquad \mathrm{u}_{F,\partial} \equiv 0 \quad \forall F \in \mathcal{F}_h \setminus \mathcal{F}_h^{\mathrm{d}}.$$

3.2.2 Stability

Let us introduce, for all $T \in \mathcal{T}_h$, the following diffusion-dependent seminorm on $\underline{\mathrm{U}}_T^k$:

$$\|\underline{\mathrm{v}}_T\|_{U,T}^2 := \rho_T^{-1} \left(\|\mathbb{M}_T^{1/2} \nabla v_T\|_T^2 + \sum_{F \in \mathcal{F}_T} \frac{\mu_{T,F}}{h_F} \|v_T - \mathrm{v}_F\|_F^2 \right). \tag{16}$$

It can be proved that the map

$$\|\underline{\mathrm{v}}_h\|_{U,h}^2 := \sum_{T \in \mathcal{T}_h} \|\underline{\mathrm{v}}_T\|_{U,T}^2,$$

defines a norm on $\underline{\mathrm{U}}_{h,0}^k$. Stability for problem (14) is expressed by the following result (cf. [29, Lemma 3.1]).

Lemma 3.3 (Stability) *For all $T \in \mathcal{T}_h$ and all $\underline{\mathrm{v}}_T \in \underline{\mathrm{U}}_T^k$, the following holds:*

$$\|\underline{\mathrm{v}}_T\|_{U,T} \lesssim a_T(\underline{\mathrm{v}}_T, \underline{\mathrm{v}}_T)^{1/2} \lesssim \rho_T \|\underline{\mathrm{v}}_T\|_{U,T}. \tag{17}$$

As a consequence, we infer that

$$\|\underline{\mathrm{v}}_h\|_{U,h}^2 \lesssim a_h(\underline{\mathrm{v}}_h, \underline{\mathrm{v}}_h) \quad \forall \underline{\mathrm{v}}_h \in \underline{\mathrm{U}}_h^k, \tag{18}$$

implying that problem (14) *is well-posed.*

3.2.3 Error Estimates

Let $u \in U$ be such that $u = u_0 + u_\partial$, where $u_0 \in U_0$ is the (unique) solution to (3), and $u_\partial \in U$ is defined in Sect. 2.1. Let $\underline{\mathrm{u}}_h \in \underline{\mathrm{U}}_h^k$ be such that $\underline{\mathrm{u}}_h = \underline{\mathrm{u}}_{h,0} + \underline{\mathrm{u}}_{h,\partial}$, where $\underline{\mathrm{u}}_{h,0} \in \underline{\mathrm{U}}_{h,0}^k$ is the (unique) solution to (14), and $\underline{\mathrm{u}}_{h,\partial} \in \underline{\mathrm{U}}_h^k$ is defined in Sect. 3.2.1. Finally, let us introduce the notation $\|\cdot\|_h := a_h(\cdot,\cdot)^{1/2}$. Then, we can state the following result, which slightly improves on [29, Theorem 4.1] [where the norm $\|\cdot\|_h$ is to be used under the supremum in Eq. (12)]. Note that the constants in the error bounds can depend on the polynomial degree following the use of discrete trace and inverse inequalities.

Theorem 3.1 (Energy-Norm Error Estimate) *Assume that u further belongs to $H^{k+2}(P_\Omega)$ (so that, by Definition 2.2, $u \in H^{k+2}(\mathcal{T}_h)$). Then, the following holds:*

$$\|\underline{\mathrm{I}}_h^k u - \underline{\mathrm{u}}_h\|_{U,h} \lesssim \|\underline{\mathrm{I}}_h^k u - \underline{\mathrm{u}}_h\|_h \lesssim \left\{ \sum_{T\in\mathcal{T}_h} \mu_{\sharp,T}\rho_T h_T^{2(k+1)} \|u\|_{H^{k+2}(T)}^2 \right\}^{1/2}, \tag{19}$$

which implies, by an additional use of Lemma 3.2, that

$$\|\mathbb{M}^{1/2}(\nabla u - \nabla_h p_h^{k+1}\underline{\mathrm{u}}_h)\| \lesssim \left\{ \sum_{T\in\mathcal{T}_h} \mu_{\sharp,T}\rho_T h_T^{2(k+1)} \|u\|_{H^{k+2}(T)}^2 \right\}^{1/2}. \tag{20}$$

In the more general case of a piecewise (non-constant) polynomial diffusivity, estimates (19) and (20) still hold with a factor ρ_T^2 instead of ρ_T (cf. [29]).

Whenever elliptic regularity holds, a L^2-norm error estimate of order h^{k+2} can be established, which slightly improves on [29, Theorem 4.2] (where the assumption of piecewise constant diffusivity is to be added).

Theorem 3.2 (**L^2-Norm Error Estimate**) *Assume elliptic regularity for problem* (3) *under the form* $\|z\|_{H^2(P_\Omega)} \lesssim \mu_\flat^{-1}\|g\|$ *for all* $g \in L^2(\Omega)$ *and* $z \in U_0$ *solving* (3) *with data* g *and homogeneous (mixed Dirichlet-Neumann) boundary conditions. Assume* $f \in H^{k+\delta}(\Omega), \phi_\partial \in W^{k+\delta,\infty}(\Gamma_{\rm n})$*, with* $\delta = 0$ *for* $k \geq 1$ *and* $\delta = 1$ *for* $k = 0$*. Then, under the same assumption on* u *as in Theorem 3.1, the following holds:*

$$\mu_\flat\|I_h^k u - u_h\| \lesssim \mu_\sharp^{1/2}\rho^{1/2}h\left\{\sum_{T\in\mathcal{T}_h}\mu_{\sharp,T}\rho_T h_T^{2(k+1)}\|u\|_{H^{k+2}(T)}^2\right\}^{1/2} + h^{k+2}\left\{\|f\|_{H^{k+\delta}(\Omega)} + \|\phi_\partial\|_{W^{k+\delta,\infty}(\Gamma_{\rm n})}\right\}. \tag{21}$$

3.2.4 Local Conservativity

For all $T \in \mathcal{T}_h$, let us first introduce the local bilinear form $\hat{a}_T : \underline{\mathrm{U}}_T^k \times \underline{\mathrm{U}}_T^k \to \mathbb{R}$ such that, for all $\underline{\mathrm{w}}_T, \underline{\mathrm{v}}_T \in \underline{\mathrm{U}}_T^k$,

$$\hat{a}_T(\underline{\mathrm{w}}_T, \underline{\mathrm{v}}_T) := (\mathbb{M}_T\nabla p_T^{k+1}\underline{\mathrm{w}}_T, \nabla p_T^{k+1}\underline{\mathrm{v}}_T)_T + \sum_{F\in\mathcal{F}_T}\frac{\mu_{T,F}}{h_F}(w_T - \mathfrak{w}_F, v_T - \mathfrak{v}_F)_F. \tag{22}$$

Then, we use (22) to define the local isomorphism $\underline{\mathrm{c}}_T^k : \underline{\mathrm{U}}_T^k \to \underline{\mathrm{U}}_T^k$ such that, for all $\underline{\mathrm{w}}_T \in \underline{\mathrm{U}}_T^k$, $\underline{\mathrm{c}}_T^k\underline{\mathrm{w}}_T$ is uniquely defined from the following local problem:

$$\hat{a}_T(\underline{\mathrm{c}}_T^k\underline{\mathrm{w}}_T, \underline{\mathrm{v}}_T) = a_T(\underline{\mathrm{w}}_T, \underline{\mathrm{v}}_T) + \sum_{F\in\mathcal{F}_T}\frac{\mu_{T,F}}{h_F}(w_T - \mathfrak{w}_F, v_T - \mathfrak{v}_F)_F \qquad \forall\, \underline{\mathrm{v}}_T \in \underline{\mathrm{U}}_T^k,$$

and $\int_T c_T^k\underline{\mathrm{w}}_T = \int_T w_T$. Finally, we define the local gradient reconstruction operator $\boldsymbol{G}_T^{k+1} : \underline{\mathrm{U}}_T^k \to \nabla\mathbb{P}_d^{k+1}(T)$ such that

$$\boldsymbol{G}_T^{k+1} := \nabla(p_T^{k+1} \circ \underline{\mathrm{c}}_T^k).$$

Adapting the arguments of [27, Lemmata 2 and 3], one can show the following result.

Lemma 3.4 (Local Conservativity) *Let* $\underline{\mathrm{u}}_h \in \underline{\mathrm{U}}_h^k$ *be defined as in* (15) *from the solution of problem* (14)*. Then, for all* $T \in \mathcal{T}_h$*, the following local equilibrium relation holds:*

$$(\mathbb{M}_T\boldsymbol{G}_T^{k+1}\underline{\mathrm{u}}_T, \nabla v_T)_T - \sum_{F\in\mathcal{F}_T}(\Phi_{T,F}(\underline{\mathrm{u}}_T), v_T)_F = (f, v_T)_T \qquad \forall\, v_T \in \mathbb{P}_d^k(T), \tag{23}$$

where the numerical flux operator $\Phi_{T,F} : \underline{\mathrm{U}}_T^k \to \mathbb{P}_{d-1}^k(F)$ *is such that, for all* $\underline{\mathrm{v}}_T \in \underline{\mathrm{U}}_T^k$,

$$\Phi_{T,F}(\underline{\mathrm{v}}_T) := \mathbb{M}_T \boldsymbol{G}_T^{k+1} \underline{\mathrm{v}}_T \cdot \boldsymbol{n}_{T,F} - \frac{\mu_{T,F}}{h_F} \left[(c_T^k \underline{\mathrm{v}}_T - v_T) - (\mathfrak{c}_F^k \underline{\mathrm{v}}_T - \mathfrak{v}_F) \right]. \tag{24}$$

In addition, the numerical fluxes are equilibrated in the following sense: For all $F \in \mathcal{F}_h^{\mathrm{i}}$ *such that* $F \subseteq \partial T_1 \cap \partial T_2$,

$$\Phi_{T_1,F}(\underline{\mathrm{u}}_T) + \Phi_{T_2,F}(\underline{\mathrm{u}}_T) = 0, \tag{25}$$

and $\Phi_{T,F}(\underline{\mathrm{u}}_T) = \pi_F^k \phi_\partial$ *for all* $F \in \mathcal{F}_h^{\mathrm{n}}$ *such that* $F \subseteq \partial T \cap \partial \Omega$.

Numerical fluxes can thus be computed by local element-by-element post-processing.

3.3 Computational Aspects

This section discusses various relevant computational aspects: the elimination of cell-based unknowns by static condensation, the offline/online decomposition of the computations, and the choice of polynomial bases.

3.3.1 Static Condensation

Following [20, Sect. 2.5], we show how cell-based unknowns can be locally eliminated from problem (14), thereby leading to a global system in terms of face-based unknowns only.

Introducing the notation $f_T := f_{|T}$ for all $T \in \mathcal{T}_h$, we begin by observing that problem (14) can be equivalently rewritten using (15) as follows:

$$a_T((u_T, 0), (v_T, 0)) = (f_T, v_T)_T - a_T((0, \mathfrak{u}_{\partial T}), (v_T, 0)) \qquad \forall v_T \in U_T^k,\ \forall T \in \mathcal{T}_h, \tag{26a}$$

$$a_h(\underline{\mathrm{u}}_h, (0, \mathfrak{v}_h)) = \sum_{F \in \mathcal{F}_h^{\mathrm{n}}} (\phi_\partial, \mathfrak{v}_F)_F \qquad \forall \mathfrak{v}_h \in \mathfrak{U}_{h,0}^k, \tag{26b}$$

that is to say, problem (14) can be split into card($\mathcal{T}_h$) local problems (26a) that allow one to express, for all $T \in \mathcal{T}_h$, u_T in terms of $\mathfrak{u}_{\partial T}$ and f_T, and one global problem (26b) written in terms of face-based unknowns only.

We now introduce two local cell-based potential lifting operators:

- a trace-based lifting $t_T^k : \mathfrak{U}_{\partial T}^k \to U_T^k$ such that, for all $\mathfrak{w}_{\partial T} \in \mathfrak{U}_{\partial T}^k$, $t_T^k \mathfrak{w}_{\partial T} \in U_T^k$ solves

$$a_T((t_T^k \mathfrak{w}_{\partial T}, 0), (v_T, 0)) = -a_T((0, \mathfrak{w}_{\partial T}), (v_T, 0)) \qquad \forall v_T \in U_T^k; \tag{27}$$

- a datum-based lifting $d_T^k : L^2(T) \to U_T^k$ such that, for all $\varphi_T \in L^2(T)$, $d_T^k \varphi_T \in U_T^k$ solves

$$a_T((d_T^k \varphi_T, 0), (v_T, 0)) = (\varphi_T, v_T)_T \qquad \forall v_T \in U_T^k. \tag{28}$$

Problems (27) and (28) are well-posed owing to the first inequality in (17) and the fact that $\|\cdot\|_{U,T}$ is a norm on the zero-trace subspace of $\underline{U}_T^k$, cf. (16). Problem (27) can be rewritten as

$$a_T((t_T^k \mathfrak{w}_{\partial T}, \mathfrak{w}_{\partial T}), (v_T, 0)) = 0 \qquad \forall v_T \in U_T^k. \tag{29}$$

Using (26a), (29), and (28), we infer that

$$\underline{u}_T = (t_T^k \mathfrak{u}_{\partial T} + d_T^k f_T, \mathfrak{u}_{\partial T}). \tag{30}$$

Introducing the global operators $t_h^k : \mathfrak{U}_h^k \to U_h^k$ and $d_h^k : L^2(\Omega) \to U_h^k$ such that, for all $\mathfrak{w}_h \in \mathfrak{U}_h^k$, all $\varphi \in L^2(\Omega)$, and all $T \in \mathcal{T}_h$, $(t_h^k \mathfrak{w}_h)_{|T} := t_T^k \mathfrak{w}_{\partial T}$ and $(d_h^k \varphi)_{|T} := d_T^k \varphi_{|T}$, we can rewrite (30) globally as follows:

$$\underline{u}_h = (t_h^k \mathfrak{u}_h + d_h^k f, \mathfrak{u}_h). \tag{31}$$

Finally, we reformulate the global problem (26b) under an equivalent form. We remark, using (31), that

$$\begin{aligned} a_h(\underline{u}_h, (0, \mathfrak{v}_h)) &= a_h(\underline{u}_h, (t_h^k \mathfrak{v}_h, \mathfrak{v}_h)) - a_h(\underline{u}_h, (t_h^k \mathfrak{v}_h, 0)) \\ &= a_h((t_h^k \mathfrak{u}_h, \mathfrak{u}_h), (t_h^k \mathfrak{v}_h, \mathfrak{v}_h)) + a_h((d_h^k f, 0), (t_h^k \mathfrak{v}_h, \mathfrak{v}_h)) \\ &\quad - a_h((t_h^k \mathfrak{u}_h, \mathfrak{u}_h), (t_h^k \mathfrak{v}_h, 0)) - a_h((d_h^k f, 0), (t_h^k \mathfrak{v}_h, 0)) \\ &:= \mathfrak{T}_1 + \mathfrak{T}_2 - \mathfrak{T}_3 - \mathfrak{T}_4, \end{aligned}$$

where $\mathfrak{T}_2 = \mathfrak{T}_3 = 0$ owing to (29) and to the symmetry of a_h, while $\mathfrak{T}_4 = (f, t_h^k \mathfrak{v}_h)$ owing to (28). Introducing for all $\mathfrak{w}_h \in \mathfrak{U}_h^k$ the notation $\underline{t}_h^k \mathfrak{w}_h := (t_h^k \mathfrak{w}_h, \mathfrak{w}_h)$ and the decomposition $\mathfrak{u}_h = \mathfrak{u}_{h,0} + \mathfrak{u}_{h,\partial}$ for the face-based unknowns, the previous relation enables us to rewrite the global problem (26b) as follows: Find $\mathfrak{u}_{h,0} \in \mathfrak{U}_{h,0}^k$ such that

$$a_h(\underline{t}_h^k \mathfrak{u}_{h,0}, \underline{t}_h^k \mathfrak{v}_h) = (f, t_h^k \mathfrak{v}_h) - a_h(\underline{t}_h^k \mathfrak{u}_{h,\partial}, \underline{t}_h^k \mathfrak{v}_h) + \sum_{F \in \mathcal{F}_h^{\mathrm{n}}} (\phi_\partial, \mathfrak{v}_F)_F \qquad \forall \mathfrak{v}_h \in \mathfrak{U}_{h,0}^k. \tag{32}$$

Problem (32) is well-posed owing to (18) and to the fact that $\|\cdot\|_{U,h}$ defines a norm on $\underline{\mathsf{U}}_{h,0}^k$. The following proposition summarizes the above considerations.

Proposition 3.1 (Characterization of the Approximate Solution) *The solution* $\underline{\mathsf{U}}_h^k \ni \underline{\mathsf{u}}_h = \underline{\mathsf{u}}_{h,0} + \underline{\mathsf{u}}_{h,\partial}$ *with* $\underline{\mathsf{u}}_{h,0} \in \underline{\mathsf{U}}_{h,0}^k$ *solving* (14) *can be expressed as* (31), *where the operator* t_h^k *and the vector of cell-based unknowns* $d_h^k f$ *are defined cell-wise as the solutions of the local problems* (27) *and* (28), *respectively, and where* $\mathfrak{u}_h \in \mathfrak{U}_h^k$ *is such that* $\mathfrak{u}_h = \mathfrak{u}_{h,0} + \mathfrak{u}_{h,\partial}$ *with* $\mathfrak{u}_{h,0} \in \mathfrak{U}_{h,0}^k$ *the unique solution of the global problem* (32).

3.3.2 Offline/Online Solution Strategy

Static condensation naturally points to an offline/online decomposition of the computations.

In the offline step, we begin by solving, for all $T \in \mathcal{T}_h$, the local problems (6), in order to compute the operator p_h^{k+1}. This first substep essentially requires to invert card$(\mathcal{T}_h)$ symmetric positive-definite matrices of size $N_{k+1,d}$. This can be done effectively using Cholesky factorization. Then, for all $T \in \mathcal{T}_h$, we solve the local problems (27) and (28). As both problems involve the same matrix, this second substep essentially requires the inversion of card$(\mathcal{T}_h)$ symmetric positive-definite matrices of size $N_{k,d}$. Note that both substeps are fully parallelizable. At the end of the offline step, one has computed the trace-based lifting t_h^k, and the restriction of the datum-based lifting d_h^k to $U_h^k = \mathbb{P}_d^k(\mathcal{T}_h)$. This fully determines d_h^k since the right-hand side of (28) only requires the projection of the datum onto U_h^k.

In the online step, given a right-hand side $f \in L^2(\Omega)$, we compute its L^2-orthogonal projection onto U_h^k, and we solve the global problem (32); the size of this problem is approximately equal to card$(\mathcal{F}_h) \times N_{k,d-1}$. The approximate solution is finally computed applying (31). A modification of the right-hand side (or of the boundary conditions) only requires to perform again the online step.

The offline/online solution strategy is particularly attractive in a multi-query context where one wants to compute the solution of problem (14) for a large number of right-hand sides $f \in L^2(\Omega)$.

3.3.3 Implementation

An important step in the implementation consists in selecting bases for the polynomial spaces on elements and faces that appear in the construction [cf. (6), (27), (28), (32)]. For $T \in \mathcal{T}_h$, we denote by $\boldsymbol{x}_T$ a point in T (typically the barycenter of T). One possibility leading to a hierarchical basis for $\mathbb{P}_d^l(T)$, $l \in \{k, k+1\}$, is to choose the following family of monomial functions:

$$\left\{\prod_{i=1}^{d} \xi_{T,i}^{\alpha_i} \mid \xi_{T,i} := \frac{x_i - x_{T,i}}{h_T} \; \forall\, 1 \le i \le d,\; \boldsymbol{\alpha} = (\alpha_i)_{1\le i\le d} \in \mathbb{N}^d,\; \|\boldsymbol{\alpha}\|_{l^1} \le l\right\}.$$

Similarly, for all $F \in \mathcal{F}_h$, we can define a basis for $\mathbb{P}^k_{d-1}(F)$ spanned by monomials with respect to a local frame scaled using the face diameter and, say, the barycenter of F.

3.3.4 Cost Assessment

Another important question linked to implementation is the scaling of the time devoted to the assembly (computation of the local contributions, static condensation, and matrix/right-hand side assembly) with respect to the time devoted to the solution (solving of the global problem), and how this scaling depends on the meshsize and on the order of approximation. Let us assume a naive implementation that does not exploit parallelism, and let us focus on problem (14) for a given right-hand side in two space dimensions. On Fig. 2, we plot, for polynomial degrees up to 5, the assembly/solution time ratio as a function of the number of mesh faces for two families of meshes corresponding, respectively, to the triangular (first) mesh family of the FVCA5 benchmark [41] and to the (predominantly) hexagonal mesh family introduced in [31, Sect. 4.2.3]. The global system is solved using the sparse direct solver of Eigen v3. This way, both the assembly and solution times are only marginally influenced by the problem data (right-hand side, boundary conditions). As illustrated in Fig. 2, the overall cost of the assembly time becomes quickly negligible in comparison with the solution time with mesh refinement (except for $k = 0$). This can be dramatically improved, e.g., using thread-based parallelism to solve the (independent) local problems for both the computation of the potential reconstructions and the static condensation inside each element.

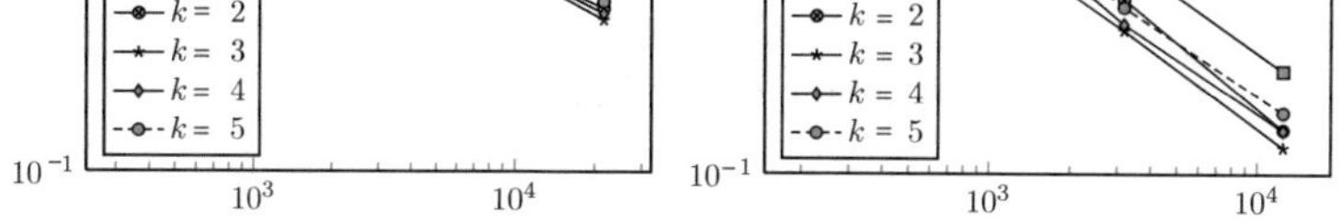

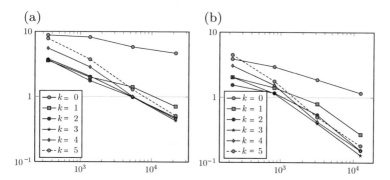

Fig. 2 Assembly time divided by the solution time as a function of card($\mathcal{F}_h$) for a triangular mesh family (**a**, *left panel*) and a (predominantly) hexagonal mesh family (**b**, *right panel*); the *symbols* indicate in both panels the polynomial degree that is being used

4 The HHO Method in Mixed Form

In this section, we study the HHO method in mixed formulation. The starting point is the following mixed form of the model problem (1): Find $\boldsymbol{s} : \Omega \to \mathbb{R}^d$, $u : \Omega \to \mathbb{R}$, such that

$$\begin{aligned} \boldsymbol{s} &= \mathbb{M}\nabla u && \text{in } \Omega, \\ -\mathrm{div}\boldsymbol{s} &= f && \text{in } \Omega, \\ u &= \psi_\partial && \text{on } \Gamma_{\mathrm{d}}, \\ \boldsymbol{s}{\cdot}\boldsymbol{n} &= \phi_\partial && \text{on } \Gamma_{\mathrm{n}}. \end{aligned} \tag{33}$$

To write this problem in weak form, we introduce the functional spaces

$$\boldsymbol{S} := \boldsymbol{H}(\mathrm{div}, \Omega), \qquad \boldsymbol{S}_0 := \left\{\boldsymbol{t} \in \boldsymbol{S} \mid \boldsymbol{t}{\cdot}\boldsymbol{n}_{|\Gamma_{\mathrm{n}}} = 0\right\}, \qquad V := L^2(\Omega),$$

so that the weak problem reads: Find $(\boldsymbol{s}_0, u) \in \boldsymbol{S}_0 \times V$ such that

$$\begin{aligned} (\mathbb{M}^{-1}\boldsymbol{s}_0, \boldsymbol{t}) + (u, \mathrm{div}\,\boldsymbol{t}) &= \langle \boldsymbol{t}{\cdot}\boldsymbol{n}, (u_\partial)_{|\Gamma}\rangle_\Gamma - (\mathbb{M}^{-1}\boldsymbol{s}_\partial, \boldsymbol{t}) && \forall \boldsymbol{t} \in \boldsymbol{S}_0, \\ -(\mathrm{div}\boldsymbol{s}_0, v) &= (f, v) + (\mathrm{div}\boldsymbol{s}_\partial, v) && \forall v \in V, \end{aligned} \tag{34}$$

where $\boldsymbol{s}_\partial \in \boldsymbol{S}$ is a lifting of the Neumann datum such that $(\boldsymbol{s}_\partial{\cdot}\boldsymbol{n})_{|\Gamma_{\mathrm{n}}} = \phi_\partial$ (which can be taken to be $\boldsymbol{s}_\partial = \nabla\theta$ where $\theta \in H^1(\Omega)$ solves $\theta - \Delta\theta = 0$ in Ω with $\nabla\theta{\cdot}\boldsymbol{n} = \phi_\Gamma$ on Γ where ϕ_Γ is the zero-extension of ϕ_∂ to Γ), and $\langle\cdot,\cdot\rangle_\Gamma$ denotes the duality pairing between $H^{-1/2}(\Gamma)$ and $H^{1/2}(\Gamma)$ (note that, owing to the fact that $\boldsymbol{t} \in \boldsymbol{S}_0$, $\langle \boldsymbol{t}{\cdot}\boldsymbol{n}, (u_\partial)_{|\Gamma}\rangle_\Gamma$ does not depend on the choice of the lifting u_∂ of ψ_∂). The solution $(\boldsymbol{s}, u) \in \boldsymbol{S} \times V$ is then computed as $(\boldsymbol{s}, u) = (\boldsymbol{s}_0 + \boldsymbol{s}_\partial, u)$.

4.1 Discrete Setting

Let us fix an integer $k \geq 0$ and consider an admissible and compatible mesh sequence $(\mathcal{T}_h)_{h\in\mathcal{H}}$ in the sense of Definitions 2.1 and 2.2. We suppose that the assumptions of Sect. 2.4 concerning the diffusivity hold.

4.1.1 Discrete Unknowns

We adopt the same notation as in Sect. 3.1.1, to which we add the use of boldface to denote vector-valued quantities. We introduce, first locally then globally, the discrete unknowns associated with the flux and with the potential. For the flux, we consider hybrid unknowns, in the sense that they consist of both cell- and face-based contributions. The cell-based flux unknowns are vector-valued while the

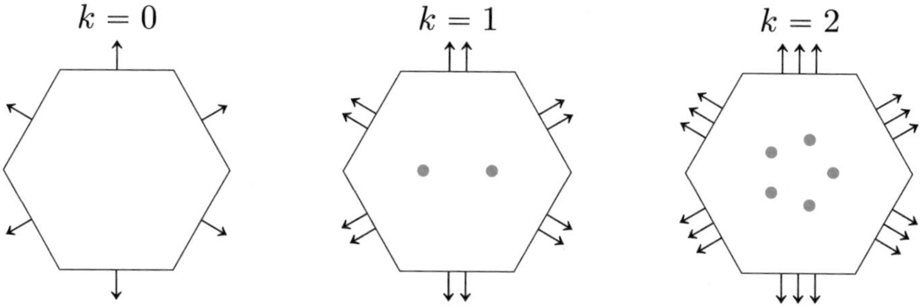

Fig. 3 Degrees of freedom associated with hybrid flux discrete unknowns, $d = 2$, $k \in \{0, 1, 2\}$

face-based ones are scalar-valued. For the potential, we consider scalar-valued cell-based unknowns.

Local Definition Let $T \in \mathcal{T}_h$. Setting

$$\boldsymbol{S}_T^k := \mathbb{M}_T \boldsymbol{\nabla} \mathbb{P}_d^k(T), \qquad \mathfrak{S}_F^k := \mathbb{P}_{d-1}^k(F) \text{ for all } F \in \mathcal{F}_T,$$

we define the local set of hybrid flux unknowns, cf. Fig. 3, as

$$\underline{\mathbf{S}}_T^k := \boldsymbol{S}_T^k \times \mathfrak{S}_{\partial T}^k, \qquad \text{where } \mathfrak{S}_{\partial T}^k := \bigtimes_{F \in \mathcal{F}_T} \mathfrak{S}_F^k.$$

In the lowest-order case $k = 0$, cell-based flux unknowns are unnecessary and $\boldsymbol{S}_T^k$ has dimension zero. Any element $\underline{\mathbf{t}}_T \in \underline{\mathbf{S}}_T^k$ can be decomposed as $\underline{\mathbf{t}}_T := (\boldsymbol{t}_T \in \boldsymbol{S}_T^k, \mathrm{t}_{\partial T} \in \mathfrak{S}_{\partial T}^k)$, with $\mathrm{t}_{\partial T} := (\mathrm{t}_F \in \mathfrak{S}_F^k)_{F \in \mathcal{F}_T}$. Letting, for $q > 2$,

$$\boldsymbol{S}^+(T) := \{\boldsymbol{t} \in \boldsymbol{L}^q(T) \mid \operatorname{div} \boldsymbol{t} \in L^2(T)\},$$

and recalling that the normal component of functions in this space can act against polynomial functions on all faces of T, we introduce the local reduction operator $\underline{\mathbf{I}}_T^k : \boldsymbol{S}^+(T) \to \underline{\mathbf{S}}_T^k$ such that, for all $\boldsymbol{t} \in \boldsymbol{S}^+(T)$,

$$\underline{\mathbf{I}}_T^k \boldsymbol{t} := \left(\mathbb{M}_T \boldsymbol{\nabla} y, (\pi_F^k(\boldsymbol{t} \cdot \boldsymbol{n}_F))_{F \in \mathcal{F}_T} \right),$$

where $y \in \mathbb{P}_d^k(T)$ is a solution (defined up to an additive constant) of the Neumann problem

$$(\mathbb{M}_T \boldsymbol{\nabla} y, \boldsymbol{\nabla} w)_T = (\boldsymbol{t}, \boldsymbol{\nabla} w)_T \qquad \forall w \in \mathbb{P}_d^k(T), \tag{35}$$

observing that the required compatibility condition on the right-hand side is verified.

As far as the potential is concerned, we let U_T^k, introduced in (4), be the associated local set of (cell-based) discrete unknowns.

Global Definition We define the global set of hybrid flux unknowns as

$$\underline{\mathbf{S}}_h^k := \boldsymbol{S}_h^k \times \left\{ \bigtimes_{F\in\mathcal{F}_h} \mathfrak{S}_F^k \right\},$$

where $\boldsymbol{S}_h^k := \bigtimes_{T\in\mathcal{T}_h} \boldsymbol{S}_T^k$. Observe that the flux unknowns attached to interfaces are single-valued. Given an element $\underline{\mathbf{t}}_h \in \underline{\mathbf{S}}_h^k$, for any $T \in \mathcal{T}_h$, we denote by $\underline{\mathbf{t}}_T = (\boldsymbol{t}_T, \mathrm{t}_{\partial T}) \in \underline{\mathbf{S}}_T^k$ its restriction to the element T. We introduce the following subspace of $\underline{\mathbf{S}}_h^k$, that allows one to account for (homogeneous) Neumann boundary conditions in a strong manner:

$$\underline{\mathbf{S}}_{h,0}^k := \left\{ \underline{\mathbf{t}}_h \in \underline{\mathbf{S}}_h^k \mid \mathrm{t}_F \equiv 0, \forall F \in \mathcal{F}_h^{\mathrm{n}} \right\}.$$

We also define the global reduction operator $\underline{\mathbf{I}}_h^k : \boldsymbol{S} \cap \boldsymbol{S}^+(\mathcal{T}_h) \to \underline{\mathbf{S}}_h^k$ such that, for all $\boldsymbol{t} \in \boldsymbol{S} \cap \boldsymbol{S}^+(\mathcal{T}_h)$, and for all $T \in \mathcal{T}_h$, $(\underline{\mathbf{I}}_h^k \boldsymbol{t})_{|T} := \underline{\mathbf{I}}_T^k \boldsymbol{t}_{|T}$. Single-valuedness at interfaces is ensured by the regularity of functions in $\boldsymbol{S} \cap \boldsymbol{S}^+(\mathcal{T}_h)$.

We finally define U_h^k, cf. (5), as the global set of discrete (cell-based) potential unknowns, and we denote by $v_T \in U_T^k$ the restriction of any $v_h \in U_h^k$ to the element $T \in \mathcal{T}_h$.

4.1.2 Divergence Reconstruction Operator

Let $T \in \mathcal{T}_h$. We define the local divergence reconstruction operator $D_T^k : \underline{\mathbf{S}}_T^k \to U_T^k$ as the operator such that, for all $\underline{\mathbf{t}}_T = (\boldsymbol{t}_T, \mathrm{t}_{\partial T}) \in \underline{\mathbf{S}}_T^k$,

$$(D_T^k \underline{\mathbf{t}}_T, v_T)_T = -(\boldsymbol{t}_T, \nabla v_T)_T + \sum_{F\in\mathcal{F}_T} (\mathrm{t}_F \varepsilon_{T,F}, v_T)_F \qquad \forall v_T \in U_T^k, \tag{36}$$

where $\varepsilon_{T,F} := \boldsymbol{n}_F \cdot \boldsymbol{n}_{T,F}$ for all $T \in \mathcal{T}_h$ and $F \in \mathcal{F}_T$. This definition reproduces at the discrete level an integration by parts formula, that brings into action the local hybrid flux unknowns. The following property is crucial for inf-sup stability, cf. [26, Lemmas 2 and 5].

Lemma 4.1 (Commuting Property) *The following holds for all* $\boldsymbol{t} \in \boldsymbol{S}^+(T)$:

$$D_T^k \underline{\mathbf{I}}_T^k \boldsymbol{t} = \pi_T^k (\operatorname{div} \boldsymbol{t}). \tag{37}$$

Proof For $\boldsymbol{t} \in \boldsymbol{S}^+(T)$, let us plug the quantity $\underline{\mathbf{t}}_T := \underline{\mathbf{I}}_T^k \boldsymbol{t} = \left(\mathbb{M}_T \nabla y, (\pi_F^k(\boldsymbol{t}\cdot\boldsymbol{n}_F))_{F\in\mathcal{F}_T} \right)$ into (36), where $y \in \mathbb{P}_d^k(T)$ is a solution to (35). Let $v_T \in U_T^k$, and observe that $v_T \in \mathbb{P}_d^k(T)$ and $v_{T|F} \in \mathbb{P}_{d-1}^k(F)$. Hence,

$$(D_T^k \underline{\mathbf{I}}_T^k \boldsymbol{t}, v_T)_T = -(\boldsymbol{t}, \nabla v_T)_T + \sum_{F\in\mathcal{F}_T} (\boldsymbol{t}\cdot\boldsymbol{n}_{T,F}, v_T)_F = (\operatorname{div} \boldsymbol{t}, v_T)_T,$$

which concludes the proof.

For further use, we introduce the global divergence reconstruction operator $D_h^k : \underline{\mathbf{S}}_h^k \to U_h^k$ such that, for all $\underline{\mathbf{t}}_h \in \underline{\mathbf{S}}_h^k$, and all $T \in \mathcal{T}_h$, $(D_h^k \underline{\mathbf{t}}_h)_{|T} := D_T^k \underline{\mathbf{t}}_T$.

4.1.3 Flux Reconstruction Operator

Let $T \in \mathcal{T}_h$. The local flux reconstruction operator $\boldsymbol{F}_T^{k+1} : \underline{\mathbf{S}}_T^k \to \boldsymbol{S}_T^{k+1}$ is defined, for all $\underline{\mathbf{t}}_T = (\boldsymbol{t}_T, \mathrm{t}_{\partial T}) \in \underline{\mathbf{S}}_T^k$, as $\boldsymbol{F}_T^{k+1} \underline{\mathbf{t}}_T := \mathbb{M}_T \nabla z$, where $z \in \mathbb{P}_d^{k+1}(T)$ is a solution (defined up to an additive constant) of the Neumann problem

$$(\mathbb{M}_T \nabla z, \nabla w)_T = (\boldsymbol{t}_T, \nabla \pi_T^k w)_T + \sum_{F \in \mathcal{F}_T} (\mathrm{t}_F \varepsilon_{T,F}, \pi_F^k w - \pi_T^k w)_F \qquad \forall w \in \mathbb{P}_d^{k+1}(T), \tag{38}$$

observing that the required compatibility condition on the right-hand side is verified. The definition of $\boldsymbol{F}_T^{k+1} \underline{\mathbf{t}}_T$ is motivated by the following link between $\boldsymbol{F}_T^{k+1} \underline{\mathbf{t}}_T$ and the divergence reconstruction operator defined in (36): For all $\underline{\mathbf{t}}_T = (\boldsymbol{t}_T, \mathrm{t}_{\partial T}) \in \underline{\mathbf{S}}_T^k$,

$$(\boldsymbol{F}_T^{k+1} \underline{\mathbf{t}}_T, \nabla w)_T = -(D_T^k \underline{\mathbf{t}}_T, w)_T + \sum_{F \in \mathcal{F}_T} (\mathrm{t}_F \varepsilon_{T,F}, w)_F \qquad \forall w \in \mathbb{P}_d^{k+1}(T). \tag{39}$$

As in Sect. 3.1.2, computing the operator $\boldsymbol{F}_T^{k+1}$ using (38) or (39) requires to invert a symmetric positive-definite matrix of size $N_{k+1,d}$, cf. (2), which can be performed effectively via Cholesky factorization. The following result can be found in [26, Lemma 3] (and requires, as its primal counterpart (8), that the diffusion tensor be piecewise constant).

Lemma 4.2 (Polynomial Consistency) *The following holds for all $\boldsymbol{t} \in \boldsymbol{S}_T^{k+1}$:*

$$\boldsymbol{F}_T^{k+1} \underline{\mathbf{I}}_T^k \boldsymbol{t} = \boldsymbol{t}. \tag{40}$$

Proof Let $\boldsymbol{t} \in \boldsymbol{S}_T^{k+1}$ and plug $\underline{\mathbf{t}}_T := \underline{\mathbf{I}}_T^k \boldsymbol{t}$ into (39). Using the commuting property (37) leads to $D_T^k \underline{\mathbf{I}}_T^k \boldsymbol{t} = \pi_T^k(\operatorname{div} \boldsymbol{t}) = \operatorname{div} \boldsymbol{t}$ since $\boldsymbol{t} \in \boldsymbol{S}_T^{k+1} \subset \mathbb{P}_d^k(T)$ ($\mathbb{M}_T$ is a constant tensor), which combined with the fact that $\pi_F^k(\boldsymbol{t} \cdot \boldsymbol{n}_F) = \boldsymbol{t} \cdot \boldsymbol{n}_F$ (since faces are planar), allows us to infer that, for all $w \in \mathbb{P}_d^{k+1}(T)$,

$$(\boldsymbol{F}_T^{k+1} \underline{\mathbf{I}}_T^k \boldsymbol{t}, \nabla w)_T = -(\operatorname{div} \boldsymbol{t}, w)_T + \sum_{F \in \mathcal{F}_T} (\boldsymbol{t} \cdot \boldsymbol{n}_{T,F}, w)_F = (\boldsymbol{t}, \nabla w)_T.$$

This last relation proves (40) since $(\boldsymbol{F}_T^{k+1} \underline{\mathbf{I}}_T^k \boldsymbol{t} - \boldsymbol{t}) \in \boldsymbol{S}_T^{k+1} = \mathbb{M}_T \nabla \mathbb{P}_d^{k+1}(T)$.

The next result is adapted from [26, Lemma 9], and is related, in the light of Lemma 5.1 below, to Lemma 3.2.

Lemma 4.3 (Approximation) *For all $v \in H^{k+2}(T)$, letting $\boldsymbol{t} := \mathbb{M}_T \nabla v$, the following holds for all $F \in \mathcal{F}_T$:*

$$\|\mathbb{M}_T^{-1/2}(\boldsymbol{t} - \boldsymbol{F}_T^{k+1}\underline{\mathbf{I}}_T^k \boldsymbol{t})\|_T + h_F^{1/2}\mu_{T,F}^{-1/2}\|(\boldsymbol{t} - \boldsymbol{F}_T^{k+1}\underline{\mathbf{I}}_T^k \boldsymbol{t})\cdot\boldsymbol{n}_F\|_F \lesssim \rho_T^{1/2}\mu_{\sharp,T}^{1/2}h_T^{k+1}\|v\|_{H^{k+2}(T)}. \tag{41}$$

For further use, we define the global flux reconstruction operator $\boldsymbol{F}_h^{k+1} : \underline{\mathbf{S}}_h^k \to \boldsymbol{S}_h^{k+1}$ such that, for all $\underline{\mathbf{t}}_h \in \underline{\mathbf{S}}_h^k$, and all $T \in \mathcal{T}_h$, $(\boldsymbol{F}_h^{k+1}\underline{\mathbf{t}}_h)_{|T} := \boldsymbol{F}_T^{k+1}\underline{\mathbf{t}}_T$.

4.1.4 Stabilization

For all $T \in \mathcal{T}_h$, we define the stabilization bilinear form $J_T : \underline{\mathbf{S}}_T^k \times \underline{\mathbf{S}}_T^k \to \mathbb{R}$ such that

$$J_T(\underline{\mathbf{s}}_T, \underline{\mathbf{t}}_T) := \sum_{F\in\mathcal{F}_T} \frac{h_F}{\mu_{T,F}}((\boldsymbol{F}_T^{k+1}\underline{\mathbf{s}}_T)\cdot\boldsymbol{n}_F - \mathfrak{s}_F, (\boldsymbol{F}_T^{k+1}\underline{\mathbf{t}}_T)\cdot\boldsymbol{n}_F - \mathfrak{t}_F)_F.$$

Notice that J_T is symmetric, positive semi-definite, and polynomially consistent (as a consequence of Lemma 4.2) in the sense that, for all $\boldsymbol{t} \in \boldsymbol{S}_T^{k+1}$,

$$J_T(\underline{\mathbf{I}}_T^k \boldsymbol{t}, \underline{\mathbf{r}}_T) = 0 \qquad \forall \underline{\mathbf{r}}_T \in \underline{\mathbf{S}}_T^k. \tag{42}$$

This result can be found in [26, Eq. (18)]. Another important property of J_T is the following approximation property (see [26, Lemma 9] and Lemma 4.3 above): For all $v \in H^{k+2}(T)$, the following holds with $\boldsymbol{t} := \mathbb{M}_T \nabla v$:

$$J_T(\underline{\mathbf{I}}_T^k \boldsymbol{t}, \underline{\mathbf{I}}_T^k \boldsymbol{t})^{1/2} \lesssim \rho_T^{1/2}\mu_{\sharp,T}^{1/2}h_T^{k+1}\|v\|_{H^{k+2}(T)}. \tag{43}$$

4.2 Discrete Problem: Formulation and Key Properties

4.2.1 Formulation

For all $T \in \mathcal{T}_h$, we define the following local bilinear form:

$$H_T : \underline{\mathbf{S}}_T^k \times \underline{\mathbf{S}}_T^k \to \mathbb{R}; \qquad (\underline{\mathbf{s}}_T, \underline{\mathbf{t}}_T) \mapsto (\mathbb{M}_T^{-1}\boldsymbol{F}_T^{k+1}\underline{\mathbf{s}}_T, \boldsymbol{F}_T^{k+1}\underline{\mathbf{t}}_T)_T + J_T(\underline{\mathbf{s}}_T, \underline{\mathbf{t}}_T), \tag{44}$$

where the notation H_T is reminiscent of the similarity with the discrete Hodge operator considered in the CDO framework in the lowest-order case [10]. Introduce now the following global bilinear form:

$$H_h : \underline{\mathbf{S}}_h^k \times \underline{\mathbf{S}}_h^k \to \mathbb{R}; \qquad (\underline{\mathbf{s}}_h, \underline{\mathbf{t}}_h) \mapsto \sum_{T\in\mathcal{T}_h} H_T(\underline{\mathbf{s}}_T, \underline{\mathbf{t}}_T). \tag{45}$$

The mixed form of the HHO method for problem (34) reads: Find $(\underline{\mathbf{s}}_{h,0}, u_h) \in \underline{\mathbf{S}}_{h,0}^k \times U_h^k$ such that

$$
\begin{aligned}
H_h(\underline{\mathbf{s}}_{h,0}, \underline{\mathbf{t}}_h) + (u_h, D_h^k \underline{\mathbf{t}}_h) &= \sum_{F \in \mathcal{F}_h^{\mathrm{d}}} (\mathrm{t}_F, \psi_\partial)_F - H_h(\underline{\mathbf{s}}_{h,\partial}, \underline{\mathbf{t}}_h) \quad && \forall \underline{\mathbf{t}}_h \in \underline{\mathbf{S}}_{h,0}^k, \\
-(D_h^k \underline{\mathbf{s}}_{h,0}, v_h) &= (f, v_h) + (D_h^k \underline{\mathbf{s}}_{h,\partial}, v_h) && \forall v_h \in U_h^k,
\end{aligned}
\tag{46}
$$

where $\underline{\mathbf{s}}_{h,\partial} := \underline{\mathbf{I}}_h^k s_\partial \in \underline{\mathbf{S}}_h^k$ is the reduction of the lifting s_∂ of the Neumann datum ϕ_∂. The discrete solution $(\underline{\mathbf{s}}_h, u_h) \in \underline{\mathbf{S}}_h^k \times U_h^k$ is finally computed as

$$
(\underline{\mathbf{s}}_h, u_h) = (\underline{\mathbf{s}}_{h,0} + \underline{\mathbf{s}}_{h,\partial}, u_h). \tag{47}
$$

Remark 4.1 (Discrete Neumann Datum) Similarly to Remark 3.2, the discrete lifting $\underline{\mathbf{s}}_{h,\partial}$ of the Neumann datum can be obtained without explicitly knowing s_∂ by setting

$$
s_{T,\partial} \equiv \mathbf{0} \quad \forall T \in \mathcal{T}_h, \qquad \mathfrak{s}_{F,\partial} = \pi_F^k \phi_\partial \quad \forall F \in \mathcal{F}_h^{\mathrm{n}}, \qquad \mathfrak{s}_{F,\partial} \equiv 0 \quad \forall F \in \mathcal{F}_h \setminus \mathcal{F}_h^{\mathrm{n}}.
$$

4.2.2 Stability

Let us introduce, for all $T \in \mathcal{T}_h$, the following norm on $\underline{\mathbf{S}}_T^k$:

$$
\|\underline{\mathbf{t}}_T\|_{S,T}^2 := \mu_{\sharp,T}^{-1} \left(\|\mathbf{t}_T\|_T^2 + \sum_{F \in \mathcal{F}_T} h_F \|\mathrm{t}_F\|_F^2 \right). \tag{48}
$$

Setting $\|\underline{\mathbf{t}}_h\|_{S,h}^2 := \sum_{T \in \mathcal{T}_h} \|\underline{\mathbf{t}}_T\|_{S,T}^2$ for all $\underline{\mathbf{t}}_h \in \underline{\mathbf{S}}_h^k$, it follows that $\|\cdot\|_{S,h}$ defines a norm on $\underline{\mathbf{S}}_h^k$. The coercivity of H_h can be expressed in terms of this norm, cf. [26, Lemma 4].

Lemma 4.4 (Stability for H_h**)** *For all $T \in \mathcal{T}_h$, and for all $\underline{\mathbf{t}}_T \in \underline{\mathbf{S}}_T^k$, the following holds:*

$$
\|\underline{\mathbf{t}}_T\|_{S,T} \lesssim H_T(\underline{\mathbf{t}}_T, \underline{\mathbf{t}}_T)^{1/2} \lesssim \rho_T^{1/2} \|\underline{\mathbf{t}}_T\|_{S,T}. \tag{49}
$$

Consequently, we infer that

$$
\|\underline{\mathbf{t}}_h\|_{S,h}^2 \lesssim H_h(\underline{\mathbf{t}}_h, \underline{\mathbf{t}}_h) \qquad \forall \underline{\mathbf{t}}_h \in \underline{\mathbf{S}}_h^k. \tag{50}
$$

We can then state the following result, whose proof hinges on Lemma 4.1, and which is a slightly modified version of [26, Lemma 5].

Lemma 4.5 (Well-Posedness of (46)**)** *For all $v_h \in U_h^k$, the following holds:*

$$\mu_\flat^{1/2}\|v_h\| \lesssim \sup_{\underline{\mathbf{t}}_h \in \underline{\mathbf{S}}_{h,0}^k, \|\underline{\mathbf{t}}_h\|_{S,h}=1} (D_h^k \underline{\mathbf{t}}_h, v_h). \tag{51}$$

Combining (51) *with Lemma 4.4, we infer that problem* (46) *is well-posed.*

4.2.3 Error Estimates

Let $(s, u) \in S \times V$ be such that $(s, u) = (s_0 + s_\partial, u)$, where $(s_0, u) \in S_0 \times V$ is the (unique) solution to (34), and $s_\partial \in S$ is defined above. We further assume that $s \in S$ fulfills the additional regularity $s \in S^+(\mathcal{T}_h)$. Similarly, let $(\underline{\mathbf{s}}_h, u_h) \in \underline{\mathbf{S}}_h^k \times U_h^k$ be such that $(\underline{\mathbf{s}}_h, u_h) = (\underline{\mathbf{s}}_{h,0} + \underline{\mathbf{s}}_{h,\partial}, u_h)$, where $(\underline{\mathbf{s}}_{h,0}, u_h) \in \underline{\mathbf{S}}_{h,0}^k \times U_h^k$ is the (unique) solution to (46), and $\underline{\mathbf{s}}_{h,\partial} \in \underline{\mathbf{S}}_h^k$ is defined in Sect. 4.2.1. Finally, let us introduce the notation $\|\cdot\|_h := H_h(\cdot,\cdot)^{1/2}$. Then, we can state the following result, whose proof can be easily adapted from the one of [26, Theorem 6]. Note that, here again, the constants in the error bounds can depend on the polynomial degree following the use of discrete trace and inverse inequalities.

Theorem 4.1 (Error Estimate for the Flux) *Assume the additional regularity $u \in H^{k+2}(P_\Omega)$ (so that, by Definition 2.2, $u \in H^{k+2}(\mathcal{T}_h)$). Then, the following holds:*

$$\|\underline{\mathbf{I}}_h^k s - \underline{\mathbf{s}}_h\|_{S,h} \lesssim \|\underline{\mathbf{I}}_h^k s - \underline{\mathbf{s}}_h\|_h \lesssim \left\{ \sum_{T \in \mathcal{T}_h} \mu_{\sharp,T} \rho_T h_T^{2(k+1)} \|u\|_{H^{k+2}(T)}^2 \right\}^{1/2}, \tag{52}$$

which implies, by an additional use of Lemma 4.3,

$$\|\mathbb{M}^{-1/2}(s - \boldsymbol{F}_h^{k+1}\underline{\mathbf{s}}_h)\| \lesssim \left\{ \sum_{T \in \mathcal{T}_h} \mu_{\sharp,T} \rho_T h_T^{2(k+1)} \|u\|_{H^{k+2}(T)}^2 \right\}^{1/2}. \tag{53}$$

Whenever elliptic regularity holds, a supercloseness result for the potential can be established, as an adaptation of [26, Theorem 7].

Theorem 4.2 (Supercloseness of the Potential) *Assume elliptic regularity for problem* (3) *under the form $\|z\|_{H^2(P_\Omega)} \lesssim \mu_\flat^{-1}\|g\|$ for all $g \in L^2(\Omega)$ and $z \in U_0$ solving* (3) *with data g and homogeneous (mixed Dirichlet-Neumann) boundary conditions. Assume $f \in H^{k+\delta}(\Omega)$, $\phi_\partial \in W^{k+\delta,\infty}(\Gamma_n)$, with $\delta = 0$ for $k \geq 1$ and*

$\delta = 1$ *for* $k = 0$. *Then, under the same assumption on* u *as in Theorem 4.1, the following holds:*

$$\mu_\flat \|I_h^k u - u_h\| \lesssim \mu_\sharp^{1/2} \rho^{1/2} h \left\{ \sum_{T \in \mathcal{T}_h} \mu_{\sharp,T} \rho_T h_T^{2(k+1)} \|u\|_{H^{k+2}(T)}^2 \right\}^{1/2} + h^{k+2} \left\{ \|f\|_{H^{k+\delta}(\Omega)} + \|\phi_\partial\|_{W^{k+\delta,\infty}(\Gamma_n)} \right\}. \quad (54)$$

4.3 Static Condensation

There are two ways of reducing the size of the discrete problem (46).

First, as exposed in [26, Sect. 3.4], it is possible to eliminate locally the cell-based flux unknowns and the potential unknowns, up to one constant value per element. Thus, the global system to solve only writes in terms of the face-based flux unknowns and of the mean value of the potential in each element. For all $T \in \mathcal{T}_h$, let $U_T^{k,0}$ be the space of d-variate polynomials of degree at most k having zero mean value in T, so that $U_T^k = U_T^0 \oplus U_T^{k,0}$. Hence, any function $v_T \in U_T^k$ can be written $v_T = v_T^0 + \hat{v}_T$ with $v_T^0 \in U_T^0$ and $\hat{v}_T \in U_T^{k,0}$. Then, we infer from (46) that, for all $T \in \mathcal{T}_h$, $(\boldsymbol{s}_T, \hat{u}_T) \in \boldsymbol{S}_T^k \times U_T^{k,0}$ can be eliminated locally by solving the following saddle point problem with right-hand side depending on $\mathfrak{s}_{\partial T} \in \mathfrak{S}_{\partial T}^k$ and $f_T := f_{|T}$:

$$\begin{aligned} \hat{H}_T(\boldsymbol{s}_T, \boldsymbol{t}_T) - (\boldsymbol{t}_T, \nabla \hat{u}_T)_T &= -H_T((\boldsymbol{0}, \mathfrak{s}_{\partial T}), (\boldsymbol{t}_T, 0)) && \forall\, \boldsymbol{t}_T \in \boldsymbol{S}_T^k, \\ (\boldsymbol{s}_T, \nabla \hat{v}_T)_T &= (f_T, \hat{v}_T)_T + \sum_{F \in \mathcal{F}_T} (\mathfrak{s}_F \varepsilon_{T,F}, \hat{v}_T)_F && \forall\, \hat{v}_T \in U_T^{k,0}, \end{aligned} \quad (55)$$

where $\hat{H}_T(\boldsymbol{s}_T, \boldsymbol{t}_T) := H_T((\boldsymbol{s}_T, 0), (\boldsymbol{t}_T, 0))$. Problem (55) is the counterpart in a mixed context of problem (26a) obtained in the primal context; the further splitting of (55) leading to datum- and trace-based lifting operators is omitted for brevity. Problem (55) is well-posed, since, according to (49) and (48), $\hat{H}_T(\boldsymbol{t}_T, \boldsymbol{t}_T)$ is uniformly equivalent to $\|\boldsymbol{t}_T\|_T^2$ and the inf-sup condition holds. The global (saddle point) problem resulting from the local elimination (55) has the same size and structure as that obtained with the Multiscale Hybrid-Mixed (MHM) method derived in [3, 40] on simplicial meshes.

The second static condensation approach is based on a reformulation of the mixed problem (46) into a primal problem. Following [1, Sect. 3.3], the reformulation is based on the introduction of Lagrange multipliers that enforce the continuity of interface-based flux unknowns and that can be interpreted as potential traces on mesh faces. One can eliminate the cell- and face-based flux unknowns, and, once the reformulation has been performed, one can adapt the arguments of Sect. 3.3.1 to further eliminate locally the cell-based potential unknowns, ending up with a global system only depending on the Lagrange multipliers (face-based potential unknowns). This static condensation approach has the double advantage that it requires to solve local coercive problems (as opposed to local saddle point problems) and that it yields a coercive global problem. For this reason, we discuss it in more detail in Sect. 5.

5 Bridging the Primal and Mixed Forms of the HHO Method

The goal of this section is to bridge the primal and mixed forms of the HHO method studied in Sects. 3 and 4, respectively. As discussed in the previous section, this can be exploited in practice to implement the mixed form of the HHO method in terms of a coercive problem posed on the Lagrange multipliers only.

5.1 Unpatching Interface-Based Flux Unknowns

We introduce a global set of hybrid flux unknowns where interface-based unknowns are two-valued; we refer to these unknowns as unpatched. The unpatched global set of hybrid flux unknowns is defined as

$$\check{\underline{\mathbf{S}}}_h^k := \bigtimes_{T\in\mathcal{T}_h} \underline{\mathbf{S}}_T^k,$$

with subset

$$\check{\underline{\mathbf{S}}}_{h,0}^k := \left\{ \check{\underline{\mathbf{t}}}_h \in \check{\underline{\mathbf{S}}}_h^k \mid \check{\mathfrak{t}}_F \equiv 0, \forall F \in \mathcal{F}_h^{\mathrm{n}} \right\}. \tag{56}$$

Given an element $\check{\underline{\mathbf{t}}}_h \in \check{\underline{\mathbf{S}}}_h^k$, for any $T \in \mathcal{T}_h$, we denote by $\check{\underline{\mathbf{t}}}_T := (\check{\boldsymbol{t}}_T, (\check{\mathfrak{t}}_{T,F})_{F\in\mathcal{F}_T}) \in \underline{\mathbf{S}}_T^k$ its restriction to the element T. For boundary faces $F \in \mathcal{F}_h^{\mathrm{b}}$, the subscript T in $\check{\mathfrak{t}}_{T,F}$ can be omitted, and we simply write $\check{\mathfrak{t}}_F$, as we already did in (56).

Let us introduce the following subspace of $\check{\underline{\mathbf{S}}}_h^k$ (respectively, $\check{\underline{\mathbf{S}}}_{h,0}^k$):

$$\check{\underline{\mathbf{Z}}}_{h(,0)}^k := \left\{ \check{\underline{\mathbf{t}}}_h \in \check{\underline{\mathbf{S}}}_{h(,0)}^k \mid \sum_{T\in\mathcal{T}_F} \check{\mathfrak{t}}_{T,F} = 0, \forall F \in \mathcal{F}_h^{\mathrm{i}} \right\}.$$

It can be easily seen that there exists a natural isomorphism $\underline{\mathbf{J}}_h^k$ from $\check{\underline{\mathbf{Z}}}_h^k$ onto the space $\underline{\mathbf{S}}_h^k$. Note that the restriction of $\underline{\mathbf{J}}_h^k$ to $\check{\underline{\mathbf{Z}}}_{h,0}^k$ defines an isomorphism onto $\underline{\mathbf{S}}_{h,0}^k$.

5.2 Unpatched Mixed Formulation

We begin by extending to $\check{\underline{\mathbf{S}}}_h^k$ the definitions, respectively built from (36) and (38), of the divergence reconstruction operator D_h^k and of the flux reconstruction operator $\boldsymbol{F}_h^{k+1}$, for which we keep the same notation (locally, the definitions are unchanged up to the replacement of $\mathrm{t}_F\varepsilon_{T,F}$ by $\check{\mathrm{t}}_{T,F}$ in face terms). We can then naturally extend the bilinear form H_h, defined in (45) and built from (44), to the product space $\check{\underline{\mathbf{S}}}_h^k \times \check{\underline{\mathbf{S}}}_h^k$.

We next introduce, for all $T \in \mathcal{T}_h$, the additional bilinear form

$$B_T : \underline{\mathbf{S}}_T^k \times \underline{\mathrm{U}}_T^k \to \mathbb{R}; \qquad (\check{\underline{\mathbf{t}}}_T, \underline{\mathrm{v}}_T) \mapsto (v_T, D_T^k \check{\underline{\mathbf{t}}}_T)_T - \sum_{F \in \mathcal{F}_T \cap \mathcal{F}_h^{\mathrm{i}}} (\mathfrak{v}_F, \check{\mathrm{t}}_{T,F})_F, \tag{57}$$

whose global version is as usual obtained by element-by-element assembly:

$$B_h : \check{\underline{\mathbf{S}}}_h^k \times \underline{\mathrm{U}}_h^k \to \mathbb{R}; \qquad (\check{\underline{\mathbf{t}}}_h, \underline{\mathrm{v}}_h) \mapsto \sum_{T \in \mathcal{T}_h} B_T(\check{\underline{\mathbf{t}}}_T, \underline{\mathrm{v}}_T).$$

This bilinear form includes interface terms that enforce the single-valuedness constraints for interface-based flux unknowns. In that vision, the face-based potential unknowns can be seen as Lagrange multipliers.

The unpatched (mixed) HHO discretization of problem (34) then reads: Find $(\check{\underline{\mathbf{s}}}_{h,0}, \check{\underline{\mathrm{u}}}_{h,0}) \in \check{\underline{\mathbf{S}}}_{h,0}^k \times \underline{\mathrm{U}}_{h,0}^k$ such that, for all $(\check{\underline{\mathbf{t}}}_h, \underline{\mathrm{v}}_h) \in \check{\underline{\mathbf{S}}}_{h,0}^k \times \underline{\mathrm{U}}_{h,0}^k$,

$$H_h(\check{\underline{\mathbf{s}}}_{h,0}, \check{\underline{\mathbf{t}}}_h) + B_h(\check{\underline{\mathbf{t}}}_h, \check{\underline{\mathrm{u}}}_{h,0}) = \sum_{F \in \mathcal{F}_h^{\mathrm{d}}} (\check{\mathrm{t}}_F, \psi_\partial)_F - H_h(\check{\underline{\mathbf{s}}}_{h,\partial}, \check{\underline{\mathbf{t}}}_h) - B_h(\check{\underline{\mathbf{t}}}_h, \underline{\mathrm{u}}_{h,\partial}), \tag{58a}$$

$$-B_h(\check{\underline{\mathbf{s}}}_{h,0}, \underline{\mathrm{v}}_h) = (f, v_h) + B_h(\check{\underline{\mathbf{s}}}_{h,\partial}, \underline{\mathrm{v}}_h), \tag{58b}$$

where $\check{\underline{\mathbf{s}}}_{h,\partial} := (\underline{\mathbf{J}}_h^k)^{-1}(\underline{\mathbf{s}}_{h,\partial}) \in \check{\underline{\mathbf{Z}}}_h^k$ is such that, for all $T \in \mathcal{T}_h$, $\check{\underline{\mathbf{s}}}_{T,\partial} = (\mathbf{s}_{T,\partial}, (\mathfrak{s}_{F,\partial}\varepsilon_{T,F})_{F \in \mathcal{F}_T})$, with $\underline{\mathbf{s}}_{h,\partial} \in \underline{\mathbf{S}}_h^k$ defined in Sect. 4.2.1, and where $\underline{\mathrm{u}}_{h,\partial}$ is defined in Sect. 3.2.1. Finally, we define

$$(\check{\underline{\mathbf{s}}}_h, \check{\underline{\mathrm{u}}}_h) := (\check{\underline{\mathbf{s}}}_{h,0} + \check{\underline{\mathbf{s}}}_{h,\partial}, \check{\underline{\mathrm{u}}}_{h,0} + \underline{\mathrm{u}}_{h,\partial}) \in \check{\underline{\mathbf{S}}}_h^k \times \underline{\mathrm{U}}_h^k. \tag{59}$$

5.3 Equivalence Between Primal and Mixed Formulations

The bridge between primal- and mixed-form HHO methods is built in two steps: first, we prove the equivalence between the mixed and unpatched mixed formulations; then, we prove that the unpatched mixed formulation can be recast into a primal formulation.

The following result is an adaptation of [1, Lemma 3.3].

Theorem 5.1 (Equivalence (46)–(58)**)** *Denote by* $(\underline{\mathbf{s}}_{h,0}, u_h) \in \underline{\mathbf{S}}_{h,0}^k \times U_h^k$ *and* $(\underline{\check{\mathbf{s}}}_{h,0}, \underline{\check{\mathrm{u}}}_{h,0}) \in \underline{\check{\mathbf{S}}}_{h,0}^k \times \underline{\mathrm{U}}_{h,0}^k$ *the solutions to* (46) *and* (58)*, respectively. Then,* $\underline{\check{\mathbf{s}}}_{h,0} \in \underline{\check{\mathbf{Z}}}_{h,0}^k$ *and* $\underline{\mathbf{s}}_{h,0} = \underline{\mathbf{J}}_h^k(\underline{\check{\mathbf{s}}}_{h,0})$*, so that* $\underline{\mathbf{s}}_h = \underline{\mathbf{J}}_h^k(\underline{\check{\mathbf{s}}}_h)$ *(*$\underline{\mathbf{s}}_h$ *and* $\underline{\check{\mathbf{s}}}_h$ *are defined in* (47) *and* (59)*, respectively); furthermore,* $u_h = \check{u}_h$ *[recall that* $\check{u}_h$ *denotes the cell-based part of* $\underline{\check{\mathrm{u}}}_h$*, defined in* (59)*].*

Following [1, Sect. 3.3], let us now introduce, for all $T \in \mathcal{T}_h$, the local potential-to-flux mapping operator $\check{\underline{\boldsymbol{\varsigma}}}_T^k : \underline{\mathrm{U}}_T^k \to \underline{\mathbf{S}}_T^k$ such that, for all $\underline{\mathrm{v}}_T \in \underline{\mathrm{U}}_T^k$,

$$H_T(\check{\underline{\boldsymbol{\varsigma}}}_T^k \underline{\mathrm{v}}_T, \underline{\check{\mathbf{t}}}_T) = -B_T(\underline{\check{\mathbf{t}}}_T, \underline{\mathrm{v}}_T) + \sum_{F \in \mathcal{F}_T \cap \mathcal{F}_h^{\mathrm{b}}} (\check{\mathrm{t}}_F, \mathfrak{v}_F)_F \qquad \forall\, \underline{\check{\mathbf{t}}}_T \in \underline{\mathbf{S}}_T^k. \tag{60}$$

This yields a well-posed problem owing to the first inequality in (49). Defining next another local flux reconstruction operator $\check{\boldsymbol{F}}_T^{k+1} : \underline{\mathrm{U}}_T^k \to \boldsymbol{S}_T^{k+1}$ such that

$$\check{\boldsymbol{F}}_T^{k+1} := \boldsymbol{F}_T^{k+1} \circ \check{\underline{\boldsymbol{\varsigma}}}_T^k, \tag{61}$$

one can prove the following result.

Lemma 5.1 (Link Between $\boldsymbol{F}_T^{k+1}$ and p_T^{k+1}) *For all* $\underline{\mathrm{v}}_T \in \underline{\mathrm{U}}_T^k$*, the following holds:*

$$\check{\boldsymbol{F}}_T^{k+1} \underline{\mathrm{v}}_T = \mathbb{M}_T \boldsymbol{\nabla} p_T^{k+1} \underline{\mathrm{v}}_T. \tag{62}$$

Proof Let $\underline{\mathrm{v}}_T \in \underline{\mathrm{U}}_T^k$, and let us plug, for $w \in \mathbb{P}_d^{k+1}(T)$, $\underline{\check{\mathbf{t}}}_T := \underline{\mathbf{I}}_T^k(\mathbb{M}_T \boldsymbol{\nabla} w)$ into (60). Using (57), (36), the polynomial consistency property of Lemma 4.2 coupled to (61), and the one of (42), we get

$$(\check{\boldsymbol{F}}_T^{k+1} \underline{\mathrm{v}}_T, \boldsymbol{\nabla} w)_T = (\boldsymbol{\nabla} v_T, \mathbb{M}_T \boldsymbol{\nabla} w)_T + \sum_{F \in \mathcal{F}_T} (\mathfrak{v}_F - v_T, \mathbb{M}_T \boldsymbol{\nabla} w \cdot \boldsymbol{n}_{T,F})_F, \tag{63}$$

where we have used that $(\check{\boldsymbol{t}}_T, \boldsymbol{\nabla} v_T)_T = (\mathbb{M}_T \boldsymbol{\nabla} w, \boldsymbol{\nabla} v_T)_T$ and $\check{\mathrm{t}}_{T,F} = \mathbb{M}_T \boldsymbol{\nabla} w \cdot \boldsymbol{n}_{T,F}$, owing to (35) and to the fact that $w \in \mathbb{P}_d^{k+1}(T)$. Finally, performing a last integration by parts in (63), and comparing to the definition (6) of p_T^{k+1}, we prove (62).

Now, defining $\check{\underline{\boldsymbol{\varsigma}}}_h^k : \underline{\mathrm{U}}_h^k \to \underline{\check{\mathbf{S}}}_h^k$ such that, for all $\underline{\mathrm{v}}_h \in \underline{\mathrm{U}}_h^k$, and for all $T \in \mathcal{T}_h$, $(\check{\underline{\boldsymbol{\varsigma}}}_h^k \underline{\mathrm{v}}_h)_{|T} := \check{\underline{\boldsymbol{\varsigma}}}_T^k \underline{\mathrm{v}}_T$, we infer from (60) that

$$H_h(\check{\underline{\boldsymbol{\varsigma}}}_h^k \underline{\check{\mathrm{u}}}_h, \underline{\mathbf{t}}_h) = -B_h(\underline{\mathbf{t}}_h, \underline{\check{\mathrm{u}}}_h) + \sum_{F \in \mathcal{F}_h^{\mathrm{d}}} (\check{\mathrm{t}}_F, \psi_\partial)_F \qquad \forall\, \underline{\mathbf{t}}_h \in \underline{\check{\mathbf{S}}}_{h,0}^k, \tag{64}$$

where we have used the fact that $\check{\mathrm{t}}_F \equiv 0$ for all $F \in \mathcal{F}_h^{\mathrm{n}}$ and that $\check{\mathrm{u}}_F = \pi_F^k \psi_\partial$ for all $F \in \mathcal{F}_h^{\mathrm{d}}$. Comparing (64) with (58a), it is readily inferred that $\underline{\check{\mathbf{s}}}_h = \check{\underline{\boldsymbol{\varsigma}}}_h^k \underline{\check{\mathrm{u}}}_h$. Plugging this relation into (58b), we get that

$$-B_h(\check{\underline{\varsigma}}_h^k \check{\underline{u}}_h, \underline{v}_h) = (f, v_h) \qquad \forall \underline{v}_h \in \underline{U}_{h,0}^k. \tag{65}$$

Using again (60), we additionally prove that

$$H_h(\check{\underline{\varsigma}}_h^k \underline{v}_h, \check{\underline{\varsigma}}_h^k \check{\underline{u}}_h) = -B_h(\check{\underline{\varsigma}}_h^k \check{\underline{u}}_h, \underline{v}_h) + \sum_{F \in \mathcal{F}_h^{\mathrm{n}}} (\check{\mathfrak{s}}_F, \mathfrak{v}_F)_F \qquad \forall \underline{v}_h \in \underline{U}_{h,0}^k, \tag{66}$$

where we have used the fact that $\mathfrak{v}_F \equiv 0$ for all $F \in \mathcal{F}_h^{\mathrm{d}}$. Plugging (66) into (65), using the symmetry of H_h, the decomposition $\check{\underline{u}}_h = \check{\underline{u}}_{h,0} + \underline{u}_{h,\partial}$, and the fact that $\check{\mathfrak{s}}_F = \pi_F^k \phi_\partial$ for all $F \in \mathcal{F}_h^{\mathrm{n}}$, we obtain

$$H_h(\check{\underline{\varsigma}}_h^k \check{\underline{u}}_{h,0}, \check{\underline{\varsigma}}_h^k \underline{v}_h) = (f, v_h) - H_h(\check{\underline{\varsigma}}_h^k \underline{u}_{h,\partial}, \check{\underline{\varsigma}}_h^k \underline{v}_h) + \sum_{F \in \mathcal{F}_h^{\mathrm{n}}} (\phi_\partial, \mathfrak{v}_F)_F \qquad \forall \underline{v}_h \in \underline{U}_{h,0}^k. \tag{67}$$

Finally, introducing the global bilinear form $A_h : \underline{U}_h^k \times \underline{U}_h^k \to \mathbb{R}$ such that $A_h(\underline{u}_h, \underline{v}_h) := H_h(\check{\underline{\varsigma}}_h^k \underline{u}_h, \check{\underline{\varsigma}}_h^k \underline{v}_h)$, problem (67) can be rewritten under the form

$$A_h(\check{\underline{u}}_{h,0}, \underline{v}_h) = (f, v_h) - A_h(\underline{u}_{h,\partial}, \underline{v}_h) + \sum_{F \in \mathcal{F}_h^{\mathrm{n}}} (\phi_\partial, \mathfrak{v}_F)_F \qquad \forall \underline{v}_h \in \underline{U}_{h,0}^k. \tag{68}$$

Using (45), (44), (61), and (62), we also infer that

$$A_h(\underline{u}_h, \underline{v}_h) = \sum_{T \in \mathcal{T}_h} (\mathbb{M}_T \nabla p_T^{k+1} \underline{u}_T, \nabla p_T^{k+1} \underline{v}_T)_T + \sum_{T \in \mathcal{T}_h} J_T(\check{\underline{\varsigma}}_T^k \underline{u}_T, \check{\underline{\varsigma}}_T^k \underline{v}_T). \tag{69}$$

Finally, owing to (69), the comparison of problem (68) to problem (14) allows to infer the following result, cf. [1, Sect. 3.3.4].

Theorem 5.2 (Equivalence (14)–(58)) *Let us denote by* $\underline{u}_{h,0} \in \underline{U}_{h,0}^k$ *and* $(\check{\underline{s}}_{h,0}, \check{\underline{u}}_{h,0}) \in \check{\underline{S}}_{h,0}^k \times \underline{U}_{h,0}^k$ *the solutions to* (14) *and* (58), *respectively. Then, up to a choice of stabilization* $j_T(\cdot,\cdot) := J_T(\check{\underline{\varsigma}}_T^k \cdot, \check{\underline{\varsigma}}_T^k \cdot)$ *in* (13) *for problem* (14), $\underline{u}_{h,0} = \check{\underline{u}}_{h,0}$, *so that* $\underline{u}_h = \check{\underline{u}}_h$ *(*$\underline{u}_h$ *and* $\check{\underline{u}}_h$ *are defined in* (15) *and* (59), *respectively).*

The combination of Theorems 5.1 and 5.2 states the equivalence between primal- and mixed-form HHO methods, up to an appropriate choice of stabilization.

From a practical point of view, to compute the solution $(\underline{s}_{h,0}, u_h)$ of the mixed problem (46), it suffices to solve the coercive global problem (68) (once the operator $\check{\underline{\varsigma}}_h^k$ has been computed solving (60) locally in each element) and to use the relation $(\underline{s}_{h,0} + \underline{s}_{h,\partial}, u_h) = (\mathbf{J}_h^k(\check{\underline{\varsigma}}_h^k \check{\underline{u}}_h), \check{u}_h)$ combined with $\check{\underline{u}}_h = \check{\underline{u}}_{h,0} + \underline{u}_{h,\partial}$. Adapting the arguments of Sect. 3.3.1, static condensation can be performed on problem (68), hence leading to a global problem expressed in terms of Lagrange multipliers (face-based potential unknowns) only.

6 Conclusion and Perspectives

HHO methods are very recent polytopal discretization methods which, by now, rest on a firm theoretical basis for elliptic PDEs in primal and mixed forms. Advantages offered by HHO methods are a dimension-independent construction, local conservativity, the possibility to consider an arbitrary polynomial order, a natural treatment of variable diffusion coefficients, and tight computational costs in particular owing to static condensation and an offline/online decomposition of the solution procedure. The price to pay is, on the one hand, the need to solve local problems in the assembly phase (numerical experiments indicate, however, that the relative cost with respect to solving the global problem swiftly decreases as mesh resolution increases). On the other hand, HHO methods are essentially nonconforming (as DG methods) so that some post-processing of the discrete solution may be useful when visualizing the solution on coarse meshes (on fine meshes, the jumps swiftly converge to zero). Note, however, that contrary to interior penalty DG methods, the stabilization does not require user-dependent parameters that must be large enough. Expanding the HHO methodology to systems of quasi-linear or even nonlinear PDEs poses new challenges. Encouraging results (in the linear case) include the robustness with respect to Péclet number in case of advection-diffusion and with respect to incompressibility in linear elasticity, while a nonlinear Leray–Lions problem is addressed in [24]. Another attractive potential application of HHO methods is in the context of multiscale problems, where adequate local problems that take into account the small scales of the problem can be coupled through a global problem posed on a coarse mesh.

References

1. J. Aghili, S. Boyaval, D.A. Di Pietro, Hybridization of mixed high-order methods on general meshes and application to the Stokes equations. Comput. Methods Appl. Math. **15**(2), 111–134 (2015)
2. P.F. Antonietti, S. Giani, P. Houston, *hp*-version composite discontinuous Galerkin methods for elliptic problems on complicated domains. SIAM J. Sci. Comput. **35**(3), A1417–A1439 (2013)
3. R. Araya, C. Harder, D. Paredes, F. Valentin, Multiscale hybrid-mixed method. SIAM J. Numer. Anal. **51**(6), 3505–3531 (2013)
4. D.N. Arnold, An interior penalty finite element method with discontinuous elements. SIAM J. Numer. Anal. **19**, 742–760 (1982)
5. D.N. Arnold, F. Brezzi, B. Cockburn, L.D. Marini, Unified analysis of discontinuous Galerkin methods for elliptic problems. SIAM J. Numer. Anal. **39**(5), 1749–1779 (2002)
6. B. Ayuso de Dios, K. Lipnikov, G. Manzini, The nonconforming virtual element method. ESAIM: Math. Model Numer. Anal. (M2AN) **50**(3), 879–904 (2016)
7. F. Bassi, L. Botti, A. Colombo, D.A. Di Pietro, P. Tesini, On the flexibility of agglomeration based physical space discontinuous Galerkin discretizations. J. Comput. Phys. **231**(1), 45–65 (2012)

8. L. Beirão da Veiga, F. Brezzi, L.D. Marini, Virtual elements for linear elasticity problems. SIAM J. Numer. Anal. **51**(2), 794–812 (2013)
9. L. Beirão da Veiga, K. Lipnikov, G. Manzini, *The Mimetic Finite Difference Method for Elliptic Problems*. MS&A, vol. 11 (Springer, New York, 2014)
10. J. Bonelle, A. Ern, Analysis of compatible discrete operator schemes for elliptic problems on polyhedral meshes. Math. Model. Numer. Anal. **48**, 553–581 (2014)
11. J. Bonelle, A. Ern. Analysis of compatible discrete operator schemes for the Stokes equations on polyhedral meshes. IMA J. Numer. Anal. **35**, 1672–1697 (2015)
12. J. Bonelle, D.A. Di Pietro, A. Ern, Low-order reconstruction operators on polyhedral meshes: application to compatible discrete operator schemes. Comput. Aided Geom. Des. **35–36**, 27–41 (2015)
13. A. Bossavit, Computational electromagnetism and geometry. J. Jpn. Soc. Appl. Electromagn. Mech. **7–8**, 150–159 (no. 1), 294–301 (no. 2), 401–408 (no. 3), 102–109 (no. 4), 203–209 (no. 5), 372–377 (no. 6) (1999–2000)
14. F. Brezzi, L.D. Marini, Virtual elements for plate bending problems. Comput. Methods Appl. Mech. Eng. **253**, 455–462 (2013)
15. F. Brezzi, G. Manzini, L.D. Marini, P. Pietra, A. Russo, Discontinuous Galerkin approximations for elliptic problems. Numer. Methods Partial Differ. Equ. **16**(4), 365–378 (2000)
16. F. Brezzi, K. Lipnikov, M. Shashkov, Convergence of the mimetic finite difference method for diffusion problems on polyhedral meshes. SIAM J. Numer. Anal. **43**(5), 1872–1896 (2005)
17. F. Brezzi, K. Lipnikov, V. Simoncini, A family of mimetic finite difference methods on polygonal and polyhedral meshes. Math. Models Methods Appl. Sci. **15**(10), 1533–1551 (2005)
18. A. Cangiani, E.H. Georgoulis, P. Houston, *hp*-Version discontinuous Galerkin methods on polygonal and polyhedral meshes. Math. Models Methods Appl. Sci. **24**(10), 2009–2041 (2014)
19. P. Castillo, B. Cockburn, I. Perugia, D. Schötzau, An a priori error analysis of the local discontinuous Galerkin method for elliptic problems. SIAM J. Numer. Anal. **38**, 1676–1706 (2000)
20. B. Cockburn, D.A. Di Pietro, A. Ern, Bridging the hybrid high-order and hybridizable discontinuous Galerkin methods. ESAIM: Math. Model Numer. Anal. (M2AN) **50**(3), 635–650 (2016)
21. B. Cockburn, J. Gopalakrishnan, R. Lazarov, Unified hybridization of discontinuous Galerkin, mixed, and continuous Galerkin methods for second-order elliptic problems. SIAM J. Numer. Anal. **47**(2), 1319–1365 (2009)
22. L. Codecasa, R. Specogna, F. Trevisan, A new set of basis functions for the discrete geometric approach. J. Comput. Phys. **19**(299), 7401–7410 (2010)
23. D.A. Di Pietro, Cell-centered Galerkin methods for diffusive problems. Math. Model. Numer. Anal. **46**(1), 111–144 (2012)
24. D.A. Di Pietro, J. Droniou. A hybrid high-order method for Leray-Lions elliptic equations on general meshes. Math. Comp. Accepted for publication. Preprint, arXiv:1508.01918 [math.NA]
25. D.A. Di Pietro, A. Ern, *Mathematical Aspects of Discontinuous Galerkin Methods*. Mathématiques & Applications, vol. 69 (Springer, Berlin/Heidelberg, 2012)
26. D.A. Di Pietro, A. Ern. Arbitrary-order mixed methods for heterogeneous anisotropic diffusion on general meshes. IMA J. Numer. Anal. (2016). Published online. doi:10.1093/imanum/drw003
27. D.A. Di Pietro, A. Ern, Equilibrated tractions for the Hybrid High-Order method. C. R. Acad. Sci. Paris Ser. I **353**, 279–282 (2015)
28. D.A. Di Pietro, A. Ern, A hybrid high-order locking-free method for linear elasticity on general meshes. Comput. Methods Appl. Mech. Eng. **283**, 1–21 (2015)
29. D.A. Di Pietro, A. Ern, Hybrid high-order methods for variable-diffusion problems on general meshes. C. R. Acad. Sci Paris Ser. I **353**, 31–34 (2015)

30. D.A. Di Pietro, J. Droniou, A. Ern, A discontinuous-skeletal method for advection-diffusionreaction on general meshes. SIAM J. Numer. Anal. **53**(5), 2135–2157 (2015)
31. D.A. Di Pietro, S. Lemaire, An extension of the Crouzeix–Raviart space to general meshes with application to quasi-incompressible linear elasticity and Stokes flow. Math. Comput. **84**, 1–31 (2015)
32. D.A. Di Pietro, A. Ern, S. Lemaire, An arbitrary-order and compact-stencil discretization of diffusion on general meshes based on local reconstruction operators. Comput. Methods Appl. Math. **14**(4), 461–472 (2014)
33. K. Domelevo, P. Omnes, A finite volume method for the Laplace equation on almost arbitrary two-dimensional grids. ESAIM Math. Model. Numer. Anal. **39**(6), 1203–1249 (2005)
34. J. Droniou, R. Eymard, A mixed finite volume scheme for anisotropic diffusion problems on any grid. Numer. Math. **105**, 35–71 (2006)
35. J. Droniou, R. Eymard, T. Gallouët, R. Herbin, A unified approach to mimetic finite difference, hybrid finite volume and mixed finite volume methods. Math. Models Methods Appl. Sci. **20**(2), 1–31 (2010)
36. J. Droniou, R. Eymard, T. Gallouët, R. Herbin, Gradient schemes: a generic framework for the discretisation of linear, nonlinear and nonlocal elliptic and parabolic equations. Math. Models Methods Appl. Sci. **23**, 2395–2432 (2013)
37. A. Ern, J.-L. Guermond, Discontinuous Galerkin methods for Friedrichs' systems. I. General theory. SIAM J. Numer. Anal. **44**(2), 753–778 (2006)
38. R. Eymard, T. Gallouët, R. Herbin, Finite volume methods, in *Techniques of Scientific Computing (Part III)*, ed. by P.G. Ciarlet, J.-L. Lions. Handbook of Numerical Analysis, vol. 7 (North-Holland, Amsterdam, 2000), pp. 713–1020
39. R. Eymard, T. Gallouët, R. Herbin, Discretization of heterogeneous and anisotropic diffusion problems on general nonconforming meshes. SUSHI: a scheme using stabilization and hybrid interfaces. IMA J. Numer. Anal. **30**(4), 1009–1043 (2010)
40. C. Harder, D. Paredes, F. Valentin, A family of multiscale hybrid-mixed finite element methods for the Darcy equation with rough coefficients. J. Comput. Phys. **245**, 107–130 (2013)
41. R. Herbin, F. Hubert, Benchmark on discretization schemes for anisotropic diffusion problems on general grids, in *Finite Volumes for Complex Applications V*, ed. by R. Eymard, J.-M. Hérard (Wiley, London, 2008), pp. 659–692
42. C. Lehrenfeld, Hybrid discontinuous Galerkin methods for solving incompressible flow problems. Ph.D. thesis, Rheinisch-Westfälischen Technischen Hochschule Aachen, 2010
43. K. Lipnikov, G. Manzini, A high-order mimetic method on unstructured polyhedral meshes for the diffusion equation. J. Comput. Phys. **272**, 360–385 (2014)
44. E. Tonti, On the formal structure of physical theories. *Quaderni dei Gruppi di Ricerca Matematica del CNR* (1975)
45. J. Wang, X. Ye, A weak Galerkin element method for second-order elliptic problems. J. Comput. Appl. Math. **241**, 103–115 (2013)
46. J. Wang, X. Ye, A weak Galerkin mixed finite element method for second-order elliptic problems. Math. Comput. **83**(289), 2101–2126 (2014)

A Survey of Trefftz Methods for the Helmholtz Equation

Ralf Hiptmair, Andrea Moiola, and Ilaria Perugia

Abstract Trefftz methods are finite element-type schemes whose test and trial functions are (locally) solutions of the targeted differential equation. They are particularly popular for time-harmonic wave problems, as their trial spaces contain oscillating basis functions and may achieve better approximation properties than classical piecewise-polynomial spaces.

We review the construction and properties of several Trefftz variational formulations developed for the Helmholtz equation, including least squares, discontinuous Galerkin, ultra weak variational formulation, variational theory of complex rays and wave based methods. The most common discrete Trefftz spaces used for this equation employ generalised harmonic polynomials (circular and spherical waves), plane and evanescent waves, fundamental solutions and multipoles as basis functions; we describe theoretical and computational aspects of these spaces, focusing in particular on their approximation properties.

One of the most promising, but not yet well developed, features of Trefftz methods is the use of adaptivity in the choice of the propagation directions for the basis functions. The main difficulties encountered in the implementation are the assembly and the ill-conditioning of linear systems, we briefly survey some strategies that have been proposed to cope with these problems.

The original version of this chapter was revised. An erratum to this chapter can be found at DOI 10.1007/978-3-319-41640-3_14.

R. Hiptmair (✉)
Seminar for Applied Mathematics, ETH Zürich, 8092 Zürich, Switzerland
e-mail: hiptmair@sam.math.ethz.ch; ralfh@ethz.ch

A. Moiola
Department of Mathematics and Statistics, University of Reading,
Whiteknights, PO Box 220, Reading RG6 6AX, UK
e-mail: a.moiola@reading.ac.uk

I. Perugia
Faculty of Mathematics, University of Vienna, 1090 Vienna, Austria

Department of Mathematics, University of Pavia, 27100 Pavia, Italy
e-mail: ilaria.perugia@univie.ac.at

G.R. Barrenechea et al. (eds.), *Building Bridges: Connections and Challenges in Modern Approaches to Numerical Partial Differential Equations*,
Lecture Notes in Computational Science and Engineering 114,
DOI 10.1007/978-3-319-41640-3_8

1 Introduction

Given a linear PDE, a Trefftz method is a volume-oriented discretisation scheme, for which all trial and test functions, when restricted to any element of a given mesh, are solutions of the PDE under consideration. The name comes from the work [112] of Trefftz, dating back to 1926, where this idea was applied to the Laplace equation. Since then, several versions of Trefftz methods have been proposed and applied to a range of PDEs by different groups of mathematicians, engineers and computational scientists, often unaware of each other. Typical PDEs addressed are *linear*, with *piecewise-constant coefficients* and *homogeneous*, i.e. with vanishing volume source term.

Trefftz methods are related to both finite element (FEM) and boundary element methods (BEM). With the former they have in common that they provide a discretisation in the volume. With the latter they share some characteristics such as the need of integration on lower-dimensional manifolds only. Compared to conventional FEMs, Trefftz methods have attracted attention mainly for two reasons: (i) they may need much fewer degrees of freedom than standard schemes to achieve the same accuracy, and (ii) they incorporate some properties of the problem's solution (such as oscillatory character, wavelength, maximum principle, boundary layers) in the trial spaces, and thus also in the discrete solution. In addition, compared to BEMs, an advantage of Trefftz schemes is that they do not require the evaluation of singular integrals.

Comparing with finite and boundary elements, in 1997 Zienkiewicz [121] stated: "*...it seems without doubt that in the future Trefftz type elements will frequently be encountered in general finite element codes.... It is the author's belief that the simple Trefftz approach will in the future displace much of the boundary type analysis with singular kernels.*" While this prediction has not yet come true, in the last years more and more work has been devoted to the formulation, the analysis and the validation of these methods and substantial progress has been accomplished.

In this chapter we survey Trefftz finite element methods for the *homogeneous Helmholtz equation* ($-\Delta u - k^2 u = 0$), which models acoustic wave propagation in time-harmonic regime. For medium and high frequencies, i.e. for values of kL in a range of 10^2 to 10^4, where $k > 0$ is the wavenumber, and L a characteristic length of the region of interest, the numerical solution of the Helmholtz equation in 2D and 3D is particularly challenging. A main reason is that Helmholtz solutions oscillate with a wavelength proportional to the inverse of k. Hence, piecewise polynomials do not provide efficient approximation. Trefftz schemes are thus particularly relevant as they can improve on the point where (polynomial) FEMs fail: the approximation properties of the basis functions. Moreover, some Trefftz methods can remedy other shortcomings that often haunt discretisations of time-harmonic problems, such as the lack of coercivity and the presence of minimal resolution conditions to guarantee unique solvability. Theorem 2 in this chapter is an example. Earlier overviews of Trefftz schemes for the Helmholtz equation, together with numerous references, can be found in [98], [85, Chap. 1] and [76, Chap. 3]. Surveys of Trefftz schemes for other equations are in [67, 75, 99, 121].

For most of the Trefftz spaces used, continuity across interfaces separating mesh elements cannot be enforced strongly, as Trefftz functions are not as "flexible" as piecewise polynomials. As a consequence, the standard Helmholtz variational formulation posed in subspaces of the Sobolev space H^1 is not applicable and discretisations must be used that can accommodate discontinuous trial functions. A wide array of different variational formulations has been proposed and in Sect. 2 we attempt a classification and a comparison of the best known. We identify three main classes of formulations: (i) *least squares* (LS, Sect. 2.1), where squares of suitable norms of residuals are minimised; (ii) *discontinuous Galerkin* (DG, Sect. 2.2), whose formulations arise from local integration by parts and which may or may not use Lagrange multipliers on mesh interfaces; (iii) *weighted residual* (Sect. 2.3), which are defined by testing residuals against suitable traces of test functions. The methods discussed include: the Trefftz-discontinuous Galerkin (TDG), the ultra weak variational formulation (UWVF), the discontinuous enrichment method (DEM), the variational theory of complex rays (VTCR) and the wave based method (WBM). Moreover, in the spirit of the symposium that led up to the present volume, to "build bridges" with a wider portion of the literature and of the computational PDE community, in Sect. 2.4 we describe some older Trefftz schemes defined on a single element and in Sect. 2.5 we consider some methods that are not Trefftz but use oscillating basis functions that are "approximately Trefftz", such as the partition of unity method (PUM). To easily compare them, we write all formulations for the same Robin–Dirichlet model boundary value problem (see Sect. 1.1).

In Sect. 2 we completely gloss over the choice of *basis functions and discrete spaces* employed, whose description is postponed to Sect. 3. This is because, apart from few exceptions such as unbounded elements, any Trefftz discrete space can be employed in any Trefftz variational formulation. We believe that separating the discussion of the two main components in the definition of a Trefftz method, i.e. variational formulations and discrete spaces, will make the presentation clearer. The most common basis functions for Trefftz methods are plane waves ($\mathbf{x} \mapsto e^{ik\mathbf{d}\cdot\mathbf{x}}$ for a fixed unit vector $\mathbf{d}$) and generalised harmonic polynomials (i.e. circular/spherical waves, products of circular/spherical harmonics and Bessel functions), for which quite a complete approximation theory exists, see Sects. 3.1 and 3.2. Other basis functions include fundamental solutions, multipoles, evanescent waves and corner waves. We note that, since the Helmholtz operator is the sum of a second- and a zero-order term, no non-vanishing piecewise-polynomial Trefftz function is possible.

In this chapter we state a few theorems, none of them is entirely new. Lemma 1 exemplifies the technique of [88] to control the L^2 norm of Trefftz functions with mesh-dependent norms containing interface jumps. If a Trefftz method is well-posed in a suitable skeleton norm, this allows to control the error in the volume; we do this for the LS method in Theorem 1 and for the TDG method (well-posed by Theorem 2) in Corollary 1. This can be combined with the approximation results for circular/spherical and plane waves in Sects. 3.1 and 3.2. In brief: we provide the tools to derive stability and orders of L^2-convergence in the volume for all Trefftz methods that are well-posed in suitable skeleton norms.

Trefftz methods suffer from two main problems: *ill-conditioning* due to the poor linear independence of the basis functions, and the need for *numerical quadrature for oscillating integrands*. On the other hand, since the PDE is solved exactly in each element, only low-dimensional integrals on the mesh skeleton need to be evaluated, leading to massively reduced computational cost for the assembly of the linear systems. Moreover, if plane wave bases are used, on any polygonal/polyhedral mesh the integrals can be computed analytically in a cheap way. In Sect. 4 we briefly review strategies developed to deal with the computation of matrix entries and to cope with ill-conditioning.

Some Trefftz methods also provide an attractive framework for implementing non-standard adaptive policies, like directional adaptivity following dominant wave directions. This is made possible, because plane wave-type Trefftz functions naturally encode a direction of propagation. More details are given in Sect. 4.2.

As mentioned, in this chapter we only discuss the Helmholtz equation, i.e. acoustic problems, and constant material parameters. The discrete Trefftz spaces used for the Helmholtz equation with variable coefficients are briefly addressed in Sect. 3.4. Other time-harmonic wave problems that have been tackled with Trefftz methods include electromagnetism (Maxwell equations) [18, 85], linearised Euler equation and general hyperbolic systems [37], linear elasticity (Navier equation) [76], (fourth order) Kirchhoff–Love plates [27, 71, 76, 102], Koiter's linear shell theory [102], poro-elasticity [27, Sect. 5.4], coupled vibro-acoustic problems [27]. A list of applications and references can be found in [25, Sect. 5.1] (with a focus in vibrational mechanics) and in [76, 85]. A related application is tackled by the *method of particular solutions* (MPS) of [15, 36], which uses Helmholtz solutions to approximate Laplace eigenvalue problems; in this setting the wavenumber is part of the unknowns. For recent work on space–time Trefftz methods for wave propagation in time-domain see [69] and references therein.

Several comparisons of the numerical performances of different Trefftz schemes for simple model problems have been published, e.g. [7] (PUM, DEM, generalised FEM), [40] (LS, UWVF), [61] (PUM, UWVF), [39] (DG, UWVF, LS), [115] (DEM, UWVF, PUM), [59] (LS, UWVF, VTCR), where we have included the PUM even if strictly speaking it is not a Trefftz method. However, from these results it is difficult to conclude that any formulation is clearly preferable from a computational point of view. A general conclusion might be that, in order to achieve the best accuracy and conditioning, the choice of the approximation space matters more than that of the variational formulation. We reiterate that these two choices are mutually independent: any Trefftz discrete space might be used in any Trefftz variational formulation. We make some further concluding remarks in Sect. 5.

1.1 Model Boundary Value Problem

We rely on a simple model boundary value problem (BVP) for the Helmholtz equation that will be used to describe and compare the different Trefftz methods. Let

$\Omega \subset \mathbb{R}^n, n = 2, 3$, be a bounded, Lipschitz, connected domain, with $\partial\Omega = \Gamma_D \cup \Gamma_R$, where Γ_D and Γ_R are disjoint components of $\partial\Omega$; $\Gamma_R \neq \emptyset$ while Γ_D might be empty. Denote by $\mathbf{n}$ the outward-pointing unit normal vector field on $\partial\Omega$. We consider the homogeneous Robin–Dirichlet BVP

$$\begin{aligned} -\Delta u - k^2 u &= 0 && \text{in } \Omega, \\ u &= g_D && \text{on } \Gamma_D, \\ \frac{\partial u}{\partial \mathbf{n}} + \mathrm{i}k\vartheta u &= g_R && \text{on } \Gamma_R. \end{aligned} \tag{1}$$

Here g_D and g_R are the boundary data, i is the imaginary unit, $k \in \mathbb{R}$ (the wavenumber) and ϑ (the impedance parameter) are positive constants. We assume that Ω, g_D and g_R are such that $u \in H^{3/2+s}(\Omega)$, for some $s > 0$. In typical sound-soft acoustic scattering problems, Γ_D represents the boundary of the scatterer, and Γ_R stands for an artificial truncation of the unbounded region where waves propagate; see e.g. [55, Sect. 2].

Simple generalisations of the BVP (1) that can be tackled by Trefftz methods are:

- Neumann boundary conditions $\partial u/\partial \mathbf{n} = g_N$ on Γ_D;
- discontinuous and piecewise-constant wavenumber k;
- piecewise constant and discontinuous tensor coefficient $\boldsymbol{A}$ in the more general Helmholtz equation $-\nabla \cdot (\boldsymbol{A}\nabla u) - k^2 u = 0$, e.g. [60] and [18, Chap. I.5];
- spatially varying impedance $0 < \vartheta \in L^\infty(\Gamma_R)$;
- absorbing media $k \in \mathbb{C}$;
- inhomogeneous Helmholtz equation $-\Delta u - k^2 u = f$, where the source term f might be either localised [37, Sect. 5], [25, 57, 58], or not [1, Sect. 2.2];
- scattering in unbounded domains;
- scattering by periodic diffraction gratings in [20, 119];
- scattering by screens (i.e. manifolds with boundary, leading to non-Lipschitz computational domains) in [120].

The presence of smoothly varying coefficients is more challenging for Trefftz methods, as in general no Trefftz functions in analytical form are available; this extension is briefly addressed in Sect. 3.4.

1.2 Notation

We introduce a finite element partition $\mathcal{T}_h = \{K\}$ of Ω, not necessarily conforming. We write $\mathbf{n}_K$ for the outward-pointing unit normal vector on ∂K, and h for the mesh width of $\mathcal{T}_h$, i.e. $h := \max_{K \in \mathcal{T}_h} h_K$, with $h_K := \operatorname{diam} K$. We denote by $\mathcal{F}_h := \bigcup_{K \in \mathcal{T}_h} \partial K$ and $\mathcal{F}_h^I := \mathcal{F}_h \setminus \partial\Omega$ the skeleton of the mesh and its inner part.

We also introduce some standard DG notation. Given two elements $K_1, K_2 \in \mathcal{T}_h$, a piecewise-smooth function v and vector field $\boldsymbol{\tau}$ on $\mathcal{T}_h$, we define on $\partial K_1 \cap \partial K_2$

$$\begin{aligned} &\text{the averages:} && \{\!\{v\}\!\} := \tfrac{1}{2}(v_{|K_1} + v_{|K_2}), && \{\!\{\boldsymbol{\tau}\}\!\} := \tfrac{1}{2}(\boldsymbol{\tau}_{|K_1} + \boldsymbol{\tau}_{|K_2}), \\ &\text{the normal jumps:} && [\![v]\!]_N := v_{|K_1}\mathbf{n}_{K_1} + v_{|K_2}\mathbf{n}_{K_2}, && [\![\boldsymbol{\tau}]\!]_N := \boldsymbol{\tau}_{|K_1}\cdot\mathbf{n}_{K_1} + \boldsymbol{\tau}_{|K_2}\cdot\mathbf{n}_{K_2}. \end{aligned}$$

We denote by ∇_h the element-wise application of the gradient ∇, and write $\partial_{\mathbf{n}} = \mathbf{n}\cdot\nabla_h$ on $\partial\Omega$ and $\partial_{\mathbf{n}_K} = \mathbf{n}_K \cdot \nabla_h$ on ∂K for the normal derivatives.

For $s > 0$, define the broken Sobolev space $H^s(\mathcal{T}_h)$ and the *Trefftz space* $T(\mathcal{T}_h)$:

$$\begin{aligned} H^s(\mathcal{T}_h) &:= \big\{v \in L^2(\Omega) : \; v_{|K} \in H^s(K) \; \forall K \in \mathcal{T}_h\big\}, \\ T(\mathcal{T}_h) &:= \big\{v \in H^1(\mathcal{T}_h) : \; -\Delta v - k^2 v = 0 \text{ in } K \text{ and } \partial_{\mathbf{n}_K} v \in L^2(\partial K) \; \forall K \in \mathcal{T}_h\big\}. \end{aligned}$$

The discrete Trefftz space $V_p(\mathcal{T}_h)$ is a finite-dimensional subspace of $T(\mathcal{T}_h)$ and can be represented as $V_p(\mathcal{T}_h) = \bigoplus_{K\in\mathcal{T}_h} V_{p_K}(K)$, where $V_{p_K}(K)$ is a p_K-dimensional subspace of $T(\mathcal{T}_h)$ of functions supported in K. We use the terms h-convergence to mean the convergence of a sequence of numerical solutions to u when the mesh $\mathcal{T}_h$ is refined, i.e. $h \to 0$, p-convergence to designate the convergence when the local spaces are enriched, i.e. $p := \min_{K\in\mathcal{T}_h} p_K \to \infty$, and hp-convergence to mean the convergence for a suitable combination of the two refinement strategies. We remark that when non-polynomial spaces are used, as it is the case for Trefftz methods in frequency domain, it is not obvious how to define the "degree" of a space, thus p_K denotes the local number of degrees of freedom. Finally, we denote by Re $\{\cdot\}$, Im $\{\cdot\}$ and $\bar{\cdot}$ the real part, the imaginary part and the conjugate of a complex value.

We note that some of the methods in Sect. 2, such as the TDG, the UWVF and the VTCR, involve sesquilinear forms (i.e. test functions are conjugated) while others, such as the DEM and the WBM, involve bilinear forms (test functions are not conjugated). Any method (if no unbounded elements are used) can be modified to either form, even though sesquilinear forms are more amenable to stability and error analysis; for each method we follow the conventions of the references we cite.

1.3 Estimation of the $L^2(\Omega)$ Norm of (Piecewise) Trefftz Functions

Given two uniformly positive functions $\lambda \in L^\infty(\mathcal{F}_h^I \cup \Gamma_D)$ and $\sigma \in L^\infty(\mathcal{F}_h^I \cup \Gamma_R)$, we introduce the following *skeleton* seminorm (defined e.g. on $H^{3/2+\varepsilon}(\mathcal{T}_h)$, $\varepsilon > 0$):

$$\begin{aligned} |||v|||_{\lambda,\sigma}^2 := \|\sigma[\![\nabla_h v]\!]_N\|_{L^2(\mathcal{F}_h^I)}^2 + \|\lambda[\![v]\!]_N\|_{L^2(\mathcal{F}_h^I)}^2 \\ + \|\sigma(\partial_{\mathbf{n}} v + \mathrm{i}k\vartheta v)\|_{L^2(\Gamma_R)}^2 + \|\lambda v\|_{L^2(\Gamma_D)}^2 \,. \end{aligned} \quad (2)$$

A special property of the Trefftz space $T(\mathcal{T}_h)$ is that this seminorm is actually a norm for it, and that it controls the $L^2(\Omega)$ norm, as it was first proved by P. Monk and D.Q. Wang using a special duality technique in [88, Theorem 3.1].

Lemma 1 *$|||\cdot|||_{\lambda,\sigma}$ is a norm in $T(\mathcal{T}_h)$. Moreover, all Trefftz functions $v \in T(\mathcal{T}_h) \cap H^{3/2+\varepsilon}(\mathcal{T}_h)$, $\varepsilon > 0$, satisfy the estimate*

$$\|v\|_{L^2(\Omega)} \le C_* |||v|||_{\lambda,\sigma},$$

with a constant $C_* > 0$ depending only on $k, \lambda, \sigma, \vartheta, \Omega$ and $\mathcal{T}_h$. Setting

$$\sigma_K := \operatorname{ess\,inf}_{\mathbf{x}\in\partial K\setminus\Gamma_D} \sigma(\mathbf{x}), \quad \lambda_K := \operatorname{ess\,inf}_{\mathbf{x}\in\partial K\setminus\Gamma_R} \lambda(\mathbf{x}) \qquad \forall K \in \mathcal{T}_K,$$

we can express the dependence of C_* on the relevant parameters in the following situations:

(i) If $\partial\Omega = \Gamma_R$ and Ω is either convex or smooth and star-shaped with respect to a ball, then

$$\|v\|_{L^2(\Omega)} \le C_1 \operatorname{diam}\Omega \max_{K\in\mathcal{T}_h} \left(\left(\frac{1}{\sigma_K^2 k} + \frac{k}{\lambda_K^2} \right) \left(1 + \frac{1}{kh_K} \right) \right)^{1/2} |||v|||_{\lambda,\sigma},$$

where $C_1 > 0$ depends on ϑ, the shape-regularity of the mesh and the shape of Ω.

(ii) If $k > 1$, $\Omega \subset \mathbb{R}^2$ has diameter $\operatorname{diam}\Omega = 1$ and satisfies

$$\mathbf{x}\cdot\mathbf{n} \ge \gamma > 0 \quad \text{a.e. on } \Gamma_R \text{ and} \quad \mathbf{x}\cdot\mathbf{n} \le 0 \quad \text{a.e. on } \Gamma_D, \tag{3}$$

and each element K is star-shaped with respect to a ball of radius $\rho_K h_K$, we have

$$\|v\|_{L^2(\Omega)} \le C_2 \max_{K\in\mathcal{T}_h} \left(\left(\frac{1}{\sigma_K^2 k} + \frac{k}{\lambda_K^2} \right) \left((kh_K)^{2t} + \frac{1}{kh_K} \right) \right)^{1/2} |||v|||_{\lambda,\sigma},$$

where $0 < t < s_\Omega \le 1/2$, s_Ω being the "elliptic regularity parameter" of [55, Eq. (6)], and $C_2 > 0$ depends only on Ω, ϑ, t, and on the shape-regularity $\inf_{K\in\mathcal{T}_h} \rho_K$ of the mesh.

The bound in part (i) of Lemma 1 can be verified following the proof of [85, Lemma 4.3.7], while that in part (ii) requires also the stability and trace estimates of [56, Eq. (7), (20)] (see also [56, Lemma 4.5] and a weaker but more general bound in [55, Lemma 4.4]). Conditions (3) on the shape of Ω are satisfied if Γ_R is boundary of a domain star-shaped with respect to a ball centred at $\mathbf{0}$ and Γ_D is boundary of a smaller domain (a scatterer, or a "hole" in Ω) star-shaped with respect to $\mathbf{0}$, see [55, Sect. 2, Fig. 2]. The value of the bounding constants arise only from (a) trace estimates for mesh elements, and (b) stability bounds for an inhomogeneous Helmholtz BVP on Ω, thus more general shapes of Ω give different dependencies on k (using e.g. the k-explicit $H^1(\Omega)$ bounds in [30, Theorem 2.4], [105, Theorem 1.6], and bounds in higher-order norms as in [41, Lemma 2.12]). This result is relevant because, for Trefftz methods that allow a priori stability or error estimates, these are typically in a skeleton norm similar to $|||\cdot|||_{\lambda,\sigma}$. Thus Lemma 1 can lead to error estimates in the mesh- and parameter-independent $L^2(\Omega)$ norm; we pursue this in Sects. 2.1 and 2.2.1.

2 Trefftz Variational Formulations

2.1 Least Squares (LS) Methods

Least squares methods are perhaps the simplest kind of Trefftz formulations. They allow simple error and stability analysis, are easy to implement, lead to sign-definite Hermitian (or symmetric) linear systems, at the price of a possibly worse conditioning. A description of Trefftz LS schemes for the Helmholtz equation with numerous references is given by Stojek in [107]. The same method is named *frameless Trefftz elements* in [99, Sect. 3.6] and *weighted variational formulation* (WVF) in [59]. In [88], Monk and Wang proposed the following Trefftz LS method for the BVP (1):

$$\text{find} \qquad u_{\text{LS}} = \operatorname*{arg\,min}_{v_{hp}\in V_p(\mathcal{T}_h)} J(v_{hp}; g_R, g_D), \qquad \text{where}$$

$$J(v; g_R, g_D) := \int_{\mathcal{F}_h^I} \left(\lambda^2 \big|[\![v]\!]_N\big|^2 + \sigma^2 \big|[\![\nabla_h v]\!]\big|^2\right) \mathrm{d}S + \int_{\Gamma_R} \sigma^2 \big|\partial_{\mathbf{n}} v + \mathrm{i}k\vartheta\, v - g_R\big|^2 \,\mathrm{d}S + \int_{\Gamma_D} \lambda^2 \big|v - g_D\big|^2 \,\mathrm{d}S, \tag{4}$$

where $[\![\nabla v]\!] := \nabla_h v_{|K_1} - \nabla_h v_{|K_2}$ on $\partial K_1 \cap \partial K_2$ is the jump of the complete gradient (whose "sign" depends on a choice of the ordering of the elements in $\mathcal{F}_h$). The LS methods in [107, Eq. (7)] and [75, Chap. 10] differ from (4) (apart from the use of different boundary conditions) in that only the normal component of the jump of the gradient $[\![\nabla_h v]\!]_N$ is penalised on $\mathcal{F}_h^I$, as opposed to the entire jump $[\![\nabla_h v]\!]$. Obviously, every Galerkin discretisation of the variational problem arising from (4) will give rise to a Hermitian linear system, which is a clear advantage of LS methods.

The choice of the relative weights $0 < \lambda, \sigma \in L^\infty(\mathcal{F}_h)$ between the terms in (4) is a crucial point for the conditioning and the accuracy of LS methods. Different choices have been proposed (for 2D problems): $\sigma = 1$ and $\lambda = k$ or $\lambda_{|e} = 1/h_e$ in [88, Sect. 2]; $\lambda = 1$ and $\sigma_{|e} = h_e/(p_{K_1} + p_{K_2})$ in [107, Sect. 3.2]; $\lambda = 1$ and $\sigma_{|e} = \mathcal{O}(\max\{p_{K_1}, p_{K_2}\}^{-1/2})$ in [75, Theorem 10.3.4]. Here, $e = \partial K_1 \cap \partial K_2$ denotes a mesh interface, h_e its length, p_{K_1} and p_{K_2} the dimensions of the local

Trefftz spaces $V_{p_{K_1}}(K_1)$ and $V_{p_{K_2}}(K_2)$ on the adjacent elements K_1 and K_2. In 2D and 3D, [59] suggests to choose $\sigma = 1$ and $\lambda = k$ and, for BVPs with singular solutions, $\sigma_{|\Gamma_R} = k^{1/2}$.

The LS method computes the element $u_{\rm LS}$ in $V_p(\mathcal{T}_h)$ that minimises the error $u - u_{\rm LS}$ measured in the skeleton norm $\|v\|_{\rm LS}^2 := J(v; 0, 0)$, thus orders of converge in this norm follow immediately from approximation bounds for the specific discrete Trefftz space $V_p(\mathcal{T}_h)$ chosen, see e.g. Sect. 3 below or [88]. Since $|||v|||_{\lambda,\sigma} \leq \|v\|_{\rm LS}$ (with equality if J in (4) is defined with $[\![\nabla_h v]\!]_N$ instead of $[\![\nabla_h v]\!]$), Lemma 1, following [88, Theorem 3.1], guarantees that the $L^2(\Omega)$ norm of the error of the LS solution is controlled by the value of the LS functional, thus convergence follows also in Ω. This is summarised in Theorem 1, see Sect. 1.3 for the extension to different domains.

Theorem 1 *Let u be the solution of (1) and $u_{\rm LS} \in V_p(\mathcal{T}_h)$ the discrete LS solution of (4). Then, for $C_* > 0$ depending only on $k, \lambda, \sigma, \vartheta, \Omega$ and $\mathcal{T}_h$,*

$$\|u - u_{\rm LS}\|_{\rm LS} = \inf_{v_{hp} \in V_p(\mathcal{T}_h)} \left\|u - v_{hp}\right\|_{\rm LS},$$

$$\|u - u_{\rm LS}\|_{L^2(\Omega)} \leq C_* \inf_{v_{hp} \in V_p(\mathcal{T}_h)} \left\|u - v_{hp}\right\|_{\rm LS}.$$

If $\lambda = k$, $\sigma = 1$, $\partial\Omega = \Gamma_R$ and Ω is either convex or smooth and star-shaped, then

$$\|u - u_{\rm LS}\|_{L^2(\Omega)} \leq C_0 \operatorname{diam} \Omega \, k^{-1/2} \left(1 + \left(k \min_{K \in \mathcal{T}_h} h_K\right)^{-1/2}\right) \inf_{v_{hp} \in V_p(\mathcal{T}_h)} \left\|u - v_{hp}\right\|_{\rm LS},$$

where $C_0 > 0$ depends only on ϑ, the shape of Ω and the shape-regularity of $\mathcal{T}_h$.

The *hp*-convergence theory of [56] easily extends to the LS method. In 2D, if the LS parameters are defined as $\lambda_{|e}^2 = kh/\min\{h_{K_1}, h_{K_2}\}$ for $e = \partial K_1 \cap \partial K_2$, $\lambda_{|e}^2 = kh/h_K$ for $e \subset \partial K \cap \Gamma_D$, and $\sigma^2 = 1/k$, under the assumptions on Ω and on the discretisation stipulated in [56], then the $\|\cdot\|_{\rm LS}$ norm of the LS error is estimated as in [56, Eq. (48)] and the $L^2(\Omega)$ norm of the same error converges to zero exponentially in the square root of the total number of degrees of freedom used.

In [75, Chap. 10], the Trefftz LS scheme is analysed for pure Dirichlet boundary conditions ($\Gamma_R = \emptyset$); the crucial parameter in the analysis is the relative distance between k^2 and the closest Dirichlet eigenvalue of $-\Delta$. Error bounds in the broken Sobolev norm $H^1(\mathcal{T}_h)$ are derived.

In the numerical tests in [39, 40], the LS method appears to be slightly less accurate than the UWVF (see Sect. 2.2.2 below) and a DG method, all employed with the same discrete space. On the other hand, in the examples in [59], the performance of the LS method is comparable to that of the UWVF and considerably better than that of the VTCR.

2.1.1 The Method of Fundamental Solutions (MFS)

A popular class of LS Trefftz methods is the method of fundamental solutions. A lucid introduction to the MFS for Helmholtz problems, together with numerous references, is in [31]. The MFS is considered a special case of *source simulation technique* in [92]. The characteristic features of the most common form of the MFS are: (i) the domain is not meshed; (ii) the N basis functions are fundamental solutions ($H_0^{(1)}(k|\mathbf{x}-\mathbf{y}_\ell|)$ in 2D, $\ell = 1,\dots,N$, where $H_0^{(1)}$ is a Hankel function of the first kind and order zero and $\mathbf{y}_\ell \in \mathbb{R}^2 \setminus \overline{\Omega}$, see Sect. 3.3); (iii) the minimisation of the $L^2(\partial\Omega)$ norm of the error is substituted by the minimisation of the squared error over $M \geq N$ points $\mathbf{x}_j \in \partial\Omega, j = 1,\dots,M$. If $M = N$, the MFS is not an LS method but it simply interpolates the boundary conditions with Trefftz functions.

The same method with plane wave bases is compared to the MFS in [1]. A variant that is popular in acoustics is the *Helmholtz equation least-squares* (HELS) method, which uses spherical-wave and multipole basis functions, see the recent book [117] and references therein. LS variants of MFS relying on higher order multipoles in addition to simple Hankel functions have a long history in wave simulations [90, Sect. 2].

The locations $\mathbf{y}_\ell$ of the basis singularities are either obtained numerically together with the coefficients multiplying the basis functions using non-linear LS solvers [31, Eq. (7)] (leading to a highly adaptive method), or can be fixed a priori on a smooth boundary in $\mathbb{R}^n \setminus \overline{\Omega}$, e.g. using complex analysis techniques (in 2D) as in [9], or are determined based on heuristic criteria [90, Sect. 3].

The MFS with fixed nodes can be interpreted as a discretisation of a compact transfer operator related to a single layer potential representation. For this reason it yields ill-conditioned linear systems; however this does not rule out efficient computations as demonstrated and analysed in [9] and in [10, Sect. 7]. According to [31, p. 766], the larger the distance between the nodes and Ω, the more ill-conditioned the linear system and the more accurate the solution (though this might seem counter-intuitive).

A strength of the MFS is its simplicity of implementation, as no mesh is needed and all geometric information is contained in only two point sets $\{\mathbf{y}_\ell\}_{\ell=1}^N \subset \mathbb{R}^n \setminus \overline{\Omega}$, $\{\mathbf{x}_j\}_{j=1}^M \subset \partial\Omega$. Since fundamental solutions satisfy the Sommerfeld radiation conditions, the MFS is often used for scattering problems in unbounded domains.

In [9], the convergence of the MFS for Dirichlet problems on a circular domain is analysed in great detail, and a special design of the curve supporting the fundamental solutions is proposed for general domains with analytic boundaries. With this choice, extremely accurate and cheap computations are possible.

In [10], Barnett and Betcke present a finite element scheme that couples the LS formulation of [107] with the MFS in 2D. They consider the scattering by sound-soft (non-convex) polygons; the total field is approximated inside an artificial boundary and the scattered field outside of it. Singular Fourier–Bessel basis functions depending on the scatterer's corners (see Sect. 3.4) are used on all elements adjacent to the scatterer, strongly enforcing the (homogeneous) Dirichlet boundary conditions; due to this, no terms on $\partial\Omega$ appear in the method formulation. Exponential orders of convergence are proved. The strong enforcement of boundary conditions may be substituted by an LS approach to deal with more general linear boundary conditions, curved boundaries and transmission problems.

2.2 *Discontinuous Galerkin (DG) Methods*

The discontinuous Galerkin (DG) methods constitute a wide class of numerical schemes for the approximation of PDEs, employing discontinuous test and trial functions [6]. A great number of tools for their design, implementation and error analysis have been devised, so they are a natural setting for Trefftz methods. In [54] we showed that when the interior penalty (IP) method, one of most common DG schemes, is applied to the Laplace equation, the use of Trefftz spaces (made of harmonic polynomials) offers better accuracy than standard spaces also in an *hp*-context. Similar considerations were made in [74] for the h-convergence of the local DG (LDG) method. To our knowledge, no *standard* DG variational formulation (e.g. any of those in [6]) has been proposed in the literature to discretise time-harmonic problems with Trefftz basis functions. Possible reasons for this are that the error analysis of standard DG schemes requires inverse estimates, which are well-known for polynomial spaces but harder in the Trefftz case (however, see [46, Sect. 3.2] for h-explicit inverse estimates for plane waves in 2D), and that the application of formulations designed for the Laplace equation to the Helmholtz case requires some problematic minimal resolution condition to ensure unique solvability [83].

In the next sections we outline some DG formulations that have been designed specifically for Trefftz discretisations; some of these have later been employed also with polynomial approximating spaces, e.g. [83, 89].

A note on terminology: all Trefftz methods presented in this survey involve the discretisation of variational formulations based on discontinuous functions, however with "DG" we denote only those methods that arrive at local variational formulations by applying integration by parts to the PDE to be approximated. On the contrary, least squares and weighted residual methods simply enforce (weakly) continuity and boundary conditions, irrespectively of the considered PDE.

2.2.1 The Trefftz-DG (TDG) Method

Originally, Trefftz-discontinuous Galerkin (TDG) methods (or plane wave DG, PWDG, when used in combination with plane wave basis functions) were introduced as a way of recasting the ultra weak variational formulation (UWVF) of [18, 19] (see Sect. 2.2.2 below) in a framework that would facilitate its theoretical analysis [17, 46]. A similar, but more general, Trefftz-DG framework was proposed in [37, 39], arising from methods for hyperbolic equations; see Remark 1 below.

We first derive the TDG formulation as in [55]. We multiply the Helmholtz equation (1) by a test function v and integrate by parts twice on each $K \in \mathcal{T}_h$:

$$\begin{aligned} 0 = \int_K (-\Delta u - k^2 u)\overline{v}\,\mathrm{d}V &= \int_K (\nabla u \cdot \overline{\nabla v} - k^2 u\overline{v})\,\mathrm{d}V - \int_{\partial K} \nabla u \cdot \mathbf{n}_K\,\overline{v}\,\mathrm{d}S \\ &= \int_K u\,(-\Delta\overline{v} - k^2\overline{v})\,\mathrm{d}V + \int_{\partial K} u\,\overline{\partial_{\mathbf{n}_K} v}\,\mathrm{d}S - \int_{\partial K} \partial_{\mathbf{n}_K} u\,\overline{v}\,\mathrm{d}S. \end{aligned}$$

We then replace u and v by discrete functions $u_{hp}, v_{hp} \in V_p(\mathcal{T}_h)$, the trace of u on ∂K by the numerical flux $\hat{u}_{hp}$, and the trace of ∇u by the numerical flux $\mathrm{i}k\widehat{\boldsymbol{\sigma}}_{hp}$ (both defined below), obtaining the elemental TDG formulation:

$$\int_{\partial K} \hat{u}_{hp}\,\overline{\partial_{\mathbf{n}_K} v_{hp}}\,\mathrm{d}S - \int_{\partial K} \mathrm{i}k\widehat{\boldsymbol{\sigma}}_{hp} \cdot \mathbf{n}_K\,\overline{v}_{hp}\,\mathrm{d}S = 0, \tag{5}$$

where the volume integral vanishes as the test function $v_{hp} \in V_P(\mathcal{T}_h) \subset T(\mathcal{T}_h)$ is a Trefftz function. Variants of DG methods are distinguished by the underlying numerical fluxes. Here we opt for the *primal fluxes*:

$$\mathrm{i}k\widehat{\boldsymbol{\sigma}}_{hp} = \begin{cases} \{\!\{\nabla_h u_{hp}\}\!\} - \alpha\,\mathrm{i}k\,[\![u_{hp}]\!]_N & \text{on faces in } \mathcal{F}_h^I, \\ \nabla_h u_{hp} - (1-\delta)\left(\nabla_h u_{hp} + \mathrm{i}k\vartheta\,u_{hp}\mathbf{n} - g_R\mathbf{n}\right) & \text{on faces in } \Gamma_R, \\ \nabla_h u_{hp} - \alpha\,\mathrm{i}k\,(u_{hp} - g_D)\mathbf{n} & \text{on faces in } \Gamma_D, \end{cases} \tag{6}$$

$$\hat{u}_{hp} = \begin{cases} \{\!\{u_{hp}\}\!\} - \beta\,(\mathrm{i}k)^{-1}[\![\nabla_h u_{hp}]\!]_N & \text{on faces in } \mathcal{F}_h^I, \\ u_{hp} - \delta\left((\mathrm{i}k\vartheta)^{-1}\nabla_h u_{hp} \cdot \mathbf{n} + u_{hp} - (\mathrm{i}k\vartheta)^{-1} g_R\right) & \text{on faces in } \Gamma_R, \\ g_D & \text{on faces in } \Gamma_D, \end{cases} \tag{7}$$

where the flux parameters $\alpha > 0$, $\beta > 0$, $0 < \delta \le 1/2$, are bounded functions defined on suitable unions of edges/faces (see also Table 1). Adding over all elements, we obtain the following formulation of the TDG method:

$$\begin{aligned} &\text{find } u_{\mathrm{TDG}} \in V_p(\mathcal{T}_h) \text{ s.t.} \quad \mathcal{A}_{\mathrm{TDG}}(u_{\mathrm{TDG}}, v_{hp}) = \ell_{\mathrm{TDG}}(v_{hp}) \quad \forall v_{hp} \in V_p(\mathcal{T}_h), \quad \text{where} \\ &\mathcal{A}_{\mathrm{TDG}}(u, v) := \qquad (8) \\ &\int_{\mathcal{F}_h^I} \Big(\{\!\{u\}\!\}[\![\overline{\nabla_h v}]\!]_N - \{\!\{\nabla_h u\}\!\} \cdot [\![\overline{v}]\!]_N + \alpha\mathrm{i}k[\![u]\!]_N \cdot [\![\overline{v}]\!]_N - \beta(\mathrm{i}k)^{-1}[\![\nabla_h u]\!]_N[\![\overline{\nabla_h v}]\!]_N \Big)\,\mathrm{d}S \\ &+ \int_{\Gamma_R} \Big((1-\delta)\mathrm{i}k\vartheta u\overline{v} + (1-\delta)u\overline{\partial_{\mathbf{n}} v} - \delta\partial_{\mathbf{n}} u\,\overline{v} - \delta(\mathrm{i}k\vartheta)^{-1}\partial_{\mathbf{n}} u\overline{\partial_{\mathbf{n}} v} \Big)\,\mathrm{d}S \\ &+ \int_{\Gamma_D} \Big(-\partial_{\mathbf{n}} u\,\overline{v} + \alpha\,\mathrm{i}k\,u\,\overline{v} \Big)\,\mathrm{d}S, \\ &\ell_{\mathrm{TDG}}(v) := \int_{\Gamma_R} g_R\Big((1-\delta)\overline{v} - \delta(\mathrm{i}k\vartheta)^{-1}\overline{\partial_{\mathbf{n}} v}\Big)\,\mathrm{d}S + \int_{\Gamma_D} g_D\Big(\alpha\mathrm{i}k\overline{v} - \overline{\partial_{\mathbf{n}} v}\Big)\,\mathrm{d}S. \end{aligned}$$

Table 1 Different TDG flux parameters in (6) and (7) that have been considered

		α	β	δ
Quasi-uniform meshes, h-convergence	Gittelson et al. [46]	$\mathtt{a}/kh_K$	$\mathtt{b}kh_K$	$\mathtt{d}kh_K$
Quasi-uniform meshes, p-convergence	Hiptmair et al. [53]	$\mathtt{a}$	$\mathtt{b}$	$\mathtt{d}$
UWVF (see Sect. 2.2.2)	Cessenat and Després [19]	$1/2$	$1/2$	$1/2$
Locally refined meshes, hp-convergence	Hiptmair et al. [55]	$\mathtt{a}h/h_K$	$\mathtt{b}h/h_K$	$\mathtt{d}h/h_K$
Geometrically graded meshes, exponential hp-convergence	Hiptmair et al. [56]	$\mathtt{a}h/h_K$	$\mathtt{b}$	$\mathtt{d}$
Polynomial (non Trefftz) basis, hp-convergence	Melenk et al. [83]	$\mathtt{a}q_K^2/kh_K$	$\mathtt{b}kh_K/q_K$	$\mathtt{d}kh_K/q_K$

Here $\mathtt{a}, \mathtt{b}, \mathtt{d}$ are positive functions independent of the other parameters; k is the wavenumber; h_K is the local meshwidth; $h = \max_{K\in\mathcal{T}_h} h_K$ is the global meshwidth; q_K is the local polynomial degree (for the non-Trefftz version)

The TDG method was introduced in the primal form described here in [44, 46] and in mixed form in [52], under the name of *plane wave DG (PWDG) method*, following the derivation of [6] of general DG schemes for elliptic equations. In [46], first-order convergence in the meshwidth was established, using Schatz' argument, for 2D Robin problems with source term $f \in L^2(\Omega)$, plane wave discrete spaces and quasi-uniform families of meshes. This was extended to higher orders in h in [84], p-convergence in [53], three dimensions in [85], locally-refined meshes in [55], and finally the exponential convergence in the number of degrees of freedom of its hp-version was proved in [56]. Its dispersion analysis was performed in [44, 45].

For polynomial discrete spaces, the advantages of using the formulation underlying the TDG method, compared to standard DG schemes, were analysed in [83]. In [14], the TDG formulation was utilised with (non-Trefftz) bases defined from oscillating functions from high-frequency asymptotics modulated with polynomials; problems with varying coefficients were also considered.

The TDG formulation (8) can be seen as a modification of either the *interior penalty* method, or of the *local DG* (LDG) method (see e.g. [6]): with respect to the *interior penalty* method, the stabilisation term multiplied by β is added in the TDG fluxes (7), while with respect to the LDG method, in the TDG fluxes (6), the consistency term is written in terms of the primal variable ($\{\nabla_h u_{hp}\}$) instead of in terms of the auxiliary variable ($\{ik\boldsymbol{\sigma}_{hp}\}$) and the additional stabilisation of the jumps of $\boldsymbol{\sigma}_{hp}$ is removed. In [106], the TDG and the UWVF are seen as special instances of a family of methods arising from integration by parts.

The a priori error analysis of the TDG relies on Theorem 2 below (e.g. [55, Sect. 4]), which makes use of the following mesh- and flux-dependent seminorms:

$$
\begin{aligned}
|||v|||_{\mathrm{TDG}}^2 &:= k^{-1}\left\|\beta^{\frac{1}{2}}[\![\nabla_h v]\!]_N\right\|_{L^2(\mathcal{F}_h^I)}^2 + k\left\|\alpha^{\frac{1}{2}}[\![v]\!]_N\right\|_{L^2(\mathcal{F}_h^I)}^2 \\
&\quad + k^{-1}\left\|\delta^{\frac{1}{2}}\vartheta^{-\frac{1}{2}}\partial_{\mathbf{n}} v\right\|_{L^2(\Gamma_R)}^2 + k\left\|(1-\delta)^{\frac{1}{2}}\vartheta^{\frac{1}{2}} v\right\|_{L^2(\Gamma_R)}^2 + k\left\|\alpha^{\frac{1}{2}} v\right\|_{L^2(\Gamma_D)}^2; \\
|||v|||_{\mathrm{TDG+}}^2 &:= |||v|||_{\mathrm{TDG}}^2 + k\left\|\beta^{-\frac{1}{2}}\{\!\{v\}\!\}\right\|_{L^2(\mathcal{F}_h^I)}^2 + k^{-1}\left\|\alpha^{-\frac{1}{2}}\{\!\{\nabla_h v\}\!\}\right\|_{L^2(\mathcal{F}_h^I)}^2 \\
&\quad + k\left\|\delta^{-\frac{1}{2}}\vartheta^{\frac{1}{2}} v\right\|_{L^2(\Gamma_R)}^2 + k^{-1}\left\|\alpha^{-\frac{1}{2}}\partial_{\mathbf{n}} v\right\|_{L^2(\Gamma_D)}^2.
\end{aligned}
$$

Theorem 2 *The seminorms $|||\cdot|||_{\mathrm{TDG}}$ and $|||\cdot|||_{\mathrm{TDG+}}$ are norms in the Trefftz space $T(\mathcal{T}_h)$. The TDG sesquilinear form is continuous and coercive:*

$$
|\mathcal{A}_{\mathrm{TDG}}(v,w)| \le 2|||v|||_{\mathrm{TDG+}}|||w|||_{\mathrm{TDG}}, \quad \mathrm{Im}\left\{\mathcal{A}_{\mathrm{TDG}}(v,v)\right\} = |||v|||_{\mathrm{TDG}}^2
$$

for all $v, w \in T(\mathcal{T}_h)$, thus there exists a unique solution $u_{\mathrm{TDG}} \in V_p(\mathcal{T}_h)$ to the TDG formulation (8) *and the quasi-optimality bound holds:*

$$
|||u - u_{\mathrm{TDG}}|||_{\mathrm{TDG}} \le 3 \inf_{v_{hp}\in V_p(\mathcal{T}_h)} |||u - v_{hp}|||_{\mathrm{TDG+}}.
$$

Choosing $\lambda^2 = \alpha k$ on $\mathcal{F}_h^I \cup \Gamma_D$, $\sigma^2 = \beta/k$ on $\mathcal{F}_h^I$ and $\sigma^2 = \min\{\delta, 1-\delta\}/2k\vartheta$ on Γ_R, the norm (2) is controlled as $|||v|||_{\lambda,\sigma} \le |||v|||_{\mathrm{TDG}}$ for all $v \in T(\mathcal{T}_h)$. Thus, by Lemma 1, the $L^2(\Omega)$ norm of the TDG error can be controlled by its $|||\cdot|||_{\mathrm{TDG}}$ norm, and so by the discrete space approximation properties. This result has been stated in several slightly different forms, depending on the regularity of the solution u, the type of mesh used, the choice of the numerical flux parameters α, β, δ; see [85, Lemma 4.3.7], [55, Lemma 4.4] and [56, Lemma 4.5]. To strike a balance between the size of the constants arising from the duality argument of Lemma 1 and approximation errors, different flux parameters have been chosen on different meshes and aiming at different types of convergence estimates, see Table 1. For illustration, we state the result in the case of constant flux parameters, quasi-uniform meshes, and domains that guarantee sufficiently smooth solutions for the dual problems; this follows from Lemma 1 and Theorem 2 (cf. [85, Corollary 4.3.8]).

Corollary 1 *Let u be the solution of* (1)*, where Ω is either convex or smooth and star-shaped, and let $u_{\mathrm{TDG}} \in V_p(\mathcal{T}_h)$ be the solution of the TDG method with flux parameters chosen as in the second row of Table 1. Then*

$$
\|u - u_{\mathrm{TDG}}\|_{L^2(\Omega)} \le C_0 \operatorname{diam}\Omega \left(1 + \left(k \min_{K\in\mathcal{T}_h} h_K\right)^{-1/2}\right) \inf_{v_{hp}\in V_p(\mathcal{T}_h)} |||u - v_{hp}|||_{\mathrm{TDG+}},
$$

where $C_0 > 0$ depends only on ϑ, the shape of Ω and the shape-regularity of the mesh, but is independent of k and $V_p(\mathcal{T}_h)$.

The combination of the abstract error analysis outlined above and approximation estimates for plane, circular and spherical waves (see Sect. 3) leads to a priori h-, p- and hp-convergence estimates in $|||\cdot|||_{\mathrm{TDG}}$ and L^2 norms, see [46, 53, 55, 56, 85]. The dependence of the error bounds on the wavenumber k is explicit, as in Corollary 1.

Remark 1 The Helmholtz equation may be written as the first order hyperbolic system $-ik\mathbf{u} + \sum_{j=1}^{n} \partial_{x_j}(\mathbf{A}^{(j)}\mathbf{u}) = \mathbf{0}$, where $\mathbf{u} := (u; \nabla u/(ik))$ and $\mathbf{A}^{(j)}$ are the $(1+n) \times (1+n)$ symmetric matrices whose only non-zero elements are $A^{(j)}_{1,j+1} = A^{(j)}_{j+1,1} = 1$, for $1 \le j \le n$. Then, similarly to [37, Eq. (22)] or [39, Eq. (5)], a general Trefftz-DG method can be written as:

$$\text{seek } \mathbf{u} \in \mathbf{V}_p(\mathcal{T}_h) := \big\{(u,\boldsymbol{\sigma}) : u \in V_p(\mathcal{T}_h), \boldsymbol{\sigma} = \nabla u/(ik)\big\} \text{ s.t. } \forall \mathbf{v} \in \mathbf{V}_p(\mathcal{T}_h)$$

$$\sum_{\substack{K_1,K_2\in\mathcal{T}_h,\\ K_1\neq K_2}} \int_{\partial K_1\cap\partial K_2} \big(\mathbf{F}^{\mathrm{in}}_{|K_1}\mathbf{u}_{|K_1} - \mathbf{F}^{\mathrm{in}}_{|K_2}\mathbf{u}_{|K_2}\big)\cdot\big(\overline{\mathbf{v}_{|K_1}} - \overline{\mathbf{v}_{|K_2}}\big)\,\mathrm{d}S + \int_{\partial\Omega} (\mathbf{F}^{\mathrm{in}}\mathbf{u} - \mathbf{g})\cdot\overline{\mathbf{v}}\,\mathrm{d}S = 0$$

where the flux-splitting matrices $\mathbf{F}^{\mathrm{in}}$, $\mathbf{F}^{\mathrm{out}}$ are defined on $\prod_{K\in\mathcal{T}_h}\partial K$ and satisfy $\mathbf{F}^{\mathrm{in}} \le 0$, $\mathbf{F}^{\mathrm{out}} \ge 0$ (i.e. are negative and positive semi-definite, respectively), $\mathbf{F}^{\mathrm{in}} + \mathbf{F}^{\mathrm{out}} = \begin{pmatrix} 0 & \mathbf{n}_K^\top \\ \mathbf{n}_K & \mathbf{0}\end{pmatrix}$ on ∂K, and $\mathbf{F}^{\mathrm{in}}_{K_1} = -\mathbf{F}^{\mathrm{out}}_{K_2}$ on $\partial K_1 \cap \partial K_2$. The boundary data are represented by a suitable vector field $\mathbf{g} = -\mathbf{F}^{\mathrm{out}}\mathbf{u}$. The TDG in (8) (up to a factor $-ik$) is obtained by choosing:

$$\mathbf{F}^{\mathrm{in}}_K = \begin{cases} \begin{pmatrix} -\alpha & \frac{1}{2}\mathbf{n}_K^\top \\ \frac{1}{2}\mathbf{n}_K & -\beta\mathbf{n}\otimes\mathbf{n}^\top \end{pmatrix} \\ \begin{pmatrix} -(1-\delta)\vartheta & \delta\mathbf{n}_K^\top \\ (1-\delta)\mathbf{n} & -\frac{\delta}{\vartheta}\mathbf{n}\otimes\mathbf{n}^\top \end{pmatrix} \\ \begin{pmatrix} -\alpha & \mathbf{n}_K^\top \\ \mathbf{0} & \mathbf{0} \end{pmatrix} \end{cases} \qquad \mathbf{F}^{\mathrm{out}}_K = \begin{cases} \begin{pmatrix} \alpha & \frac{1}{2}\mathbf{n}_K^\top \\ \frac{1}{2}\mathbf{n}_K & \beta\mathbf{n}\otimes\mathbf{n}^\top \end{pmatrix} & \text{on } \partial K \cap \mathcal{F}_h^I, \\ \begin{pmatrix} (1-\delta)\vartheta & (1-\delta)\mathbf{n}_K^\top \\ \delta\mathbf{n} & \frac{\delta}{\vartheta}\mathbf{n}\otimes\mathbf{n}^\top \end{pmatrix} & \text{on } \partial K \cap \Gamma_R, \\ \begin{pmatrix} \alpha & \mathbf{0}^\top \\ \mathbf{n}_K & \mathbf{0} \end{pmatrix} & \text{on } \partial K \cap \Gamma_D. \end{cases}$$

The right-hand side is represented by the vector $\mathbf{g} = -\frac{1}{ik}\begin{pmatrix} 1-\delta \\ \delta\vartheta^{-1}\mathbf{n}_K \end{pmatrix} g_R$ on Γ_R and $\mathbf{g} = -\begin{pmatrix} \alpha \\ \mathbf{n}_K \end{pmatrix} g_D$ on Γ_D.

2.2.2 The Ultra Weak Variational Formulation (UWVF)

The ultra weak variational formulation (UWVF) has been introduced in the 1990s by Cessenat and Després in [18, 19]. Since then it has received a great deal of attention and has been applied to numerous PDEs and BVPs; we refer to [60] for a description of its computational aspects and to [76, Sect. 3.5.2] for an extensive bibliography. Different derivations can be found e.g. in [17, 19, 37, 39, 46]; in particular [17, 46] obtain the UWVF in the setting of DG schemes for elliptic problems of [6], while

[37, 39] derive it for general first-order hyperbolic systems using a flux-splitting approach as we did for the TDG in Remark 1. Note that different papers use different sign conventions. The extension of the UWVF to problems with smooth coefficients has been tackled in [65].

To write its formulation for the BVP (1) in the Robin case, i.e. $\Gamma_D = \emptyset$, we first define the trace space $X := \prod_{K\in\mathcal{T}_h} L^2(\partial K)$, and the operators $F_K : L^2(\partial K) \to L^2(\partial K)$, mapping the boundary datum y_K of a local adjoint-impedance Helmholtz BVP into the impedance trace of the BVP solution e_K itself:

$$F_K(y_K) := (\partial_{\mathbf{n}_K} + \mathrm{i}k)e_K, \qquad \text{where} \qquad \begin{cases} -\Delta e_K - k^2 e_K = 0 & \text{in } K, \\ (-\partial_{\mathbf{n}_K} + \mathrm{i}k)e_K = y_K & \text{on } \partial K. \end{cases}$$

The Helmholtz BVP is written as a transmission problem across the mesh interfaces, i.c., for all $K, K' \in \mathcal{T}_h$,

$$\begin{aligned} -\Delta u - k^2 u &= 0 && \text{in } K, \\ \partial_{\mathbf{n}_K} u + \mathrm{i}ku &= -\partial_{\mathbf{n}_{K'}} u + \mathrm{i}ku && \text{on } \partial K \cap \partial K', \\ \partial_{\mathbf{n}_K} u + \mathrm{i}k\vartheta u &= g_R && \text{on } \partial K \cap \Gamma_R. \end{aligned}$$

Then, after multiplying the first equation by $e_{|K}$, $e \in T(\mathcal{T}_h)$, integrating by parts twice, taking into account transmission and boundary conditions, and introducing $x, y \in X$ defined as $x_{|\partial K} = -\partial_{\mathbf{n}_K} u + \mathrm{i}ku$ and $y_{|\partial K} = -\partial_{\mathbf{n}_K} e + \mathrm{i}ke$, the UWVF of problem (1) [19, (1.4)] reads: find $x \in X$ such that, for every $y \in X$,

$$\begin{aligned} &\sum_{K\in\mathcal{T}_h} \int_{\partial K} x_{|\partial K}\, \overline{y_{|\partial K}}\, \mathrm{d}S - \sum_{K,K'\in\mathcal{T}_h} \int_{\partial K\cap\partial K'} x_{|\partial K'}\, \overline{F_K(y_{|\partial K})}\, \mathrm{d}S \\ &\quad - \sum_{K\in\mathcal{T}_h} \int_{\partial K\cap\Gamma_R} \frac{1-\vartheta}{1+\vartheta} x_{|\partial K}\, \overline{F_K(y_{|\partial K})}\, \mathrm{d}S = \sum_{K\in\mathcal{T}_h} \int_{\partial K\cap\Gamma_R} \frac{2}{1+\vartheta} g_R\, \overline{F_K(y_{|\partial K})}\, \mathrm{d}S. \end{aligned} \tag{9}$$

(Note that for $\vartheta = 1$ the term on $\partial K \cap \Gamma_R$ at left-hand side vanishes and $2/(1+\vartheta) = 1$.) The expression (9) is a variational formulation for the skeleton unknown x; after the equation is solved for x, the Helmholtz solution $u_{|K}$ can be recovered in the interior of each element by solving a local (in K) adjoint-impedance Helmholtz BVP with datum $(-\partial_{\mathbf{n}_K} + \mathrm{i}k)u_{|K} = x_{|\partial K}$. If the formulation is discretised choosing a finite dimensional subspace X_h of X corresponding to the impedance traces of a Trefftz space, namely

$$X_h := \{x_h \in X : x_{h|\partial K} = (-\partial_{\mathbf{n}_K} + \mathrm{i}k)v_{|K}\ \forall K \in \mathcal{T}_h,\ v \in V_p(\mathcal{T}_h)\},$$

then the action of F_K and the reconstruction of u_K in K are immediately computed.

Theorem 2.1 of [19] states that the discrete problem obtained by substituting X_h to X in (9) is solvable, independently of the meshsize h; Corollary 3.8 shows that, for plane wave discrete spaces, the Dirichlet and Robin traces of the UWVF solution converge to the corresponding traces of u with algebraic orders of convergence in $L^2(\Gamma_R)$. In [17, Sect. 4], these results have been used together with the duality technique of [88] to prove orders of convergence for the $L^2(\Omega)$ norm of the error.

The UWVF has been recast as a DG method with Trefftz basis functions in several different ways in [17, 37, 39, 46]. In particular, [46, Remark 2.1] shows that *the UWVF is a special case of the TDG* formulation (8) for flux parameters $\alpha = \beta = \delta = 1/2$. As a consequence, the orders of convergence in h and p proved for the TDG on quasi-uniform meshes in [46, 53] carry over to the UWVF (with suboptimal orders in h); on the other hand, the *hp*-type results of [55, 56] require variable numerical flux parameters to cope with elements of different sizes (see Table 1), so they do not apply to the UWVF. Thus, the TDG can be understood as the extension of the UWVF to non quasi-uniform meshes. Alternatively, in [89, Sect. 4.3, 5.2], the UWVF is employed on meshes refined towards solution singularities by choosing Trefftz spaces on large elements and polynomial spaces on small ones. No applications of the TDG combining mesh-dependent parameters and polynomial spaces in small elements have been documented.

2.2.3 DG Schemes with Lagrange Multipliers

The DG schemes described so far enforce weak continuity between elements using numerical fluxes, in the spirit of [6]. A different approach is to enforce continuity using Lagrange multipliers. This was probably first proposed for Trefftz methods in [63, Sect. 2.3], for the 1D Helmholtz equation.

This strategy has been followed in the *discontinuous enrichment method* (DEM), introduced by Farhat et al. in [32], combining a space of piecewise-constant Lagrange multipliers on mesh interfaces with a discrete space composed by sums of continuous piecewise polynomials and discontinuous plane waves. Subsequently, in [33], the polynomial part of the trial space was dropped, leaving a plane wave trial space and thus reducing to a Trefftz method; in this version, the DEM was renamed *discontinuous Galerkin method* (DGM) and the Lagrange multipliers were approximated by oscillatory functions. This formulation performed very well for test cases and was later extended to "higher order elements" (i.e. elements containing more plane waves) and other PDEs. We refer again to [76, Sect. 3.5.3] for a comprehensive bibliography.

Here we briefly describe the formulation of the DGM following [33, Sect. 2]:

$$\text{find } (u,\lambda) \in H^1(\mathcal{T}_h) \times W(\mathcal{T}_h) \text{ s.t.}$$

$$\begin{cases} \mathcal{A}_{\mathrm{DGM}}(u,v) + \mathcal{B}_{\mathrm{DGM}}(\lambda,v) = \displaystyle\int_{\Gamma_R} g_R\, v \,\mathrm{d}S & \forall v \in H^1(\mathcal{T}_h), \\ \mathcal{B}_{\mathrm{DGM}}(\mu,u) = \displaystyle\int_{\Gamma_D} \mu\, g_D \,\mathrm{d}S & \forall \mu \in W(\mathcal{T}_h), \end{cases}$$

where

$$\mathcal{A}_{\mathrm{DGM}}(w,v) := \sum_{K\in\mathcal{T}_h}\int_K (\nabla w\cdot\nabla v - k^2 u\, v)\,\mathrm{d}V + \int_{\Gamma_R} \mathrm{i}k\vartheta\, w\, v\,\mathrm{d}S,$$

$$\mathcal{B}_{\mathrm{DGM}}(\mu,w) := \sum_{K,K'\in\mathcal{T}_h}\int_{\partial K\cap\partial K'} \mu(w_{|K'} - w_{|K})\,\mathrm{d}S + \int_{\Gamma_D} \mu\, w\,\mathrm{d}S,$$

$$W(\mathcal{T}_h) := \Bigg(\prod_{K,K'\in\mathcal{T}_h} \tilde{H}^{-1/2}(\partial K\cap\partial K')\Bigg)\times H^{-1/2}(\Gamma_D).$$

It is immediate to verify that the solution u to BVP (1) satisfies this formulation, and that the multiplier λ represents the normal derivative of u on the mesh interfaces and on Γ_D. This formulation is then discretised by restricting it to finite dimensional spaces $V_p(\mathcal{T}_h)\subset H^1(\mathcal{T}_h)$ and $W_p(\mathcal{T}_h)\subset W(\mathcal{T}_h)$. In the DEM of [32], $V_p(\mathcal{T}_h)$ is the direct sum of a continuous polynomial and a plane wave space, in the DGM of [33] and subsequent papers only the plane wave part is retained, so $V_p(\mathcal{T}_h)\subset T(\mathcal{T}_h)$. The volume degrees of freedom, i.e. those corresponding to $V_p(\mathcal{T}_h)$, are then eliminated by static condensation in order to reduce the computational cost of the scheme.

A stability and convergence analysis of the simplest version of the DGM (four plane waves per element and piecewise-constant multipliers) is attempted in [2]: for a Robin–Neumann BVP on a domain decomposed in rectangles, under a mesh resolution condition, the scheme is shown to be well-posed, and a priori orders of convergence are proved (in $H^1(\mathcal{T}_h)$ norm for the primal variable and in $L^2(\mathcal{F}_h)$ for the multipliers), along with residual-type a posteriori error bounds.

We are not aware of any error analysis for the DGM method holding in more general situations (e.g. more than four plane waves per elements, propagation directions not aligned to the mesh, non-rectangular mesh elements).

A similar formulation, named *hybrid-Trefftz finite element method*, is described in [99, Sect. 3.5] (deriving the functional in Eq. (65) therein): the same form $\mathcal{A}_{\mathrm{DGM}}$ above is used, while $\mathcal{B}_{\mathrm{DGM}}$ is substituted by $\mathcal{B}_{\mathrm{HT}}(\mu,w) := -\int_{\mathcal{F}_h^I}\mu\,[\![\nabla_h w]\!]_N\,\mathrm{d}S - \int_{\Gamma_N}\mu\,\partial_{\mathbf{n}}w\,\mathrm{d}S$, where now the multiplier μ approximates the Dirichlet trace of u, the right-hand sides and the space $W(\mathcal{T}_h)$ are changed accordingly. A further variant of hybrid-Trefftz methods is presented in [109] and related papers.

Another DG method with Trefftz basis, called *modified DG method* (mDGM), has been proposed in [48]. The Lagrange multipliers are double-valued on the interfaces (differently from the DEM/DGM of [32, 33]) and belong to $\prod_{K\in\mathcal{T}_h} L^2(\partial K\setminus\Gamma_R)$. A two-step procedure is adopted. First, for each basis element $\lambda\in L^2(\partial K\setminus\Gamma_R)$ of the discrete Lagrange multiplier space, a well-posed Helmholtz BVP on K with impedance datum λ is solved in the local Trefftz space $V_{p_K}(K)$ using the classical $H^1(K)$-conforming variational formulation. Second, these local solutions are combined in a global LS formulation leading to a positive semi-definite system whose unknowns are the Lagrange multipliers themselves. The mDGM was further improved in [3] leading to the *stable DG method* (SDGM), which differs from the mDGM in that the local impedance problems are solved with a least squares formulation posed on ∂K, which gives local Hermitian matrices.

Lagrange multipliers are also used to tackle problems with discontinuous coefficients by means of the partition of unity method, see [73] and Sect. 2.5 below.

2.3 Weighted Residual Methods

Trefftz discretisations lend themselves well to weighted residual formulations: the discrete solution is automatically a local solution of the PDE, only the residual on interfaces (the jumps) and on the boundary (the mismatch with boundary conditions) need to be enforced by multiplying them to suitable traces of test functions. The choice of these traces leads to different variational formulations, the most developed of which are the VTCR and the WBM described in the following. While it is simple to design weighted residual methods, their error analysis is by no means easy, as they arise neither from integration by parts, nor from a minimisation principle.

An earlier weighted-residual Trefftz formulation is the *weak element method* of [47], where the integral averages of Dirichlet and Neumann jumps on mesh faces are set to zero (equivalently, test functions are constant on each mesh face).

We note that some of the earliest Trefftz schemes, e.g. the *indirect approximation* of [22, Eq. (35)], are of weighted-residual type, even though testing was confined to the boundary of the domain only, see Sect. 2.4 below.

2.3.1 The Variational Theory of Complex Rays (VTCR)

The VTCR is a weighted residual Trefftz method introduced in the 1990s by P. Ladevéze and coworkers for problems arising in computational mechanics and later extended to the Helmholtz case in [100]. Recent surveys are [70, 71, 102].

Several VTCR formulations, slightly different from each other, have been presented. A general VTCR formulation for the BVP (1) can be written as:

$$\text{find } u_{\mathrm{VTCR}} \in V_p(\mathcal{T}_h) \text{ s.t. } \quad \mathcal{A}_{\mathrm{VTCR}}(u_{\mathrm{VTCR}}, v_{hp}) = \ell_{\mathrm{VTCR}}(v_{hp}) \quad \forall v_{hp} \in V_p(\mathcal{T}_h), \text{ where}$$

$$\mathcal{A}_{\mathrm{VTCR}}(u,v) := \mathrm{Im}\left\{ \int_{\mathcal{F}_h^I} \Big([\![u]\!]_N \cdot \{\!\{\overline{\nabla_h v}\}\!\} - [\![\nabla_h u]\!]_N \{\!\{\overline{v}\}\!\} \Big) \,\mathrm{d}S \right. \tag{10}$$

$$\left. + \int_{\Gamma_D} u\, \overline{\partial_{\mathbf{n}} v} \,\mathrm{d}S + \int_{\Gamma_R} \Big(\frac{C_1}{ik\vartheta} (\partial_{\mathbf{n}} u + ik\vartheta u)\overline{\partial_{\mathbf{n}} v} + C_2 (\partial_{\mathbf{n}} u + ik\vartheta u)\overline{v} \Big) \,\mathrm{d}S \right\},$$

$$\ell_{\mathrm{VTCR}}(v) := \mathrm{Im}\left\{ \int_{\Gamma_D} g_D \overline{\partial_{\mathbf{n}} v} \,\mathrm{d}S + \int_{\Gamma_R} \Big(\frac{C_1}{ik\vartheta} g_R\, \overline{\partial_{\mathbf{n}} v} + C_2\, g_R\, \overline{v} \Big) \,\mathrm{d}S \right\},$$

where we have reported the formulation with only the imaginary part of the left- and right-hand side, following the VTCR convention; however dropping "Im" does not modify the method.

The formulations in [102, Eq. (21)] and in [70, Eq. (5)] correspond to the choice of coupling parameters $C_1 = 1/2$ and $C_2 = -1/2$ (up to an overall factor k and using $\operatorname{Re}\{-iz\} = \operatorname{Im}\{z\}$); that in [101, Eq. (6)] to $C_1 = 1/2$ and $C_2 = 1/2$; that in [68, Eq. (4)] to $C_1 = 1$ and $C_2 = 0$. The choice of the coupling parameters does not affect the consistency of the method as all terms in (10) are products of residuals (internal jumps and boundary conditions) and traces of test functions. In some of the papers cited, using $\operatorname{Im}\{a\overline{b}\} = -\operatorname{Im}\{\overline{a}b\}\ \forall a, b \in \mathbb{C}$, the conjugation is written on the trial, rather than test, functions in some of the terms, without affecting the formulation.

The VTCR (and similarly the WBM) does not correspond to any of the classical DG schemes listed in [6]. Indeed, to derive it from the elemental DG equation (5), one would need to choose numerical fluxes that, in the terminology of [6], are neither consistent (they do not equal the fields ∇u and u when applied to the exact solution u itself) nor conservative (they are not single-valued on the interfaces).

Following [68, Sect. 2.2], it is possible to show that if absorption is present then the VTCR is well-posed. More precisely, provided that $C_1 = 1$, $C_2 = 0$, $\operatorname{Re} k > 0$ and $\operatorname{Im}\{k^2\} > 0$, the VTCR bilinear form satisfies

$$\mathcal{A}_{\mathrm{VTCR}}(v, v) = -\operatorname{Im}\{k^2\}\, \|v\|^2_{L^2(\Omega)} - \frac{\operatorname{Re} k}{|k|^2}\, \|\vartheta^{-1/2}\partial_{\mathbf{n}} v\|^2_{L^2(\Gamma_R)} \qquad \forall v \in T(\mathcal{T}_h),$$

thus the VTCR solution is unique in the Trefftz space and coercivity in $L^2(\Omega)$ norm holds (the analogous result for $C_1 = -C_2 = 1/2$ is proved in [70, Proposition 2]). However, this does not extend to the setting we considered so far, i.e. propagating waves with $k \in \mathbb{R}$: in this case it can easily be shown that $\mathcal{A}_{\mathrm{VTCR}}(v, v) = 0$ for all $v \in T(\mathcal{T}_h)$ such that $v = 0$ on all elements adjacent to the Robin boundary Γ_R and for any choice $C_1, C_2 \in \mathbb{C}$, thus well-posedness can not be ensured using a coercivity argument. Following [70, Proposition 2], for $C_1 = 1/2, C_2 = -1/2, k \in \mathbb{R}$, we have:

$$A_{\mathrm{VTCR}}(v, v) = -\frac{1}{2}\Big(\frac{1}{k}\, \|\vartheta^{-1/2}\partial_{\mathbf{n}} u\|^2_{L^2(\Gamma_R)} + k\, \|\vartheta^{1/2} u\|^2_{L^2(\Gamma_R)}\Big) \qquad \forall v \in T(\mathcal{T}_h),$$

thus (using Holmgren's theorem [21, Theorem 2.4]) uniqueness of the solution of (10) is proved if all mesh elements are adjacent to Γ_R. For more general cases, coercivity appears to be too strong an argument. We conjecture that discrete inf-sup conditions might be a more viable way for proving well-posedness of the VTCR.

Sect. 3 of [70] considers the application of the VTCR formulation, corrected with suitable volume terms, with non-Trefftz (piecewise-polynomial) discrete spaces. This variation is termed *weak Trefftz* and analysed therein.

2.3.2 The Wave Based Method (WBM)

The WBM is a weighted residual Trefftz method, analogous to the VTCR, first introduced in the dissertation of Desmet [26] and later extended to a wide variety of engineering applications, mainly in the realm of vibro-acoustics. Recent reviews of the state of the art of the research on the WBM can be found in [25, 27]. The discrete space typically used together with the WBM is composed of propagating and evanescent plane waves, as outlined in Sect. 3.2.

The basic variational formulation of the WBM applied to BVP (1), translating Sect. 4.1.4 of [27] to our notation and multiplying all terms by $(-ik)$, reads

$$\text{find } u_{\mathrm{WBM}} \in V_p(\mathcal{T}_h) \text{ s.t. } \mathcal{A}_{\mathrm{WBM}}(u_{\mathrm{WBM}}, v_{hp}) = \ell_{\mathrm{WBM}}(v_{hp}) \quad \forall v_{hp} \in V_p(\mathcal{T}_h), \text{ where}$$

$$\begin{aligned}
\mathcal{A}_{\mathrm{WBM}}(u, v) &:= \int_{\mathcal{F}_h^I} \left(2[\![\nabla_h u]\!]_N \{v\} + \frac{\mathrm{i}k}{Z_{int}} [\![u]\!]_N \cdot [\![v]\!]_N \right) \mathrm{d}S \\
&\quad + \int_{\Gamma_R} (\partial_{\mathbf{n}} u + \mathrm{i}k\vartheta u)\, v \,\mathrm{d}S - \int_{\Gamma_D} u\, \partial_{\mathbf{n}} v \,\mathrm{d}S \\
\ell_{\mathrm{WBM}}(v) &:= \int_{\Gamma_R} g_R\, v \,\mathrm{d}S - \int_{\Gamma_D} g_D\, \partial_{\mathbf{n}} v \,\mathrm{d}S,
\end{aligned}$$

where Z_{int} is an interior coupling factor. In some works, a slightly different formulation is used, e.g. in [98, Eq. (81)] different terms are used on the internal interfaces. We are not aware of any rigorous stability or error analysis of the WBM formulation.

2.4 *Single-Element Direct and Indirect Trefftz Methods*

Most schemes described so far were introduced not earlier than mid 1990s, but a lot of research on Trefftz methods has been carried out since the late 1970s by I. Herrera, J. Jirousek, A.P. Zieliński, O.C. Zienkiewicz and numerous co-workers, mainly for static elasticity problems. General reviews of these works are in [67, 121]; the Helmholtz case is described in detail in [22]. A major difference between these methods and those we described in the previous sections is that in many instances of the former ones no mesh is introduced on the domain Ω, so that the unknowns are defined on $\partial\Omega$ only. For this reason, these Trefftz methods more closely resemble standard boundary element methods rather than finite element schemes.

There are two main classes of these Trefftz methods: direct and indirect. (We use the terms "direct" and "indirect" as in [22, 67] and [98, Sect. 5.1].) We describe them for a modification of BVP (1) where we drop the Robin boundary Γ_R and we consider instead a Neumann boundary portion Γ_N with boundary condition $\partial_{\mathbf{n}} u = g_N$.

The *indirect method* is the simplest kind of weighted residual scheme:

$$\int_{\Gamma_D} u\,\overline{\partial_{\mathbf{n}} v}\,\mathrm{d}S - \int_{\Gamma_N} \partial_{\mathbf{n}} u\,\overline{v}\,\mathrm{d}S = \int_{\Gamma_D} g_D\,\overline{\partial_{\mathbf{n}} v}\,\mathrm{d}S - \int_{\Gamma_N} g_N \overline{v}\,\mathrm{d}S, \tag{11}$$

(see [22, Eq. (35)] for sound-hard scattering problems in unbounded domains, [98, Eq. (47)], [121, Eq. (16)], [67, Eq. (16), (26)]). For Dirichlet exterior problems this is also the method of [8, Sect. 3]. In most references the test function is not conjugated. We note that the indirect method is nothing else than the WBM of Sect. 2.3.2 posed on a single element, i.e. $\mathcal{T}_h = \{\Omega\}$ and $\mathcal{F}_h^I = \emptyset$. In the indirect method, the trial functions approximating u are global solutions of the Helmholtz equation on the whole of Ω; on the other hand the test function v only needs to be defined on $\partial\Omega$. If the Trefftz test and trial spaces coincide, then the obtained stiffness matrix is symmetric (by Green's second identity). If the signs of the terms on Γ_N are changed, as in [67, Eq. (22)], a non-symmetric formulation is obtained.

Subtracting from (11) the second Green's identity $\int_{\partial\Omega}(u\,\overline{\partial_{\mathbf{n}} v} - \partial_{\mathbf{n}} u\,\overline{v})\,\mathrm{d}S = 0$, which holds for all Helmholtz solutions u and v in Ω, we derive the *direct method*:

$$\int_{\Gamma_D} \partial_{\mathbf{n}} u\,\overline{v}\,\mathrm{d}S - \int_{\Gamma_N} u\,\overline{\partial_{\mathbf{n}} v}\,\mathrm{d}S = \int_{\Gamma_D} g_D\,\overline{\partial_{\mathbf{n}} v}\,\mathrm{d}S - \int_{\Gamma_N} g_N\,\overline{v}\,\mathrm{d}S, \tag{12}$$

(see [22, Eq. (42)], [98, Eq. (50)]). The direct method for the Dirichlet problem may be viewed as the TDG of Sect. 2.2.1 with $\alpha = 0$ posed on a single element $K = \Omega$. Conversely to the indirect method, consistency of (12) is guaranteed only if the test functions are Helmholtz solutions in Ω, while the trial functions might be defined (and often are) on $\partial\Omega$ only, for better computational efficiency; the solution is then evaluated in Ω with a representation formula in a post-processing step as for BEMs. The stiffness matrix arising from the direct formulation (12) is the transpose to that of the indirect method (11). Theorem 6.44 in [106] gives sufficient conditions for the well-posedness of the direct method. Theorem 7.19 in [20] proves that, for well-posed Dirichlet problems with $H^1(\partial\Omega)$ data, if the Neumann traces of the trial space coincide with the Dirichlet traces of the test space, then the direct method is well-posed and computes the best approximation of the exact solution in $L^2(\partial\Omega)$ norm. If Ω is unbounded, the direct and the indirect methods can still be used choosing discrete functions that satisfy Sommerfeld radiation condition; however in (12) the conjugation on the test function must be dropped to preserve consistency. In this case, if a multipole basis is used, Waterman's *null-field* method is obtained, see [78, Chap. 7], which is a special instance of the *T-matrix* method [78, Sect. 7.9]. (Note that [92] uses the name *null-field method* for the indirect method with non-conjugated test functions, and *Cremer equations* for the same with conjugated test functions.)

For a special choice of Trefftz test functions v indexed by a complex parameter (see the last paragraph of Sect. 3.2), method (12) is called "*global relation*" and is the variational formulation at the heart of the *Fokas transform method*, see [23, Eq. (2)], [106, Eq. (6.142–143)] or [20, Eq. (7.156)]. In this context, this formulation is typically discretised using piecewise-polynomial (on $\partial\Omega$) trial functions, even though Trefftz functions may be used as well.

2.5 Non-Trefftz Methods with Oscillatory Basis Functions

The main reason for the success of Trefftz methods in the context of time-harmonic wave problems is that the oscillatory basis functions may offer much better approximation properties than piecewise polynomials used in standard FEMs. On the other hand, similar approximation can also be achieved if the discrete functions are not exact local solution of the PDE to be discretised, but are only "approximate solutions". If basis functions of this kind are used, the Trefftz formulations described in the previous sections cannot be employed as they stand, because the residual in the elements will not vanish any more and consistency will fail.

Approximate Trefftz functions are especially attractive for problems with smoothly varying material parameters, where no analytic Trefftz function might be known. Trefftz formulations, possibly with additional volume terms, can be used with basis functions that are solutions of the equation only up to a certain order; see [14, 65, 111], where this idea is pursued for DG, UWVF and DEM formulations.

In the following we briefly discuss a few methods that have been proposed employing oscillatory and k-dependent basis functions that are not Trefftz.

A very well-known scheme of this kind is the *partition of unity method* (PUM or PUFEM), introduced by I. Babuška and J.M. Melenk in the mid 1990s, see e.g. [81]. The PUM combines the approximation properties of Trefftz functions with the standard variational formulation of the problem, e.g. for the BVP (1) with $\Gamma_D = \emptyset$

$$\int_\Omega \left(\nabla_h u \cdot \overline{\nabla_h v} - k^2 u\,\overline{v}\right) \mathrm{d}V + \int_{\Gamma_R} \mathrm{i}k\vartheta\, u\,\overline{v}\,\mathrm{d}S = \int_{\Gamma_R} g_R\,\overline{v}\,\mathrm{d}S \qquad \forall v \in H^1(\Omega). \tag{13}$$

This requires the use of $H^1(\Omega)$-conforming trial and test functions, thus continuity on interfaces needs to be enforced strongly, which is not viable in Trefftz spaces. The PUM uses as basis a set of Trefftz functions multiplied to a partition of unity defined on a FEM mesh, e.g. piecewise linear/multilinear polynomial FEMs on simplicial/tensor elements. Theorem 2.1 in [81] ensures that the trial space obtained enjoys the same approximation properties of the Trefftz space employed. If a p-dimensional local Trefftz space is used in each element, together with a piecewise linear/multilinear partition of unity, the total number of degrees of freedom used equals p times the number of mesh vertices, while for a similar Trefftz method on the same mesh (providing comparable accuracy) it would equal p times the number of mesh elements; this means that on tensor meshes almost the same number of DOFs would be employed by the two methods, while on triangles and tetrahedra a saving of a factor up to two or six, respectively, can be achieved by the PUM. A shortcoming of the PUM is that the formulation (13) is not sign-definite and its well-posedness requires a scale resolution condition, while this is not needed for some Trefftz schemes such as the TDG/UWVF presented in Sects. 2.2.1 and 2.2.2. Differently from Trefftz schemes, the implementation of the PUM requires the computation of volume integrals; moreover, the numerical integration of the PUM basis functions may be more expensive than that of genuine Trefftz functions, see Sect. 4.1.

The PUM for the Helmholtz and other frequency-domain equations was further developed by R.J. Astley, P. Bettes, A. El Kacimi, O. Laghrouche, M.S. Mohamed, E. Perrey-Debain, J. Trevelyan and collaborators, see e.g. [72, 96]. When a PUM and a standard FEM discrete spaces are combined, e.g. using formulation (13), the method obtained is termed *generalised finite element method* (GFEM); e.g. [108] employs high-order tensor-product polynomials summed to products of plane waves and bilinear functions. In problems with discontinuous wavenumber k, the PUM can be applied by coupling the homogeneous regions by means of Lagrange multipliers as in [73]; this is not necessary as formulation (13) holds on the whole domain, but enhance the accuracy as in each subdomain only basis functions oscillating with the correct local wavelength are used. In [51] and related papers, the *trigonometric finite wave elements* (TFWE) is described: the PUM is used with special basis functions adapted to waveguides, lasers and geometries with a single dominant wave propagation direction. The *finite ray element method* of [79] consists in the use of a PUM basis in a *first order system of least squares* (FOSLS) formulation; as the unknown is constituted by both u and its gradient, more unknowns are needed but the system matrix is Hermitian. Finally, in the *hybrid numerical asymptotic method* of [42], the PUM space is constructed by multiplying nodal finite elements to oscillating functions whose phases are derived from geometrical optics (GO) or geometrical theory of diffraction (GTD), e.g. by solving the eikonal equation, cf. Sect. 4.2.

Plane wave bases have been combined in [97] with the *virtual element method* (VEM) framework [11], in order to design a high-order, conforming method for the Helmholtz problem, in the spirit of the PUM, but allowing for general polytopic meshes. The main ingredients of the resulting PW-VEM are (i) a low frequency space made of low order VEM functions, which do not need to be explicitly computed in the element interiors, (ii) a proper local projection operator onto a high-frequency space made of plane waves, and (iii) an approximate stabilisation term. The implementation of the PW-VEM does not require computation of volume integrals, and no quadrature formulas are required for the assembly of the stiffness matrix, for meshes with flat interelement boundaries.

The *hybridizable DG* method of [91] employs two discontinuous discrete spaces (one scalar and one vector) and a space of Lagrange multipliers on the mesh interfaces. Though Trefftz spaces might be used with this formulation, the authors consider basis functions constructed as products of polynomials and geometrical optics-based oscillating functions, similar to those in [42] but discontinuous.

A Trefftz approach has been proposed in the context of finite difference schemes in the *flexible local approximation method* (FLAME) by I. Tsukerman, see e.g. the comprehensive review [113]. In the FLAME, the Taylor expansion of the solution to be approximated used to define classical finite difference schemes is substituted by an expansion in a series of Trefftz basis functions, leading to better accuracy.

Oscillatory basis functions have been successfully used in boundary element methods, in particular for scattering problems, see the review on the *hybrid numerical-asymptotic BEM* (HNA-BEM) [21], the *plane-wave basis boundary elements* [96, Sect. 3] and the *extended isogeometric boundary element method* (XIBEM) [93].

3 Trefftz Discrete Spaces and Approximation

Given a Trefftz variational formulation of a BVP, as those in Sect. 2, the definition of a Trefftz finite element method is completed by the choice of a discrete space

$$V_p(\mathcal{T}_h) = \big\{v \in T(\mathcal{T}_h) : \ v_{|K} \in V_{p_K}(K)\big\} \subset T(\mathcal{T}_h),$$

where $V_{p_K}(K)$ is a p_K-dimensional space of functions v on K such that $\Delta v + k^2 v = 0$. We describe next the main features of the most common local Trefftz spaces $V_{p_K}(K)$; we do not consider Lagrange multiplier spaces on mesh faces for the methods in Sect. 2.2.3. The discussion of the conditioning properties of the basis functions described and of the techniques for their numerical integration is postponed to Sect. 4.

3.1 Generalised Harmonic Polynomials (GHPs)

Generalised harmonic polynomials are smooth Helmholtz solutions that are separable in polar and spherical coordinates in 2D and 3D, respectively, i.e. *circular and spherical waves* (also called Fourier–Bessel functions or Fourier basis). The local spaces $V_{p_K}(K)$ are defined as follows:

$$\text{2D:}\quad V_{p_K}(K) = \Big\{v : \ v(\mathbf{x}) = \sum_{\ell=-q_K}^{q_K} \alpha_\ell\, J_\ell(k\,|\mathbf{x}-\mathbf{x}_K|)\, e^{i\ell\theta}, \ \alpha_\ell \in \mathbb{C}\Big\},$$

$$\text{3D:}\quad V_{p_K}(K) = \Big\{v : \ v(\mathbf{x}) = \sum_{\ell=0}^{q_K}\sum_{m=-\ell}^{\ell} \alpha_{\ell,m}\, j_\ell(k\,|\mathbf{x}-\mathbf{x}_K|)\, Y_\ell^m\Big(\frac{\mathbf{x}-\mathbf{x}_K}{|\mathbf{x}-\mathbf{x}_K|}\Big), \ \alpha_{\ell,m} \in \mathbb{C}\Big\},$$

where $\mathbf{x}_K \in K$ (e.g. is the mass centre of K), θ is the angle of $\mathbf{x}$ in the local polar coordinate system centred at $\mathbf{x}_K$, J_ℓ is the Bessel function of the first kind and order ℓ, $\{Y_\ell^m\}_{m=-\ell}^{\ell}$ is a basis of spherical harmonics of order ℓ (see e.g. [85, Eq. (B.30)]), and j_ℓ is the spherical Bessel function defined by $j_\ell(z) = \sqrt{\frac{\pi}{2z}}\, J_{\ell+\frac{1}{2}}(z)$. The space dimension p_K is given by $p_K = 2q_K + 1$ in 2D and by $p_K = (q_K+1)^2$ in 3D. We call

q_K, the maximal index of the (spherical) Bessel functions used, the "degree" of the GHP space, as it plays the same role of the polynomial degree in the approximation theory. A particular feature of GHP spaces is that they are hierarchical.

The name "generalised harmonic polynomials" was coined in [80] and comes from the fact that they are images of *harmonic polynomials* under the operator that maps harmonic functions into Helmholtz solutions, in the framework of Vekua–Bergman's theory [12, 114] (see also [50, 87]). The same theory allows to transfer approximation results for harmonic functions by spaces of harmonic polynomials into results on the approximation of Helmholtz solutions by GHPs. The density of GHPs in a space of Helmholtz solutions was proved in [50, Theorem 4.8] and [114, Sect. 22.8]. Approximation estimates in two dimensions were first proved in [28, Theorem 6.2] (in L^∞ norm) and in [80] (in Sobolev norms), and later sharpened and extended to three dimensions in [86]. We summarise here the estimates of [86, Theorem 3.2].

Let $D \in \mathbb{R}^n$, $n = 2, 3$, be a bounded, open set with Lipschitz boundary and diameter h_D, containing $B_{\rho h_D}(\mathbf{x}_D)$ (the ball centred at some $\mathbf{x}_D \in D$ and with radius ρh_D), and star-shaped with respect to $B_{\rho_0 h_D}(\mathbf{x}_D)$, where $0 < \rho_0 \leq \rho \leq 1/2$. Assume that $u \in H^{s+1}(D)$, $s \in \mathbb{N}$, satisfies $\Delta u + k^2 u = 0$ in D and define the k-weighted Sobolev norm $\|u\|_{j,k,D} := (\sum_{m=0}^{j} k^{2(j-m)} |u|^2_{m,D})^{1/2}$, $j \in \mathbb{N}$, where $|\cdot|_{m,D}$ is the Sobolev seminorm of order m on D.

(i) If $n = 2$ and D satisfies the exterior cone condition with angle $\lambda_D \pi$ [86, Definition 3.1] ($\lambda_D = 1$ if D is convex), then for every $L \geq s$ there exists a GHP Q_L of degree at most L such that, for every $j \leq s + 1$, it holds

$$\|u - Q_L\|_{j,k,D} \leq C\big(1 + (h_D k)^{j+6}\big) e^{\frac{3}{4}(1-\rho)h_D k} \left(\left(\frac{\log(L+2)}{L+2} \right)^{\lambda_D} h_D \right)^{s+1-j} \|u\|_{s+1,k,D},$$

where the constant $C > 0$ depends only on the shape of D, j and s, but is independent of h_D, k, L and u.

(ii) If $n = 3$, there exists a constant $\lambda_D > 0$ depending only on the shape of D, such that for every $L \geq \max\{s, 2^{1/\lambda_D}\}$ there exists a GHP Q_L of degree at most L such that, for every $j \leq s + 1$, it holds

$$\|u - Q_L\|_{j,k,D} \leq C\big(1 + (h_D k)^{j+6}\big) e^{\frac{3}{4}(1-\rho)h_D k} L^{-\lambda_D(s+1-j)} h_D^{s+1-j} \|u\|_{s+1,k,D},$$

where the constant $C > 0$ depends only on the shape of D, j and s, but is independent of h_D, k, L and u.

The main difference between the two results is that the positive shape-dependent parameter λ_D entering the exponent of L (thus the p-convergence order) is explicitly known in 2D (it depends on the largest non-convex corner of D) but not in 3D.

Exponential convergence of the GHP approximation of Helmholtz solutions that possess analytic extension outside D were proved in [85, Proposition 3.3.3] and improved in 2D in [56], based upon the corresponding result for harmonic functions of [54]. Roughly speaking, the error is bounded by a negative exponential of the form $C\exp(-bL) \sim C\exp(-bp_D^{1/(n-1)})$, while classical bounds for polynomials achieve at most $C\exp(-bp_D^{1/n})$, since the dimension p_D of the GHP space of order L is $\mathcal{O}(L^{n-1})$, while the dimension p_D of the polynomial space of degree L is $\mathcal{O}(L^n)$. Thus, Trefftz methods based on GHPs (and similarly on PWs) can achieve better asymptotic order than standard schemes; however the value of the positive coefficients b, C and their dependence on the BVP and discretisation are not entirely clear.

Approximation estimates in the (discontinuous) spaces $V_p(\mathcal{T}_h)$ immediately follow from the local approximation estimates with $\overline{D} = K$, for all $K \in \mathcal{T}_h$. In case of (H^1-conforming) partition of unity spaces enriched with GHPs, global estimates follow from combining the local estimates with [81, Theorem 2.1].

GHPs have been proposed in numerous Trefftz formulations: LS [88, 107], UWVF [77], VTCR [68], hybrid-Trefftz [99, Eq. (62)], direct and indirect single-element schemes [22, 121], HELS [117], MPS [15, 36].

3.2 *Plane Waves (PWs)*

Plane waves probably constitute the most common choice of Trefftz basis functions. In this case, the local space $V_{p_K}(K)$ is defined by

$$V_{p_K}(K) = \left\{v : \ v(\mathbf{x}) = \sum_{\ell=1}^{p_K} \alpha_\ell \, e^{ik\mathbf{d}_\ell\cdot(\mathbf{x}-\mathbf{x}_K)}, \ \alpha_\ell \in \mathbb{C}\right\}, \tag{14}$$

where $\{\mathbf{d}_\ell\}_{\ell=1}^{p_K} \subset \mathbb{R}^n$, $|\mathbf{d}_\ell| = 1$, are distinct propagation directions. To obtain isotropic approximations, in 2D, uniformly-spaced directions on the unit circle can be chosen (i.e. $\mathbf{d}_\ell = (\cos(2\pi\ell/p_K), \sin(2\pi\ell/p_K))$); in 3D, [103] and [94] provide directions that are "almost equally spaced" (see [1, Sect. 3.4] for a simpler version). In these cases, the PW spaces are not hierarchical. However, one of the potential benefits of PW approximations is the possibility to depart from the isotropic case and to adapt the basis propagation directions to the specific BVP at hand and to different elements, either a priori or a posteriori, see Sect. 4.2.

The linear independence of arbitrary sets of plane waves (and of their traces) is proved in [1, 20]. PW bases whose linear independence does not degenerate for small values of kh_K were introduced in [46, Sect. 3.1] in 2D and in [86, Sect. 4.1] in 3D (see also [85, Sect. 3.4.1]) for analysis purposes. These stable PW bases converge to GHP bases in the low-frequency limit [86, p. 815]. The existence of these stable bases, which is instrumental to the derivation of approximation estimates for Helmholtz solutions in PW spaces in [86], is guaranteed, provided that the set of directions $\{\mathbf{d}_\ell\}_{\ell=1}^{p_K}$ constitutes a fundamental system for certain harmonic polynomials. In 2D, any choice of $p_K = 2q_K + 1$ distinct directions, q_K being the maximal degree of the considered harmonic polynomials, guarantees this property. In 3D, sufficient conditions on $p_K = (q_K + 1)^2$ directions are stated in [86, Lemma 4.2].

Approximation estimates in PW spaces can be derived from similar bounds for GHPs such as those in Sect. 3.1. In [80, Chap. 8], GHPs were approximated by PWs by approximating their smooth Herglotz kernel with delta functions, leading to p-estimates in 2D, while in [86] the Jacobi–Anger expansion was used to link PWs and GHPs in 2D and 3D. Theorems 5.2 and 5.3 of [86] (see also [85, Sect. 3.5]) show that Helmholtz solutions of given Sobolev regularity can be approximated in PW spaces with hp-estimates similar to those shown in Sect. 3.1 for GHPs. For PWs, these estimates hold with $L = q_K$, so that q_K plays the role of a "degree" for the considered PW space. As mentioned, for these bounds to hold in 3D, the PW directions have to satisfy some further conditions. A different derivation of h-approximation estimates based on a Taylor argument can be found in [19, Theorem 3.7]. In [95], the PW approximation of Helmholtz solutions on the unit disc is analysed in detail, together with the conditioning of different linear systems used for its computation (least squares and collocation for a Dirichlet problem on the disc) and the implications on the accuracy of the approximation computed in finite-precision arithmetic. We refer again to [56, Sect. 5.2] for the exponential convergence in 2D of PW approximations of analytic Helmholtz solutions (see also [85, Remark 3.5.8] which holds in 2D and 3D).

Similar to PWs are the *evanescent waves:* the basis elements have the same expression $v(\mathbf{x}) = e^{ik\mathbf{d}\cdot\mathbf{x}}$ but with a more general $\mathbf{d} \in \mathbb{C}^n$, $\mathbf{d}\cdot\mathbf{d} = 1$. If $\mathbf{d} = \mathbf{d}_R + i\mathbf{d}_I$, with $\mathbf{d}_R, \mathbf{d}_I \in \mathbb{R}^n$, then v oscillates in the direction $\mathbf{d}_R$ (with wavenumber $k|\mathbf{d}_R| \geq k$) and decays exponentially in the orthogonal direction $\mathbf{d}_I$ (i.e. $|v(\mathbf{x})| = e^{-k\mathbf{d}_I\cdot\mathbf{x}}$). Evanescent waves are used in combination with plane waves to approximate interface problems in the DEM [110] and the UWVF [77], and to represent outgoing waves in a 2D unbounded half-strip of the form $\{a < x < b, y > c\}$ in [20, 119].

A special combination of propagative and evanescent waves is typically used in the WBM. We describe a 2D version of this space as in [25, Eqs. (14)–(21)] (see [27, Sect. 4.1] for 3D). This space is not invariant under rotation but depends on the choice of the Cartesian axes. For a mesh element K, we fix a truncation parameter $N > 0$ (typically $1 \leq N \leq 6$) and define $L_x := \sup_{(x_1,y_1),(x_2,y_2)\in K} |x_1 - x_2|$ and $L_y := \sup_{(x_1,y_1),(x_2,y_2)\in K} |y_1 - y_2|$ as the edge lengths of the smallest rectangle containing K and aligned to the Cartesian axes. Two sets of basis functions are used:

$$\cos(k_{xj}x)\, e^{\pm i\sqrt{k^2-k_{xj}^2}\, y}, \qquad k_{xj} := \frac{j\pi}{L_x^K}, \quad j = 0, \dots, \lfloor NkL_x^K/\pi \rfloor,$$

$$e^{\pm i\sqrt{k^2-k_{yj}^2}\, x} \cos(k_{yj}y), \qquad k_{yj} := \frac{j\pi}{L_y^K}, \quad j = 0, \dots, \lfloor NkL_y^K/\pi \rfloor,$$

for a total dimension $p_K = 4 + 2(\lfloor NkL_x/\pi \rfloor + \lfloor NkL_y/\pi \rfloor)$. Each basis function is half the sum of two plane (or evanescent) waves, symmetric to one another with respect to the x or y axis: e.g. $\cos(k_{xj}x)\exp(i\sqrt{k^2 - k_{xj}^2}y) = \frac{1}{2}(e^{ik\mathbf{d}_{xj}^{+}\cdot\mathbf{x}} + e^{ik\mathbf{d}_{xj}^{-}\cdot\mathbf{x}})$, with $\mathbf{d}_{xj}^{\pm} := (\pm k_{xj}/k, \sqrt{1-(k_{xj}/k)^2})$. A maximum of $4 + 2(\lfloor kL_x/\pi \rfloor + \lfloor kL_y/\pi \rfloor)$ basis functions are propagative PWs, this number designed to keep the conditioning under control. If $N > 1$, then roughly a fraction $(N-1)/N$ of the total basis functions are evanescent waves decaying in a direction parallel to one of the Cartesian axes. Refinement is obtained by increasing N: for $N \leq 1$ only propagative waves are present, for higher values evanescent waves are introduced.

In 2D, both evanescent and plane waves may be written as $\exp\{\frac{k}{2}(\mathrm{i}(\nu + 1/\nu)x + (\nu - 1/\nu)y\} = \exp\{ik(x\sin\theta + y\cos\theta\}$, parametrised by $\nu \in \mathbb{C}$ or $\theta \in \mathbb{C}$ with $\nu = \mathrm{e}^{\mathrm{i}\theta}$; these waves constitute the test space (but usually not the trial) for the Fokas method in [23, 106] and [20, Sect. 7.3.4] (see also Sect. 2.4).

3.3 Fundamental Solutions and Multipoles

Fundamental solutions and multipoles are Helmholtz solution in the complement of a point and satisfy Sommerfeld radiation condition ($\lim_{r\to\infty} r^{\frac{n-1}{2}}(\frac{\partial u}{\partial r} - iku) = 0$, where $r = |\mathbf{x}|$). Thcy arc particularly uscful to define Trefftz spaces on unbounded elements, e.g. for scattering problems.

If the local spaces are spanned by fundamental solutions, simple sources are located at distinct poles $\mathbf{x}_\ell$ in the complement of K:

$$2D: \quad V_{p_K}(K) = \Big\{v : v(\mathbf{x}) = \sum_{\ell=1}^{p_K} \alpha_\ell H_0^{(1)}(k|\mathbf{x}-\mathbf{x}_\ell|),\ \alpha_\ell \in \mathbb{C}\Big\},$$

$$3D: \quad V_{p_K}(K) = \Big\{v : v(\mathbf{x}) = \sum_{\ell=1}^{p_K} \alpha_\ell \frac{\mathrm{e}^{-\mathrm{i}k|\mathbf{x}-\mathbf{x}_\ell|}}{|\mathbf{x}-\mathbf{x}_\ell|},\ \alpha_\ell \in \mathbb{C}\Big\},$$

where $H_0^{(1)}$ is the Hankel function of the first kind and of order 0. Different a priori or a posteriori strategies are used to fix the location of the poles, see Sect. 2.1.1 and the references cited therein. As the distance of the points $\mathbf{x}_\ell$ from K increases, these basis functions approach plane waves, so they permit flexibility not only in the choice of the propagation directions but also in the wavefront curvature.

Apart from the MFS and its modifications (see Sect. 2.1.1 and [1, 9, 10, 31, 92, 120]), spaces of fundamental solutions have been used in connection to the UWVF (see [57], where ray-tracing is used to determine the poles, and [58]).

Theorem 6 of [104] ensures that Helmholtz solutions in K can be approximated in Hölder norms by fundamental solutions centred at any "embracing boundary" in 2D and 3D, under weak assumptions on the regularity of ∂K. We are not aware of any result providing orders of convergence.

An alternative approach consists in choosing local spaces generated by multipole expansions, where multiple sources with increasing order are located at a single pole

$\mathbf{x}_0$ (or only at few poles):

$$2D: \; V_{p_K}(K) = \Big\{ v : \; v(\mathbf{x}) = \sum_{\ell=-q_K}^{q_K} \alpha_\ell \, H_\ell^{(1)}(k\,|\mathbf{x}-\mathbf{x}_0|)\, \mathrm{e}^{i\ell\theta}, \; \alpha_\ell \in \mathbb{C} \Big\},$$

$$3D: \; V_{p_K}(K) = \Big\{ v : \; v(\mathbf{x}) = \sum_{\ell=0}^{q_K} \sum_{m=-\ell}^{\ell} \alpha_{\ell,m} \, h_\ell^{(1)}(k\,|\mathbf{x}-\mathbf{x}_0|)\, Y_\ell^m\Big(\frac{\mathbf{x}-\mathbf{x}_0}{|\mathbf{x}-\mathbf{x}_0|}\Big), \; \alpha_{\ell m} \in \mathbb{C} \Big\},$$

where $H_\ell^{(1)}$ ($h_\ell^{(1)}$) are Hankel functions (spherical Hankel functions, respectively) of the first kind and order ℓ. As for the GHPs in Sect. 3.1, θ is the angle of $\mathbf{x}$ in the local coordinate system centred at $\mathbf{x}_0$, which is located in the complement of K, and the space dimension is $p_K = 2q_K+1$ in 2D and $p_K = (q_K+1)^2$ in 3D. According to [10, Remark 2.2], fundamental solutions lead to more stable methods than multipoles.

Multipole spaces have been used in connection to LS schemes [90, 107], WBM [25, Eq. (23)], [27, Sect. 4.1.2], hybrid-Trefftz [99, Eq. (63)], HELS [117], source simulation techniques [92], null-field [78] and single-element schemes [8, 22, 121]. In [49] and related papers, some 2D multipoles with suitably chosen index ℓ (not necessarily integer) are used on infinite sectors, in such a way to ensure continuity of discrete functions across rays; this might be more efficient than full multipole spaces for solutions with a preferred propagation direction.

3.4 Other Basis Functions

Other discrete Trefftz spaces have been proposed in literature for use with the various approaches covered in Sect. 2.

In 2D, *corner waves* such as $J_{\ell/\alpha}(k|\mathbf{x}|)\sin(\ell\theta/\alpha)$, with $\ell \in \mathbb{N}$ and $0 < \alpha < 2$, capture the behaviour of Helmholtz solutions near a domain corner of angle $\pi\alpha$. They have been used e.g. in the WBM [24], in LS methods [10, 107, 119] and in the MPS [15, 36]. In [120], they are used with $\alpha = 2$ on tips of 1D screens to represent the strong singularities of the solution in a non-Lipschitz domain. Theorem 6.3 of [28] uses Vekua–Bergman theory to give orders of convergence for the approximation of singular functions by spaces of corner waves and GHPs (see also [10, Sect. 5] and references therein). We are not aware of any use of similar functions in 3D.

The *wave band functions*, introduced in the VTCR context [100], are Herglotz functions with piecewise-constant kernel, e.g. $\int_a^b \mathrm{e}^{ik(x\cos\theta+y\sin\theta)}\,\mathrm{d}\theta$ in 2D.

In the presence of a circular hole, suitable combinations of *Hankel and Bessel functions* a priori fulfil homogeneous boundary conditions [107, Eq. (13)].

If the wavenumber varies inside an element, the basis functions described so far do not lead to Trefftz methods. In case of linearly variable wavenumber, *Airy functions* can be used to construct Trefftz spaces [111]. In [64, 65] *generalised plane waves* in the form $\mathrm{e}^{P(\mathbf{x})}$, for suitable polynomials P, are introduced and analysed in a UWVF setting: they solve a perturbed Helmholtz problem and converge with high orders in h_K. Similar "almost-Trefftz" waves are used in [43] and named *oscillated polynomials*. *Modulated plane waves*, i.e. products of PWs and polynomials, are the basis functions of the DG method of [13, 14]; as they are only "approximately Trefftz", volume terms appear in the formulation.

Products of (continuous) low-order polynomials and PWs or GHPs constitute the basis of the PUM [51, 73, 81, 96, 108], while products of polynomials and oscillating functions derived from high-frequency asymptotics are basis elements in [42, 91].

4 Further Topics

4.1 Assembly of Linear Systems

All the Trefftz finite element methods for (1) discussed in Sect. 2 give rise to dense or sparse linear systems of equations. Entries of coefficient matrices are obtained by integrating products of (derivatives of) trial and test functions over bounded d-dimensional sub-manifolds of Ω, $d < n$. The stable and accurate (approximate) evaluation of these integrals is a key implementation issue.

Among all Trefftz approximation spaces and associated bases presented in Sect. 3, plane waves (PWs) $e^{ik\mathbf{d}\cdot\mathbf{x}}$ (either propagative with $\mathbf{d} \in \mathbb{R}^n$ or evanescent with $\mathbf{d} \in \mathbb{C}^n$) are exceptional, because they allow a closed-form evaluation of their integrals over any flat sub-manifold with piecewise flat/straight boundary. For instance, in all variants of PW-based Trefftz methods on polyhedral meshes in 3D, expressing mesh faces by 2D parametrisations, we eventually encounter integrals of the form

$$\int_F \exp(\mathbf{w}\cdot\mathbf{x})\,\mathrm{d}V, \qquad F \subset \mathbb{R}^2 \text{ a bounded polygon, } \mathbf{w} \in \mathbb{C}^2 \text{ constant.} \tag{15}$$

Then we can take the cue from [38, Sect. 2.1] or [29, Sect. 4] and apply integration by parts in order to reduce (15) to integrals over the straight edges $e_1, e_2, \ldots e_q$, $q \in \mathbb{N}$ of F:

$$\int_F \exp(\mathbf{w}\cdot\mathbf{x})\,\mathrm{d}V = \frac{1}{\mathbf{w}\cdot\mathbf{w}}\int_F \mathbf{w}\cdot\nabla\exp(\mathbf{w}\cdot\mathbf{x})\,\mathrm{d}V = \sum_{\ell=1}^{q}\frac{\mathbf{w}\cdot\mathbf{n}_\ell}{\mathbf{w}\cdot\mathbf{w}}\int_{e_\ell}\exp(\mathbf{w}\cdot\mathbf{x})\,\mathrm{d}s,$$

where $\mathbf{n}_\ell$ is the exterior normal at e_ℓ. Then, as in [44, Chap. 2], if $e_\ell = [\mathbf{a}, \mathbf{b}]$, $\mathbf{a}, \mathbf{b} \in \mathbb{R}^2$, we find, $\int_{e_\ell} \exp(\mathbf{w}\cdot\mathbf{x})\,\mathrm{d}s = \exp(\mathbf{w}\cdot\mathbf{a})|\mathbf{b}-\mathbf{a}|\psi(\mathbf{w}\cdot(\mathbf{b}-\mathbf{a}))$, where $\psi(z) := (\exp(z)-1)/z$. Of course, a numerically stable implementation of this function for small arguments is essential.[1] This approach can be generalised to yield

[1] A stable algorithm for point evaluations of ψ even for arguments close to 0 is provided by the MATLAB function `expm1`.

analytic formulas for computing integrals of products of PWs times polynomials, see [29, 38], with increased computational effort, however.

Approximate evaluation of the integrals becomes inevitable for all choices of Trefftz basis functions other than PWs, and even for a PW basis on meshes with curved elements. Then Gauss–Legendre numerical quadrature seems to be the most widely used option. However, the integrands may be oscillatory, which delays the onset of (exponential) convergence of the quadrature error until the number of quadrature points surpasses a threshold roughly proportional to the ratio of the local mesh size and the wavelength. This leads to higher computational cost per degree of freedom for larger values of kh_K. One may think of using special quadrature rules for oscillatory integrals, as derived, for instance, in [62]. Those avoid an increase in the number of quadrature points for growing spatial frequency of the oscillations, but unfortunately require precise knowledge of the oscillatory term in the integrand.

4.2 Adaptive Trefftz Methods

Besides classical *h*-, *p*- or *hp*-adaptivity, Trefftz methods offer scope for more sophisticated adaptive strategies consisting in the choice of specific basis functions for different BVPs and in different mesh elements, either a priori or a posteriori.

The main strand of **a priori adaptive** Trefftz methods falls into the category of *hybrid numerical-asymptotic* methods. High-frequency limit models, such as ray optics or geometric theory of diffraction (GTD), guide the selection of local Trefftz spaces in the individual cells of a mesh. In a non-Trefftz PUM framework this idea was pursued in [42], and within the hybridizable DG method in [91], in both cases for 2D acoustic scattering at a smooth sound-soft object. In these works, local phase factors $\mathbf{x} \mapsto \exp(ikS(\mathbf{x}))$ derived from reflected and diffracted waves multiply standard continuous nodal basis functions, in [42], or local polynomials, in [91], thus generating a basis for (local) trial spaces.

The policy of incorporating local directions of rays is particularly attractive for PW-based methods, because PW basis functions naturally encode a direction of propagation. For problems where excitation is due to an incident PW and material properties are piecewise constant, ray tracing and related techniques [91, Sect. 3.2] based on geometric optics (specular reflection and Snell's law of refraction at material interfaces) can provide information about the local orientation of wave fronts for $k \to \infty$. PWs matching the found ray directions are then used to build local bases, either exclusively or augmented by a reduced set of generalised harmonic polynomials (GHPs) or "equi-spaced" PWs.

This idea for TDG was first outlined and tested in [13] and further elaborated and extended in [57, Chap. 5] (for UWVF). In the latter work, in an attempt to resolve curved wave fronts and take into account diffracted waves from corners, also Hankel functions $\mathbf{x} \mapsto H_0^{(1)}(k|\mathbf{x} - \mathbf{y}_*|)$ with $\mathbf{y}_*$ outside a mesh cell have been proposed as local basis functions. Approximation of curved wave fronts deduced from GTD corrections is also attempted in [14]. There the authors move beyond

Trefftz methods and use DG with trial spaces of polynomially modulated PWs, which are more suitable for approximating propagating circular waves.

In simple 2D situations with convex smooth or polygonal scatterers and incident plane wave, overall accuracy seems to benefit substantially from a priori directional adaptivity. However, if there are more than only a few dominant wave directions as in the case of more complicated geometries, trapping of waves, dark zones and shadow boundaries, current directional adaptivity soon meets its limitations. On the other hand, this strategy appears as the most promising way to achieve k-uniform accuracy with numbers of degrees of freedom that remain k-uniformly bounded or display only moderate growth as $k \to \infty$. The potential of this idea has been strikingly demonstrated in the case of BEM for 2D scattering problems [21].

A Posteriori Directional Adaptivity seeks to extract information about dominant wave directions from intermediate approximations of u. A refine-and-coarsen strategy is embraced in [13]. In each step of the adaptive cycle it first computes a PWDG solution u of the scattering problem based on a relatively large number of local Trefftz basis functions (GHPs and PWs). Subsequently, by solving local non-linear L^2-least squares problems, the directions of fewer PWs are determined so that u can still be well approximated locally.

A p-hierarchical error indicator is studied in [44]. In a step of the adaptive scheme starting from the approximate solution u a presumably improved solution $\hat{u}$ is computed using double the number of local PWs. Then a single local plane wave direction $\mathbf{d}_K$ on a mesh element K is extracted from the error $e(\mathbf{x}) := \hat{u}(\mathbf{x}) - u(\mathbf{x})$ through the projection formula

$$\widetilde{\mathbf{d}}_K := \operatorname{Re} \int_K \frac{\nabla e(\mathbf{x})}{ike(\mathbf{x})}\, \mathrm{d}V, \qquad \mathbf{d}_K := \frac{\widetilde{\mathbf{d}}_K}{|\widetilde{\mathbf{d}}_K|}.$$

Detailed numerical experiments are reported in [44, Chap. 6]. In the pre-asymptotic regime, when the resolution of the trial spaces is still rather low, one observes a pronounced gain in accuracy in the case of the adaptive approach compared to approximation with the same total number of equi-spaced PWs.

Directional adaptivity for Trefftz methods has also been tried in other flavours. In the context of least squares methods as discussed in Sect. 2.1 an offset angle for the sets of local equi-spaced PWs is introduced as another degree of freedom in [4], aiming to align them with a local dominant wave direction. For the VTCR method presented in Sect. 2.3.1, an error indicator based on local wave energy is used in [101] to steer angular refinement of local Trefftz spaces.

A Posteriori Mesh Adaptivity is considered in [66], where classical "elliptic" error estimation and mesh refinement strategies are adapted for the h-version of TDG. In a low-frequency setting, the method inherits the good performance of the underlying adaptive mesh refinement algorithms for polynomial DG for the Poisson equation. However, there is little hope that this carries over to larger wavenumbers k. A similar error estimator, aimed at adaptive mesh refinement, has been described in [2, Sect. 3.2] for the DEM/DGM presented in Sect. 2.2.3.

4.3 Ill-Conditioning and Solvers

The linear systems of equations spawned by PW-based finite element methods are highly prone to ill-conditioning, when high resolution trial spaces are used, see e.g. [60, Sect. 5], [37, 40, Sect. 4.3], and [72] for a PUM setting. This is largely caused by an inherent instability of the PW basis on cells, whose size is relatively small compared to the wavelength. Intuitively, for $|\mathbf{x}| \ll k^{-1}$, the functions $\mathbf{x} \mapsto \mathrm{e}^{ik\mathbf{d}_\ell \cdot \mathbf{x}}$ from (14) are almost constant, hence, nearly linearly dependent, cf. [72, Sect. 4.2]. The same heuristics applies, when their density increases; even for cell sizes comparable to the wavelength, PWs are hardly distinct when their directions are close, cf. [72, Sect. 4.3].

Empirically, for the local PW Galerkin matrix $\mathbf{M}_K$ associated with the L^2 inner product on a single mesh cell K, we find that its spectral condition number grows like $\sim h_K^{-q}$ for cell size $h_K \to 0$, where $q > 0$ is proportional to the number p_K of (approximately uniformly spaced) PWs in 2D, and to the square root of p_K in 3D. Essentially, q is related to the "degree" of the considered set of p_K PWs; see Sect. 3.2. Even worse, the condition number soars exponentially in q: $\mathrm{cond}_2(\mathbf{M}_K) \sim \mathrm{e}^{\alpha q}$ for $q \to \infty$ and $\alpha > 0$; see Appendix. A similar explosion of condition numbers is observed for the full systems matrices as meshes are refined or more PW basis functions per element are used.

There is circumstantial evidence that direct sparse elimination can cope fairly well with the ill-conditioned linear systems arising from UWVF or PUM, see [40, Sect. 5.3.3], [77]. Yet, eventually the instability of the basis will impact the quality of the solution [108, Sect. 5.4]. A remedy proposed in [60] for the UWVF is to limit p_K based on monitoring condition numbers of element matrices. Apparently, this also curbs the condition number of the global system matrix. Alternatively, there exist different heuristic recipes for choosing a priori the number of PWs per element to balance accuracy and conditioning: in 2D, the widely cited [61, Eq. (14)] suggests $p_K = \mathrm{round}(kh_K + C(kh_K)^{1/3})$ with $3 \le C \le 14$ for the UWVF, while [70, Sect. 5.1.1] proposes $p_K = \lfloor 2kh_K \rfloor$ for the VTCR. For the WBM, [25, Sect. 3.2] proposes a rule to balance propagative and evanescent basis functions, see Sect. 3.2.

The most straightforward cure for instability would trade the PW basis of $V_{p_K}(K)$ from (14) for a more stable basis, found by local orthonormalisation as in the case of polynomial FEM, cf. the approach from [91, Sect. 3.1]. However, instability may sneak in through the back door and manifest itself in severe impact of round-off errors during orthonormalisation and recombination of element matrices. The use of high-precision arithmetic may be advisable, but has never been documented.

For the sake of stability, PWs may be replaced by the generalised harmonics polynomials introduced in Sect. 3.1. In 2D, a scaling of the GHPs has been devised in [77], in order to lower the condition number of the resulting UWVF:

$$\frac{J_\ell(k\,|\mathbf{x} - \mathbf{x}_K|)\,\mathrm{e}^{i\ell\theta}}{k\sqrt{\left|J'_\ell(kh_K)\right|^2 + |J_\ell(kh_K)|^2}}.$$

In [77], it is also shown that the conditioning of GHP-based UWVF schemes is better than for methods based on PWs, and that it improves on regular meshes. This might be related to the orthogonality of GHPs on balls.

The numerical experiments in [57, Sect. 3.7] suggest that the use of fundamental solutions as basis functions may considerably reduce the conditioning of UWVF matrices, at the expense of accuracy. Both accuracy and conditioning increase the further the centres of the fundamental solutions are from the element.

The use of iterative solvers for linear systems generated by Trefftz methods entails preconditioning. For PW basis functions, the first proposal in [19, Sect. 2.4] for the UWVF was a local preconditioner, equivalent to an orthonormalisation of the PW basis with respect to an L^2 inner product on the boundary of mesh cells. An interesting connection of the local preconditioner with non-overlapping optimised Schwarz domain decomposition methods was discovered in [16]. The local preconditioner was used in conjunction with a BiCGStab Krylov subspace solver in [60] and augmented by a coarse-grid correction in the spirit of non-overlapping domain decomposition in [59, 118]. The coarse space is again spanned by PWs. This is also true for the two-level sub-structuring preconditioner proposed for DEM/DGM (see Sect. 2.2.3) in [34]. Two-level, non-overlapping Schwarz domain decomposition preconditioners for PWDG (essentially UWVF) have been tested in [5]; these preconditioners seem to be robust with respect to the wavenumber k and the local number of PW directions, although they do not seem to be perfectly scalable with respect to the number of subdomains.

5 Assessment and Conclusion

Faced with a flurry of different Trefftz methods and a wealth of numerical data, we feel at a loss about making unequivocal statements about the merits of Trefftz methods, let alone ranking them according to some undisputed criteria. Rigorous theory is available for LS methods (Sect. 2.1), TDG (Sect. 2.2.1), and PUM (Sect. 2.5). Combined with approximation results for suitable Trefftz bases, this leads to better asymptotic estimates in terms of orders of convergence in the number of degrees of freedom to what is available for polynomial FEM (e.g. [53, 56]). The dependence of crucial constants on the wavenumber k is explicit in several cases, but the orders in k are usually not better than for polynomial methods. Thus theory fails to provide information about the key issue of "k-robust" accuracy with "k-independent" cost. Moreover, numerical dispersion will also haunt local Trefftz methods in the case of h-refinement; thus they provide no escape from the pollution error.

We also advise caution when reading numerical experiments, because they may be tarnished by selection bias, making authors subliminally pick test cases matching the intended message of an article. Disregarding this, even "objective" comparisons are inevitably confined to a few simple model problems. This is problematic, because different model problems sometimes seem to support opposite conclusions.

From our experience, the power of Trefftz methods can best harnessed by p-refinement using approximation by Trefftz functions in regions as large as possible. In the presence of singularities we recommend either the use of corner basis functions (Sect. 3.4) in 2D, or hp-refinement, maybe using standard polynomial approximation on small elements as in [89]. There is a solid theoretical foundation, when this is done in the LS, TDG, or PUM framework. The resulting methods should be able to compete successfully with polynomial FEM even in their more sophisticated versions tailored to wave propagation problems [30, 35, 82].

The discussion of adaptive approaches in Sect. 4.2 hints that some Trefftz trial spaces have approximation capabilities well beyond the reach of polynomials. Directional adaptivity seems to be very promising, but much research will still be required to convert it into a reliable practical algorithm. The same applies to iterative solvers and preconditioners for Trefftz schemes, see Sect. 4.3, which might also benefit considerably from the extra information contained in Trefftz trial spaces. Hence, we believe that many exciting possibilities offered by the idea of Trefftz approximation still await discovery and that the full potential of Trefftz methods is only gradually being realised.

Appendix: Condition Numbers of Plane Wave Mass Matrices

Given a wave number $k > 0$ and $p \in \mathbb{N}$ distinct unit vectors $\mathbf{d}_\ell \in \mathbb{R}^n$, $\ell = 1, \dots, p$, and a domain $K \subset \mathbb{R}^n$ with barycentre $\mathbf{x}_K$, the symmetric positive definite plane wave element mass matrix $\mathbf{M}_K$ on K is defined as

$$\mathbf{M}_K := \left(\int_K e^{\mathrm{i}k\mathbf{d}_\ell \cdot (\mathbf{x}-\mathbf{x}_K)} \cdot e^{-\mathrm{i}k\mathbf{d}_m \cdot (\mathbf{x}-\mathbf{x}_K)} \, \mathrm{d}V \right)_{\ell,m=1}^{p} .$$

For $n = 2$ we computed spectral condition numbers of $\mathbf{M}_K$ for equi-spaced directions $\mathbf{d}_\ell = (\cos(2\pi\ell/p), \sin(2\pi\ell/p))$, $\ell = 0, \dots, p-1$. For $n = 3$ we choose the directions $\mathbf{d}_\ell$ as the "minimum norm points" according to Sloan and Womersley [103, 116]. These points are indexed by a level $q \in \mathbb{N}$ and $p = (q+1)^2$. The spectral condition numbers are plotted in Fig. 1 for $n = 2$, $K = (-1, 1)^2$, and Fig. 2 for $n = 3$, $K = (-1, 1)^3$. They have been computed with MATLAB using the high-precision arithmetic (200 decimal digits) provided by the Advanpix Multi-Precision Toolbox.[2]

[2]http://www.advanpix.com/.

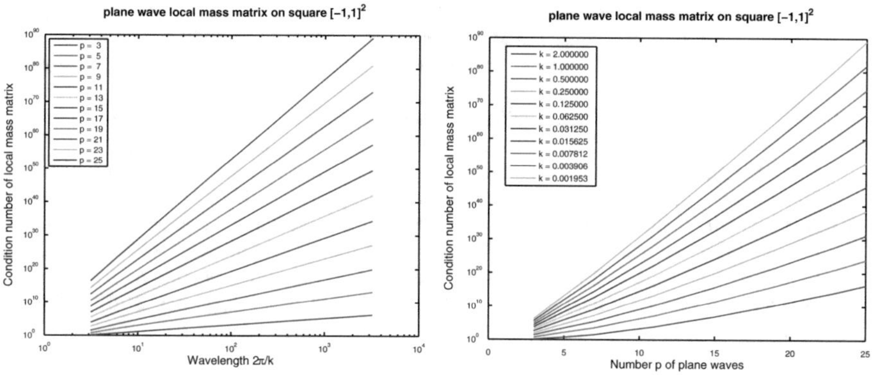

Fig. 1 Condition numbers of element mass matrices on the square $(-1, 1)^2$

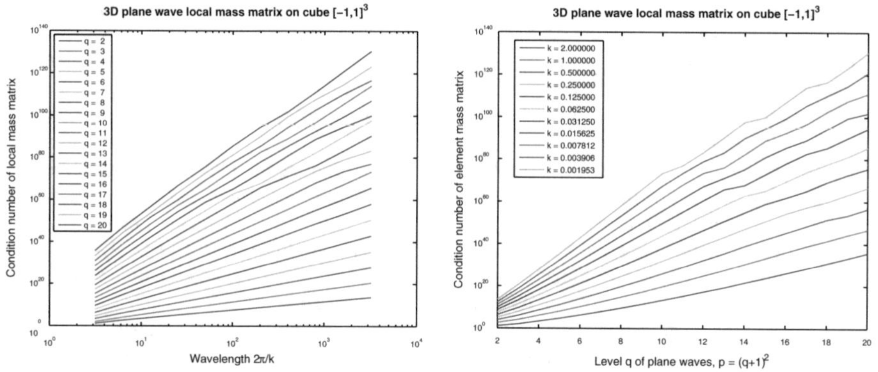

Fig. 2 Condition numbers of element mass matrices on the cube $(-1, 1)^3$

References

1. C.J. Alves, S.S. Valtchev, Numerical comparison of two meshfree methods for acoustic wave scattering. Eng. Anal. Boundary Elem. **29**(4), 371–382 (2005)
2. M. Amara, R. Djellouli, C. Farhat, Convergence analysis of a discontinuous Galerkin method with plane waves and Lagrange multipliers for the solution of Helmholtz problems. SIAM J. Numer. Anal. **47**(2), 1038–1066 (2009)
3. M. Amara, H. Calandra, R. Dejllouli, M. Grigoroscuta-Strugaru, A stable discontinuous Galerkin-type method for solving efficiently Helmholtz problems. Comput. Struct. **106–107**, 258–272 (2012)
4. M. Amara, S. Chaudhry, J. Diaz, R. Djellouli, S.L. Fiedler, A local wave tracking strategy for efficiently solving mid- and high-frequency Helmholtz problems. Comput. Methods Appl. Mech. Eng. **276**, 473–508 (2014)
5. P.F. Antonietti, I. Perugia, D. Zaliani, Schwarz domain decomposition preconditioners for plane wave discontinuous Galerkin methods, in *Numerical Mathematics and Advanced Applications - ENUMATH 2013*, ed. by A. Abdulle, S. Deparis, D. Kressner, F. Nobile, M. Picasso. Lecture Notes in Computational Science and Engineering, vol. 103 (Springer, Berlin, 2015), pp. 557–572

6. D.N. Arnold, F. Brezzi, B. Cockburn, L.D. Marini, Unified analysis of discontinuous Galerkin methods for elliptic problems. SIAM J. Numer. Anal. **39**(5), 1749–1779 (2002)
7. R.J. Astley, P. Gamallo, Special short wave elements for flow acoustics. Comput. Methods Appl. Mech. Eng. **194**(2–5), 341–353 (2005)
8. A.K. Aziz, M.R. Dorr, R.B. Kellogg, A new approximation method for the Helmholtz equation in an exterior domain. SIAM J. Numer. Anal. **19**(5), 899–908 (1982)
9. A.H. Barnett, T. Betcke, Stability and convergence of the method of fundamental solutions for Helmholtz problems on analytic domains. J. Comput. Phys. **227**(14), 7003–7026 (2008)
10. A.H. Barnett, T. Betcke, An exponentially convergent nonpolynomial finite element method for time-harmonic scattering from polygons. SIAM J. Sci. Comput. **32**(3), 1417–1441 (2010)
11. L. Beirão da Veiga, F. Brezzi, A. Cangiani, G. Manzini, L.D. Marini, A. Russo, Basic principles of virtual element methods. Math. Models Methods Appl. Sci. **23**(01), 199–214 (2013)
12. S. Bergman, *Integral Operators in the Theory of Linear Partial Differential Equations*. Ergebnisse der Mathematik und ihrer Grenzgebiete, vol. 23 (Springer, New York, 1969); Second revised printing
13. T. Betcke, J. Phillips, Adaptive plane wave discontinuous Galerkin method for Helmholtz problems, in *Proceedings of the 10th International Conference on the Mathematical and Numerical Aspects of Waves*, Vancouver, 2011, pp. 261–264
14. T. Betcke, J. Phillips, Approximation by dominant wave directions in plane wave methods. Technical Report, UCL (2012). Available at http://discovery.ucl.ac.uk/1342769/
15. T. Betcke, L.N. Trefethen, Reviving the method of particular solutions. SIAM Rev. **47**(3), 469–491 (2005)
16. T. Betcke, M. Gander, J. Phillips, Block Jacobi relaxation for plane wave discontinuous Galerkin methods, in *Domain Decomposition Methods in Science and Engineering XXI*, ed. by J. Erhel, M.J. Gander, L. Halpern, G. Pichot, T. Sassi, O. Widlund. Lecture Notes in Computational Science and Engineering, vol. 98 (Springer, Berlin, 2014), pp. 577–585
17. A. Buffa, P. Monk, Error estimates for the ultra weak variational formulation of the Helmholtz equation. M2AN, Math. Model. Numer. Anal. **42**(6), 925–940 (2008)
18. O. Cessenat, Application d'une nouvelle formulation variationnelle aux équations d'ondes harmoniques. Problèmes de Helmholtz 2D et de Maxwell 3D. Ph.D. thesis, Université Paris IX Dauphine, 1996
19. O. Cessenat, B. Després, Application of an ultra weak variational formulation of elliptic PDEs to the two-dimensional Helmholtz equation. SIAM J. Numer. Anal. **35**(1), 255–299 (1998)
20. S.N. Chandler-Wilde, S. Langdon, Acoustic scattering: high-frequency boundary element methods and unified transform methods, in *Unified Transform Method for Boundary Value Problems: Applications and Advances*, ed. by A. Fokas, B. Pelloni (SIAM, Philadelphia, 2015), pp. 181–226
21. S.N. Chandler-Wilde, I.G. Graham, S. Langdon, E. Spence, Numerical-asymptotic boundary integral methods in high-frequency acoustic scattering. Acta Numer. **21**, 89–305 (2012)
22. Y.K. Cheung, W.G. Jin, O.C. Zienkiewicz, Solution of Helmholtz equation by Trefftz method. Int. J. Numer. Methods Eng. **32**(1), 63–78 (1991)
23. C.I.R. Davis, B. Fornberg, A spectrally accurate numerical implementation of the Fokas transform method for Helmholtz-type PDEs. Complex Var. Elliptic Equ. **59**, 564–577 (2014)
24. E. Deckers, B. Bergen, B. Van Genechten, D. Vandepitte, W. Desmet, An efficient wave based method for 2D acoustic problems containing corner singularities. Comput. Methods Appl. Mech. Eng. **241–244**, 286–301 (2012)
25. E. Deckers et al., The wave based method: an overview of 15 years of research. Wave Motion **51**(4), 550–565 (2014); Innovations in Wave Modelling
26. W. Desmet, A wave based prediction technique for coupled vibro-acoustic analysis. Ph.D. thesis, KU Leuven, Belgium, 1998
27. W. Desmet et al., The wave based method, in *"Mid-Frequency" CAE Methodologies for Mid-Frequency Analysis in Vibration and Acoustics* (KU Leuven, Belgium, 2012), pp. 1–60

28. S.C. Eisenstat, On the rate of convergence of the Bergman-Vekua method for the numerical solution of elliptic boundary value problems. SIAM J. Numer. Anal. **11**, 654–680 (1974)
29. A. El Kacimi, O. Laghrouche, Improvement of PUFEM for the numerical solution of high-frequency elastic wave scattering on unstructured triangular mesh grids. Int. J. Numer. Methods Eng. **84**(3), 330–350 (2010)
30. S. Esterhazy, J. Melenk, On stability of discretizations of the Helmholtz equation, in *Numerical Analysis of Multiscale Problems*, ed. by I. Graham, T. Hou, O. Lakkis, R. Scheichl. Lecture Notes in Computational Science and Engineering, vol. 83 (Springer, Berlin, 2011), pp. 285–324
31. G. Fairweather, A. Karageorghis, P.A. Martin, The method of fundamental solutions for scattering and radiation problems. Eng. Anal. Boundary Elem. **27**(7), 759–769 (2003)
32. C. Farhat, I. Harari, L. Franca, The discontinuous enrichment method. Comput. Methods Appl. Mech. Eng. **190**(48), 6455–6479 (2001)
33. C. Farhat, I. Harari, U. Hetmaniuk, A discontinuous Galerkin method with Lagrange multipliers for the solution of Helmholtz problems in the mid-frequency regime. Comput. Methods Appl. Mech. Eng. **192**(11), 1389–1419 (2003)
34. C. Farhat, R. Tezaur, J. Toivanen, A domain decomposition method for discontinuous Galerkin discretizations of Helmholtz problems with plane waves and Lagrange multipliers. Int. J. Numer. Methods Eng. **78**(13), 1513–1531 (2009)
35. X.B. Feng, H.J. Wu, *hp*-Discontinuous Galerkin methods for the Helmholtz equation with large wave number. Math. Comput. **80**(4), 1997–2024 (2011)
36. L. Fox, P. Henrici, C. Moler, Approximations and bounds for eigenvalues of elliptic operators. SIAM J. Numer. Anal. **4**, 89–102 (1967)
37. G. Gabard, Discontinuous Galerkin methods with plane waves for time-harmonic problems. J. Comput. Phys. **225**, 1961–1984 (2007)
38. G. Gabard, Exact integration of polynomial-exponential products with application to wave-based numerical methods. Commun. Numer. Methods Eng. **25**(3), 237–246 (2009)
39. G. Gabard, P. Gamallo, T. Huttunen, A comparison of wave-based discontinuous Galerkin, ultra-weak and least-square methods for wave problems. Int. J. Numer. Methods Eng. **85**(3), 380–402 (2011)
40. P. Gamallo, R.J. Astley, A comparison of two Trefftz-type methods: the ultra-weak variational formulation and the least squares method for solving shortwave 2D Helmholtz problems. Int. J. Numer. Methods Eng. **71**, 406–432 (2007)
41. M. Gander, I. Graham, E. Spence, Applying GMRES to the Helmholtz equation with shifted Laplacian preconditioning: what is the largest shift for which wavenumber-independent convergence is guaranteed? Numer. Math. **131**(3), 567–614 (2015). doi:10.1007/s00211-015-0700-2. http://dx.doi.org/10.1007/s00211-015-0700-2
42. E. Giladi, J.B. Keller, A hybrid numerical asymptotic method for scattering problems. J. Comput. Phys. **174**(1), 226–247 (2001)
43. A. Gillman, R. Djellouli, M. Amara, A mixed hybrid formulation based on oscillated finite element polynomials for solving Helmholtz problems. J. Comput. Appl. Math. **204**(2), 515–525 (2007)
44. C.J. Gittelson, Plane wave discontinuous Galerkin methods. Master's thesis, SAM, ETH Zürich, Switzerland, 2008. Available at http://www.sam.math.ethz.ch/~hiptmair/StudentProjects/Gittelson/thesis.pdf
45. C.J. Gittelson, R. Hiptmair, Dispersion analysis of plane wave discontinuous Galerkin methods. Int. J. Numer. Methods Eng. **98**(5), 313–323 (2014)
46. C.J. Gittelson, R. Hiptmair, I. Perugia, Plane wave discontinuous Galerkin methods: analysis of the *h*-version. M2AN, Math. Model. Numer. Anal. **43**(2), 297–332 (2009)
47. C.I. Goldstein, The weak element method applied to Helmholtz type equations. Appl. Numer. Math. **2**(3–5), 409–426 (1986)
48. M. Grigoroscuta-Strugaru, M. Amara, H. Calandra, R. Djellouli, A modified discontinuous Galerkin method for solving efficiently Helmholtz problems. Commun. Comput. Phys. **11**(2), 335–350 (2012)

49. I. Harari, P. Barai, P.E. Barbone, Numerical and spectral investigations of Trefftz infinite elements. Int. J. Numer. Methods Eng. **46**(4), 553–577 (1999)
50. P. Henrici, A survey of I. N. Vekua's theory of elliptic partial differential equations with analytic coefficients. Z. Angew. Math. Phys. **8**, 169–202 (1957)
51. B. Heubeck, C. Pflaum, G. Steinle, New finite elements for large-scale simulation of optical waves. SIAM J. Sci. Comput. **31**(2), 1063–1081 (2008/09)
52. R. Hiptmair, I. Perugia, Mixed plane wave DG methods, in *Domain Decomposition Methods in Science and Engineering XVIII*, ed. by M. Bercovier, M.J. Gander, R. Kornhuber, O. Widlund. Lecture Notes in Computational Science and Engineering (Springer, Berlin, 2008), pp. 51–62
53. R. Hiptmair, A. Moiola, I. Perugia, Plane wave discontinuous Galerkin methods for the 2D Helmholtz equation: analysis of the *p*-version. SIAM J. Numer. Anal. **49**, 264–284 (2011)
54. R. Hiptmair, A. Moiola, I. Perugia, C. Schwab, Approximation by harmonic polynomials in star-shaped domains and exponential convergence of Trefftz *hp*-dGFEM. Math. Model. Numer. Anal. **48**, 727–752 (2014)
55. R. Hiptmair, A. Moiola, I. Perugia, Trefftz discontinuous Galerkin methods for acoustic scattering on locally refined meshes. Appl. Numer. Math. **79**, 79–91 (2014)
56. R. Hiptmair, A. Moiola, I. Perugia, Plane wave discontinuous Galerkin methods: exponential convergence of the *hp*-version. Found. Comput. Math. (2015). doi:10.1007/s10208-015-9260-1
57. C.J. Howarth, New generation finite element methods for forward seismic modelling. Ph.D. thesis, University of Reading, UK, 2014. Available at http://www.reading.ac.uk/maths-and-stats/research/theses/maths-phdtheses.aspx
58. C. Howarth, P. Childs, A. Moiola, Implementation of an interior point source in the ultra weak variational formulation through source extraction. J. Comput. Appl. Math. **271**, 295–306 (2014)
59. Q. Hu, L. Yuan, A weighted variational formulation based on plane wave basis for discretization of Helmholtz equations. Int. J. Numer. Anal. Model. **11**(3), 587–607 (2014)
60. T. Huttunen, P. Monk, J.P. Kaipio, Computational aspects of the ultra-weak variational formulation. J. Comput. Phys. **182**(1), 27–46 (2002)
61. T. Huttunen, P. Gamallo, R. Astley, A comparison of two wave element methods for the Helmholtz problem. Commun. Numer. Methods Eng. **25**(1), 35–52 (2009)
62. D. Huybrechs, S. Olver, Highly oscillatory quadrature, in *Highly Oscillatory Problems*. London Mathematical Society Lecture Note Series, vol. 366 (Cambridge University Press, Cambridge, 2009), pp. 25–50
63. F. Ihlenburg, I. Babuška, Solution of Helmholtz problems by knowledge-based FEM. Comput. Assist. Mech. Eng. Sci. **4**, 397–416 (1997)
64. L.M. Imbert-Gérard, Interpolation properties of generalized plane waves. Numer. Math. (2015). doi:10.1007/s00211-015-0704-y
65. L.M. Imbert-Gérard, B. Després, A generalized plane-wave numerical method for smooth nonconstant coefficients. IMA J. Numer. Anal. **34**(3), 1072–1103 (2014)
66. S. Kapita, P. Monk, T. Warburton, Residual based adaptivity and PWDG methods for the Helmholtz equation. arXiv:1405.1957v1 (2014)
67. E. Kita, N. Kamiya, Trefftz method: an overview. Adv. Eng. Softw. **24**(1–3), 3–12 (1995)
68. L. Kovalevsky, P. Ladevéze, H. Riou, The Fourier version of the variational theory of complex rays for medium-frequency acoustics. Comput. Methods Appl. Mech. Eng. **225/228**, 142–153 (2012)
69. F. Kretzschmar, A. Moiola, I. Perugia, S.M. Schnepp, A priori error analysis of space-time Trefftz discontinuous Galerkin methods for wave problems. arXiv:1501.05253v2 (2015)
70. P. Ladevéze, H. Riou, On Trefftz and weak Trefftz discontinuous Galerkin approaches for medium-frequency acoustics. Comput. Methods Appl. Mech. Eng. **278**, 729–743 (2014)
71. P. Ladevéze, A. Barbarulo, H. Riou, L. Kovalevsky, The variational theory of complex rays, in *"Mid-Frequency" CAE Methodologies for Mid-Frequency Analysis in Vibration and Acoustics* (KU Leuven, Belgium, 2012), pp. 155–217

72. O. Laghrouche, P. Bettes, R.J. Astley, Modelling of short wave diffraction problems using approximating systems of plane waves. Int. J. Numer. Methods Eng. **54**, 1501–1533 (2002)
73. O. Laghrouche, P. Bettess, E. Perrey-Debain, J. Trevelyan, Wave interpolation finite elements for Helmholtz problems with jumps in the wave speed. Comput. Methods Appl. Mech. Eng. **194**(2–5), 367–381 (2005)
74. F. Li, C.W. Shu, A local-structure-preserving local discontinuous Galerkin method for the Laplace equation. Methods Appl. Anal. **13**(2), 215–233 (2006)
75. Z.C. Li, T.T. Lu, H.Y. Hu, A.H.D. Cheng, *Trefftz and Collocation Methods* (WIT Press, Southampton, 2008)
76. T. Luostari, Non-polynomial approximation methods in acoustics and elasticity. Ph.D. thesis, University of Eastern Finland, 2013. Available at http://venda.uef.fi/inverse/Frontpage/Publications/Theses
77. T. Luostari, T. Huttunen, P. Monk, Improvements for the ultra weak variational formulation. Int. J. Numer. Methods Eng. **94**(6), 598–624 (2013)
78. P.A. Martin, Multiple scattering, *Encyclopedia of Mathematics and Its Applications*, vol. 107 (Cambridge University Press, Cambridge, 2006); Interaction of time-harmonic waves with N obstacles
79. P. Mayer, J. Mandel, The finite ray element method for the Helmholtz equation of scattering: first numerical experiments. Technical Report 111, Center for Computational Mathematics, UC Denver, 1997. Available at http://ccm.ucdenver.edu/reports/
80. J.M. Melenk, On generalized finite element methods. Ph.D. thesis, University of Maryland, 1995
81. J.M. Melenk, I. Babuška, The partition of unity finite element method: basic theory and applications. Comput. Methods Appl. Mech. Eng. **139**(1–4), 289–314 (1996)
82. J.M. Melenk, S. Sauter, Wavenumber explicit convergence analysis for Galerkin discretizations of the Helmholtz equation. SIAM J. Numer. Anal. **49**(3), 1210–1243 (2011)
83. J.M. Melenk, A. Parsania, S. Sauter, General DG-methods for highly indefinite Helmholtz problems. J. Sci. Comput. **57**(3), 536–581 (2013)
84. A. Moiola, Approximation properties of plane wave spaces and application to the analysis of the plane wave discontinuous Galerkin method. Report 2009-06, SAM, ETH Zürich, 2009
85. A. Moiola, Trefftz-discontinuous Galerkin methods for time-harmonic wave problems. Ph.D. thesis, Seminar for Applied Mathematics, ETH Zürich, 2011. Available at http://e-collection.library.ethz.ch/view/eth:4515
86. A. Moiola, R. Hiptmair, I. Perugia, Plane wave approximation of homogeneous Helmholtz solutions. Z. Angew. Math. Phys. **62**, 809–837 (2011)
87. A. Moiola, R. Hiptmair, I. Perugia, Vekua theory for the Helmholtz operator. Z. Angew. Math. Phys. **62**, 779–807 (2011)
88. P. Monk, D. Wang, A least squares method for the Helmholtz equation. Comput. Methods Appl. Mech. Eng. **175**(1/2), 121–136 (1999)
89. P. Monk, J. Schöberl, A. Sinwel, Hybridizing Raviart-Thomas elements for the Helmholtz equation. Electromagnetics **30**, 149–176 (2010)
90. E. Moreno, D. Erni, C. Hafner, R. Vahldieck, Multiple multipole method with automatic multipole setting applied to the simulation of surface plasmons in metallic nanostructures. J. Opt. Soc. Am. A **19**(1), 101–111 (2002)
91. N. Nguyen, J. Peraire, F. Reitich, B. Cockburn, A phase-based hybridizable discontinuous Galerkin method for the numerical solution of the Helmholtz equation. J. Comput. Phys. **290**, 318–335 (2015)
92. M. Ochmann, The source simulation technique for acoustic radiation problems. Acta Acustica united with Acustica **81**(6), 512–527 (1995)
93. M.J. Peake, J. Trevelyan, G. Coates, Extended isogeometric boundary element method (XIBEM) for two-dimensional Helmholtz problems. Comput. Methods Appl. Mech. Eng. **259**, 93–102 (2013)

94. M.J. Peake, J. Trevelyan, G. Coates, The equal spacing of N points on a sphere with application to partition-of-unity wave diffraction problems. Eng. Anal. Boundary Elem. **40**, 114–122 (2014)
95. E. Perrey-Debain, Plane wave decomposition in the unit disc: convergence estimates and computational aspects. J. Comput. Appl. Math. **193**(1), 140–156 (2006)
96. E. Perrey-Debain, O. Laghrouche, P. Bettess, Plane-wave basis finite elements and boundary elements for three-dimensional wave scattering. Philos. Trans. R. Soc. Lond. Ser. A Math. Phys. Eng. Sci. **362**(1816), 561–577 (2004)
97. I. Perugia, P. Pietra, A. Russo, A plane wave virtual element method for the Helmholtz problem. arXiv:1505.04965v1 (2015)
98. B. Pluymers, B. van Hal, D. Vandepitte, W. Desmet, Trefftz-based methods for time-harmonic acoustics. Arch. Comput. Methods Eng. **14**(4), 343–381 (2007)
99. Q.H. Qin, Trefftz finite element method and its applications. Appl. Mech. Rev. **58**(5), 316–337 (2005)
100. H. Riou, P. Ladevéze, B. Sourcis, The multiscale VTCR approach applied to acoustics problems. J. Comput. Acoust. **16**(4), 487–505 (2008)
101. H. Riou, P. Ladevéze, B. Sourcis, B. Faverjon, L. Kovalevsky, An adaptive numerical strategy for the medium-frequency analysis of Helmholtz's problem. J. Comput. Acoust. **20**(01), 1250001 (2012)
102. H. Riou, P. Ladevéze, L. Kovalevsky, The variational theory of complex rays: an answer to the resolution of mid-frequency 3d engineering problems. J. Sound Vib. **332**(8), 1947–1960 (2013)
103. I.H. Sloan, R.S. Womersley, Extremal systems of points and numerical integration on the sphere. Adv. Comput. Math. **21**(1–2), 107–125 (2004)
104. Y.S. Smyrlis, Density results with linear combinations of translates of fundamental solutions. J. Approx. Theory **161**(2), 617–633 (2009)
105. E.A. Spence, Wavenumber-explicit bounds in time-harmonic acoustic scattering. SIAM J. Math. Anal. **46**(4), 2987–3024 (2014)
106. E. Spence, "When all else fails, integrate by parts": an overview of new and old variational formulations for linear elliptic PDEs, in *Unified Transform Method for Boundary Value Problems: Applications and Advances*, ed. by A. Fokas, B. Pelloni (SIAM, Philadelphia, 2015), pp. 93–159
107. M. Stojek, Least-squares Trefftz-type elements for the Helmholtz equation. Int. J. Numer. Methods Eng. **41**(5), 831–849 (1998)
108. T. Strouboulis, I. Babuška, R. Hidajat, The generalized finite element method for Helmholtz equation: theory, computation, and open problems. Comput. Methods Appl. Mech. Eng. **37–40**, 4711–4731 (2006)
109. K.Y. Sze, G.H. Liu, H. Fan, Four- and eight-node hybrid-Trefftz quadrilateral finite element models for Helmholtz problem. Comput. Methods Appl. Mech. Eng. **199**, 598–614 (2010)
110. R. Tezaur, L. Zhang, C. Farhat, A discontinuous enrichment method for capturing evanescent waves in multiscale fluid and fluid/solid problems. Comput. Methods Appl. Mech. Eng. **197**(19–20), 1680–1698 (2008)
111. R. Tezaur, I. Kalashnikova, C. Farhat, The discontinuous enrichment method for medium-frequency Helmholtz problems with a spatially variable wavenumber. Comput. Methods Appl. Mech. Eng. **268**, 126–140 (2014)
112. E. Trefftz, Ein Gegenstuck zum Ritzschen Verfahren, in *Proceedings of the 2nd International Congress for Applied Mechanics*, Zurich, 1926, pp. 131–137
113. I. Tsukerman, A class of difference schemes with flexible local approximation. J. Comput. Phys. **211**(2), 659–699 (2006)
114. I.N. Vekua, *New Methods for Solving Elliptic Equations* (North Holland, Amsterdam, 1967); Translation from Russian edition (1948)

115. D. Wang, R. Tezaur, J. Toivanen, C. Farhat, Overview of the discontinuous enrichment method, the ultra-weak variational formulation, and the partition of unity method for acoustic scattering in the medium frequency regime and performance comparisons. Int. J. Numer. Methods Eng. **89**(4), 403–417 (2012)
116. R.S. Womersley, I.H. Sloan, Interpolation and cubature on the sphere. http://web.maths.unsw.edu.au/~rsw/Sphere
117. S.F. Wu, *The Helmholtz Equation Least Squares Method.* Modern Acoustics and Signal Processing (Springer, New York, 2015)
118. L. Yuan, Q. Hu, A solver for Helmholtz system generated by the discretization of wave shape functions. Adv. Appl. Math. Mech. **5**(6), 791–808 (2013)
119. E. Zheng, F. Ma, D. Zhang, A least-squares non-polynomial finite element method for solving the polygonal-line grating problem. J. Math. Anal. Appl. **397**(2), 550–560 (2013)
120. E. Zheng, F. Ma, D. Zhang, A least-squares finite element method for solving the polygonal-line arc-scattering problem. Appl. Anal. **93**(6), 1164–1177 (2014)
121. O. Zienkiewicz, Trefftz type approximation and the generalized finite element method- history and development. Comput. Assist. Mech. Eng. Sci. **4**(3), 305–316 (1997)

Review of Discontinuous Galerkin Finite Element Methods for Partial Differential Equations on Complicated Domains

Paola F. Antonietti, Andrea Cangiani, Joe Collis, Zhaonan Dong, Emmanuil H. Georgoulis, Stefano Giani, and Paul Houston

Abstract The numerical approximation of partial differential equations (PDEs) posed on complicated geometries, which include a large number of small geometrical features or microstructures, represents a challenging computational problem. Indeed, the use of standard mesh generators, employing simplices or tensor product elements, for example, naturally leads to very fine finite element meshes, and hence the computational effort required to numerically approximate the underlying PDE problem may be prohibitively expensive. As an alternative approach, in this article we present a review of composite/agglomerated discontinuous Galerkin finite element methods (DGFEMs) which employ general polytopic elements. Here, the elements are typically constructed as the union of standard element shapes; in this way, the minimal dimension of the underlying composite finite element space is

P.F. Antonietti
MOX, Dipartimento di Matematica, Politecnico di Milano, Piazza Leonardo da Vinci 32, 20133 Milano, Italy
e-mail: paola.antonietti@polimi.it

A. Cangiani • Z. Dong
Department of Mathematics, University of Leicester, Leicester LE1 7RH, UK
e-mail: Andrea.Cangiani@le.ac.uk; zd14@le.ac.uk

J. Collis • P. Houston (✉)
School of Mathematical Sciences, University of Nottingham, University Park, Nottingham, NG7 2RD, UK
e-mail: Joe.Collis@nottingham.ac.uk; Paul.Houston@nottingham.ac.uk

E.H. Georgoulis
Department of Mathematics, University of Leicester, Leicester LE1 7RH, UK

School of Applied Mathematical and Physical Sciences, National Technical University of Athens, Zografou 15780, Greece
e-mail: Emmanuil.Georgoulis@le.ac.uk

S. Giani
School of Engineering and Computing Sciences, Durham University, South Road, Durham, DH1 3LE, UK
e-mail: Stefano.Giani@durham.ac.uk

G.R. Barrenechea et al. (eds.), *Building Bridges: Connections and Challenges in Modern Approaches to Numerical Partial Differential Equations*, Lecture Notes in Computational Science and Engineering 114,
DOI 10.1007/978-3-319-41640-3_9

independent of the number of geometrical features. In particular, we provide an overview of *hp*-version inverse estimates and approximation results for general polytopic elements, which are sharp with respect to element facet degeneration. On the basis of these results, a priori error bounds for the *hp*-DGFEM approximation of both second-order elliptic and first-order hyperbolic PDEs will be derived. Finally, we present numerical experiments which highlight the practical application of DGFEMs on meshes consisting of general polytopic elements.

1 Introduction

In many application areas arising in engineering and biological sciences, for example, one is often required to numerically approximate partial differential equations (PDEs) posed on complicated domains which contain small (relative to the size of the overall domain) geometrical features, or so-called microstructures. The key underlying issue for all classes of finite element/finite volume methods is the design of a suitable computational mesh upon which the underlying PDE problem will be discretized. On the one hand, the mesh should provide an accurate description of the given geometry with a granularity sufficient to compute numerical approximations to within desired engineering accuracy constraints. On the other hand, the mesh should not be so fine that the computational time required to compute the desired solution is too high for practical turn-around times. These issues are particularly pertinent when high-order methods are employed, since in this setting it is desirable to employ relatively coarse meshes, so that the polynomial degree may be suitably enriched.

Standard mesh generators typically generate grids consisting of triangular/quadrilateral elements in two-dimensions and tetrahedral/hexahedral/prismatic/pyramidal elements in three-dimensions. On the basis of the mesh, in the traditional finite element setting, the underlying finite element space, consisting of (continuous/discontinuous) piecewise polynomials, is then constructed based on mapping polynomial bases defined on a canonical/reference element to the physical domain. In the presence of boundary layers, anisotropic meshing may be exploited; however, in areas of high curvature the use of such highly-stretched elements may lead to element self-intersection, unless the curvature of the geometry is carefully 'propagated' into the interior of the mesh through the use of isoparametric element mappings. The use of what we shall refer to as standard element shapes necessitates the exploitation of very fine computational meshes when the geometry possesses small details or microstructures. Indeed, in such situations, an extremely large number of elements may be required for a given mesh generator to produce even a 'coarse' mesh which adequately describes the underlying geometry. Thereby, the solution of the resulting system of equations emanating, for example, from a finite element discretization of the underlying PDE on the resulting coarse mesh, may be impractical due to the large numbers of degrees of freedom involved. Moreover, since this initial coarse mesh already contains such a large number of elements,

Fig. 1 Example of a porous scaffold used for in vitro bone tissue growth, cf. [4, 5]

the use of efficient multilevel solvers may be difficult, as an adequate sequence of coarser grids which represent the geometry is unavailable. As an example arising in biological applications, in Fig. 1, we show a finite element mesh of a porous scaffold employed for in vitro bone tissue growth, cf. [4, 5]. Here, the mesh, consisting of 3.2 million elements, has been generated based on μCT image data represented in the form of voxels.

From the above discussion, we naturally conclude that, when standard element shapes are employed, the dimension of the underlying finite element space is proportional to the complexity of the given computational geometry. A natural alternative is to consider the exploitation of computational meshes consisting of general polytopic elements, i.e., polygons in two-dimensions and polyhedra in three-dimensions. In the context of discretizing PDEs in complicated geometries, Composite Finite Elements (CFEs) have been developed in the articles [32, 33] and [1, 31] for both conforming finite element and discontinuous Galerkin (DGFEM) methods, respectively, which exploit general meshes consisting of agglomerated elements, where each element is generated from a collection of neighbouring elements present within a standard finite element method. A closely related technique based on employing the so-called agglomerated DGFEM has also been considered in [7–9]. We point out that the DGFEM CFE approach developed in [1, 31] is essentially identical to the agglomerated DGFEM considered in [7–9]; however, the CFE methodology is more general in the sense that, depending on the selection of the underlying prolongation operator employed to construct the coarse finite element space, cf. Remark 2 below, the resulting elemental basis functions may only be locally piecewise smooth on each polytopic element, cf. [32, 33]. From a meshing point of view, the exploitation of general polytopic elements provides enormous flexibility. Indeed, in addition to meshing complicated geometries using a minimal number of elements, they are naturally suited to applications in complicated/moving domains, such as in solid mechanics, fluid structure interaction, geophysical problems, including earthquake engineering and flows in fractured porous media, and mathematical biology, for example. Indeed, general element shapes are often exploited as transitional elements in finite element meshes, for example, when fictitious domain methods, unfitted methods or overlapping meshes are employed, cf. [16–18, 36, 39], for example. The use of similar techniques in the context of characteristic-based/Lagrange–Galerkin methods is also highly relevant. The practical relevance and potential impact of employing such general computational meshes is an extremely exciting topic which has witnessed a vast amount of intensive research in recent years by a number of leading research groups. In the conforming setting, we mention the CFE method [32, 33], the Polygonal Finite Element Method [45], and the Extended Finite Element Method [27]. These latter two approaches achieve conformity by enriching/modifying the standard polynomial finite element spaces, in the spirit of the Generalized Finite Element framework of Babuška and Osborn in [6]. Typically, the handling of non-standard shape functions carries an increase in computational effort. The recently proposed Virtual Element Method [12], overcomes this difficulty, achieving the extension of conforming finite element methods to polytopic elements while maintaining the ease of implementation of these schemes; see also the closely related Mimetic Finite Difference method, cf. [11, 14, 19], for example.

In this article we present an overview of CFEs, and in particular consider their construction and analysis within the *hp*-version DGFEM setting. With this in mind, we follow the work presented in [1, 32, 33]; the inclusion of general polytopic meshes which admit arbitrarily small/degenerate $(d-k)$-dimensional element facets, $k = 1, \ldots, d-1$, where d denotes the spatial dimension, will also be discussed, following [21, 22]. The structure of this article is as follows. In Sect. 2, we introduce composite/agglomerated DGFEMs for the numerical approximation of second-order elliptic PDEs. Section 3 is devoted to the stability and a priori analysis of the proposed method; in particular, we derive *hp*-version inverse estimates and approximation results which are sharp with respect to element facet degeneration. In Sect. 4 we analyze the *hp*-version DGFEM discretization of first-order hyperbolic PDEs on polytopic meshes. The practical performance of the proposed DGFEMs for application to incompressible fluid flow problems is studied in Sect. 5. Finally, in Sect. 6 we summarize the work presented in this article and draw some conclusions.

2 Construction of Composite Finite Element Methods

The original idea behind the construction of CFEs, as presented in [32, 33] for conforming finite element methods, is to exploit general shaped element domains upon which elemental basis functions may only be locally piecewise smooth. In particular, an element domain within a CFE may consist of a collection of neighbouring elements present within a standard finite element method, with the basis function of the CFE being constructed as a linear combination of those defined on the standard finite element subdomains. The extension of this general approach to the DGFEM setting has been considered in the series of articles [1, 30, 31]; see also [2, 29] for their application within Schwarz-type domain decomposition preconditioners. For related work on the application of DGFEMs on meshes consisting of agglomerated elements, we refer to the articles [7–9]. We note that in the context of DGFEMs, the elemental finite element bases simply consist of polynomial functions, since inter-element conformity is not required.

For generality, we introduce CFE methods based on the construction proposed in [33] and [1]. Here, the philosophy underlying CFE methods is to construct finite element spaces based on first generating a hierarchy of meshes, such that the finest mesh does indeed provide an accurate representation of the underlying computational domain, followed by the introduction of appropriate prolongation operators which determine how the finite element basis functions on the coarse mesh are defined in terms of those on the fine grid. In this manner, CFEs naturally lend themselves to adaptive enrichment of the finite element space by locally varying the hierarchical level from which an element belongs, cf. [8, 31].

For concreteness, throughout this section, we concentrate on the numerical approximation of the Poisson equation. However, we stress that this class of methods naturally extends to a wide range of PDEs; indeed, it is the treatment of the underlying second-order PDE operator which gives rise to a number of theoretical and practical difficulties which we will address Sect. 3. With this in mind, given that Ω is a bounded, connected Lipschitz domain in $\mathbb{R}^d$, $d > 1$, with boundary $\partial\Omega$, consider the following PDE problem: find u such that

$$-\Delta u = f \quad \text{in } \Omega, \tag{1}$$

$$u = g \quad \text{on } \partial\Omega, \tag{2}$$

where $f \in L_2(\Omega)$ and g is a sufficiently regular boundary datum. In particular, it is assumed that Ω is a 'complicated' domain, in the sense that it contains small details or microstructures.

2.1 Composite/Agglomerated Meshes

The approach developed in [33], cf. also [1], is to construct the underlying physical/agglomerated meshes by first introducing a hierarchy of overlapping reference and logical meshes, from which a very fine geometry-conforming mesh, consisting of standard-shaped elements, may be defined, based on possibly moving nodes in the finest logical mesh onto the boundary $\partial\Omega$ of the computational domain. The coarse mesh, consisting of polytopic elements, is then constructed based on agglomerating elements which share the same parent within the underlying refinement tree.

More precisely, given an open bounded Lipschitz domain Ω, which potentially contains small features/microstructures, we first define the coarsest *reference* mesh $\mathcal{R}_H \equiv \mathcal{R}_{h_1}$ to be an overlapping grid in the sense that it does not resolve the boundary $\partial\Omega$ of the domain Ω. In particular, we let $\mathcal{R}_H = \{\hat{\kappa}\}$ be a coarse conforming shape-regular mesh consisting of (closed) standard element domains

$\hat{\kappa}$, cf. above, whose open intersection is empty such that

$$\Omega \subset \Omega_H = \left(\bigcup_{\hat{\kappa} \in \mathcal{R}_H} \hat{\kappa} \right)^{\circ} \quad \text{and} \quad \hat{\kappa}^{\circ} \cap \Omega \neq \emptyset \;\; \forall \hat{\kappa} \in \mathcal{R}_H,$$

where, for a closed set $D \subset \mathbb{R}^d$, D° denotes the interior of D.

On the basis of the coarse mesh $\mathcal{R}_H$, a hierarchy of reference meshes $\mathcal{R}_{h_i}$, $i = 2, 3, \ldots, \ell$, are now constructed based on adaptively refining the coarse mesh $\mathcal{R}_H$ with a view to improving the approximation of the boundary of Ω. With this in mind, given an input tolerance TOL, we proceed as follows:

1. Set $\mathcal{R}_{h_1} = \mathcal{R}_H$, the mesh counter $i = 1$, and store the elements $\hat{\kappa} \in \mathcal{R}_{h_1}$ as the root nodes of the refinement tree $\hat{\mathfrak{T}}$; we assign these elements with a level number $L = 1$.
2. Writing children($\hat{\kappa}$) to denote the number of children that element $\hat{\kappa}$ possesses, construct the refinement set $\mathfrak{R}$:

$$\mathfrak{R} = \left\{ \hat{\kappa} \in \hat{\mathfrak{T}} : \text{ children}(\hat{\kappa}) = 0 \;\; \wedge \;\; \hat{\kappa}^{\circ} \cap \partial\Omega \neq \emptyset \;\; \wedge \;\; h_{\hat{\kappa}} > \text{TOL} \right\}, \qquad (3)$$

   where $h_{\hat{\kappa}} = \text{diam}(\hat{\kappa})$.
3. If $\mathfrak{R} = \emptyset$, then `STOP`. Otherwise, for each $\hat{\kappa} \in \mathfrak{R}$, refine the element $\hat{\kappa} = \bigcup_{i=1}^{n_{\hat{\kappa}}} \hat{\kappa}_i$. Here, we store the child elements $\hat{\kappa}_i$, $i = 1, \ldots, n_{\hat{\kappa}}$, within the tree $\hat{\mathfrak{T}}$, where $\hat{\kappa}$ is their parent, level($\hat{\kappa}_i$) $=$ level($\hat{\kappa}$) $+ 1$, $i = 1, \ldots, n_{\hat{\kappa}}$, and level($\hat{\kappa}$) denotes the level of the element $\hat{\kappa}$ in $\hat{\mathfrak{T}}$. We point out that $n_{\hat{\kappa}}$ will depend on both the type of element to be refined, and the type of refinement, i.e., isotropic/anisotropic. For isotropic refinement of a quadrilateral element $\hat{\kappa}$ in two-dimensions, we have that $n_{\hat{\kappa}} = 4$.
4. Perform any additional refinements to undertake necessary mesh smoothing, for example, to ensure that the resulting mesh is 1-irregular, cf. [1].
5. Update mesh counter $i = i + 1$ and construct the reference mesh $\mathcal{R}_{h_i}$ from the tree structure $\hat{\mathfrak{T}}$ in the following manner:

$$\mathcal{R}_{h_i} = \left\{ \hat{\kappa} \in \hat{\mathfrak{T}} : \text{level}(\hat{\kappa}) = i \vee (\text{level}(\hat{\kappa}) \leq i \wedge \text{children}(\hat{\kappa}) = 0) \right\}.$$

6. Return to step 2 and continue to iterate until either the condition in 3 is satisfied, or a maximum number of allowable refinements have been undertaken.

Remark 1 We point out that the above procedure provides a generic refinement algorithm which may be employed to generate the sequence of reference meshes $\{\mathcal{R}_{h_i}\}_{i=1}^{\ell}$, though alternative sequences of hierarchical meshes may be exploited within the CFE framework.

On the basis of the reference meshes $\{\mathcal{R}_{h_i}\}_{i=1}^{\ell}$, we now define the corresponding sequences of logical and physical meshes $\{\mathcal{L}_{h_i}\}_{i=1}^{\ell}$ and $\{\mathcal{M}_{h_i}\}_{i=1}^{\ell}$, respectively. To this end, we first consider the finest reference mesh $\mathcal{R}_{h_\ell}$: given that the stopping criterion in step 2 above, cf. (3), is satisfied, then vertex nodes $\hat{x}_v \in \hat{\kappa}$, $\hat{\kappa} \in \mathcal{R}_{h_\ell}$, which are close to the boundary $\partial\Omega$ in the sense that

$$\mathrm{dist}(\hat{x}_v, \partial\Omega) \ll h_{\hat{\kappa}},$$

are moved onto the boundary of the computational domain. As a result of this node movement procedure, some of the elements stored in the tree $\hat{\mathfrak{T}}$ may end up lying outside of Ω; these are subsequently removed from $\hat{\mathfrak{T}}$ to yield the cropped tree $\mathfrak{T}$. On the basis of the cropped tree data structure $\mathfrak{T}$, the logical meshes are constructed based on agglomerating elements which share a common parent within a given level of the mesh tree hierarchy $\mathfrak{T}$. More precisely, following [30], we introduce the following notation: for $\tilde{\kappa}_{\mathcal{C}} \in \mathfrak{T}$, with $\mathrm{level}(\tilde{\kappa}_{\mathcal{C}}) = j$, we write $\mathfrak{F}_i^j(\tilde{\kappa}_{\mathcal{C}})$, $j \geq i$, to denote the unique element $\tilde{\kappa}_{\mathcal{P}} \in \mathfrak{T}$ with $\mathrm{level}(\tilde{\kappa}_{\mathcal{P}}) = i$ who is directly related to $\tilde{\kappa}_{\mathcal{C}}$ in the sense that $\tilde{\kappa}_{\mathcal{C}} \subset \tilde{\kappa}_{\mathcal{P}}$; i.e., $\tilde{\kappa}_{\mathcal{C}}$ has resulted from subsequent refinement of $\tilde{\kappa}_{\mathcal{P}}$. In the trivial case when $j = i$, $\mathfrak{F}_i^j(\tilde{\kappa}_{\mathcal{C}}) = \tilde{\kappa}_{\mathcal{C}}$. Thereby, the logical meshes $\{\mathcal{L}_{h_i}\}_{i=1}^{\ell}$ may be constructed from $\mathfrak{T}$ as follows:

$$\begin{aligned}\mathcal{L}_{h_i} = \{\tilde{\kappa} : &(\tilde{\kappa} \in \mathfrak{T} \wedge \mathrm{level}(\tilde{\kappa}) \leq i \wedge \mathrm{children}(\tilde{\kappa}) = 0)\\ &\vee(\tilde{\kappa} = \cup_{\tilde{\kappa}' \in \mathfrak{T}} \tilde{\kappa}' : \mathrm{children}(\tilde{\kappa}') = 0 \ \wedge \mathfrak{F}_i^j(\tilde{\kappa}') = P, j = \mathrm{level}(\tilde{\kappa}')\\ &\ \wedge P \text{ is identical for all members of this set})\}\,.\end{aligned}$$

We point out that in the absence of any node movement the finest reference and logical meshes $\mathcal{R}_{h_\ell}$ and $\mathcal{L}_{h_\ell}$, respectively, are identical.

Finally, the set of physical meshes $\{\mathcal{M}_{h_i}\}_{i=1}^{\ell}$ are defined based on moving the nodes in the respective logical meshes $\{\mathcal{L}_{h_i}\}_{i=1}^{\ell}$. More precisely, writing $\hat{\mathcal{N}}_\ell$ to denote the set of nodal points which define the finest logical mesh $\mathcal{L}_{h_\ell}$, the process of node movement naturally defines a bijective mapping

$$\Phi : \hat{\mathcal{N}}_\ell \rightarrow \mathcal{N}_\ell,$$

where $\mathcal{N}_\ell$ denotes the set of mapped vertex nodes. The mapping Φ can then be employed to map an element $\tilde{\kappa} \in \mathcal{L}_{h_\ell}$ to the physical element κ. For simplicity, we denote this mapping by Φ also; hence, we write

$$\Phi(\tilde{\kappa}) = \kappa.$$

With this notation, the physical meshes $\{\mathcal{M}_{h_i}\}_{i=1}^{\ell}$ may be defined as follows:

$$\mathcal{M}_{h_i} = \{\kappa : \kappa = \Phi(\tilde{\kappa}) \ \text{ for some } \tilde{\kappa} \in \mathcal{L}_{h_i}\},$$

$i = 1, \ldots, \ell$. We point out that both the logical and physical meshes $\{\mathcal{L}_{h_i}\}_{i=1}^{\ell}$ and $\{\mathcal{M}_{h_i}\}_{i=1}^{\ell}$, respectively, may consist of general polygonal/polyhedral element

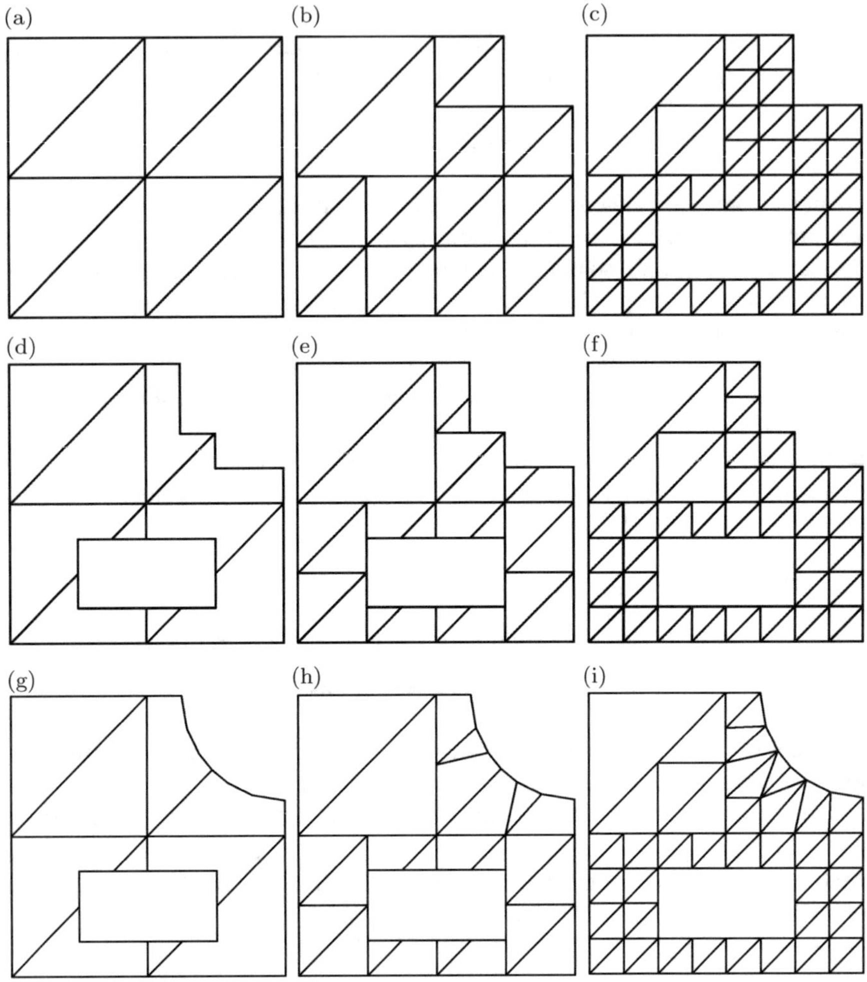

Fig. 2 Hierarchy of meshes: (**a**)–(**c**) Reference meshes; (**d**)–(**f**) Logical meshes; (**g**)–(**i**) Corresponding physical meshes. (**a**) $\mathcal{R}_H = \mathcal{R}_{h_1}$. (**b**) $\mathcal{R}_{h_2}$. (**c**) $\mathcal{R}_{h_3}$. (**d**) $\mathcal{L}_H = \mathcal{L}_{h_1}$. (**e**) $\mathcal{L}_{h_2}$. (**f**) $\mathcal{L}_{h_3}$. (**g**) $\mathcal{M}_{\mathrm{CFE}} \equiv \mathcal{M}_{h_1}$. (**h**) $\mathcal{M}_{h_2}$. (**i**) $\mathcal{M}_{h_3}$

domains. We refer to the coarsest physical mesh $\mathcal{M}_{h_1}$ as the CFE mesh, and accordingly write $\mathcal{M}_{\mathrm{CFE}} \equiv \mathcal{M}_{h_1}$. As a simple example, in Fig. 2, we consider the case when Ω is the unit square, which has had both the rectangular region $(1/4, 3/4) \times (1/8, 3/8)$ and the circular region enclosed by $r < 3/8$, where $r^2 = (x-1)^2 + (y-1)^2$, removed. Here, we show the reference, logical, and physical meshes $\{\mathcal{R}_{h_i}\}_{i=1}^{\ell}$, $\{\mathcal{L}_{h_i}\}_{i=1}^{\ell}$, and $\{\mathcal{M}_{h_i}\}_{i=1}^{\ell}$, respectively, when $\ell = 3$.

2.2 Finite Element Spaces

Given the set of physical (polytopic) meshes $\{\mathcal{M}_{h_i}\}_{i=1}^{\ell}$, constructed in the previous section, we introduce the corresponding sequence of DGFEM finite element spaces $V(\mathcal{M}_{h_i}, \mathbf{p}_i)$, $i = 1, \ldots, \ell$, respectively, consisting of piecewise discontinuous polynomials. To this end, for each element $\kappa \in \mathcal{M}_{\mathrm{CFE}} (\equiv \mathcal{M}_{h_1})$, we associate a positive integer p_κ, henceforth referred to as the polynomial degree of the element $\kappa \in \mathcal{M}_{\mathrm{CFE}}$, and collect the p_κ in the vector $\mathbf{p}_1 = (p_\kappa : \kappa \in \mathcal{M}_{\mathrm{CFE}})$. The polynomial degree vectors $\mathbf{p}_i$, $i = 2, \ldots, \ell$, associated with the respective meshes $\mathcal{M}_{h_i}$, $i = 2, \ldots, \ell$, are then defined in such a manner that the polynomial degree of the child element contained within the refinement tree $\mathfrak{T}$ is directly inherited from its parent element. More precisely,

$$\mathbf{p}_i = (p_\kappa,\ \kappa \in \mathcal{M}_{h_i} : p_\kappa = p_{\kappa'},\ \text{where } \kappa' = \mathfrak{F}_1^j(\kappa)\ \wedge \mathrm{level}(\kappa) = j,\ \kappa' \in \mathcal{M}_{\mathrm{CFE}}).$$

With this in mind, we write

$$V(\mathcal{M}_{h_i}, \mathbf{p}_i) = \{u \in L_2(\Omega) : u|_\kappa \in \mathcal{P}_{p_\kappa}(\kappa)\ \forall \kappa \in \mathcal{M}_{h_i}\},$$

$i = 1, \ldots, \ell$, where $\mathcal{P}_p(\kappa)$ denotes the set of polynomials of degree at most $p \geq 1$ defined over the general polytope κ.

With this construction, noting that the meshes $\{\mathcal{M}_{h_i}\}_{i=1}^{\ell}$ are nested, we deduce that

$$V(\mathcal{M}_{h_1}, \mathbf{p}_1) \subset V(\mathcal{M}_{h_2}, \mathbf{p}_2) \subset \ldots \subset V(\mathcal{M}_{h_\ell}, \mathbf{p}_\ell).$$

We now introduce the classical prolongation (injection) operator from $V(\mathcal{M}_{h_i}, p)$ to $V(\mathcal{M}_{h_{i+1}}, p)$, $1 \leq i \leq \ell - 1$, given by

$$P_i^{i+1} : V(\mathcal{M}_{h_i}, \mathbf{p}_i) \to V(\mathcal{M}_{h_{i+1}}, \mathbf{p}_{i+1}), \quad i = 1, \ldots, \ell - 1.$$

Hence, the prolongation operator from $V(\mathcal{M}_{h_i}, \mathbf{p}_i)$ to $V(\mathcal{M}_{h_\ell}, \mathbf{p}_\ell)$, $1 \leq i \leq \ell - 1$, is defined by

$$P_i = P_{\ell-1}^{\ell} P_{\ell-2}^{\ell-1} \ldots P_i^{i+1}.$$

With this notation, we may write $V(\mathcal{M}_{h_i}, \mathbf{p}_i)$, $1 \leq i \leq \ell - 1$, in the following alternative manner

$$V(\mathcal{M}_{h_i}, \mathbf{p}_i) = \{u \in L_2(\Omega) : u = P_i^\top \phi,\ \phi \in V(\mathcal{M}_{h_\ell}, \mathbf{p}_\ell)\}, \tag{4}$$

where the restriction operator $P_i^\top$ is defined as the transpose of P_i, with respect to the standard $L_2(\Omega)$-inner product.

Remark 2 The exploitation of the prolongation operator P_i within the definition of the finite element spaces $V(\mathcal{M}_{h_i}, \mathbf{p}_i)$, $i = 1, \ldots, \ell$, stated in (4) allows for the introduction of different spaces, depending on the specific choice of P_i. Here, cf. also [1], the finite element spaces are constructed so that on each (composite) element $\kappa \in \mathcal{M}_{h_i}$, $i = 1, \ldots, \ell$, the restriction of a function $v \in V(\mathcal{M}_{h_i}, \mathbf{p}_i)$ to κ is a polynomial of degree p_κ. In the case when the finite element spaces consist of continuous piecewise polynomials, cf. [33], for example, alternative prolongation operators are employed which leads to basis functions which are *piecewise* polynomials on each composite/polytopic element domain.

The space $V(\mathcal{M}_{h_1}, \mathbf{p}_1) \equiv V(\mathcal{M}_{\mathrm{CFE}}, \boldsymbol{p})$ is referred to as the composite finite element space. We stress that the dimension of $V(\mathcal{M}_{\mathrm{CFE}}, \boldsymbol{p})$ is *independent* of the underlying domain Ω in the sense that it does not directly depend on the number of microstructures contained in Ω. Indeed, the dimension of $V(\mathcal{M}_{\mathrm{CFE}}, \boldsymbol{p})$ can be chosen by the user; of course, if $V(\mathcal{M}_{\mathrm{CFE}}, \boldsymbol{p})$ is not sufficiently rich, then the accuracy of any computed finite element approximation $u_h \in V(\mathcal{M}_{\mathrm{CFE}}, \boldsymbol{p})$ may be low. However, given the construction of the composite finite element mesh $\mathcal{M}_{\mathrm{CFE}}$, the underlying numerical scheme naturally lends itself to adaptive enrichment of the finite element space $V(\mathcal{M}_{\mathrm{CFE}}, \boldsymbol{p})$, cf. [30, 31].

Remark 3 As a final remark, we note that an alternative approach for the construction of the composite finite element mesh $\mathcal{M}_{\mathrm{CFE}}$ is to simply employ a standard mesh generator to produce a fine mesh $\mathcal{M}_{\mathrm{fine}}$ which accurately describes the domain Ω. Then coarse agglomerated meshes may be constructed based on employing graph partitioning algorithms. One of the most popular software packages employed for this purpose is METIS [37], cf. [21, 29]. From a theoretical point of view, this setting is more difficult to analyse; we shall return to this issue in Sect. 3.

To define the forthcoming DGFEM, cf. Sect. 2.3, we define the broken Sobolev space $H^{\mathbf{s}}(\Omega, \mathcal{M}_{\mathrm{CFE}})$ with respect to the subdivision $\mathcal{M}_{\mathrm{CFE}}$ up to composite order $\mathbf{s}$ in the standard fashion:

$$H^{\mathbf{s}}(\Omega, \mathcal{M}_{\mathrm{CFE}}) = \{u \in L_2(\Omega) : u|_\kappa \in H^{s_\kappa}(\kappa) \quad \forall \kappa \in \mathcal{M}_{\mathrm{CFE}}\}.$$

Moreover, for $u \in H^1(\Omega, \mathcal{M}_{\mathrm{CFE}})$, we define the broken gradient $\nabla_h u$ by $(\nabla_h u)|_\kappa = \nabla(u|_\kappa)$, $\kappa \in \mathcal{M}_{\mathrm{CFE}}$.

2.3 Discontinuous Galerkin Methods on Polytopic Meshes

In this section, we consider the DGFEM discretization of the second-order elliptic PDE model problem (1)–(2). For concreteness, we focus our attention on the *hp*-version of the (symmetric) interior penalty DGFEM.

For the proceeding analysis, we introduce the concept of mesh interfaces and faces, cf. [22]. In order to admit hanging nodes/edges, which are permitted in $\mathcal{M}_{\text{CFE}}$, the interfaces of $\mathcal{M}_{\text{CFE}}$ are defined to be the intersection of the $(d-1)$-dimensional facets of neighbouring elements; on the boundary an interface is simply a $(d-1)$-dimensional facet of $\kappa \in \mathcal{M}_{\text{CFE}}$. In the two-dimensional setting, i.e., $d = 2$, the interfaces of a given element $\kappa \in \mathcal{M}_{\text{CFE}}$ simply consists of line segments ($(d-1)$-dimensional simplices). For $d = 3$, we assume that each interface of an element $\kappa \in \mathcal{M}_{\text{CFE}}$ may be subdivided into a set of co-planar triangles; we use the terminology 'face' to refer to a $(d-1)$-dimensional simplex (line segment or triangle for $d = 2$ or 3, respectively), which forms part of the boundary (interface) of an element $\kappa \in \mathcal{M}_{\text{CFE}}$. For $d = 2$, the face and interface of an element $\kappa \in \mathcal{M}_{\text{CFE}}$ coincide.

Following [21, 22], we assume that a sub-triangulation into faces of each mesh interface is given if $d = 3$, and denote by $\mathcal{F}_{\text{CFE}}$ the union of all open mesh interfaces if $d = 2$ and the union of all open triangles belonging to the sub-triangulation of all mesh interfaces if $d = 3$. In this way, $\mathcal{F}_{\text{CFE}}$ is always defined as a set of $(d-1)$-dimensional simplices. Further, we write $\mathcal{F}_{\text{CFE}} = \mathcal{F}^{\mathcal{I}_{\text{CFE}}} \cup \mathcal{F}^{\mathcal{B}_{\text{CFE}}}$, where $\mathcal{F}^{\mathcal{I}_{\text{CFE}}}$ denotes the union of all open $(d-1)$-dimensional element faces $F \subset \mathcal{F}_{\text{CFE}}$ that are contained in Ω, and $\mathcal{F}^{\mathcal{B}_{\text{CFE}}}$ is the union of element boundary faces, i.e., $F \subset \partial\Omega$ for $F \in \mathcal{F}^{\mathcal{B}_{\text{CFE}}}$. The boundary $\partial\kappa$ of an element κ and the sets $\partial\kappa \setminus \partial\Omega$ and $\partial\kappa \cap \partial\Omega$ will be identified in a natural way with the corresponding subsets of $\mathcal{F}_{\text{CFE}}$.

Given $\kappa \in \mathcal{M}_{\text{CFE}}$, the trace of a function $v \in H^1(\Omega, \mathcal{M}_{\text{CFE}})$ on $\partial\kappa$, relative to κ, is denoted by v_κ^+. Then for almost every $\mathbf{x} \in \partial\kappa\setminus\partial\Omega$, there exists a unique $\kappa' \in \mathcal{M}_{\text{CFE}}$ such that $\mathbf{x} \in \partial\kappa'$; with this notation, the outer/exterior trace v_κ^- of v on $\partial\kappa\setminus\partial\Omega$, relative to κ, is defined as the inner trace $v_{\kappa'}^+$ relative to the element(s) κ' such that the intersection of $\partial\kappa'$ with $\partial\kappa\setminus\partial\Omega$ has positive $(d-1)$-dimensional measure.

Next, we introduce some additional trace operators. Let κ_i and κ_j be two adjacent elements of $\mathcal{M}_{\text{CFE}}$ and let $\mathbf{x}$ be an arbitrary point on the interior face $F \in \mathcal{F}^{\mathcal{I}_{\text{CFE}}}$ given by $F = \partial\kappa_i \cap \partial\kappa_j$. We write $\mathbf{n}_{\kappa_i}$ and $\mathbf{n}_{\kappa_j}$ to denote the outward unit normal vectors on F, relative to $\partial\kappa_i$ and $\partial\kappa_j$, respectively. Furthermore, let v and $\mathbf{q}$ be scalar- and vector-valued functions, which are smooth inside each element κ_i and κ_j. Using the above notation, we write $(v_{\kappa_i}^+, \mathbf{q}_{\kappa_i}^+)$ and $(v_{\kappa_j}^+, \mathbf{q}_{\kappa_j}^+)$, we denote the traces of $(v, \mathbf{q})$ on F taken from within the interior of κ_i and κ_j, respectively. The averages of v and $\mathbf{q}$ at $\mathbf{x} \in F \in \mathcal{F}^{\mathcal{I}_{\text{CFE}}}$ are given by

$$\{\!\{v\}\!\} = \frac{1}{2}(v_{\kappa_i}^+ + v_{\kappa_j}^+), \quad \{\!\{\mathbf{q}\}\!\} = \frac{1}{2}(\mathbf{q}_{\kappa_i}^+ + \mathbf{q}_{\kappa_j}^+),$$

respectively. Similarly, the jumps of v and $\mathbf{q}$ at $\mathbf{x} \in F \in \mathcal{F}^{\mathcal{I}_{\text{CFE}}}$ are given by

$$[\![v]\!] = v_{\kappa_i}^+ \mathbf{n}_{\kappa_i} + v_{\kappa_j}^+ \mathbf{n}_{\kappa_j}, \quad [\![\mathbf{q}]\!] = \mathbf{q}_{\kappa_i}^+ \cdot \mathbf{n}_{\kappa_i} + \mathbf{q}_{\kappa_j}^+ \cdot \mathbf{n}_{\kappa_j},$$

respectively. On a boundary face $F \in \mathcal{F}^{\mathcal{B}_{\text{CFE}}}$, such that $F \subset \partial\kappa_i$, $\kappa_i \in \mathcal{M}_{\text{CFE}}$, we set

$$\{\!\{v\}\!\} = v_{\kappa_i}^+, \quad \{\!\{\mathbf{q}\}\!\} = \mathbf{q}_{\kappa_i}^+, \quad [\![v]\!] = v_{\kappa_i}^+ \mathbf{n}_{\kappa_i} \quad [\![\mathbf{q}]\!] = \mathbf{q}_{\kappa_i}^+ \cdot \mathbf{n}_{\kappa_i},$$

with $\mathbf{n}_{\kappa_i}$ denoting the unit outward normal vector on the boundary $\partial\Omega$. Since below it will always be clear from the context to which element κ in the subdivision $\mathcal{M}_{\mathrm{CFE}}$ the quantities $v_\kappa^\pm$, and so on, correspond to, for the sake of notational simplicity we shall suppress the letter κ in the subscript and write, respectively, $v^\pm$ instead.

With this notation, the symmetric interior penalty DGFEM for the numerical approximation of (1)–(2) is given by: find $u_h \in V(\mathcal{M}_{\mathrm{CFE}}, \boldsymbol{p})$ such that

$$B_{\mathtt{Diff}}(u_h, v_h) = F_{\mathtt{Diff}}(v_h) \tag{5}$$

for all $v_h \in V(\mathcal{M}_{\mathrm{CFE}}, \boldsymbol{p})$, where

$$\begin{aligned} B_{\mathtt{Diff}}(w, v) = & \sum_{\kappa \in \mathcal{M}_{\mathrm{CFE}}} \int_\kappa \nabla w \cdot \nabla v \, d\mathbf{x} - \sum_{F \in \mathcal{F}_{\mathrm{CFE}}} \int_F \left(\{\!\{\nabla_h v\}\!\} \cdot [\![w]\!] + \{\!\{\nabla_h w\}\!\} \cdot [\![v]\!] \right) ds \\ & + \sum_{F \in \mathcal{F}_{\mathrm{CFE}}} \int_F \sigma \, [\![w]\!] \cdot [\![v]\!] \, ds, \\ F_{\mathtt{Diff}}(v) = & \int_\Omega f v \, d\mathbf{x} - \sum_{F \in \mathcal{F}^{\mathcal{B}}{}_{\mathrm{CFE}}} \int_F g(\nabla_h v \cdot \mathbf{n} - \sigma v) \, \mathrm{d}s. \end{aligned}$$

Here, the non-negative function $\sigma \in L_\infty(\mathcal{F}_{\mathrm{CFE}})$ is the discontinuity stabilization function; the precise definition of σ is given in Lemma 4 below.

3 Stability and Approximation Results

In this section we consider the stability and error analysis of the *hp*-version DGFEM defined in (5). We point out that the original a priori error analysis of the DGFEM (5) on CFE meshes was first undertaken in the article [1], based on exploiting the work developed in both the CFE and DGFEM settings in the articles [33] and [35], respectively. Indeed, the analysis presented in [1] was based on bounding the error in terms of Sobolev norms of an extension, cf. Theorem 1 below, of the analytical solution u from an element belonging to the logical mesh to its respective element in the reference mesh, assuming the mapping Φ is sufficiently regular. This approach is advantageous since the (coarsest) reference mesh $\mathcal{R}_{h_1}$ consists of non-overlapping standard-shaped elements. In order to treat general polytopes, where an underlying reference and logical mesh may not be available, for example, on meshes generated from graph partitioning software, cf. Remark 3, we proceed based on employing the recent analysis developed in [22].

In contrast to the case when standard element domains are employed, the exploitation of general polytopic elements presents a number of key challenges for the construction and analysis of stable numerical schemes. In particular, shape-regular polytopes may admit arbitrarily small/degenerate $(d-k)$-dimensional element facets, $k = 1, \ldots, d-1$, under mesh refinement, where d denotes the spatial dimension. Thereby, standard inverse and approximation results must be carefully extended to the polytopic setting in such a manner that the resulting bounds are indeed sharp with respect to facet degeneration. With this in mind, we now summarise a number of key results derived in [22].

Firstly, we outline the key assumptions on the underlying CFE mesh $\mathcal{M}_{\mathrm{CFE}}$.

Assumption 3.1 *There exists a positive constant C_F, independent of the mesh parameters, such that*

$$\max_{\kappa\in\mathcal{M}_{\mathrm{CFE}}} (card\{F \in \mathcal{F}_{\mathrm{CFE}} : F \subset \partial\kappa\}) \leq C_F.$$

In order to deal with the case of general polytopic meshes, i.e., when reference/logical meshes are not available, we need to assume the existence of the following coverings of the mesh.

Definition 1 A *covering* $\mathcal{T}_\sharp = \{\mathcal{K}\}$ related to the polytopic mesh $\mathcal{M}_{\mathrm{CFE}}$ is a set of shape-regular d-simplices $\mathcal{K}$, such that for each $\kappa \in \mathcal{M}_{\mathrm{CFE}}$, there exists a $\mathcal{K} \in \mathcal{T}_\sharp$ such that $\kappa \subset \mathcal{K}$. Given $\mathcal{T}_\sharp$, we denote by $\Omega_\sharp$ the *covering domain* given by $\Omega_\sharp = \left(\cup_{\mathcal{K}\in\mathcal{T}_\sharp}\bar{\mathcal{K}}\right)^\circ$.

Assumption 3.2 *There exists a covering $\mathcal{T}_\sharp$ of $\mathcal{M}_{\mathrm{CFE}}$ and a positive constant $\mathcal{O}_\Omega$, independent of the mesh parameters, such that*

$$\max_{\kappa\in\mathcal{M}_{\mathrm{CFE}}} \mathcal{O}_\kappa \leq \mathcal{O}_\Omega,$$

where, for each $\kappa \in \mathcal{M}_{\mathrm{CFE}}$,

$$\mathcal{O}_\kappa = card\left\{\kappa' \in \mathcal{M}_{\mathrm{CFE}} : \kappa' \cap \mathcal{K} \neq \emptyset,\ \mathcal{K} \in \mathcal{T}_\sharp \ \text{ such that } \ \kappa \subset \mathcal{K}\right\}.$$

Thereby,

$$\mathrm{diam}(\mathcal{K}) \leq C_{\mathrm{diam}} h_\kappa,$$

for each pair $\kappa \in \mathcal{M}_{\mathrm{CFE}}$, $\mathcal{K} \in \mathcal{T}_\sharp$, with $\kappa \subset \mathcal{K}$, for a constant $C_{\mathrm{diam}} > 0$, uniformly with respect to the mesh size.

Remark 4 We note that for the classes of meshes constructed in Sect. 2.1, the coarsest reference mesh, subject to the (potential) application of the mapping Φ, may serve as the covering mesh $\mathcal{T}_\sharp$; in this setting Assumption 3.2 is trivially satisfied.

The proceeding *hp*-approximation results and inverse estimates for polytopic elements are based on considering d-dimensional simplices, where standard results can be applied. With this in mind, we introduce the following element submesh.

Definition 2 For each element κ in the computational mesh $\mathcal{M}_{\mathrm{CFE}}$, we define the family $\mathcal{F}_\flat^\kappa$ of all possible d-dimensional simplices contained in κ and having at least one face in common with κ. The notation $\kappa_\flat^F$ will be used to indicate a simplex belonging to $\mathcal{F}_\flat^\kappa$ and sharing with $\kappa \in \mathcal{M}_{\mathrm{CFE}}$ a given face F.

Equipped with these results, we first consider the derivation of *hp*-version inverse estimates on general polytopes.

3.1 Inverse Estimates

Inverse estimates, which bound a norm of a polynomial on an element face by a norm on the element itself, are fundamental for the study of the stability and error analysis of DGFEMs. In order to derive bounds which are sharp with respect to small/degenerate $(d-k)$-dimensional element facets, $k = 1, \ldots, d-1$, we first introduce the following definition.

Definition 3 Let $\tilde{\mathcal{M}}_{\mathrm{CFE}}$ denote the subset of elements κ, $\kappa \in \mathcal{M}_{\mathrm{CFE}}$, such that each $\kappa \in \tilde{\mathcal{M}}_{\mathrm{CFE}}$ can be covered by at most $m_{\mathcal{M}_{\mathrm{CFE}}}$ shape-regular simplices K_i, $i = 1, \ldots, m_{\mathcal{M}_{\mathrm{CFE}}}$, such that

$$\mathrm{dist}(\kappa, \partial K_i) < C_{as} \, \mathrm{diam}(K_i)/p_\kappa^2,$$

and

$$|K_i| \geq c_{as}|\kappa|$$

for all $i = 1, \ldots, m_{\mathcal{M}_{\mathrm{CFE}}}$, for some $m_{\mathcal{M}_{\mathrm{CFE}}} \in \mathbb{N}$ and $C_{as}, c_{as} > 0$, independent of κ and $\mathcal{M}_{\mathrm{CFE}}$.

We now state the main result of this section; see [21, 22] for details of the proof.

Lemma 1 *Let $\kappa \in \mathcal{M}_{\mathrm{CFE}}$, $F \subset \partial\kappa$ denote one of its faces, and $\tilde{\mathcal{M}}_{\mathrm{CFE}}$ be defined as in Definition 3. Then, for each $v \in \mathcal{P}_p(\kappa)$, we have the inverse estimate*

$$\|v\|_{L_2(F)}^2 \leq C_{\mathrm{INV}}(p, \kappa, F) \frac{p^2|F|}{|\kappa|} \|v\|_{L_2(\kappa)}^2, \tag{6}$$

with

$$C_{\mathrm{INV}}(p, \kappa, F) := C_{\mathrm{inv}} \begin{cases} \min\left\{ \dfrac{|\kappa|}{\sup_{\kappa_\flat^F \subset \kappa} |\kappa_\flat^F|}, p^{2(d-1)} \right\}, & \text{if } \kappa \in \tilde{\mathcal{M}}_{\mathrm{CFE}}, \\ \dfrac{|\kappa|}{\sup_{\kappa_\flat^F \subset \kappa} |\kappa_\flat^F|}, & \text{if } \kappa \in \mathcal{M}_{\mathrm{CFE}} \backslash \tilde{\mathcal{M}}_{\mathrm{CFE}}, \end{cases}$$

and $\kappa_\flat^F \in \mathcal{F}_\flat^\kappa$ as in Definition 2. Furthermore, C_{inv} is a positive constant, which if $\kappa \in \tilde{\mathcal{M}}_{\text{CFE}}$ depends on the shape regularity of the covering of κ given in Definition 3, but is always independent of $|\kappa|/\sup_{\kappa_\flat^F \subset \kappa} |\kappa_\flat^F|$ (and, therefore, of $|F|$), p, and v.

Remark 5 Loosely speaking, the proof of Lemma 1 is based on exploiting standard inverse inequalities, cf. [43], for example, together with Definition 3. Indeed, for $\kappa \in \tilde{\mathcal{M}}_{\text{CFE}}$, the essential idea is to derive two bounds, one based on extending results from [28], and one based on employing an $L_\infty(\kappa)$ bound. Taking the minimum of these two bounds gives rise to an inverse inequality which is both sharp with respect to the polynomial degree p, and moreover is sensitive with respect to the measure of the face F relative to that of the element κ.

We finish this section by recalling the inverse estimate for the H^1-(semi)norm derived in [21], cf. also [3]. In this setting, the shape regularity assumption on the covering $\mathcal{T}_\sharp$, cf. Definition 1, must be strengthened as follows.

Assumption 3.3 *The subdivision $\mathcal{M}_{\text{CFE}}$ is shape regular in the sense of [24], i.e., there exists a positive constant C_{shape}, independent of the mesh parameters, such that:*

$$\forall \kappa \in \mathcal{M}_{\text{CFE}}, \qquad \frac{h_\kappa}{\rho_\kappa} \leq C_{\text{shape}},$$

with ρ_κ denoting the diameter of the largest ball contained in κ.

Following, [21], we also require the following assumption.

Assumption 3.4 *Every polytopic element $\kappa \in \mathcal{M}_{\text{CFE}} \backslash \tilde{\mathcal{M}}_{\text{CFE}}$, admits a sub-triangulation into at most $n_{\mathcal{M}_{\text{CFE}}}$ shape-regular simplices $\mathfrak{s}_i$, $i = 1, 2, \ldots, n_{\mathcal{M}_{\text{CFE}}}$, such that $\bar{\kappa} = \cup_{i=1}^{n_{\mathcal{M}_{\text{CFE}}}} \bar{\mathfrak{s}}_i$ and*

$$|\mathfrak{s}_i| \geq \hat{c}|\kappa|$$

for all $i = 1, \ldots, n_{\mathcal{M}_{\text{CFE}}}$, for some $n_{\mathcal{M}_{\text{CFE}}} \in \mathbb{N}$ and $\hat{c} > 0$, independent of κ and $\mathcal{M}_{\text{CFE}}$.

Lemma 2 *Given Assumptions 3.3 and 3.4 are satisfied, for each $v \in \mathcal{P}_p(\kappa)$, the following inverse inequality holds*

$$\|\nabla v\|_{L_2(\kappa)}^2 \leq \tilde{C}_{\text{inv}} \frac{p^4}{h_\kappa^2} \|v\|_{L_2(\kappa)}^2, \tag{7}$$

where $\tilde{C}_{\text{inv}}$ is a positive constant, independent of the element diameter h_κ, the polynomial order p_κ, and the function v, but dependent on the shape regularity of the covering of κ, if $\kappa \in \tilde{\mathcal{M}}_{\text{CFE}}$, or the sub-triangulation of κ, if $\kappa \in \mathcal{M}_{\text{CFE}} \backslash \tilde{\mathcal{M}}_{\text{CFE}}$.

3.2 Approximation Results

Functions defined on Ω can be extended to the covering domain $\Omega_\sharp$ based on employing the following extension operator, cf. [44].

Theorem 1 *Let Ω be a domain with a Lipschitz boundary. Then there exists a linear extension operator $\mathcal{E} : H^s(\Omega) \to H^s(\mathbb{R}^d)$, $s \in \mathbb{N}_0$, such that $\mathcal{E}v|_\Omega = v$ and*

$$\|\mathcal{E}v\|_{H^s(\mathbb{R}^d)} \leq C\|v\|_{H^s(\Omega)},$$

where C is a positive constant depending only on s and Ω.

We point out that the assumptions stated in Theorem 1 on the domain Ω may be weakened. Indeed, [44] only requires that Ω is a domain with a minimally smooth boundary; the extension to domains which are simply connected, but may contain microscales, is treated in [42].

With the above notation, we now quote Lemma 4.2 from [22].

Lemma 3 *Let $\kappa \in \mathcal{M}_{\mathrm{CFE}}$, $F \subset \partial\kappa$ denote one of its faces, and $\mathcal{K} \in \mathcal{T}_\sharp$ denote the corresponding simplex such that $\kappa \subset \mathcal{K}$, cf. Definition 1. Suppose that $v \in L_2(\Omega)$ is such that $\mathcal{E}v|_\mathcal{K} \in H^{l_\kappa}(\mathcal{K})$, for some $l_\kappa \geq 0$. Then, given Assumption 3.2 is satisfied, there exists $\tilde{\Pi}v$, such that $\tilde{\Pi}v|_\kappa \in \mathcal{P}_{p_\kappa}(\kappa)$, and the following bounds hold*

$$\|v - \tilde{\Pi}v\|_{H^q(\kappa)} \leq C\frac{h_\kappa^{s_\kappa - q}}{p_\kappa^{l_\kappa - q}}\|\mathcal{E}v\|_{H^{l_\kappa}(\mathcal{K})}, \quad l_\kappa \geq 0,$$

for $0 \leq q \leq l_\kappa$, and

$$\|v - \tilde{\Pi}v\|_{L_2(F)} \leq C|F|^{1/2}\frac{h_\kappa^{s_\kappa - d/2}}{p_\kappa^{l_\kappa - 1/2}}C_m(p_\kappa,\kappa,F)^{1/2}\|\mathcal{E}v\|_{H^{l_\kappa}(\mathcal{K})}, \quad l_\kappa > d/2,$$

where

$$C_m(p_\kappa,\kappa,F) = \min\left\{\frac{h_\kappa^d}{\sup_{\kappa_\flat^F\subset\kappa}|\kappa_\flat^F|}, \frac{1}{p_\kappa^{1-d}}\right\}.$$

Here, $s_\kappa = \min\{p_\kappa + 1, l_\kappa\}$ and C is a positive constant, which depends on the shape-regularity of $\mathcal{K}$, but is independent of v, h_κ, and p_κ.

3.3 Error Analysis of the DGFEM

On the basis of the results stated in Sects. 3.1 and 3.2, we now proceed with the stability and error analysis of the DGFEM defined in (5). To this end, following the work presented in [40], we begin by defining the following extensions of the forms

$B_{\mathtt{Diff}}(\cdot,\cdot)$ and $F_{\mathtt{Diff}}(\cdot)$:

$$\tilde{B}_{\mathtt{Diff}}(w,v) = \sum_{\kappa\in\mathcal{M}_{\mathrm{CFE}}} \int_\kappa \nabla w\cdot\nabla v\, d\mathbf{x} + \sum_{F\in\mathcal{F}_{\mathrm{CFE}}} \int_F \sigma\, [\![w]\!]\cdot[\![v]\!]\, ds$$
$$- \sum_{F\in\mathcal{F}_{\mathrm{CFE}}} \int_F \left(\{\!\{\boldsymbol{\Pi}_2(\nabla_h v)\}\!\}\cdot[\![w]\!] + \{\!\{\boldsymbol{\Pi}_2(\nabla_h w)\}\!\}\cdot[\![v]\!]\right) ds,$$

$$\tilde{F}_{\mathtt{Diff}}(v) = \int_\Omega f v\, d\mathbf{x} - \sum_{F\in\mathcal{F}^{\mathcal{B}_{\mathrm{CFE}}}} \int_F g(\boldsymbol{\Pi}_2(\nabla_h v)\cdot\mathbf{n} - \sigma v)\, \mathrm{d}s,$$

respectively. Here, $\boldsymbol{\Pi}_2 : [L_2(\Omega)]^d \to [V(\mathcal{M}_{\mathrm{CFE}},\boldsymbol{p})]^d$ denotes the orthogonal L_2-projection onto the finite element space $[V(\mathcal{M}_{\mathrm{CFE}},\boldsymbol{p})]^d$. Thereby, face integrals involving the terms $\{\!\{\boldsymbol{\Pi}_2(\nabla_h w)\}\!\}$, $\{\!\{\boldsymbol{\Pi}_2(\nabla_h v)\}\!\}$ and $\boldsymbol{\Pi}_2(\nabla_h v)$ are well defined for all $v, w \in \mathcal{S} = H^1(\Omega) + V(\mathcal{M}_{\mathrm{CFE}},\boldsymbol{p})$, as these terms are now traces of elementwise polynomial functions. Moreover, it is clear that

$$\tilde{B}_{\mathtt{Diff}}(w,v) = B_{\mathtt{Diff}}(w,v) \qquad \text{for all} \quad w,v\in V(\mathcal{M}_{\mathrm{CFE}},\boldsymbol{p}),$$

and

$$\tilde{F}_{\mathtt{Diff}}(v) = F_{\mathtt{Diff}}(v) \qquad \text{for all} \quad v\in V(\mathcal{M}_{\mathrm{CFE}},\boldsymbol{p}).$$

Hence, we may rewrite the discrete problem (5) in the following equivalent manner: find $u_h \in V(\mathcal{M}_{\mathrm{CFE}},\boldsymbol{p})$ such that

$$\tilde{B}_{\mathtt{Diff}}(u_h,v_h) = \tilde{F}_{\mathtt{Diff}}(v_h) \qquad \forall v_h \in V(\mathcal{M}_{\mathrm{CFE}},\boldsymbol{p}). \tag{8}$$

Given the discrete nature of the L_2-projection operator $\boldsymbol{\Pi}_2$, the DGFEM formulation (8) is no longer consistent.

For the proceeding error analysis, we introduce the DG-norm $|||\cdot|||_{\mathtt{Diff}}$ by

$$|||w|||_{\mathtt{Diff}} = \Big(\sum_{\kappa\in\mathcal{M}_{\mathrm{CFE}}} \int_\kappa |\nabla w|^2\, \mathrm{d}\mathbf{x} + \sum_{F\in\mathcal{F}_{\mathrm{CFE}}} \int_F \sigma |[\![w]\!]|^2\, \mathrm{d}s\Big)^{1/2},$$

for $w \in \mathcal{S}$ and $\sigma > 0$.

With this notation, we recall the following coercivity and continuity properties of the bilinear form $\tilde{B}_{\mathtt{Diff}}(\cdot,\cdot)$ derived in [22].

Lemma 4 *Let $\sigma : \mathcal{F}_{\mathrm{CFE}} \to \mathbb{R}_+$ be defined facewise by*

$$\sigma(\mathbf{x}) = \begin{cases} C_\sigma \max\limits_{\kappa\in\{\kappa^+,\kappa^-\}} \left\{C_{\mathrm{INV}}(p_\kappa,\kappa,F)\dfrac{p_\kappa^2|F|}{|\kappa|}\right\}, & \mathbf{x}\in F\in\mathcal{F}^{\mathcal{I}_{\mathrm{CFE}}},\ F=\partial\kappa^+\cap\partial\kappa^-, \\ C_\sigma C_{\mathrm{INV}}(p_\kappa,\kappa,F)\dfrac{p_\kappa^2|F|}{|\kappa|}, & \mathbf{x}\in F\in\mathcal{F}^{\mathcal{B}_{\mathrm{CFE}}},\ F=\partial\kappa\cap\partial\Omega, \end{cases} \tag{9}$$

with $C_\sigma > 0$ large enough, depending on C_F, and independent of p, $|F|$, and $|\kappa|$; here C_{INV} is defined as in Lemma 1. Then, given Assumption 3.1 holds, we have that

$$\tilde{B}_{\mathtt{Diff}}(v,v) \geq C_{\mathrm{coer}} |||v|||^2_{\mathtt{Diff}} \quad \text{for all} \quad v \in \mathcal{S},$$

and

$$\tilde{B}_{\mathtt{Diff}}(w,v) \leq C_{\mathrm{cont}} |||w|||_{\mathtt{Diff}}\, |||v|||_{\mathtt{Diff}} \quad \text{for all} \quad w, v \in \mathcal{S},$$

where C_{coer} and C_{cont} are positive constants, independent of the discretization parameters.

Remark 6 We point out that Lemma 4 assumes that the number of element faces remains bounded under mesh refinement, cf. Assumption 3.1. However, based on the computations undertaken in [3], in practice we observe that C_{cocr} remains uniformly bounded on sequences of agglomerated polygons which violate this condition. Indeed, for $C_\sigma = 10$ numerical experiments suggest that $C_{\mathrm{coer}} \geq 0.8$.

Given the definition of the discontinuity stabilization function σ stated in Lemma 4, we now state the following a priori error bound.

Theorem 2 *Let $\Omega \subset \mathbb{R}^d$, $d = 2, 3$, be a bounded polyhedral domain, and let $\mathcal{M}_{\mathrm{CFE}} = \{\kappa\}$ be a subdivision of Ω consisting of general polytopic elements satisfying Assumption 3.1. Further, $\mathcal{T}_\sharp = \{\mathcal{K}\}$ denotes the associated covering of Ω consisting of shape-regular d-simplices as in Definition 1, satisfying Assumption 3.2. Let $u_h \in V(\mathcal{M}_{\mathrm{CFE}}, \boldsymbol{p})$ be the DGFEM approximation to $u \in H^1(\Omega)$ defined by* (5) *with the discontinuity stabilization parameter given by* (9)*, and suppose that $u|_\kappa \in H^{l_\kappa}(\kappa)$, $l_\kappa > 1 + d/2$, for each $\kappa \in \mathcal{M}_{\mathrm{CFE}}$, such that $\mathcal{E}u|_{\mathcal{K}} \in H^{l_\kappa}(\mathcal{K})$, where $\mathcal{K} \in \mathcal{T}_\sharp$ with $\kappa \subset \mathcal{K}$. Then, the following bound holds:*

$$|||u - u_h|||^2_{\mathtt{Diff}} \leq C \sum_{\kappa \in \mathcal{M}_{\mathrm{CFE}}} \frac{h_\kappa^{2(s_\kappa - 1)}}{p_\kappa^{2(l_\kappa - 1)}} \left(1 + \mathcal{G}_\kappa(F, C_{\mathrm{INV}}, C_m, p_\kappa)\right) \|\mathcal{E}u\|^2_{H^{l_\kappa}(\mathcal{K})},$$

where

$$\mathcal{G}_\kappa(F, C_{\mathrm{INV}}, C_m, p_\kappa) = p_\kappa h_\kappa^{-d} \sum_{F \in \mathcal{F}_{\mathrm{CFE}}} C_m(p_\kappa, \kappa, F) \sigma^{-1} |F|$$
$$+ p_\kappa^2 |\kappa|^{-1} \sum_{F \in \mathcal{F}_{\mathrm{CFE}}} C_{\mathrm{INV}}(p_\kappa, \kappa, F) \sigma^{-1} |F| + h_\kappa^{-d+2} p_\kappa^{-1} \sum_{F \in \mathcal{F}_{\mathrm{CFE}}} C_m(p_\kappa, \kappa, F) \sigma |F|,$$

with $s_\kappa = \min\{p_\kappa + 1, l_\kappa\}$ and $p_\kappa \geq 1$, where C is a positive constant which is independent of the discretization parameters. Here, we recall that C_{INV} and C_m are defined as in Lemmas 1 and 3, respectively.

Proof See [22] for details.

Remark 7 For uniform orders $p_\kappa = p \geq 1$, $h = \max_{\kappa \in \mathcal{M}_{\mathrm{CFE}}} h_\kappa$, $s_\kappa = s$, $s = \min\{p+1, l\}$, $l > 1 + d/2$, under the assumption that the diameter of the faces of each element $\kappa \in \mathcal{M}_{\mathrm{CFE}}$ is of comparable size to the diameter of the corresponding element, the a priori error bound stated in Theorem 2 coincides with the bounds derived in [35, 41], for example, for DGFEMs defined on standard element domains. In particular, this bound is optimal in h and suboptimal in p by $p^{1/2}$.

4 Hyperbolic PDEs

In this section we consider the generalization of CFE/DGFEMs posed on general polytopic meshes for the numerical approximation of first-order hyperbolic PDEs. To this end, we consider the following model problem: find u such that

$$\nabla \cdot (\mathbf{b}u) + cu = f \quad \text{in } \Omega, \tag{10}$$

$$u = g \quad \text{on } \partial_-\Omega, \tag{11}$$

where $c \in L_\infty(\Omega), f \in L_2(\Omega)$, and $\mathbf{b} = (b_1, b_2, \ldots, b_d)^\top \in [W^1_\infty(\Omega)]^d$. Here, the inflow and outflow portions of the boundary $\partial\Omega$ are defined, respectively, by

$$\partial_-\Omega = \Big\{\mathbf{x} \in \partial\Omega : \mathbf{b}(\mathbf{x}) \cdot \mathbf{n}(\mathbf{x}) < 0\Big\}, \quad \partial_+\Omega = \Big\{\mathbf{x} \in \partial\Omega : \mathbf{b}(\mathbf{x}) \cdot \mathbf{n}(\mathbf{x}) \geq 0\Big\},$$

where $\mathbf{n}$ denotes the unit outward normal vector to the boundary $\partial\Omega$. Throughout this section, we assume that the following (standard) positivity condition holds: there exists a positive constant γ_0 such that

$$c_0(x)^2 = c(x) + \frac{1}{2}\nabla \cdot \mathbf{b}(x) \geq \gamma_0 \quad \text{a.e. } \mathbf{x} \in \Omega. \tag{12}$$

The DGFEM approximation to (10)–(11) is then given by: find $u_h \in V(\mathcal{M}_{\mathrm{CFE}}, \boldsymbol{p})$ such that

$$\sum_{\kappa \in \mathcal{M}_{\mathrm{CFE}}} \Bigg\{ \int_\kappa \Big(- u_h \mathbf{b} \cdot \nabla v_h + c u_h v_h \Big)\, \mathrm{d}\mathbf{x} + \int_{\partial\kappa} \mathcal{H}(u_h^+, u_h^-, \mathbf{n}_\kappa) v_h^+ \,\mathrm{d}s \Bigg\} = \int_\Omega f v_h \,\mathrm{d}\mathbf{x} \tag{13}$$

for all $v_h \in V(\mathcal{M}_{\mathrm{CFE}}, \boldsymbol{p})$. Here, $\mathcal{H}(w_h^+, w_h^-, \mathbf{n}_\kappa)|_{\partial\kappa}$, which depends on both the inner- and outer-trace of w_h on $\partial\kappa$, $\kappa \in \mathcal{M}_{\mathrm{CFE}}$, and the unit outward normal vector $\mathbf{n}_\kappa$ to $\partial\kappa$, is a *numerical flux* function; this serves as an approximation to the normal flux $(\mathbf{b}u) \cdot \mathbf{n}_\kappa$ on the boundary of each element $\kappa \in \mathcal{M}_{\mathrm{CFE}}$. The numerical flux function $\mathcal{H}(\cdot,\cdot,\cdot)$ may be chosen to be any two-point monotone Lipschitz function which is both consistent and conservative; see [38, 46], for example. In the current setting, the most natural choice of numerical flux is the standard upwind flux given by

$$\mathcal{H}(u_h^+, u_h^-, \mathbf{n}_\kappa)|_F = \begin{cases} \mathbf{b}\cdot\mathbf{n}_\kappa \lim_{s\to 0^+} u_h(\mathbf{x}-s\mathbf{b}) & F\subset\partial\kappa\backslash\partial_-\Omega,\ \kappa\in\mathcal{M}_{\mathrm{CFE}},\\ \mathbf{b}\cdot\mathbf{n}_\kappa\, g & F\subset\partial\kappa\cap\partial_-\Omega,\ \kappa\in\mathcal{M}_{\mathrm{CFE}}, \end{cases}$$

for all $F\in\mathcal{F}_{\mathrm{CFE}}$, cf. [26].

Using the above definition of the numerical flux function $\mathcal{H}(\cdot,\cdot,\cdot)$, the DGFEM (13) can be rewritten in the following equivalent form: find $u_h \in V(\mathcal{M}_{\mathrm{CFE}},\boldsymbol{p})$ such that

$$B_{\mathrm{Hyp}}(u_h, v_h) = F_{\mathrm{Hyp}}(v_h)$$

for all $v_h\in V(\mathcal{M}_{\mathrm{CFE}},\boldsymbol{p})$, where

$$\begin{aligned} B_{\mathrm{Hyp}}(w,v) &= \sum_{\kappa\in\mathcal{M}_{\mathrm{CFE}}}\int_\kappa\Big(-w\mathbf{b}\cdot\nabla v + cwv\Big)\,\mathrm{d}\mathbf{x}\\ &\quad+\sum_{\kappa\in\mathcal{M}_{\mathrm{CFE}}}\left\{\int_{\partial_+\kappa}\mathbf{b}\cdot\mathbf{n}_\kappa\, w^+v^+\,\mathrm{d}s+\int_{\partial_-\kappa\backslash\partial_-\Omega}\mathbf{b}\cdot\mathbf{n}_\kappa\, w^-v^+\,\mathrm{d}s\right\},\\ F_{\mathrm{Hyp}}(v_h) &= \int_\Omega f v_h\,\mathrm{d}\mathbf{x}-\sum_{\kappa\in\mathcal{M}_{\mathrm{CFE}}}\int_{\partial_-\kappa\cap\partial_-\Omega}\mathbf{b}\cdot\mathbf{n}_\kappa\, g v^+\,\mathrm{d}s. \end{aligned}$$

Remark 8 We note that, upon application of integration by parts elementwise, the bilinear form $B_{\mathrm{Hyp}}(\cdot,\cdot)$ may be written in the familiar form:

$$\begin{aligned} B_{\mathrm{Hyp}}(w,v) &= \sum_{\kappa\in\mathcal{M}_{\mathrm{CFE}}}\int_\kappa\Big(\nabla\cdot(\mathbf{b}w)\,v + cwv\Big)\,\mathrm{d}\mathbf{x}\\ &\quad-\sum_{\kappa\in\mathcal{M}_{\mathrm{CFE}}}\left\{\int_{\partial_-\kappa\backslash\partial_-\Omega}\mathbf{b}\cdot\mathbf{n}_\kappa\,(w^+-w^-)v^+\,\mathrm{d}s+\int_{\partial_-\kappa\cap\partial_-\Omega}\mathbf{b}\cdot\mathbf{n}_\kappa\, w^+v^+\,\mathrm{d}s\right\}, \end{aligned}$$

cf. [21, 35], for example.

4.1 Error Analysis

The analysis of the DGFEM (13) in the *hp*-version setting may be tackled by a number of different approaches. In the articles [13, 34], additional streamline-diffusion terms are included within the underlying discretization method; in this setting, optimal *hp*-error bounds may then be derived in a straightforward manner. However, as noted in [34], the streamline-diffusion stabilization offers very little, if any, practical advantage over the standard DGFEM (with no stabilization), and is mainly employed for analysis purposes. In the absence of streamline-diffusion stabilization, under the assumption that

$$\mathbf{b} \cdot \nabla_h \xi \in V(\mathcal{M}_{\mathrm{CFE}}, \boldsymbol{p}) \quad \forall \xi \in V(\mathcal{M}_{\mathrm{CFE}}, \boldsymbol{p}), \tag{14}$$

holds, together *hp*-optimal approximation results for the local L_2-projector, optimal *hp*-bounds for (13) have been derived in the article [35] for meshes consisting of shape-regular d-parallelepipeds. For *hp*-optimal approximation results of the L_2-projector on d-simplices, we refer to [23].

Following [21], for the case when general polytopic elements are admitted, in the absence of optimal *hp*-approximation results for the local L_2-projection operator, we prove an inf-sup condition for the bilinear form $B_{\mathrm{Hyp}}(\cdot,\cdot)$, with respect to the following streamline DGFEM-norm:

$$|||v|||_{\mathrm{SD}}^2 = |||v|||_{\mathrm{Hyp}}^2 + \sum_{\kappa \in \mathcal{M}_{\mathrm{CFE}}} \tau_\kappa \|\mathbf{b} \cdot \nabla v\|_{L_2(\kappa)}^2, \tag{15}$$

where

$$|||v|||_{\mathrm{Hyp}}^2 = \sum_{\kappa \in \mathcal{M}_{\mathrm{CFE}}} \left(\|c_0 v\|_{L_2(\kappa)}^2 + \frac{1}{2}\|v^+\|_{\partial\kappa \cap \partial\Omega}^2 + \frac{1}{2}\|v^+ - v^-\|_{\partial_-\kappa \setminus \partial\Omega}^2 \right).$$

Here, c_0 is defined as in (12) and $\|\cdot\|_\tau$, $\tau \subset \partial\kappa$, denotes the (semi)norm associated with the (semi)inner product $(v, w)_\tau = \int_\tau |\mathbf{b} \cdot \mathbf{n}| v w \, \mathrm{d}s$. Finally, the streamline-diffusion parameter τ_κ, $\kappa \in \mathcal{M}_{\mathrm{CFE}}$, is given by

$$\tau_\kappa = \frac{1}{\|\mathbf{b}\|_{L_\infty(\kappa)}} \frac{1}{p_\kappa^2} \min_{F \subset \partial\kappa} \frac{\sup_{\kappa_\flat^F \subset \kappa} |\kappa_\flat^F|}{|F|} d \quad \forall \kappa \in \mathcal{M}_{\mathrm{CFE}}, \tag{16}$$

for $d = 2, 3$ and $p_\kappa \geq 1$, and $\kappa_\flat^F$ is as defined in Definition 2. In the case when $p_\kappa = 0$, τ_κ is formally defined to be zero.

Under the assumption that (14) holds, the following inf-sup condition for the bilinear form $B_{\mathrm{Hyp}}(\cdot,\cdot)$, with respect to the streamline DGFEM-norm (15), may be established, cf. [21]; this represents a generalization of the results in [15, 20].

Theorem 3 *Given Assumptions 3.1, 3.3, and 3.4 hold, there exists a positive constant Λ_s, independent of the mesh size h and the polynomial degree p, such that:*

$$\inf_{\nu \in V(\mathcal{M}_{\mathrm{CFE}}, \boldsymbol{p}) \setminus \{0\}} \sup_{\mu \in V(\mathcal{M}_{\mathrm{CFE}}, \boldsymbol{p}) \setminus \{0\}} \frac{B_{\mathrm{Hyp}}(\nu, \mu)}{|||\nu|||_{\mathrm{SD}} |||\mu|||_{\mathrm{SD}}} \geq \Lambda_s. \tag{17}$$

On the basis of the inf-sup condition stated in Theorem 3, together with the approximation results given in Lemma 3, we deduce the following a priori error bound for the DGFEM (13).

Theorem 4 *Let $\Omega \subset \mathbb{R}^d$, $d = 2, 3$, be a bounded polyhedral domain, and $\mathcal{M}_{\mathrm{CFE}} = \{\kappa\}$ be a subdivision of Ω consisting of general polytopic elements satisfying Assumptions 3.1, 3.3, and 3.4. Further, let $\mathcal{T}_\sharp = \{\mathcal{K}\}$ denote the associated covering of Ω consisting of shape-regular d-simplices as in Definition 1, which satisfies Assumption 3.2. Let $u_h \in V(\mathcal{M}_{\mathrm{CFE}}, \boldsymbol{p})$ be the DGFEM approximation to $u \in H^1(\Omega)$ defined by (13) and suppose that $u|_\kappa \in H^{l_\kappa}(\kappa)$, $l_\kappa > 1 + d/2$, for each $\kappa \in \mathcal{M}_{\mathrm{CFE}}$, such that $\mathcal{E}u|_{\mathcal{K}} \in H^{l_\kappa}(\mathcal{K})$, where $\mathcal{K} \in \mathcal{T}_\sharp$ with $\kappa \subset \mathcal{K}$. Then, the following error bound holds:*

$$|||u - u_h|||_{\mathrm{SD}}^2 \leq C \sum_{\kappa \in \mathcal{M}_{\mathrm{CFE}}} \frac{h_\kappa^{2s_\kappa}}{p_\kappa^{2l_\kappa}} \mathcal{G}_\kappa(F, C_m, p_\kappa, \tau_\kappa) \|\mathcal{E}u\|_{H^{l_\kappa}(\mathcal{K})}^2, \tag{18}$$

where

$$\mathcal{G}_\kappa(F, C_m, p_\kappa, \tau_\kappa) = \|c_0\|_{L_\infty(\kappa)}^2 + \gamma_\kappa^2 + \tau_\kappa^{-1} + \tau_\kappa \beta_\kappa^2 p_\kappa^2 h_\kappa^{-2} + \beta_\kappa p_\kappa h_\kappa^{-d} \sum_{F \subset \partial\kappa} C_m(p_\kappa, \kappa, F)|F|, \tag{19}$$

$s_\kappa = \min\{p_\kappa + 1, l_\kappa\}$ and $p_\kappa \geq 1$. Here, $\gamma_\kappa = \|c_1\|_{L_\infty(\kappa)}$, with $c_1(x) = c(x)/c_0(x)$, c_0 as in (12), $\beta_\kappa = \|\mathbf{b}\|_{L_\infty(\kappa)}$, and C_m is defined as in Lemma 3. The positive constant C is independent of the discretization parameters.

Remark 9 For uniform orders $p_\kappa = p \geq 1$, $h = \max_{\kappa \in \mathcal{M}_{\mathrm{CFE}}} h_\kappa$, $s_\kappa = s$, $s = \min\{p+1, l\}$, $l > 1 + d/2$, under the assumption that the diameter of the faces of each element $\kappa \in \mathcal{M}_{\mathrm{CFE}}$ is of comparable size to the diameter of the corresponding element, the error bound stated in Theorem 4 reduces to

$$|||u - u_h|||_{\mathrm{Hyp}} \leq |||u - u_h|||_{\mathrm{SD}} \leq C \frac{h^{s-\frac{1}{2}}}{p^{l-1}} \|u\|_{H^l(\Omega)};$$

which is optimal in h and suboptimal in p by $p^{1/2}$. This generalizes the error estimate derived in [35] to general polytopic meshes under the same assumption (14).

Remark 10 On the basis of the error analysis undertaken in both the current section and Sect. 3, a priori error bounds for the DGFEM discretization of second-order PDEs with non-negative characteristic form on general polytopic meshes may be established; for details, we refer to our recent article [21].

5 Numerical Experiments

In this section we present a series of computational examples to illustrate the performance of the DGFEM on general classes of polytopic meshes. The computational validation of the error bounds derived in Theorems 2 and 4 have been presented in

[22] and [21], respectively; cf., also, [1]. Thereby, for the purposes of this section we consider the numerical approximation of incompressible flows in complicated geometries, cf. [30]. Throughout this section, we select $C_\sigma = 10$, cf. Lemma 4.

5.1 Example 1: Flow Through a Complicated T-pipe Domain

In this first example we consider the application of goal-oriented dual-weighted-residual mesh adaptation for the DGFEM discretization of the incompressible Navier–Stokes equations, cf. [10]. To this end, the computational domain Ω is defined to be an upside-down T-shaped pipe, which has had a series of randomly located, randomly sized, holes removed from both the vertical and horizontal sections. Figure 3a depicts the initial composite mesh, constructed based on employing the algorithm outlined in Sect. 2, which consists of only 128 polygonal elements. Here, the inflow boundary is specified at the top of the vertical section of the pipe, i.e., along $y = 6$, $4 \leq x \leq 8$, where Poiseuille flow enters Ω; the left-hand and right-hand side boundaries of the horizontal portion of the pipe, located at $x = 0$, $0 \leq y \leq 3$ and $x = 12$, $0 \leq y \leq 3$, respectively, are defined to be outflow Neumann boundaries. No slip boundary conditions are imposed on the remaining walls of the T-pipe geometry, together with the boundaries of the circular holes; finally, we set $Re = 100$. This test case represents a modification of the test problem considered in [30].

Here we consider goal-oriented control of the error in the target functional J, defined by $J(\mathbf{u}, p) = p(10, 1.5) \approx 3.49924\text{E-}3$, where $\mathbf{u}$ and p denote the velocity and pressure of the underlying flow, respectively. More precisely, following the notation in [30], we may establish an (approximate) error representation formula of the form

$$J(\mathbf{u}, p) - J(\mathbf{u}_h, p_h) \approx \sum_{\kappa \in \mathcal{M}_{\mathrm{CFE}}} \eta_\kappa,$$

where $\mathbf{u}_h$ and p_h denote the DGFEM approximation to $\mathbf{u}$ and p, respectively, and η_κ, $\kappa \in \mathcal{M}_{\mathrm{CFE}}$, denote the corresponding (weighted) error indicators, which depend on both $\mathbf{u}_h$ and p_h, as well as the approximate solution of a corresponding dual problem; for full details, see [30].

In Table 1, we demonstrate the performance of exploiting an adaptive mesh refinement strategy based on marking elements for refinement according to the size of the local error indicators $|\eta_\kappa|$. Here, we set the polynomial degrees for the approximation of the velocity field equal to 2, and employ piecewise discontinuous linear polynomials for the approximation of the pressure. In Table 1 we show the number of elements in the composite mesh $\mathcal{M}_{\mathrm{CFE}}$, the number of degrees of freedom in the underlying finite element space, the true error in the functional $J(\mathbf{u}, p) - J(\mathbf{u}_h, p_h)$, the computed error representation formula $\sum_{\kappa \in \mathcal{M}_{\mathrm{CFE}}} \eta_\kappa$, and the effectivity index $\theta = \sum_{\kappa \in \mathcal{M}_{\mathrm{CFE}}} \eta_\kappa / (J(\mathbf{u}, p) - J(\mathbf{u}_h, p_h))$. Here, we see that, even on

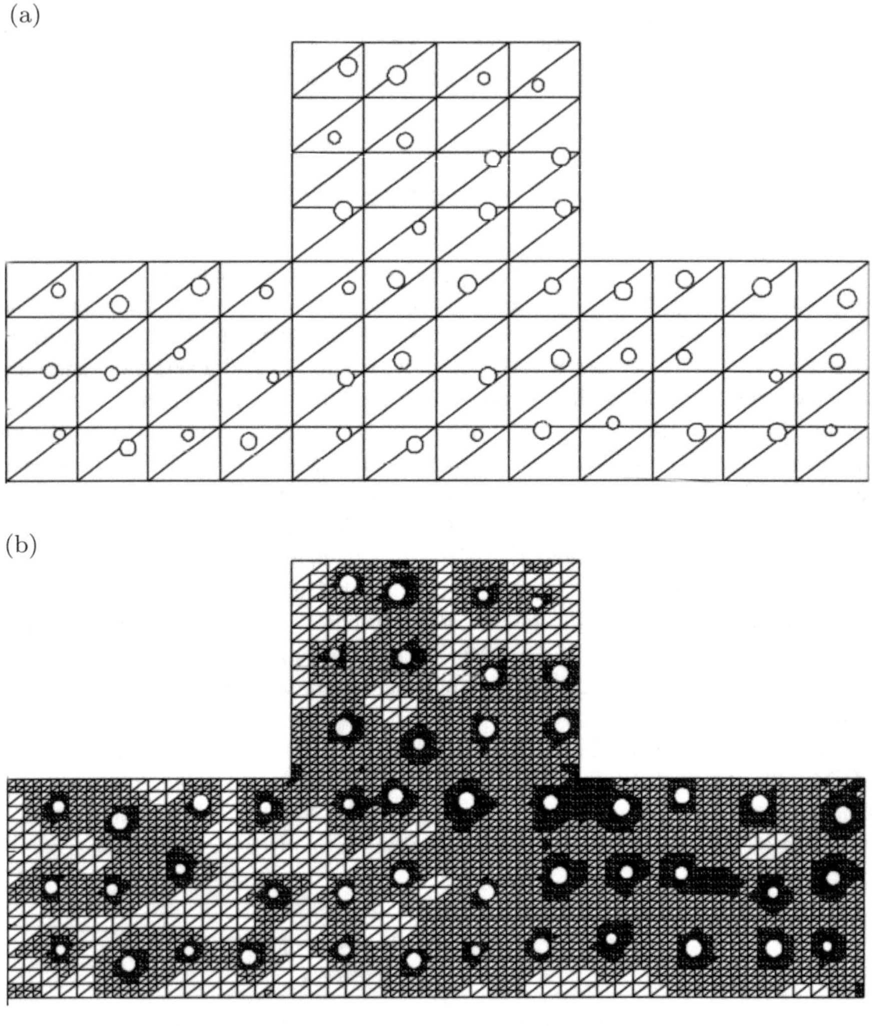

Fig. 3 Example 1. (**a**) Initial composite finite element mesh consisting of 128 polygonal elements; (**b**) Composite mesh after nine adaptive refinements with 13,356 elements

such coarse finite element meshes, the quality of the computed error representation formula is relatively good, in the sense that the effectivity indices are not too far away from unity. Indeed, as the mesh is refined, we observe that θ improves and approaches one. We note that *practical/engineering* accuracy can be attained using a very small number of degrees of freedom; indeed, fewer degrees of freedom are necessary than what would be required to accurately mesh the domain Ω using standard element shapes. The results presented in Table 1 are plotted in Fig. 4; here, we also compare the performance of the adaptive mesh refinement strategy with uniform mesh refinement. We observe that initially both strategies lead to

Table 1 Example 1: Adaptive algorithm

No of Eles	No of Dofs	$J(\mathbf{u}, p) - J(\mathbf{u}_h, p_h)$	$\sum_{\kappa \in \mathcal{M}_{\mathrm{CFE}}} \eta_\kappa$	θ
128	1920	−2.207E-2	−1.583E-2	0.72
206	3090	−4.720E-3	−2.478E-3	0.52
356	5340	−3.720E-3	−1.909E-3	0.51
618	9270	−1.620E-3	−8.014E-4	0.49
1079	16, 185	−8.216E-4	−4.427E-4	0.54
1749	26, 235	−3.929E-4	−1.965E-4	0.50
2996	44, 940	−1.707E-4	−7.457E-5	0.44
4861	72, 915	−8.728E-5	−7.197E-5	0.82
8000	120, 000	−2.164E-5	−2.324E-5	1.07
13, 356	200, 340	−5.073E-6	−5.073E-5	1.00

We present the number of elements in the composite mesh $\mathcal{M}_{\mathrm{CFE}}$ and the corresponding number of degrees of freedom in $V(\mathcal{M}_{\mathrm{CFE}}, \boldsymbol{p})$ (first two columns), the computed error in the target functional (third column), the sum of the (weighted) error indicators (fourth column), and the effectivity index (last column) at each step of the adaptive algorithm

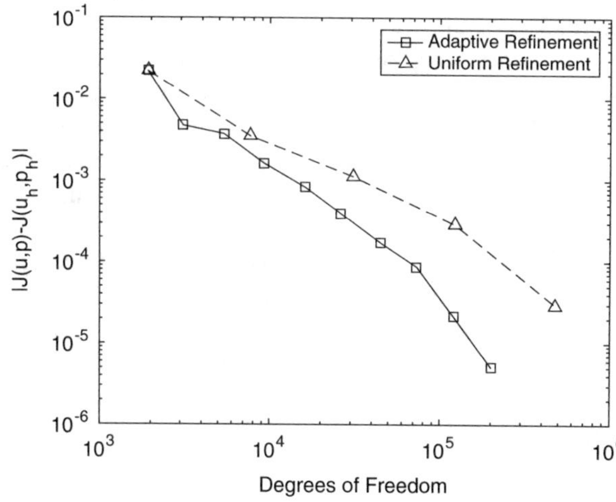

Fig. 4 Example 1: Comparison between uniform and adaptive mesh refinement

a comparable error in the computed target functional of interest J, for a given number of degrees of freedom; however, as both refinement procedures continue, the adaptive algorithm leads to over an order of magnitude improvement in the error in J for a comparable number of degrees of freedom.

5.2 *Example 2: Flow Past a 3D Scaffold Geometry*

In this final example, we consider incompressible flow past the three-dimensional scaffold geometry shown in Fig. 1. More precisely, the domain Ω is defined to be the elliptical cylinder $\{(x,y) : (x-x_0)/a^2 + (y-y_0)^2/b^2 < 1\} \times (0.015, 1.14)$, with the scaffold removed; here $(x_0, y_0) = (4.1325, 4.1625)$, $a = 4.1175$, and $b = 4.1475$. Based on the work undertaken in the article [25], we model a Newtonian fluid with density $\rho = 1000\,\text{kg/m}^3$ and viscosity $\mu = 8.1 \times 10^{-4}\,\text{Pa}\cdot\text{s}$. Prescribing a flow rate of $53\,\mu\text{ms}^{-1}$ yields a Reynolds number, $Re = 2 \times 10^{-3}$. The fine mesh which accurately describes Ω is generated based on image data supplied by Prof. El Haj and Dr. Kuiper. Here, only a coarse model has been employed; a more detailed description of the scaffold geometry is presented in the articles [4, 5]. However, even for this 'coarse' model, the underlying fine finite element mesh consists of 15.8 million elements. To demonstrate the exploitation of general polytopic elements generated by agglomeration, we employ METIS [37] to generate a very coarse mesh consisting of only 32,000 elements. We prescribe an inlet Poiseuille flow on the top of the geometry, where $z = 1.14$, together with no-slip wall boundary conditions on both the outer vertical walls of the elliptical cylinder, as well as on the scaffold itself. The bottom portion of the geometry located at $z = 0.015$ is identified as an outflow Neumann boundary. In Fig. 5 we plot the iso-surface of the magnitude of the velocity field; for the purposes of visualization, it was necessary to split the upper and lower regions of the computational domain. Clearly, by employing such a coarse agglomeration, we cannot expect that the computed DGFEM solution is sufficiently accurate, even within engineering constraints. However, this example clearly highlights a key issue we mentioned in Sect. 1: by employing polytopic elements, the dimension of the underlying finite element space is no longer proportional to the complexity of the geometry. Indeed, by exploiting *a posteriori* error estimation, cf. Example 1 above, then agglomerated elements may be marked for refinement; these can then be refined by again employing graph partitioning algorithms to the set of fine elements which form each marked (agglomerated) element. In this way, adaptive refinement of agglomerated elements, without the need to store mesh refinement trees, may be undertaken in a relatively simple manner, in order to automatically design polytopic meshes to yield reliable error control in quantities of interest. This will be investigated as part of our future programme of research.

6 Concluding Remarks

In this article, we have studied the application of DGFEMs on general finite element meshes consisting of polytopic elements. This class of methods is particularly attractive for a number of important reasons: (1) In the context of PDEs posed on complex domains Ω, the dimension of the underlying finite element space is independent of the number of small scale features/microstructures present in Ω; (2) Adaptivity can easily be employed to enhance the error in the computed numerical solution by only refining regions of the domain which directly contribute to the error in given quantities of interest; (3) High-order/*hp*-finite elements are naturally

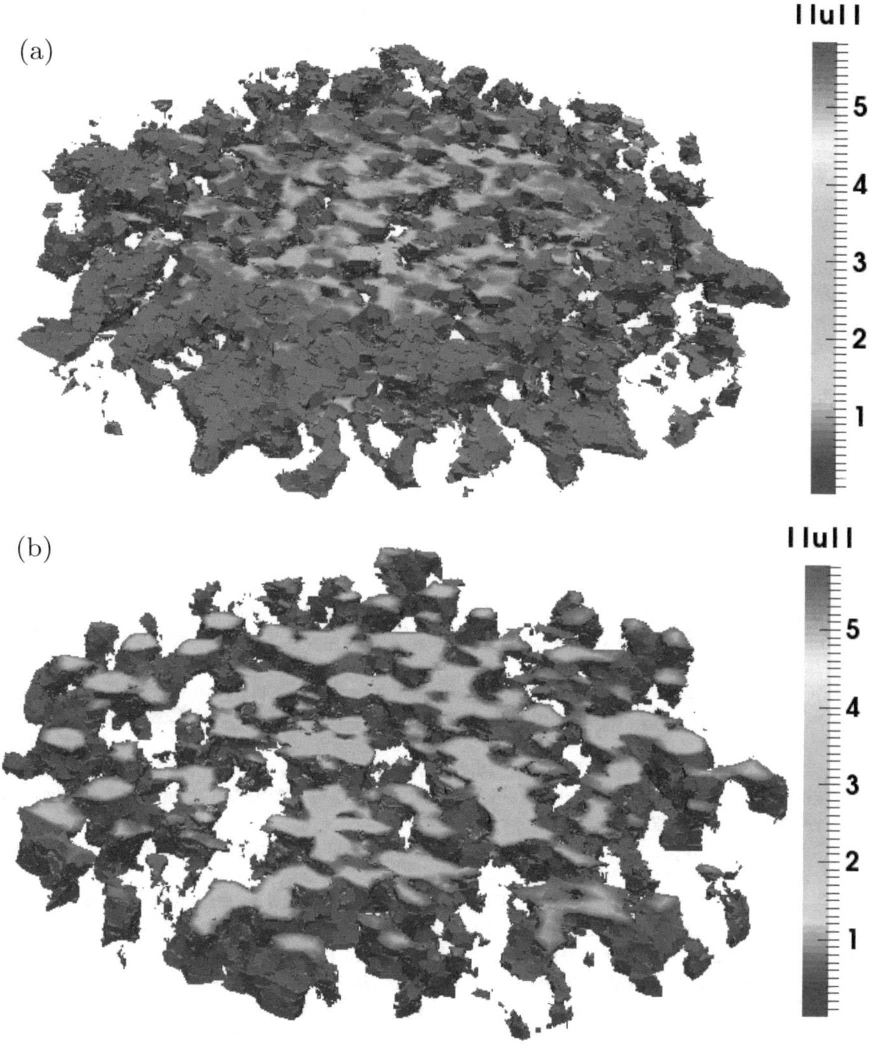

Fig. 5 Example 2. Plot of the norm of the velocity field: (**a**) upper section; (**b**) lower section

admitted; (4) The construction of coarse grid solvers for multilevel iterative solvers can easily be handled, cf. [2, 29]. In our present work, see, in particular, our recent articles [21, 22], great care has been taken to derive both inverse estimates and approximation results which are sharp with respect to element facet degeneration. This is particularly important for the definition of the interior penalty stabilization arising in the discretization of second-order elliptic PDEs. We believe this class of methods has huge potential for a wide variety of application areas, and in particular for problems arising in geophysics and biology. Indeed, as we have shown in Sect. 5, very complicated geometries can be treated, and with the use of general agglomerated refinement strategies, efficient and reliable computations may be undertaken. However, work on developing efficient quadrature and evaluation of appropriate stable polynomial bases on general polytopes still needs further work. Other future areas of research also include exploiting mesh partitioning algorithms for mesh refinement purposes, as well as the design and analysis of multilevel iterative solvers on polytopic meshes, for a wider range of application areas.

Acknowledgements We would like to thank Prof. El Haj and Dr. Kuiper for supplying the data needed to generate the scaffold geometry shown in Figs. 1 and 5. Paola Antonietti has been partially funded by SIR Project n. RBSI14VT0S *PolyPDEs: Non-conforming polyhedral finite element methods for the approximation of partial differential equations* funded by MIUR and by the Indam-GNCS Project 2015: *Non-standard numerical methods for Geophysics*. Joe Collis acknowledges the financial support of the EPSRC under the grant EP/K039342/1. Andrea Cangiani was partially supported by the EPSRC under the grant EP/L022745/1.

References

1. P.F. Antonietti, S. Giani, P. Houston, *hp*-Version composite discontinuous Galerkin methods for elliptic problems on complicated domains. SIAM J. Sci. Comput. **35**(3), A1417–A1439 (2013)
2. P.F. Antonietti, S. Giani, P. Houston, Domain decomposition preconditioners for discontinuous Galerkin methods for elliptic problems on complicated domains. J. Sci. Comput. **60**(1), 203–227 (2014)
3. P.F. Antonietti, P. Houston, M. Sarti, M. Verani, Multigrid algorithms for *hp*-version interior penalty discontinuous Galerkin methods on polygonal and polyhedral meshes. arXiv preprint arXiv:1412.0913 (2014)
4. E. Baas, J.H. Kuiper, A numerical model of heterogeneous surface strains in polymer scaffolds. J. Biomech. **41**, 1374–1378 (2008)
5. E. Baas, J.H. Kuiper, Y. Yang, M.A. Wood, A.J. El Haj, In vitro bone growth responds to local mechanical strain in three-dimensional polymer scaffolds. J. Biomech. **43**, 733–739 (2010)
6. I. Babuška, J. E. Osborn, Generalized finite element methods: their performance and their relation to mixed methods. SIAM J. Numer. Anal. **20**(3), 510–536 (1983)
7. F. Bassi, L. Botti, A. Colombo, D.A. Di Pietro, P. Tesini, On the flexibility of agglomeration based physical space discontinuous Galerkin discretizations. J. Comput. Phys. **231**(1), 45–65 (2012)
8. F. Bassi, L. Botti, A. Colombo, S. Rebay, Agglomeration based discontinuous Galerkin discretization of the Euler and Navier-Stokes equations. Comput. Fluids **61**, 77–85 (2012)
9. F. Bassi, L. Botti, A. Colombo, Agglomeration-based physical frame dG discretizations: an attempt to be mesh free. Math. Models Methods Appl. Sci. **24**(8), 1495–1539 (2014)
10. R. Becker, R. Rannacher, An optimal control approach to a-posteriori error estimation in finite element methods, in *Acta Numerica*, ed. by A. Iserles (Cambridge University Press, Cambridge, 2001), pp. 1–102
11. L. Beirão da Veiga, J. Droniou, G. Manzini, A unified approach for handling convection terms in finite volumes and mimetic discretization methods for elliptic problems. IMA J. Numer. Anal. **31**(4), 1357–1401 (2011)
12. L. Beirão da Veiga, F. Brezzi, A. Cangiani, G. Manzini, L.D. Marini, A. Russo, Basic principles of virtual element methods. Math. Models Methods Appl. Sci. **23**(1), 199–214 (2013)

13. K.S. Bey, T. Oden, *hp*-Version discontinuous Galerkin methods for hyperbolic conservation laws. Comput. Methods Appl. Mech. Eng. **133**, 259–286 (1996)
14. F. Brezzi, A. Buffa, K. Lipnikov, Mimetic finite differences for elliptic problems. ESAIM Math. Model. Numer. Anal. **43**(2), 277–295 (2009)
15. A Buffa, T.J.R. Hughes, G Sangalli, Analysis of a multiscale discontinuous Galerkin method for convection-diffusion problems. SIAM J. Numer. Anal. **44**(4), 1420–1440 (2006)
16. E. Burman, P. Hansbo, Fictitious domain finite element methods using cut elements: I. A stabilized Lagrange multiplier method. Comput. Methods Appl. Mech. Eng. **199**, 2680–2686 (2010)
17. E. Burman, P. Hansbo, An interior-penalty-stabilized Lagrange multiplier method for the finite-element solution of elliptic interface problems. IMA J. Numer. Anal. **30**, 870–885 (2010)
18. E. Burman, P. Hansbo, Fictitious domain finite element methods using cut elements: II. A stabilized Nitsche method. Appl. Numer. Math. **62**, 328–341 (2012)
19. A. Cangiani, G. Manzini, A. Russo, Convergence analysis of the mimetic finite difference method for elliptic problems. SIAM J. Numer. Anal. **47**(4), 2612–2637 (2009)
20. A. Cangiani, J. Chapman, E.H. Georgoulis, M. Jensen, On the stability of continuous–discontinuous Galerkin methods for advection–diffusion–reaction problems. J. Sci. Comput. **57**(2), 313–330 (2013)
21. A. Cangiani, Z. Dong, E.H. Georgoulis, P. Houston, *hp*-Version discontinuous Galerkin methods for advection–diffusion–reaction problems on polytopic meshes. ESAIM: M2AN **50**(3), 699–725 (2016)
22. A. Cangiani, E.H. Georgoulis, P. Houston, *hp*-Version discontinuous Galerkin methods on polygonal and polyhedral meshes. Math. Models Methods Appl. Sci. **24**(10), 2009–2041 (2014)
23. A. Chernov, Optimal convergence estimates for the trace of the polynomial L^2-projection operator on a simplex. Math. Comput. **81**(278), 765–787 (2012)
24. P.G. Ciarlet, *The Finite Element Method for Elliptic Problems*. Studies in Mathematics and Its Applications, vol. 4 (North-Holland, Amsterdam, 1978)
25. M. Cioffi, F. Boschetti, M.T. Raimondi, G. Dubini, Modeling evaluation of the fluid-dynamic microenvironment in tissue-engineered constructs: a micro-CT based model. Biotech. Bioeng. **93**(3), 500–510 (2006)
26. B. Cockburn, G.E. Karniadakis, C.-W. Shu, The development of discontinuous Galerkin methods, in *Discontinuous Galerkin Methods: Theory, Computation and Applications*, ed. by B. Cockburn, G.E. Karniadakis, C.-W. Shu. Lecture Notes in Computational Science and Engineering, vol. 11 (Springer, Heidelberg, 2000), pp. 3–50
27. T.-P. Fries, T. Belytschko, The extended/generalized finite element method: an overview of the method and its applications. Int. J. Numer. Methods Eng. **84**(3), 253–304 (2010)
28. E.H. Georgoulis, Inverse-type estimates on *hp*-finite element spaces and applications. Math. Comput. **77**(261), 201–219 (electronic) (2008)
29. S. Giani, P. Houston, Domain decomposition preconditioners for discontinuous Galerkin discretizations of compressible fluid flows. Numer. Math. Theory Methods Appl. **7**(2) (2014)
30. S. Giani, P. Houston, Goal-oriented adaptive composite discontinuous Galerkin methods for incompressible flows. J. Comput. Appl. Math. **270**, 32–42 (2014)
31. S. Giani, P. Houston, *hp*-Adaptive composite discontinuous Galerkin methods for elliptic problems on complicated domains. Numer. Methods Partial Differ. Equ. **30**(4), 1342–1367 (2014)
32. W. Hackbusch, S.A. Sauter, Composite finite elements for problems containing small geometric details. Part II: implementation and numerical results. Comput. Vis. Sci. **1**, 15–25 (1997)
33. W. Hackbusch, S.A. Sauter, Composite finite elements for the approximation of PDEs on domains with complicated micro-structures. Numer. Math. **75**, 447–472 (1997)
34. P. Houston, C. Schwab, E. Süli, Stabilized *hp*-finite element methods for first-order hyperbolic problems. SIAM J. Numer. Anal. **37**(5), 1618–1643 (electronic) (2000)

35. P. Houston, C. Schwab, E. Süli, Discontinuous *hp*-finite element methods for advection-diffusion-reaction problems. SIAM J. Numer. Anal. **39**(6), 2133–2163 (electronic) (2002)
36. A. Johansson, M.G. Larson, A high order discontinuous Galerkin Nitsche method for elliptic problems with fictitious boundary. Numer. Math. **123**(4), 607–628 (2013)
37. G. Karypis, V. Kumar, A fast and highly quality multilevel scheme for partitioning irregular graphs. SIAM J. Sci. Comput. **20**(1), 359–392 (1999)
38. D. Kröner, *Numerical Schemes for Conservation Laws*. Wiley-Teubner (Wiley, Chichester, 1997)
39. A. Massing, Analysis and implementation of finite element methods on overlapping and fictitious domains. PhD thesis, University of Oslo (2012)
40. I. Perugia, D. Schötzau, An *hp*-analysis of the local discontinuous Galerkin method for diffusion problems. J. Sci. Comput. **17**(1–4), 561–571 (2002)
41. B. Rivière, M.F. Wheeler, V. Girault, Improved energy estimates for interior penalty, constrained and discontinuous Galerkin methods for elliptic problems. I. Comput. Geosci. **3**(3–4), 337–360 (2000)
42. S.A. Sauter, R. Warnke, Extension operators and approximation on domains containing small geometric details. East West J. Numer. Math. **7**(1), 61–77 (1999)
43. C. Schwab, *p- and hp-Finite Element Methods: Theory and Applications in Solid and Fluid Mechanics*. Numerical Mathematics and Scientific Computation (Oxford University Press, New York, 1998)
44. E.M. Stein, *Singular Integrals and Differentiability Properties of Functions* (University Press, Princeton, 1970)
45. N. Sukumar, A. Tabarraei, Conforming polygonal finite elements. Int. J. Numer. Methods Eng. **61**(12), 2045–2066 (2004)
46. E.F. Toro, *Riemann Solvers and Numerical Methods for Fluid Dynamics* (Springer, Heidelberg, 1997)

Discretization of Mixed Formulations of Elliptic Problems on Polyhedral Meshes

Konstantin Lipnikov and Gianmarco Manzini

Abstract We review basic design principles underpinning the construction of the mimetic finite difference and a few finite volume and finite element schemes for mixed formulations of elliptic problems. For a class of low-order mixed-hybrid schemes, we show connections between these principles and prove that the consistency and stability conditions must lead to a member of the mimetic family of schemes regardless of the selected discretization framework. Finally, we give two examples of using flexibility of the mimetic framework: derivation of arbitrary-order schemes and inexpensive convergent schemes for nonlinear problems with small diffusion coefficients.

1 Introduction

The mixed formulation allows us to calculate simultaneously the primary solution of a PDE and its flux. For this reason, mixed formulations are very useful for numerical solution of multiphysics systems. The focus of this work is on a single diffusive process that is a part of almost any complex multiphysics system.

In this paper, we present design principles used in the derivation of mimetic finite difference (MFD) schemes on polygonal and polyhedral meshes and establish bridges to design principles used by a few other discretization frameworks (finite volumes and finite elements). The focus on the design principle allows us to avoid technical details and provide a more clear connection between different frameworks in comparison with the work performed in [32]. We also illustrate the flexibility of the mimetic framework with two challenging examples: derivation of arbitrary-order accurate schemes for linear problems and convergent schemes for nonlinear problems with degenerate diffusion coefficients.

K. Lipnikov (✉) • G. Manzini
Applied Mathematics and Plasma Physics Group, Theoretical Division, Los Alamos National Laboratory, Los Alamos, NM, USA
e-mail: lipnikov@lanl.gov; manzini@lanl.gov

G.R. Barrenechea et al. (eds.), *Building Bridges: Connections and Challenges in Modern Approaches to Numerical Partial Differential Equations*, Lecture Notes in Computational Science and Engineering 114,
DOI 10.1007/978-3-319-41640-3_10

Many ideas underpinning the MFD method were originally formulated in the sixties for orthogonal meshes using the finite difference framework from which the name of the method was derived. Over the years, the MFD method has been extensively developed for the solution of a wide range of scientific and engineering problems in continuum mechanics [47], electromagnetics [39, 43], fluid flows [13, 21–23, 41], elasticity [19, 20], obstacle and control problems [1–3], diffusion [40], discretization of differential forms [5, 10, 49], and eigenvalue analysis [14]. An extensive list of people who contributed to the development of the MFD method can be found in the recent book [24] and review paper [44]. The paper summarizes almost all known results on Cartesian and curvilinear meshes for various PDEs including the Lagrangian hydrodynamics. The book complements the paper by providing numerous examples and describing basic tools used in the convergence analysis of mimetic schemes for elliptic PDEs.

The MFD method preserves or mimics essential mathematical and physical properties of underlying partial differential equations (PDEs) on general polygonal and polyhedral meshes. For the elliptic equation, these properties include the *local flux balance* and the *duality between gradient and divergence operators*. The latter implies symmetry and positive definiteness of the resulting matrix operator and is desirable for robustness and reliability of numerical simulations. The duality of the primary and derived mimetic operators is one of the major design principles. The *definition of the primary mimetic operators is coordinate invariant*, which is another design principle that allows us to build discrete schemes for non-Cartesian coordinate systems. The discrete operators are also built to satisfy *exact identities*, the property that is critical for avoiding spurious numerical solutions, providing accurate modeling of conservation laws, and making the convergence analysis possible.

The mimetic literature mentions a variety of schemes for elliptic equations including the nodal (e.g., [9]), mixed (e.g., [5, 8, 40]) and mixed-hybrid (e.g., [42]) schemes. It is pertinent to note that these schemes have roots in only two mixed formulations for two different pairs of gradient and divergence operators. This connection was mentioned in the original work on the nodal mimetic schemes; however, it deserves a more detailed investigation in the future.

The related discretization frameworks considered in this paper include the finite volume methods [30, 36], the mixed finite element (MFE) method [50], and the virtual element method (VEM) [11]. Other finite volume frameworks exist that are based on mimetic principles such as the discrete duality finite volume methods (DDFV), see, e.g., [18, 28], but these methods do not fit in the MFD framework and will not be considered here.

The FV methods, originally introduced in [34, 35] for the heat equation and dubbed as the integrated finite difference method, form, perhaps, the largest class of schemes that can handle unstructured polygonal and polyhedral meshes, non-linear problems, and problems with anisotropic coefficients. An introduction to the finite volume methodology can be found in the recent review [29]. Almost all FV methods starts with a discrete representation of the flux balance equation. This representation is exact and this property is so important that all the methods that we consider in this paper use the same discrete form of the balance equation and the difference between them is only in the discretization of the constitutive equation.

The classical cell-centered FV scheme uses a two-point flux formula that is second-order accurate for special meshes such as the Voronoi tessellations. To overcome this limitation, a class of FV methods, consistent by design, is proposed by introducing additional unknowns on mesh faces. Examples of such methods are the *hybrid finite volume* method [36], and the *mixed finite volume* method [30]. These FV methods start with different definitions of the cell-based discrete gradient that are exact for linear solutions. The formula for the numerical flux based on this gradient needs a stabilization term. Construction of the stabilized flux uses two principles. First, the stabilization term should be zero on linear solutions. Second, the stabilized flux is defined as the solution of a certain equation with a symmetric and positive definite bilinear form. We will show that these design principles imply the duality principle in the mimetic framework.

The VEM was originally introduced as an evolution of the MFD method. In the classical finite element spirit, the duality principle is incorporated directly in the weak formulation. The exact identities are replaced by the exact sequence of virtual finite element spaces. A new design principle is the unisolvency property where the space of degrees of freedom is isomorphic to a space of finite element functions and includes polynomial as well as non-polynomial functions. The bilinear forms are split explicitly into consistency and stability forms using L^2 and H^1 projectors. The VEM literature distinguish two different methods, the nodal and mixed VEMs. The former is applied directly to the primary formulation of the elliptic problem. In contrast to the MFD method, a connection of the nodal VEM with an alternative mixed formulation has never been studied. Later, we discuss how the new design principles of the mixed VEM are connected to the stability and consistency conditions in the mimetic framework.

The recent developments of the MFD framework exploit its flexibility for selecting non-standard degrees of freedom, optimization of inner products, and non-standard approximations of primary operators to build schemes with higher order of accuracy and convergence schemes for nonlinear PDEs with degenerate coefficients (see also Sect. 4.2).

Extension to higher-order mixed scheme is almost straightforward in the mimetic framework. The key step is the proper selection of degrees of freedom that (a) simplify the discretization of the primary divergence operator and (b) allows us to formulate a computable consistency condition. A new design principle is introduced in this case, which states that a commuting relation exists between the interpolation operators defining the degrees of freedom of scalar and vector fields, and the divergence operators in the discrete and continuum settings.

All of the aforementioned methods discretize effectively the divergence operator $div\,(\cdot)$. To solve nonlinear parabolic equations, we employ new MFD schemes where the primary operator discretizes the combined operator $div\,(k\,\cdot)$, where k is the non-constant scalar diffusion coefficient. The resulting scheme uses both cell-centered and face-centered values of the diffusion coefficient. The face-centered values help to develop efficient schemes for nonlinear heat diffusion [16] and moisture transport in porous media [51]. The duality property mentioned above guarantees that the schemes can be formulated as algebraic problems with symmetric and positive definite matrices. Matrices with these properties lead to better performance of scalable iterative solvers, such as algebraic multigrid solvers and Krylov solvers such as the preconditioned conjugate gradient. A related strategy for solving nonlinear elliptic equations with degenerate coefficients is described in [4]. It uses only cell-centered values of the diffusion coefficient. We also mention the convergence analysis of gradient schemes for nonlinear parabolic problems in [31].

Finally, we mention other discretization methods that work on general meshes. Our necessarily incomplete list include the polygonal/polyhedral finite element method (PFEM) [46, 53, 54], hybrid high-order method [26], the discontinuous Galerkin (DG) method [15, 25], the hybridized discontinuous Galerkin (HDG) method [17], compatible discrete operators [6, 7], and the weak Galerkin (wG) method [55].

The outline of the paper is as follows. In Sect. 2, we review the basic discretization principles of the mimetic framework. In Sect. 3, we derive the mimetic finite difference method for elliptic problems through the consistency and stability conditions. We also prove that any mixed-hybrid method that uses the same degrees of freedom leads to a member of the mimetic family of schemes. In Sect. 4, we review the recent progress in the development of mimetic methods for mixed formulations of elliptic problems. Our final remarks and conclusions are given in Sect. 5.

2 Principles of the Mimetic Discretization Framework

The MFD method mimics important mathematical and physical properties of underlying PDEs. We start with some notation followed by the introduction of dual mimetic operators and two examples showing importance of having discrete duality and discrete exact identities in physics simulations. This section is based on the material presented in [39, 44].

Consider a polygonal or polyhedral mesh Ω_h. We denote the sets of mesh nodes, edges, faces, and cells by symbols $\mathcal{N}$, $\mathcal{E}$, $\mathcal{F}$ and $\mathcal{C}$, respectively, the set of vectors collecting the degrees of freedom associated with those mesh objects by the corresponding symbol with the subscript "h", and the restriction to cell "c" by the subscript "h, c". Each set of vectors of degrees of freedom with the (obvious) definitions of addition and multiplication by a scalar number is a linear space. For example, $\mathcal{F}_h$ is the linear space of vectors formed by the degrees of freedom located on the mesh faces, and $\mathcal{F}_{h,c}$ is its restriction to cell c. Its precise definition depends on the scheme. An illustration of particular discrete spaces restricted to a single cell is shown in Fig. 1.

The mimetic finite difference method operates with discrete analogs of the first-order operators. These operators are designed to satisfy exact identities and duality principles.

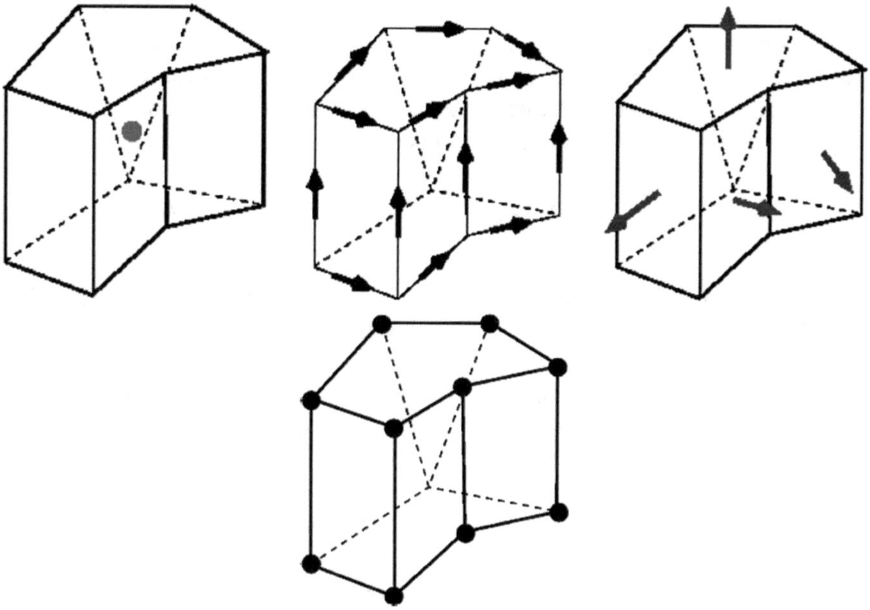

Fig. 1 Illustration of the degrees of freedom in low-order mimetic schemes. The local spaces associated with cell c from *left* to *right* are $\mathcal{C}_{h,c}$, $\mathcal{E}_{h,c}$, $\mathcal{F}_{h,c}$, and $\mathcal{N}_{h,c}$. The degrees of freedom are marked with *dots* and *arrows* and shown only on the visible objects

2.1 Global Mimetic Operators

In the mimetic framework, we usually discretize pairs of adjoint operators, such as the primary divergence $\mathcal{DIV}\colon \mathcal{F}_h \to \mathcal{C}_h$ and the derived gradient $\widetilde{\mathcal{GRAD}}\colon \mathcal{C}_h \to \mathcal{F}_h$. Hereafter, we will distinguish the derived operators from the primary operators by using a tilde on the operator's symbol. It is convenient to think about these operators as matrices acting between finite dimensional linear spaces. To discretize a large class of PDEs, we need three pairs of primary and derived operators, which are discrete analogs of gradient, curl and divergence operators. Each pair of operators satisfies a discrete integration by parts formula, e.g.

$$\big[\mathcal{DIV}\,\mathbf{u}_h,\, q_h\big]_{\mathcal{C}_h} = -\big[\mathbf{u}_h,\, \widetilde{\mathcal{GRAD}}\,q_h\big]_{\mathcal{F}_h} \qquad \forall \mathbf{u}_h \in \mathcal{F}_h,\ \forall q_h \in \mathcal{C}_h. \tag{1}$$

This formula represents one of the mimetic discretization principles as it mimics the continuum Green formula

$$\int_\Omega (\operatorname{div}\mathbf{u})\, q\, \mathrm{d}x = -\int_\Omega \mathbf{u}\cdot\nabla q\, \mathrm{d}x \qquad \forall \mathbf{u} \in H(div,\, \Omega),\ \forall q \in H^1_0(\Omega). \tag{2}$$

The brackets $[\cdot,\cdot]_{\mathcal{C}_h}$ and $[\cdot,\cdot]_{\mathcal{F}_h}$ in (1) stand for an approximation of the integrals in (2) and will be referred to as the *mimetic inner products* (or, simply *inner products*). The inner products are typically constructed from local (e.g. cell-based or node-based) inner products, which simplifies their derivation. For example, the two inner products in (1) can be reformulated as

$$\big[\mathbf{u}_h,\, \mathbf{v}_h\big]_{\mathcal{F}_h} = \sum_{c\in\Omega_h} \big[\mathbf{u}_c,\, \mathbf{v}_c\big]_{\mathcal{F}_{h,c}}, \qquad \big[p_h,\, q_h\big]_{\mathcal{C}_h} = \sum_{c\in\Omega_h} \big[p_c,\, q_c\big]_{\mathcal{C}_{h,c}} \tag{3}$$

where $\mathbf{v}_c$, $\mathbf{u}_c$, q_c and p_c denote the restriction to mesh cell c of the corresponding global vectors and $[\,\cdot,\cdot\,]_{\mathcal{F}_{h,c}}$ and $[\,\cdot,\cdot\,]_{\mathcal{C}_{h,c}}$ are the local contribution from c to the global inner products $[\,\cdot,\cdot\,]_{\mathcal{F}_h}$ and $[\,\cdot,\cdot\,]_{\mathcal{C}_h}$, respectively.

Let $\mathsf{M}_{\mathcal{C}}$ and $\mathsf{M}_{\mathcal{F}}$ be the symmetric positive definite matrices induced by the inner products $[\cdot,\cdot]_{\mathcal{C}_h}$ and $[\cdot,\cdot]_{\mathcal{F}_h}$, respectively. Then, the explicit formula for the derived gradient operator is

$$\widetilde{\mathcal{GRAD}} = -\mathsf{M}_{\mathcal{F}}^{-1}\, \mathcal{DIV}^T\, \mathsf{M}_{\mathcal{C}}.$$

This formula shows that this operator has a nonlocal stencil when matrix $\mathsf{M}_{\mathcal{F}}$ is irreducible as is typical for unstructured meshes. Note that the same property holds true for many other discretization frameworks.

Formula (1) implies that in general the discrete operators cannot be discretized independently. If we discretize one of the operators, e.g., the divergence, and select the inner products, the other operator, the gradient, must be derived from formula (1). The existing freedom is in the selection of the inner products.

The selection of the discrete spaces is often in tune with the discretization of the primary mimetic operator. For instance, for the pair of discrete spaces, $\mathcal{N}_h$ and $\mathcal{E}_h$, it is more natural to discretize the gradient operator as the primary operator. In such a case, the gradient operator $\mathcal{GRAD}\colon \mathcal{N}_h \to \mathcal{E}_h$ is the primary mimetic operator and the discrete divergence operator $\widetilde{\mathcal{DIV}}\colon \mathcal{E}_h \to \mathcal{N}_h$ is the derived operator. This pair of operators satisfies another discrete integration by parts formula that mimics (2):

$$\big[\widetilde{\mathcal{DIV}}\, \mathbf{v}_h,\, p_h\big]_{\mathcal{N}_h} = -\big[\mathbf{v}_h,\, \mathcal{GRAD}\, p_h\big]_{\mathcal{E}_h} \qquad \forall p_h \in \mathcal{N}_h,\ \forall \mathbf{v}_h \in \mathcal{E}_h. \tag{4}$$

The explicit formula for the derived divergence operator is

$$\widetilde{\mathcal{DIV}} = -\mathsf{M}_{\mathcal{N}}^{-1}\, \mathcal{GRAD}^T\, \mathsf{M}_{\mathcal{E}},$$

where $\mathsf{M}_{\mathcal{E}}$ and $\mathsf{M}_{\mathcal{N}}$ are symmetric positive definite matrices induced by the inner products. Note that in some low-order mimetic schemes, matrix $\mathsf{M}_{\mathcal{N}}$ is diagonal, so that the derived operator has a local stencil.

Remark 1 Both pairs $\mathcal{DIV}$, $\widetilde{\mathcal{GRAD}}$ and $\widetilde{\mathcal{DIV}}$, $\mathcal{GRAD}$ could be used to discretize an elliptic equation. A different representation of two similar pairs of dual operators based on the discrete Hodge operator can be found in [6, 7]. For instance, [6] gives an interesting viewpoint on the second pair of operators by linking them with a dual mesh.

The third pair of discrete operators approximates the continuum operators "*curl*". Let $\mathcal{CURL}: \mathcal{E}_h \to \mathcal{F}_h$ and $\widetilde{\mathcal{CURL}}: \mathcal{F}_h \to \mathcal{E}_h$ satisfy the discrete integration by parts formula

$$\left[\mathcal{CURL}\,\mathbf{v}_h,\,\mathbf{u}_h\right]_{\mathcal{F}_h} = \left[\mathbf{v}_h,\,\widetilde{\mathcal{CURL}}\,\mathbf{u}_h\right]_{\mathcal{E}_h} \qquad \forall \mathbf{v}_h \in \mathcal{E}_h,\ \forall \mathbf{u}_h \in \mathcal{F}_h,$$

which mimics the continuum formula

$$\int_\Omega (\mathrm{curl}\,\mathbf{v}) \cdot \mathbf{u}\,\mathrm{d}x = \int_\Omega \mathbf{v} \cdot (\mathrm{curl}\,\mathbf{u})\,\mathrm{d}x \qquad \forall \mathbf{u} \in H_0(curl,\,\Omega),\ \forall \mathbf{v} \in H(curl,\,\Omega).$$

The explicit formula for the derived curl operator is

$$\widetilde{\mathcal{CURL}} = \mathsf{M}_{\mathcal{E}}^{-1}\,\mathcal{CURL}^T\,\mathsf{M}_{\mathcal{F}},$$

so that this derived operator has typically a non-local stencil.

The spaces of discrete functions that we have introduced so far satisfy homogeneous boundary conditions. In [38], these spaces were enriched conveniently to approximate the boundary integrals that appear in general Green formulas. The resulting derived mimetic operators include an approximation of the boundary conditions. We will not follow this approach here, since the focus of this paper is on mixed-hybrid formulations, which provide another way to incorporate boundary conditions in a numerical scheme.

The duality of the discrete operators helps us to build numerical schemes that satisfy discrete conservation laws. For example, consider the Euler equations in the Lagrangian form:

$$\frac{1}{\rho}\frac{d\rho}{dt} = -\mathrm{div}\,\mathbf{u}, \qquad \rho\frac{d\mathbf{u}}{dt} = -\nabla p, \qquad \rho\frac{de}{dt} = -p\,\mathrm{div}\,\mathbf{u}, \tag{5}$$

where p is the pressure, ρ is the density, $\mathbf{u}$ is the velocity, and e is the internal energy. The system is closed by an equation of state. A mimetic discretization of (5) is given by

$$\frac{1}{\rho_h}\frac{d\rho_h}{dt} = -\mathcal{DIV}\,\mathbf{u}_h, \qquad \rho_h\frac{d\mathbf{u}_h}{dt} = -\widetilde{\mathcal{GRAD}}p_h, \qquad \rho_h\frac{de_h}{dt} = -p_h\,\mathcal{DIV}\,\mathbf{u}_h, \tag{6}$$

where p_h, ρ_h, $\mathbf{u}_h$ and e_h are the discrete analogs of the corresponding continuum quantities that appear in (5) and $\mathcal{DIV}$ and $\widetilde{\mathcal{GRAD}}$ are the mimetic operators acting, respectively, as divergence and gradient. Let us assume that no external work is done on the system, e.g., $p = 0$ of $\partial\Omega$. The integration by parts and the continuity equation from (5) lead to the conservation of the total energy E:

$$\frac{dE}{dt} = \int_{\Omega(t)} \rho\Big(\frac{d\mathbf{u}}{dt}\cdot\mathbf{u} + \frac{de}{dt}\Big)\,\mathrm{d}x = -\int_{\Omega(t)} \big(\mathbf{u}\cdot\nabla p + p\,\mathrm{div}\,\mathbf{u}\big)\,\mathrm{d}x = 0. \qquad (7)$$

To mimic this property, we need the discrete gradient and divergence operators $\widetilde{\mathcal{GRAD}}$ and $\mathcal{DIV}$ to satisfy a discrete integration by parts formula like (1). Using the same argument that leads to (7), we obtain the conservation of the total discrete energy E_h:

$$\frac{dE_h}{dt} = -\big[\mathbf{u}_h,\, \widetilde{\mathcal{GRAD}}\, p_h\big]_{\mathcal{F}_h} - \big[p_h,\, \mathcal{DIV}\mathbf{u}_h\big]_{\mathcal{C}_h} = 0.$$

We emphasize that numerical methods that conserve energy usually have other important properties such as the correct prediction of a shock position.

Another discretization principle is to derive primary operators that mimic exact identities. This is typically achieved by using the first principles (the divergence and Stokes theorems) to define the primary operators, e.g. (17). As the result, we have

$$\mathcal{DIV}\,\mathcal{CURL}\,\mathbf{v}_h = 0, \qquad \mathcal{CURL}\,\mathcal{GRAD}\,p_h = 0 \qquad \forall \mathbf{v}_h \in \mathcal{E}_h,\ \forall p_h \in \mathcal{N}_h.$$

Another consequence of the duality principle is that similar identities hold for the derived mimetic operators. Using the aforementioned explicit formulas for these operators, we immediately obtain that

$$\widetilde{\mathcal{DIV}}\,\widetilde{\mathcal{CURL}}\,\mathbf{u}_h = 0, \qquad \widetilde{\mathcal{CURL}}\,\widetilde{\mathcal{GRAD}}\,q_h = 0 \qquad \forall \mathbf{u}_h \in \mathcal{F}_h,\ \forall q_h \in \mathcal{C}_h.$$

These exact identities allows us to design numerical schemes without non-physical spurious modes. For instance, in the numerical solution of Maxwell's equations, such operators guarantee that the magnetic field $\mathbf{B}_h$ remains divergence-free for all times. Applying the primary divergence operator to a semi-discrete form of Faraday's law of induction, i.e., $\partial\mathbf{B}_h/\partial t = -\mathcal{CURL}\,\mathbf{E}_h$, we obtain

$$\frac{\partial}{\partial t}\big(\mathcal{DIV}\,\mathbf{B}_h\big) = \mathcal{DIV}\,\frac{\partial\mathbf{B}_h}{\partial t} = -\mathcal{DIV}\,\mathcal{CURL}\,\mathbf{E}_h = 0.$$

Therefore, if $\mathbf{B}_h$ is such that $\mathcal{DIV}\,\mathbf{B}_h = 0$ at time $t = 0$, this relation will be satisfied at any time $t > 0$.

2.2 Local Mimetic Operators

From this section, we limit our discussion to diffusion problem (11) and one pair of the primary and derived operators. For the practical implementation of mimetic schemes, it is convenient to write a local integration by parts formula that implies the global one. To do it, we need an additional space Λ_h of pressure unknowns defined typically on mesh faces. Its restriction to cell c is denoted by $\Lambda_{h,c}$ and consists of the vectors $\boldsymbol{\lambda}_c$. We recall that the subscript "c" is added to denote the local mimetic operators and the local discrete spaces corresponding to cell c. Also note that the introduction of $\boldsymbol{\lambda}_c$ is inspired by the classical hybridization technique in the mixed FE method.

Let $\mathcal{DIV}_c\colon \mathcal{F}_{h,c} \to \mathcal{C}_{h,c}$ be the primary divergence operator. The derived gradient operator $\widetilde{\mathcal{GRAD}}_c\colon \mathcal{C}_{h,c} \times \Lambda_{h,c} \to \mathcal{F}_{h,c}$ satisfies the discrete integration by parts formula

$$\left[\mathcal{DIV}_c\,\mathbf{u}_c,\, q_c\right]_{\mathcal{C}_{h,c}} - [\mathbf{u}_c,\, \boldsymbol{\lambda}_c]_{\Lambda_{h,c}} = -\left[\mathbf{u}_c,\, \widetilde{\mathcal{GRAD}}_c \begin{pmatrix} q_c \\ \boldsymbol{\lambda}_c \end{pmatrix}\right]_{\mathcal{F}_{h,c}}$$

$$\forall \mathbf{u}_c \in \mathcal{F}_{h,c},\ \forall q_c \in \mathcal{C}_{h,c},\ \forall \boldsymbol{\lambda}_c \in \Lambda_{h,c}, \tag{8}$$

which mimics the continuum Green formula for cell c:

$$\int_c (\mathrm{div}\,\mathbf{u})\, q\, \mathrm{d}x - \int_{\partial c} (\mathbf{u}\cdot\mathbf{n})\, q\, \mathrm{d}x = -\int_c \mathbf{u}\cdot\nabla q\, \mathrm{d}x \qquad \forall \mathbf{u} \in H(div,\, c),\ \forall q \in H^1(c).$$

In order to recover formula (1) for the global discrete gradient operator, we impose the continuity of $\boldsymbol{\lambda}_c$ and $\mathbf{u}_c$ on the mesh faces, define the local divergence operator as the restriction of the global one, define the local spaces as restrictions of global ones, require that the interface terms cancel each other,

$$\sum_{c\in\Omega_h} [\mathbf{u}_c,\, \boldsymbol{\lambda}_c]_{\Lambda_{h,c}} = 0, \tag{9}$$

and that the local inner products are summed up into global inner products as in (3). In most mimetic schemes, continuity of $\mathbf{u}_c$ and $\boldsymbol{\lambda}_c$ implies (9).

The derivation of a mimetic scheme follows three generic steps. First, we select the degrees of freedom such that the local primary operator, e.g., $\mathcal{DIV}_c$, has a simple form. Second, we define the inner products in the discrete spaces that satisfy the consistency and stability conditions. Third, we postulate the discrete integration by parts formula and obtain the derived operator, e.g., $\widetilde{\mathcal{GRAD}}_c$, from it. Note that the local derived operator is defined uniquely.

These three steps are discussed in Sect. 3 for the mixed formulation of the diffusion problem. The flexibility of the mimetic framework is exploited in Sect. 4.2, where we derive another pair of primary divergence and derived gradient operators for a nonlinear parabolic problem. More examples of mimetic schemes can be found in [24].

2.3 Material Properties

The material properties are often included in the derived mimetic operator. Indeed, the Green formula (2) can be rewritten as follows:

$$\int_{\Omega} (\operatorname{div} \mathbf{u})\, q \,\mathrm{d}x = -\int_{\Omega} \mathbb{K}^{-1}\mathbf{u} \cdot (\mathbb{K}\nabla) q \,\mathrm{d}x \qquad \forall \mathbf{u} \in H(div,\, \Omega),\ \forall q \in H^1_0(\Omega). \tag{10}$$

According to (10), we can define $\widetilde{\mathcal{GRAD}}$ as an approximation of the combined operator $\mathbb{K}\nabla(\cdot)$ and the inner product $[\mathbf{u}_h, \mathbf{v}_h]_{\mathcal{F}_h}$ as an approximation of the right-hand side integral $\int_{\Omega} \mathbb{K}^{-1}\mathbf{u}\cdot\mathbf{v}\,\mathrm{d}x$, provided that $\mathbf{u}_h$ and $\mathbf{v}_h$ are the degrees of freedom of $\mathbf{u}$ and $\mathbf{v}$.

For a perfectly conducting medium, we have the following duality relationship for the first-order curl operators:

$$\int_{\Omega} \operatorname{curl} \mathbf{E} \cdot \mu^{-1}\mathbf{B} \,\mathrm{d}x = \int_{\Omega} \varepsilon\mathbf{E} \cdot \left(\varepsilon^{-1}\operatorname{curl} \mu^{-1}\mathbf{B}\right) \mathrm{d}x.$$

In this case the inner products in spaces $\mathcal{E}_h$ and $\mathcal{F}_h$ are the weighted inner products. The weights are the magnetic permeability μ^{-1} and electric permittivity ε. The derived curl operator $\widetilde{\mathcal{CURL}}$ is an approximation of the combined operator $\varepsilon^{-1}\operatorname{curl}(\mu^{-1}\cdot)$.

3 Mixed Formulation of Diffusion Problem

Let $\Omega \in \Re^d$ be a polygonal ($d = 2$) or polyhedral ($d = 3$) domain with Lipschitz continuous boundary. Consider the mixed formulation of the diffusion problem:

$$\begin{aligned} \mathbf{u} &= -\mathbb{K}\nabla p \qquad && \text{in } \Omega, \\ \operatorname{div} \mathbf{u} &= b && \text{in } \Omega, \end{aligned} \tag{11}$$

subject to the homogeneous Dirichlet boundary conditions on $\partial\Omega$. As usual, we will refer to the scalar unknown as the *pressure* and to the vector unknown as the *flux*. We assume that the diffusion tensor is piecewise constant on mesh Ω_h and we denote its restriction to cell c by $\mathbb{K}_c$. If $\mathbb{K}_c$ is not constant on c, we can take its values at the centroids of the mesh cells without losing the approximation order.

In this section, we consider one local mimetic formulation, two FV schemes and two mixed-hybrid FE schemes with the same set of degrees of freedom. These schemes use the same discrete divergence operator and can be formally written as:

$$\mathbf{u}_c = \mathcal{L}_c(p_c, \boldsymbol{\lambda}_c), \quad \mathcal{DIV}_c\mathbf{u}_c = b^I_c, \tag{12}$$

where $\mathcal{L}_c$ is a linear operator and b_c^I is defined later. Formulas (12) subject to continuity of $\boldsymbol{\lambda}_c$ and $\mathbf{u}_c$ across mesh faces and boundary conditions define a *mixed-hybrid scheme*. A mixed-hybrid scheme is called *linearity preserving* when (12) is exact for any linear pressure p which implies constant flux $\mathbf{u}$ and zero source b.

This section is based on the material presented in [8, 24] for the MFD method, in [30, 32, 36] for the FV methods, and in [11] for the VEM.

3.1 Regular Polygonal and Polyhedral Meshes

The analysis of discretization schemes is typically conducted on a sequence of conformal meshes Ω_h where h is the diameter of the largest cell in Ω_h and $h \to 0$. A mesh is called conformal if the intersection of any two distinct cells c_1 and c_2 is either empty, or a few mesh points, or a few mesh edges, or a few mesh faces. Cell c is defined as a closed domain in $\Re^3$ (or $\Re^2$) with flat faces and straight edges.

Following [24], we make a few assumptions on the regularity of 3D meshes. Similar assumptions can be derived for 2D meshes by reducing the dimension. Let $n_\star$, $\rho_\star$ and $\gamma_\star$ denote various mesh independent constants explained below.

(M1) Every polyhedral cell c has at most $n_\star$ faces and each face f has at most $n_\star$ edges.

(M2) For every cell c with faces f and edges e, we have

$$\rho_\star \left(\mathrm{diam}(c)\right)^3 \leq |c|, \quad \rho_\star \left(\mathrm{diam}(c)\right)^2 \leq |f|, \quad \rho_\star\, \mathrm{diam}(c) \leq |e|, \tag{13}$$

where $|\cdot|$ denotes the Euclidean measure of a mesh object.

(M3) For each cell c, there exists a point $\mathbf{x}_c$ such that c is star-shaped with respect to every point in the sphere of radius $\gamma_\star\,\mathrm{diam}(c)$ centered at $\mathbf{x}_c$. For each face f, there exists a point $\mathbf{x}_f \in f$ such that f is star-shaped with respect to every point in the disk of radius $\gamma_\star\,\mathrm{diam}(c)$ centered at $\mathbf{x}_f$ as shown in Fig. 2.

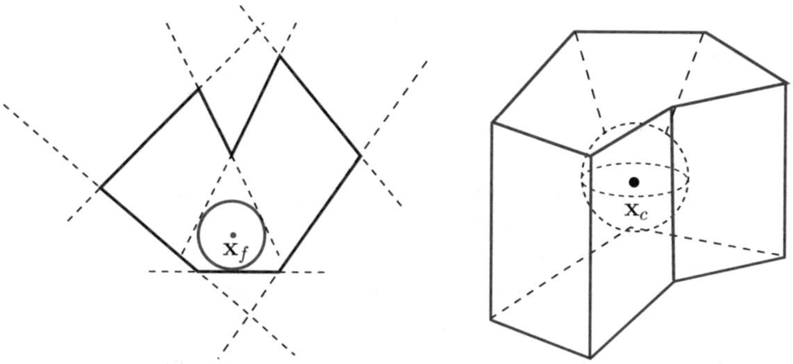

Fig. 2 Shape-regular mesh objects

(M4) For every cell c, and for every $f \in \partial c$, there exists a pyramid contained in c such that its base equals f, its height equals $\gamma_\star$ diam(c) and the projection of its vertex onto f is $\mathbf{x}_f$.

The conditions **(M1)**–**(M4)** are sufficient to develop an a priori error analysis of various discretization schemes. We recall only two results underpinning this error analysis. The first one is the Agmon inequality that uses **(M4)** and allows us to bound traces of functions:

$$\sum_{f\in\partial c} \|q\|^2_{L^2(f)} \leq C\left(\left(\mathrm{diam}(c)\right)^{-1}\|q\|^2_{L^2(c)} + \mathrm{diam}(c)\,|q|^2_{H^1(c)}\right) \qquad \forall q \in H^1(c). \tag{14}$$

The second one is the following approximation result: For any function $q \in H^2(c)$ there exists a polynomial $q^1 \in \mathcal{P}^1(c)$ such that

$$\|q - q^1\|_{L^2(c)} + \mathrm{diam}(c)\,|q - q^1|_{H^1(c)} \leq C\big(\mathrm{diam}(c)\big)^2 |q|_{H^2(c)}. \tag{15}$$

Hereafter, we will use symbols C, C_1, C_2 to denote generic constants independent of h.

Remark 2 The mesh regularity assumptions **(M1)**–**(M4)** are not optimal. They could be generalized to non-star shaped cells [24] by splitting each polygonal or polyhedral cell into finite number of shape regular simplexes. Note that an analysis of mimetic schemes for elliptic equations can be also done via the framework of gradient schemes [33] that allows us to use weaker mesh regularity assumptions in some cases. In particular, small mesh faces are allowed by this framework and optimal convergence rate was observed in many numerical experiments.

3.2 Mimetic Discretization Framework

As pointed out at the end of Sect. 2.2, the *first step* of the construction of a mimetic scheme consists in the selection of the degrees of freedom. The degrees of freedom are such that the primary divergence operator has a simple form.

The discrete space $\mathcal{C}_h$ consists of one degree of freedom per cell; its dimension equals the number of mesh cells; and for each vector $p_h \in \mathcal{C}_h$ we shall denote the value of p_h associated with cell c by $p_c \in \mathcal{C}_{h,c}$ Furthermore, we denote the vector of degrees of freedom of a smooth function p by $p_h^I \in \mathcal{C}_h$. In the MFD method p_c^I is typically the cell average of p over cell c and this definition is used in the superconvergence analysis [8].

The discrete space Λ_h consists of one degree of freedom per mesh face, e.g., λ_f; its dimension equals the number of mesh faces; and for each vector $\boldsymbol{\lambda}_h \in \Lambda_h$ we shall denote its restriction to cell c by $\boldsymbol{\lambda}_c \in \Lambda_{h,c}$. The continuity of local vectors $\boldsymbol{\lambda}_c$

across mesh cells is satisfied automatically. The value λ_f can be associated with the value of a smooth function p at the face centroid.

The discrete space $\mathcal{F}_h$ consists of one degree of freedom per boundary face and two degrees of freedom per interior face. For vector $\mathbf{u}_h \in \mathcal{F}_h$, we denote by $\mathbf{u}_c$ its restriction to cell c, and by u_f^c its component associated with face f of cell c. For a smooth function $\mathbf{u}$, we denote by $\mathbf{u}_h^I \in \mathcal{F}_h$ the vector of degrees of freedom. The value $(u_f^c)^I$ is defined as the integral average of flux $\mathbf{u} \cdot \mathbf{n}_f$ through face f, where $\mathbf{n}_f$ is the face normal fixed once and for all. Hereafter, we consider a subspace of $\mathcal{F}_h$ whose members satisfy the flux continuity constraint

$$u_f^{c_1} = u_f^{c_2} \tag{16}$$

on each interior face f shared by cells c_1 and c_2. With a slight abuse of notation, we shall refer to $\mathcal{F}_h$ as the space that satisfies condition (16).

The local primary divergence operator is defined using a straightforward discretization of the divergence theorem:

$$(\mathcal{DIV}\mathbf{u}_h)|_c \equiv \mathcal{DIV}_c\mathbf{u}_c = \frac{1}{|c|}\sum_{f\in\partial c} |f|\,\sigma_{c,f}\,u_f^c, \tag{17}$$

where $\sigma_{c,f}$ is either 1 or -1 depending on the mutual orientation of the fixed normal $\mathbf{n}_f$ and the exterior normal $\mathbf{n}_{c,f}$ to ∂c. Observe that this definition remains the same in all coordinate systems.

The *second step* of the construction of a mimetic scheme is to define accurate inner products in $\mathcal{C}_{h,c}$ and $\mathcal{F}_{h,c}$ that satisfy the consistency and stability conditions. The inner product for space $\mathcal{C}_{h,c}$ is simple:

$$\left[p_c,\, q_c\right]_{\mathcal{C}_{h,c}} = |c|\, p_c\, q_c. \tag{18}$$

Let $\mathcal{SC}_c$ be the space of constant functions. For consistency with the subsequent presentation, we use notation $\mathcal{SC}_c$ instead of $\mathcal{P}^0(c)$. Then, the above inner product implies the obvious result:

$$\left[p_c^I,\, q_c^I\right]_{\mathcal{C}_{h,c}} = \int_c p\, q\, \mathrm{d}x \qquad \forall p \in \mathcal{P}^0(c),\ \forall q \in \mathcal{SC}_c.$$

Despite its simplicity, we can use this relation to formulate the general principle that we apply to the derivation of other inner products. We define the consistency condition as the following exactness property: *the L^2 inner product of two smooth functions is equal to the mimetic inner product of their interpolants when one of the functions (p in this case) is a polynomial of a given degree and the other one belongs to a sufficiently rich space (possibly infinite dimensional) that must include polynomial functions.*

The exactness property implies that L^2 inner products of a large class of functions can be calculated exactly using the degrees of freedom. In particular, we write the consistency condition for the inner product on space $\mathcal{F}_{h,c}$ as the exactness property:

$$\left[\mathbf{u}_c^I,\, \mathbf{v}_c^I\right]_{\mathcal{F}_{h,c}} = \int_c \mathbb{K}_c^{-1}\mathbf{u}\cdot\mathbf{v}\,\mathrm{d}x \qquad \forall\mathbf{u}\in(\mathcal{P}^0(c))^d,\ \forall\mathbf{v}\in\mathcal{SF}_c, \tag{19}$$

where $\mathcal{SF}_c$ is a specially designed space containing the constant vector-functions:

$$\mathcal{SF}_c = \left\{\mathbf{v}\colon\ \operatorname{div}\mathbf{v}\in\mathcal{P}^0(c),\quad \mathbf{v}\cdot\mathbf{n}_f\in\mathcal{P}^0(f)\quad \forall f\in\partial c\right\}.$$

Let us show that the right-hand side of (19) can be calculated using the degrees of freedom. Let q^1 be a linear function such that $\mathbb{K}_c\nabla q^1 = \mathbf{u}$. Inserting $\mathbf{u}$ in (19), integrating by parts, and using the properties of $\mathcal{SF}_c$, we obtain

$$\begin{aligned}\int_c \nabla q^1\cdot\mathbf{v}\,\mathrm{d}x &= -\int_c(\operatorname{div}\mathbf{v})q^1\,\mathrm{d}x + \int_{\partial c}(\mathbf{v}\cdot\mathbf{n})\,q^1\,\mathrm{d}x\\ &= \sum_{f\in\partial c}\mathbf{v}\cdot\mathbf{n}_{c,f}\left(\int_f q^1\,\mathrm{d}x - \frac{|f|}{|c|}\int_c q^1\,\mathrm{d}x\right).\end{aligned}$$

Let $\mathsf{M}_{\mathcal{F},c}$ be the inner product matrix and $\mathbf{r}_c(q^1)\in\mathcal{F}_{h,c}$ be the vector with components $\sigma_{c,f}\left(\int_f q^1\,\mathrm{d}x - \frac{|f|}{|c|}\int_c q^1\,\mathrm{d}x\right)$. Then, combining the last formulas, we have

$$\begin{aligned}\left[(\mathbb{K}_c\nabla q^1)_c^I,\,\mathbf{v}_c^I\right]_{\mathcal{F}_{h,c}} &= \left((\mathbb{K}_c\nabla q^1)_c^I\right)^T\mathsf{M}_{\mathcal{F},c}\,\mathbf{v}_c^I = \sum_{f\in\partial c}\mathbf{v}\cdot\mathbf{n}_f\sigma_{cf}\left(\int_f q^1\,\mathrm{d}x - \frac{|f|}{|c|}\int_c q^1\,\mathrm{d}x\right)\\ &= \left(\mathbf{r}_c(q^1)\right)^T\mathbf{v}_c^I.\end{aligned}$$

Since $\mathbf{v}$ is any function in $\mathcal{SF}_c$, we can show that its interpolant $\mathbf{v}_c^I$ is any vector in $\mathcal{F}_{h,c}$. Indeed, it is sufficient to define a few functions $\mathbf{v}$ as the gradients of solutions of cell-based Poisson problems with various Neumann boundary conditions. Hence, the consistency condition gives us the following matrix equations

$$\mathsf{M}_{\mathcal{F},c}\,(\mathbb{K}_c\nabla q^1)_c^I = \mathbf{r}_c(q^1) \qquad \forall q^1\in\mathcal{P}^1(c). \tag{20}$$

Due to the linearity of these equations, it is sufficient to consider only three (two in two dimensions) linearly independent linear functions: $q_x^1 = x$, $q_y^1 = y$, and $q_z^1 = z$. Let

$$\mathsf{N}_c = [(\mathbb{K}_c\nabla x)_c^I\ (\mathbb{K}_c\nabla y)_c^I\ (\mathbb{K}_c\nabla z)_c^I],\quad \mathsf{R}_c = [\mathbf{r}_c(x)\ \mathbf{r}_c(y)\ \mathbf{r}_c(z)] \tag{21}$$

be two rectangular $n_c \times 3$ matrices where n_c is the number of faces in cell c. The Eqs. (20) are now equivalent to the matrix equation:

$$\mathsf{M}_{\mathcal{F},c}\mathsf{N}_c = \mathsf{R}_c. \tag{22}$$

Note that a symmetric positive definite solution $\mathsf{M}_{\mathcal{F},c}$ (if it exists) is not unique even for a tetrahedral cell c. The existence of solutions is based on the following result proved in [24].

Lemma 1 *Let matrices* N_c *and* R_c *be defined as in (21). Then,* $\mathsf{R}_c^T\mathsf{N}_c = |c|\,\mathbb{K}_c$.

This lemma allows us to write the explicit formula for matrix $\mathsf{M}_{\mathcal{F},c}$:

$$\mathsf{M}_{\mathcal{F},c} = \mathsf{R}_c(\mathsf{R}_c^T\mathsf{N}_c)^{-1}\mathsf{R}_c^T + \gamma_c\mathsf{P}_c, \quad \mathsf{P}_c = \mathsf{I} - \mathsf{N}_c(\mathsf{N}_c^T\mathsf{N}_c)^{-1}\mathsf{N}_c^T$$

with a positive factor γ_c in front of the projection matrix P_c. A recommended choice for γ_c is the mean trace of the first term. A family of mimetic schemes is obtained if we replace γ_c by an arbitrarily symmetric positive definite matrix G_c:

$$\mathsf{M}_{\mathcal{F},c} = \mathsf{R}_c(\mathsf{R}_c^T\mathsf{N}_c)^{-1}\mathsf{R}_c^T + \mathsf{P}_c\,\mathsf{G}_c\,\mathsf{P}_c. \tag{23}$$

The stability of the resulting mimetic method depends on the spectral bounds of matrix G_c which should be uniformly bounded by γ_c. Algebraically, it means that there exists two generic constants C_1 and C_2 such that

$$C_1|c|\sum_{f\in\partial c}|v_f^c|^2 \le \mathbf{v}_c^T\,\mathsf{M}_{\mathcal{F},c}\,\mathbf{v}_c^T \le C_2|c|\sum_{f\in\partial c}|v_f^c|^2 \tag{24}$$

holds for every $\mathbf{v}_c = \{v_f^c\}_{f\in\partial c}$. This formula is called the *stability condition* in the mimetic discretization framework. The existence of the mesh independent constants C_1 and C_2 can be shown using assumptions (**M1**)–(**M4**).

The *third* step of the construction of a mimetic scheme is to postulate either the local or global integration by parts formula and derive the gradient operator from it. For instance, given the global discrete operators $\mathcal{DIV}\colon \mathcal{F}_h \to \mathcal{C}_h$ and $\widetilde{\mathcal{GRAD}}\colon \mathcal{C}_h \to \mathcal{F}_h$, we can write the mimetic scheme as follows: Find $\mathbf{u}_h \in \mathcal{F}_h$ and $p_h \in \mathcal{C}_h$ such that

$$\mathbf{u}_h = -\widetilde{\mathcal{GRAD}}\,p_h, \qquad \mathcal{DIV}\,\mathbf{u}_h = b^I,$$

where $b^I \in \mathcal{C}_h$.

The local mimetic formulation requires us to define $[\mathbf{u}_c, \boldsymbol{\lambda}_c]_{\Lambda_{h,c}}$ which satisfies condition (9). Let

$$[\mathbf{u}_c, \boldsymbol{\lambda}_c]_{\Lambda_{h,c}} = \sum_{f\in\partial c}\sigma_{c,f}\,|f|\,\lambda_f\,u_f^c. \tag{25}$$

Then, the local formulation is to find $\mathbf{u}_c \in \mathcal{F}_{h,c}$, $p_c \in \mathcal{C}_{h,c}$ and $\boldsymbol{\lambda}_c \in \Lambda_{h,c}$ in all mesh cells such that

$$\mathbf{u}_c = -\widetilde{\mathcal{GRAD}}_c \begin{pmatrix} p_c \\ \boldsymbol{\lambda}_c \end{pmatrix}, \qquad \mathcal{DIV}_c\, \mathbf{u}_c = b_c^I$$

subject to the flux continuity conditions (16) and the homogeneous Dirichlet boundary conditions $\lambda_f = 0$ for $f \in \partial\Omega$. The local derived operator has the explicit form:

$$\widetilde{\mathcal{GRAD}}_c \begin{pmatrix} p_c \\ \boldsymbol{\lambda}_c \end{pmatrix} = -\mathsf{M}_{\mathcal{F},c}^{-1} \begin{pmatrix} \sigma_{c,f_1}\, |f_1|\, (p_c - \lambda_{f_1}) \\ \vdots \\ \sigma_{c,f_{n_c}}\, |f_{n_c}|\, (p_c - \lambda_{f_{n_c}}) \end{pmatrix}. \tag{26}$$

Lemma 2 *Under assumption (3) the local and global mimetic formulations are equivalent.*

Proof Let $\mathbf{v}_h$ be an arbitrary vector in $\mathcal{F}_h$ and $\mathbf{v}_c$ be its restriction to cell c. To show that the solution to the local mimetic formulation is the solution to the global one, we first multiply both sides of the local constitutive equations by $\mathbf{v}_c$, then sum up the results over the mesh cells, cancel all internal face terms containing λ_f, and finally use the local duality relation (8) between $\mathcal{DIV}_c$ and $\widetilde{\mathcal{GRAD}}_c$ to obtain:

$$\sum_c \left[\mathbf{u}_c,\, \mathbf{v}_c\right]_{\mathcal{F}_{h,c}} = -\sum_c \left[\widetilde{\mathcal{GRAD}}_c \begin{pmatrix} p_c \\ \boldsymbol{\lambda}_c \end{pmatrix},\, \mathbf{v}_c\right]_{\mathcal{F}_{h,c}} = \sum_c \left[\mathcal{DIV}_c\, \mathbf{v}_c,\, p_c\right]_{\mathcal{C}_{h,c}}.$$

From the additivity of the inner products and (1) we obtain that

$$\left[\mathbf{u}_h,\, \mathbf{v}_h\right]_{\mathcal{F}_h} = \left[\mathcal{DIV}\, \mathbf{v}_h,\, p_h\right]_{\mathcal{C}_h} = -\left[\widetilde{\mathcal{GRAD}}\, p_h,\, \mathbf{v}_h\right]_{\mathcal{F}_h} \qquad \forall \mathbf{v}_h \in \mathcal{F}_h,$$

which implies that $\mathbf{u}_h = -\widetilde{\mathcal{GRAD}}\, p_h$.

To show the opposite statement, we repeat the above argument in the reverse order. This proves the assertion of the lemma. □

Let us consider the matrix equation $\mathsf{N}_c = \mathsf{W}_{\mathcal{F},c}\, \mathsf{R}_c$ [compare with (22)]. To implement the mimetic scheme in a computer program, we need to know only matrix $\mathsf{W}_{\mathcal{F},c}$. The general solution to the matrix equation is

$$\mathsf{W}_{\mathcal{F},c} = \mathsf{N}_c\, (\mathsf{N}_c^T \mathsf{R}_c)^{-1}\, \mathsf{N}_c^T + \widetilde{\mathsf{G}}_c\, (\mathsf{I} - \mathsf{R}_c\, (\mathsf{R}_c^T \mathsf{R}_c)^{-1}\, \mathsf{R}_c^T),$$

where $\widetilde{\mathsf{G}}_c$ is an arbitrary $n_c \times n_c$ matrix, possibly non-symmetric. This formula leads to a large family of stable and unstable schemes that we refer to as the *extended mixed-hybrid family* of schemes. Note that non-symmetric schemes are natural for some plasma physics applications where the diffusion tensor is not symmetric.

Positive definiteness of $\mathsf{W}_{\mathcal{F},c}$ is necessary for proving the scheme's convergence following the path described in [42].

If in addition $\mathsf{W}_{\mathcal{F},c}$ is symmetric, then it is one of the matrices $\mathsf{M}_{\mathcal{F},c}^{-1}$. The classical two-point flux FV scheme is obtained when all matrices $\mathsf{W}_{\mathcal{F},c}$ are positive definite and diagonal.

3.2.1 Error Estimates

Let Ω have a Lipschitz continuous boundary. Furthermore, let every cell c be shape regular as explained in Sect. 3.1. We assume that $\mathbf{x}_c$ is the centroid of cell c. We use the triple-bar notation, e.g., $|||\cdot|||$, for the norms induced by the mimetic inner products. Then, the interpolants of the exact solution, $p^I \in \mathcal{C}_h$ and $\mathbf{u}^I \in \mathcal{F}_h$, satisfy [8]

$$|||p^I - p_h|||_{\mathcal{C}_h} + |||\mathbf{u}^I - \mathbf{u}_h|||_{\mathcal{F}_h} \leq C\,h,$$

where p_h and $\mathbf{u}_h$ are solutions of the global mimetic formulation. If in addition Ω is convex and C_1 in (24) is sufficiently large, then

$$|||p^I - p_h|||_{\mathcal{C}_h} \leq C\,h^2.$$

3.3 *Finite Volume Discretization Framework*

Two examples of FV schemes that fit within the mimetic framework are the mixed finite volume (MFV) method [30] and the hybrid finite volume (HFV) method [36]. Both methods give a family of schemes because a stabilization term, which can be suitably parameterized, appears in their formulation. Their design principles are based on conditions that imply the mimetic duality principle.

Both methods use the same degrees of freedom for pressure and flux as the local mimetic formulation and the same discrete flux balance equation, i.e., $\mathcal{DIV}_c\,\mathbf{u}_c = b_c^I$. Recall that $\mathbf{u}_c$ collects the flux unknowns associated with cell c. The local pressure unknowns are the cell pressures p_c associated with cells c and the interface pressures λ_f associated with faces f. In the formulation of these FV methods, p_c can be associated with the pressure value at any point inside the cell. As in the previous subsection, $\boldsymbol{\lambda}_c = \{\lambda_f\}_{f\in\partial c}$ is the vector whose size is equal to the number of faces in cell c.

In the HFV method, a discrete gradient in cell c is defined by applying the mid-point quadrature rule to the divergence theorem:

$$\nabla_c \begin{pmatrix} p_c \\ \boldsymbol{\lambda}_c \end{pmatrix} = \frac{1}{|c|}\sum_{f\in\partial c} |f|\,\lambda_f\,\mathbf{n}_{c,f} = \frac{1}{|c|}\sum_{f\in\partial c} |f|(\lambda_f - p_c)\mathbf{n}_{c,f}. \qquad (27)$$

This formula provides the exact value of the gradient whenever p is a linear function since for any constant vector $\mathbf{a}$ and any position vector $\mathbf{x}_c$, we have the geometric identity

$$|c|\,\mathbf{a} = \sum_{f\in\partial c} |f|\,\mathbf{a}\cdot(\mathbf{x}_f - \mathbf{x}_c)\,\mathbf{n}_{c,f},$$

where $\mathbf{x}_f$ is the face centroid.

Formula (27) is used to define the numerical scheme after an additional *stabilization term* $\mathsf{S}_{c,f}$, is included in the definition of the numerical flux u_f^c:

$$u_f^c = -\mathbf{n}_f \cdot \mathbb{K}_c \nabla_c \begin{pmatrix} p_c \\ \boldsymbol{\lambda}_c \end{pmatrix} + \mathsf{S}_{c,f}. \tag{28}$$

Like in the mimetic framework, the stabilization term is designed very carefully to preserve the consistency of the scheme. Let F_c be the diagonal matrix with entries $|f|$ on the main diagonal, $f \in \partial c$. Similarly, let Σ_c be the diagonal matrix with entries $\sigma_{c,f}$. Furthermore, let $\mathbf{1} = (1, 1, \ldots, 1)^T$, and

$$S_{c,f}(p_c, \boldsymbol{\lambda}_c) = \lambda_f - p_c - \nabla_c \begin{pmatrix} p_c \\ \boldsymbol{\lambda}_c \end{pmatrix} \cdot (\mathbf{x}_f - \mathbf{x}_c).$$

Note that the last expression is zero on linear pressure functions. The vector of numerical fluxes $\mathbf{u}_c$ is defined implicitly as the solution of

$$(\tilde{p}_c \mathbf{1} - \tilde{\boldsymbol{\lambda}}_c)^T\, \Sigma_c\, \mathsf{F}_c\, \mathbf{u}_c = |c|\,\mathbb{K}_c \nabla_c \begin{pmatrix} p_c \\ \boldsymbol{\lambda}_c \end{pmatrix} \cdot \nabla_c \begin{pmatrix} \tilde{p}_c \\ \tilde{\boldsymbol{\lambda}}_c \end{pmatrix} + \sum_{f\in\partial c} \alpha_{c,f} \frac{|f|}{d_{c,f}}\, S_{c,f}(p_c, \boldsymbol{\lambda}_c)\, S_{c,f}(\tilde{p}_c, \tilde{\boldsymbol{\lambda}}_c) \tag{29}$$

for any $\tilde{p}_c$ and $\tilde{\boldsymbol{\lambda}}_c$. Here $\alpha_{c,f}$ is a positive parameter and $d_{c,f}$ is the distance between $\mathbf{x}_c$ and the plane containing face f. After a few algebraic manipulations, we obtain:

$$\mathsf{S}_{c,f} = \sigma_{c,f} \sum_{f'\in\partial c} \alpha_{cf'} \frac{|f'|}{d_{cf'}} S_{cf'}(p_c, \boldsymbol{\lambda}_c) \left[-\frac{\delta_{ff'}}{|f|} + \frac{1}{|c|}\mathbf{n}_{c,f} \cdot (\mathbf{x}_{f'} - \mathbf{x}_c) \right]$$

where $\delta_{ff'}$ is the Kronecker symbol. It is obvious that the stabilization term is zero when it is calculated using the degrees of freedom of a linear function.

The right-hand side of (29) is a symmetric bilinear form with respect to the pressure unknowns. It is positive definite when $\alpha_{c,f} > 0$ and $\lambda_f = 0$ on $f \in \partial\Omega$. It is uniformly bounded from below when $\alpha_{c,f}$ are sufficiently large and the mesh satisfies the regularity conditions described in Sect. 3.1.

Remark 3 Equation (29) is the key design principle. When p_c is associated with the cell centroid, the left-hand side of this equation coincides with the left-hand side of (8) where the divergence is given by (17) and the interface term is defined by (25). So, Eq. (29) is also a representation of the duality principle.

Formula (27) can be rewritten using the mimetic matrix N_c as follows:

$$\nabla_c^{HFV}\begin{pmatrix} p_c \\ \boldsymbol{\lambda}_c \end{pmatrix} = \frac{\mathbb{K}_c^{-1}}{|c|}\mathsf{N}_c^T\,\Sigma_c\,\mathsf{F}_c\,\boldsymbol{\lambda}_c.$$

This formula should be compared with the formula for the discrete gradient in the MFV method. This gradient is reconstructed from face fluxes:

$$\mathcal{G}(\mathbf{u}_c) = -\frac{\mathbb{K}_c^{-1}}{|c|}\,\mathsf{R}_c^T\,\Sigma_c\,\mathbf{u}_c.$$

This formula is exact for a linear pressure function and constant flux, but it also has to be stabilized. The vector of numerical fluxes $\mathbf{u}_c$ is defined as the solution of

$$(\mathbf{v}_c)^T\mathsf{G}_c\mathbf{u}_c = (p_c\mathbf{1} - \boldsymbol{\lambda}_c)^T\,\Sigma_c\,\mathsf{F}_c\,\mathbf{v}_c \qquad \forall\mathbf{v}_c \in \mathcal{F}_{h,c} \tag{30}$$

where G_c is a symmetric positive definite matrix. This design principle shows even clear connection with the mimetic duality principle (8). Since G_c is invertible, we have that $\mathbf{u}_c$ is a linear combination of p_c and $\boldsymbol{\lambda}_c$, i.e. condition (12). We can summarize the above discussions in the following lemma.

Lemma 3 *Any linearity preserving mixed-hybrid scheme of type* (12) *with the degrees of freedom given by* $\mathcal{C}_h$, Λ_h, *and* $\mathcal{F}_h$ *is a member of the extended mixed-hybrid family of schemes. In addition, let* $\mathbf{u}_c = \mathcal{L}_c(p_c, \boldsymbol{\lambda}_c)$ *and assume that the bilinear form*

$$\mathcal{B}\big((\tilde{p}_c, \tilde{\boldsymbol{\lambda}}_c), (p_c, \boldsymbol{\lambda}_c)\big) := (\tilde{p}_c\mathbf{1} - \tilde{\boldsymbol{\lambda}}_c)^T\,\Sigma_c\,\mathsf{F}_c\,\mathcal{L}(p_c, \boldsymbol{\lambda}_c) \tag{31}$$

is symmetric, uniformly coercive and uniformly bounded (with respect to some norm that may depend on the problem). Then, the resulting scheme belongs to the mimetic family of schemes.

Proof The most general form of the constitutive equation in (12) is given by the linear relationship

$$\mathbf{u}_c = \mathcal{L}_c(p_c, \boldsymbol{\lambda}_c) = -\widetilde{\mathsf{W}}_{\mathcal{F},c}\begin{pmatrix} \boldsymbol{\lambda}_c \\ p_c \end{pmatrix}$$

with a rectangular $n_c \times (n_c + 1)$ matrix $\widetilde{\mathsf{W}}_{\mathcal{F},c} = [\widetilde{\mathsf{W}}^{(1)}_{\mathcal{F},c}, \widetilde{\mathsf{W}}^{(2)}_{\mathcal{F},c}]$. Since the scheme is exact for constant pressure functions, we have

$$0 = -\widetilde{\mathsf{W}}_{\mathcal{F},c}\begin{pmatrix} \mathbf{1} \\ 1 \end{pmatrix}.$$

Multiplying the last equation by p_c and subtracting from the previous one, we obtain

$$\mathbf{u}_c = \widetilde{\mathsf{W}}^{(1)}_{\mathcal{F},c}(p_c\mathbf{1} - \boldsymbol{\lambda}_c).$$

By our assumption, this formula is exact for all linear pressure functions and the corresponding constant flux functions. Let us take linearly independent pressure functions x, y, z and use the matrix notations introduced above to derive the following consequence of the linearity preservation property:

$$\mathsf{N}_c = \widetilde{\mathsf{W}}^{(1)}_{\mathcal{F},c}\, \Sigma_c\, \mathsf{F}_c^{-1}\, \mathsf{R}_c.$$

Hence, $\widetilde{\mathsf{W}}^{(1)}_{\mathcal{F},c} = \mathsf{W}_{\mathcal{F},c}\, \mathsf{F}_c\, \Sigma_c$ and the resulting scheme belongs to the extended mixed-hybrid family of schemes. Now

$$(\tilde{p}_c\mathbf{1} - \tilde{\boldsymbol{\lambda}}_c)^T\, \Sigma_c\, \mathsf{F}_c\, \mathbf{u}_c = (\tilde{p}_c\mathbf{1} - \tilde{\boldsymbol{\lambda}}_c)^T\, \Sigma_c\, \mathsf{F}_c\, \mathsf{W}_{\mathcal{F},c}\, \mathsf{F}_c\, \Sigma_c(p_c\mathbf{1} - \boldsymbol{\lambda}_c).$$

Our symmetry and coercivity assumptions imply that $\mathsf{W}_{\mathcal{F},c}$ is symmetric and positive definite; hence, it is invertible and $\mathsf{W}^{-1}_{\mathcal{F},c}$ coincides with one of the mimetic inner product matrices $\mathsf{M}_{\mathcal{F},c}$. Let $\tilde{\mathbf{u}}_c = \mathsf{W}^{(1)}_{\mathcal{F},c}(\tilde{p}_c\mathbf{1} - \tilde{\boldsymbol{\lambda}}_c)$. Then,

$$(\tilde{p}_c\mathbf{1} - \tilde{\boldsymbol{\lambda}}_c)^T\, \Sigma_c\, \mathsf{F}_c\, \mathbf{u}_c = \tilde{\mathbf{u}}_c^T\, \mathsf{W}^{-1}_{\mathcal{F},c}\, \mathbf{u}_c.$$

The uniform coercivity and boundness conditions imply that matrix $\mathsf{W}^{-1}_{\mathcal{F},c}$ satisfies the mimetic stability condition. This proves the assertion of the lemma. □

A detailed comparison of MFD, HFV and MFV methods is performed in [32]. In particular, the authors have shown that all schemes can be generalized to provide the identical families of numerical schemes.

3.4 Finite Element Discretization Framework

3.4.1 Raviart-Thomas FE Method on Simplexes

Let us consider the family of mimetic schemes, where the mass matrix is given in the form of Eq. (23). When the mesh is formed by simplexes, e.g., triangles in 2D and tetrahedra in 3D, there is a choice of the stabilization matrix $\mathsf{P}_c\, \mathsf{G}_c\, \mathsf{P}_c$ that provides the mass matrix from the lowest order Raviart-Thomas space [12]. For a simplex, the stabilization matrix is a rank-one matrix, $\mathsf{P}_c\, \mathsf{G}_c\, \mathsf{P}_c = \mathbf{p}_c^T\, [g_c]\, \mathbf{p}_c$, where

$$g_c = \frac{1}{d^2(d+1)} \sum_{f\in\partial c} (\mathbf{x}_f - \mathbf{x}_c)^T \mathbb{K}_c^{-1} (\mathbf{x}_f - \mathbf{x}_c)$$

and $\mathbf{p}_c^T = \big(|f_1|, |f_2|, \dots, |f_{d+1}|\big), f_i \in \partial c$.

3.4.2 Virtual Element Method

The VEM was originally introduced as an evolution of the MFD method. In the classical finite element spirit, the duality principle is now incorporated in the weak formulation: *Find* $\mathbf{u}_h \in \widetilde{\mathcal{SF}}_h$ *and* $p_h \in \widetilde{\mathcal{C}}_h$ *such that*

$$a_h(\mathbf{u}_h, \mathbf{v}_h) - (\mathrm{div}_h\, \mathbf{v}_h, p_h)_{L^2(\Omega)} = 0 \qquad \forall \mathbf{v}_h \in \widetilde{\mathcal{SF}}_h, \tag{32a}$$

$$(\mathrm{div}_h\, \mathbf{u}_h, q_h)_{L^2(\Omega)} = (b, q_h)_{L^2(\Omega)} \qquad \forall q_h \in \widetilde{\mathcal{C}}_h, \tag{32b}$$

where $\widetilde{\mathcal{C}}_h$ is the space of piecewise constant functions isometric to $\mathcal{C}_h$, $\widetilde{\mathcal{SF}}_h$ is a virtual space built from the local virtual spaces $\widetilde{\mathcal{SF}}_c$,

$$a_h(\mathbf{u}_h, \mathbf{v}_h) = \sum_{c\in\Omega_h} a_{h,c}(\mathbf{u}_h, \mathbf{v}_h), \quad a_{h,c}(\mathbf{u}_h, \mathbf{v}_h) = a^{(1)}_{h,c}(\mathbf{u}_h, \mathbf{v}_h) + a^{(2)}_{h,c}(\mathbf{u}_h, \mathbf{v}_h),$$

$a^{(1)}_{h,c}$ and $a^{(2)}_{h,c}$ are the consistency and stability terms (to be defined later in the section), and operator div_h is defined cell-by-cell as the local L^2-orthogonal projection of the continuum divergence operator onto $\widetilde{\mathcal{C}}_h$. The virtual space $\widetilde{\mathcal{SF}}_c$ includes polynomial as well as non-polynomial functions.

A new design principle is given by the unisolvency property. The virtual space $\widetilde{\mathcal{SF}}_c$ is build to be isomorphic to the mimetic space $\mathcal{F}_{h,c}$, i.e., the space of degrees of freedom. The virtual space is defined as the subspace of $\mathcal{SF}_c$:

$$\widetilde{\mathcal{SF}}_c = \left\{\mathbf{v}\colon \mathrm{div}\, \mathbf{v} \in \mathcal{P}^0(c), \quad \mathbf{v}\cdot\mathbf{n}_f \in \mathcal{P}^0(f) \quad \forall f \in \partial c, \quad \mathrm{curl}\, \mathbf{v} = 0\right\}.$$

In order to describe the splitting of the bilinear form $a_{h,c}(\mathbf{u}_h, \mathbf{v}_h)$ we need to introduce the problem-dependent L^2 projector $\Pi_c\colon \widetilde{\mathcal{SF}}_c \to \left(\mathcal{P}^0(c)\right)^d$:

$$a(\mathbf{v} - \Pi_c(\mathbf{v}), \mathbf{v}^0) = 0 \qquad \forall \mathbf{v}^0 \in \left(\mathcal{P}^0(c)\right)^d,$$

where $a(\mathbf{v}, \mathbf{u}) = (\mathbb{K}_c^{-1}\mathbf{v}, \mathbf{u})_{L^2(c)}$. The projector is computable using only the degrees of freedom [11]. We define the consistency term as

$$a^{(1)}_{h,c}(\mathbf{u}_h, \mathbf{v}_h) = a(\Pi_c(\mathbf{u}_h), \Pi_c(\mathbf{v}_h)) = \int_c \mathbb{K}_c^{-1}\Pi_c(\mathbf{u}_h)\cdot\Pi_c(\mathbf{v}_h)\,\mathrm{d}x,$$

and the stability term as

$$a^{(2)}_{h,c}(\mathbf{u}_h, \mathbf{v}_h) = (\mathbf{u}_h - \Pi_c(\mathbf{u}_h), \mathbf{v}_h - \Pi_c(\mathbf{v}_h))_{L^2(c)}.$$

To generate a family of schemes, we can replace the L^2 inner product with any other spectrally equivalent bilinear form. Like in the mimetic framework, the consistency term is the exactness property that holds when the first argument in the bilinear form is the constant function:

$$a_{h,c}^{(1)}(\mathbf{u}_h, \mathbf{v}_h) = \int_c \mathbb{K}_c^{-1} \Pi_c(\mathbf{u}_h) \cdot \Pi_c(\mathbf{v}_h)\, \mathrm{d}x = \int_c \mathbb{K}_c^{-1} \Pi_c(\mathbf{u}_h) \cdot \mathbf{v}_h\, \mathrm{d}x \qquad \forall \mathbf{v}_h \in \widetilde{\mathcal{SF}}_h.$$

Hence, the contribution of this term to the local mass matrix is like in the mimetic scheme, $\mathsf{R}_c\,(\mathsf{R}_c^T\,\mathsf{N}_c)^{-1}\,\mathsf{R}_c^T$.

The conventional FE hybridization procedure works for the VEM as well. The first equation in the variational formulation is replaced by a set of cell-based equations:

$$a_{h,c}(\mathbf{u}_{h,c}, \mathbf{v}_{h,c}) - (\mathrm{div}_{h,c}\mathbf{v}_{h,c}, p_h)_{L^2(c)} + \langle \mathbf{v}_{h,c} \cdot \mathbf{n}, \boldsymbol{\lambda}_h \rangle_{\partial c} = 0, \tag{33}$$

where $\mathbf{u}_{h,c}, \mathbf{v}_{h,c} \in \widetilde{\mathcal{SF}}_c$, div_c is the restriction of div_h to cell c, and $\boldsymbol{\lambda}_h$ is the Lagrange multiplier. It also holds that $\mathrm{div}_{h,c}\, \mathbf{v}_{h,c} = \mathcal{DIV}_c\, \mathbf{v}_c$ and

$$\langle \mathbf{v}_{h,c} \cdot \mathbf{n}, \boldsymbol{\lambda}_h \rangle_{\partial c} = \sum_{f \in \partial c} \sigma_{c,f}\, |f|\, v_f^c\, \lambda_f,$$

where λ_f is the constant value of the Lagrange multiplier on face f [compare with (25)]. The continuity equation is algebraically equivalent to (16). Introducing vector $\boldsymbol{\lambda}_c = \{\lambda_f\}_{f \in \partial c}$ and recalling that p_h has the constant value p_c over cell c, the hybridized equation can be rewritten as follows:

$$(p_c \mathbf{1} - \boldsymbol{\lambda}_c)^T\, \Sigma_c\, \mathsf{F}_c\, \mathbf{v}_c = a_{h,c}(\mathbf{u}_{h,c}, \mathbf{v}_{h,c}),$$

which also provides the relation $\mathbf{u}_c = \mathcal{L}(p_c, \boldsymbol{\lambda}_c)$. The VEM is a linearity-preserving method. According to Lemma 3, the lowest-order mixed-hybrid virtual element scheme is a member of the mimetic family of schemes.

4 Recent Developments of the Mimetic Framework

We highlight the new developments that extend the value of the mimetic framework for various physical applications. This section is based on the material presented in [37, 45] where we considered arbitrary-order mimetic schemes for linear diffusion and mimetic schemes for nonlinear elliptic equations, respectively. A different family of arbitrary-order mimetic schemes has been developed in [27]. Note that the mimetic schemes presented in Sect. 4.2 differ from the gradient schemes in [31].

4.1 High-Order Schemes

The development of a high-order mimetic scheme follows the same three steps described above for the low-order mimetic schemes. In the first step, we select the degrees of freedom that are convenient for the definition of the primary divergence operator, still denoted by $\mathcal{DIV}$. With a slight abuse of notation we still use the symbols $\mathcal{C}_h$ and $\mathcal{F}_h$ for the discrete spaces of pressure and flux unknowns, respectively.

The discrete space $\mathcal{C}_h$ contains multiple pressure unknowns that can be associated with the solution moments up to order r. The discrete space $\mathcal{F}_h$ contains multiple flux unknowns both associated with the mesh cells and the mesh faces. The cell-based degrees of freedom represent moments of the flux up to order r except for the zeroth order moment. The face-based degrees of freedom represent flux moments up to order $r+1$, see Fig. 3.

Remark 4 There exists another family of arbitrary-order mimetic schemes that coincides with the mimetic scheme described earlier when $r = 0$.

The discrete divergence operator $\mathcal{DIV}: \mathcal{F}_h \to \mathcal{C}_h$ is defined cell-wise from the commutation property:

$$\left(\mathcal{DIV}\,\mathbf{u}^I\right)_c = \mathcal{DIV}_c\,\mathbf{u}_c^I = \left(\operatorname{div}\mathbf{u}\right)_c^I,$$

which is also a useful property for the error analysis. This is the new design principle that can be generalized to other mimetic operators. The right-hand side is computable using only the degrees of freedom of $\mathbf{u}^I$. Let $\psi \in \mathcal{P}^r(c)$ be a polynomial of order at most r. Then, the definition of the moment and integration by parts give

$$\left(\operatorname{div}\mathbf{u}\right)_c^I = \frac{1}{|c|}\int_c (\operatorname{div}\mathbf{u})\,\psi\;\mathrm{d}x = -\int_c \mathbf{u}\cdot\nabla\psi\;\mathrm{d}x + \sum_{f\in\partial c}\int_f (\mathbf{u}\cdot\mathbf{n}_{c,f})\,\psi\;\mathrm{d}x.$$

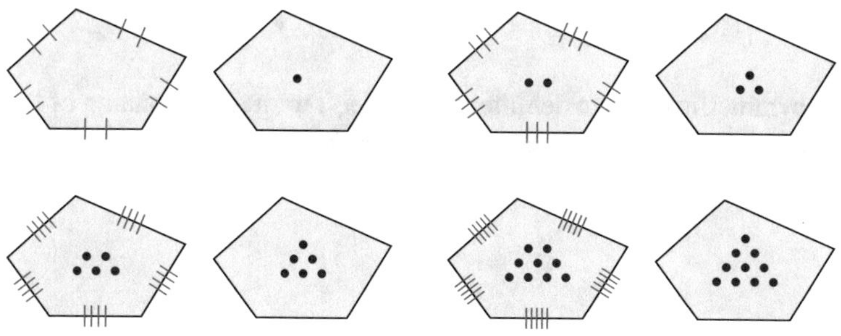

Fig. 3 Degrees of freedom for $0 \le r \le 3$ on a polygonal cell; for each polynomial degree r we show the flux degrees of freedom on the *left* and the scalar degrees of freedom on the *right*. The edge/face moments of the normal component of the flux are denoted by a *vertical line*; the cell moments are denoted by a *bullet*

In the second step, we define the mimetic inner products in spaces $\mathcal{C}_h$ and $\mathcal{F}_h$ as accurate approximations of the L^2 inner products of pressure and flux functions. The derivation is based on two high-order consistency conditions. Since the local space $\mathcal{C}_{h,c}$ is isomorphic to $\mathcal{P}^r(c)$, the first consistency condition is the obvious generalization of (18):

$$\left[p_c^I,\, q_c^I\right]_{\mathcal{C}_{h,c}} = \int_c p\, q\, \mathrm{d}x \qquad \forall p \in \mathcal{P}^r(c),\ \forall q \in \mathcal{P}^r(c).$$

The consistency condition in space $\mathcal{F}_{h,c}$ is defined as the following exactness property:

$$\left[\mathbf{u}_c^I,\, \mathbf{v}_c^I\right]_{\mathcal{F}_{h,c}} = \int_c \mathbb{K}_c^{-1}\mathbf{u}\cdot\mathbf{v}\, \mathrm{d}x \qquad \forall \mathbf{u} \in \mathbb{K}_c\, \nabla\mathcal{P}^{r+2}(c),\ \forall \mathbf{v} \in \mathcal{SF}_c,$$

where $\mathcal{SF}_c$ is a specially designed space containing the vector functions $(\mathcal{P}^{r+1}(c))^d$:

$$\mathcal{SF}_c = \left\{\mathbf{v}\colon\ \mathrm{div}\,\mathbf{v} \in \mathcal{P}^r(c), \quad \mathbf{v}\cdot\mathbf{n}_f \in \mathcal{P}^{r+1}(f) \quad \forall f \in \partial c\right\}.$$

It is easy to show that the right-hand side of the consistency condition is computable using the degrees of freedom introduced above. Let $q \in \mathcal{P}^{r+2}(c)$ be such that $\mathbf{u} = \mathbb{K}_c\, \nabla q$. Then,

$$\int_c \mathbb{K}_c^{-1}\mathbf{u}\cdot\mathbf{v}\, \mathrm{d}x = -\int_c (\mathrm{div}\mathbf{v})\, q\, \mathrm{d}x + \sum_{f\in\partial c}\int_f (\mathbf{v}\cdot\mathbf{n}_{c,f})\, q\, \mathrm{d}x.$$

As $\mathbf{v}$ is in $\mathcal{SF}_c$, the arguments of all the integrals in the right-hand side above are polynomials. Using the degrees of freedom of $\mathbf{v}$ it is possible to reconstruct $\mathrm{div}\,\mathbf{v}$ inside c and $\mathbf{v}\cdot\mathbf{n}_f$ on each $f \in \partial c$, see [37], and all these integrals are computable. Combining the last formulas, we obtain the algebraic form of the consistency condition:

$$\left[(\mathbb{K}_c\nabla q)^I,\, \mathbf{v}_c^I\right]_{\mathcal{F}_{h,c}} = \left((\mathbb{K}_c\nabla q)^I\right)^T \mathsf{M}_{\mathcal{F},c}\, \mathbf{v}_c^I = \left(\mathbf{r}_c(q)\right)^T \mathbf{v}_c^I.$$

To find a symmetric positive definite matrix $\mathsf{M}_{\mathcal{F},c}$, we need the analog of Lemma 1, which obviously holds since the function $\mathbb{K}_c\nabla\tilde{q}$ is in the space $\mathcal{SF}_c$ for any polynomial $\tilde{q} \in \mathcal{P}^{r+2}(c)$.

In the third step, we formulate the duality formula for the derived gradient operator and use it in the global mimetic formulation. The following error estimates have been shown in [37]:

$$|||p^I - p_h|||_{\mathcal{C}_h} + |||\mathbf{u}^I - \mathbf{u}_h|||_{\mathcal{F}_h} \le C\,h^{r+2}.$$

Remark 5 In the case when the diffusion tensor $\mathbb{K}$ is no longer constant, the consistency condition has to be modified. Let Π_c^r denote the local L^2 projector on the space of polynomial functions of order r. Then, the modified consistency condition reads:

$$\left[\left(\Pi_c^{r+1}(\mathbb{K}\nabla q)\right)^I,\, \mathbf{v}_c^I\right]_{\mathcal{F}_{h,c}} = \int_c \nabla q \cdot \mathbf{v}\, \mathrm{d}x \qquad \forall q \in \mathcal{P}^{r+2}(c),\ \forall \mathbf{v} \in \mathcal{SF}_c.$$

After this, the inner product matrix $\mathsf{M}_{\mathcal{F},c}$ is derived following the same steps.

4.2 Nonlinear Parabolic Problems

The consistency term in the formula for matrix $\mathsf{M}_{\mathcal{F},c}$ [see (19)] contains the inverse of the diffusion tensor. Therefore, numerical difficulties may arise in solving nonlinear parabolic problems of type

$$\frac{\partial p}{\partial t} - \mathrm{div}(k(p)\,\nabla p) = b,$$

where function $k(p)$ cannot be uniformly bounded from below. For instance, on a uniform one dimensional mesh, the numerical flux at mesh point x_i is proportional to the difference of the neighboring pressures and the transmissibility coefficient T_i:

$$u_i = -T_i \frac{p_{i+1/2} - p_{i-1/2}}{h}, \qquad T_i = \frac{2\,k_{i-1/2}\,k_{i+1/2}}{k_{i-1/2} + k_{i+1/2}}.$$

If $k_{i-1/2} \ll k_{i+1/2}$, the numerical flux goes to zero as $k_{i-1/2} \to 0$ and may lead to a nonphysical solution as shown by the numerical experiment considered in Sect. 4.2.2 (see also [48]). To obtain an accurate solution, we have to replace the harmonic average with the arithmetic average in the definition of the transmissibility coefficient. A possible strategy in the mixed finite element framework (see, e.g. [4])

consists in using two velocity variables, $\mathbf{v} = -\nabla p$ and $\mathbf{u} = k(p)\mathbf{v}$. However, the corresponding weak formulation cannot have face-based equations of type $v_c^f = k_f u_c^f$ (without breaking symmetry of the final discrete system) which are natural from a physical viewpoint. For instance, usage of face-based diffusion coefficients k_f is the well established numerical practice in modeling subsurface flows. We describe a new mimetic scheme that allows us to use different values k_f on different mesh faces. The mimetic framework always guarantees the symmetry of the resulting algebraic problem.

4.2.1 A New Pair of Primary and Derived Mimetic Operators

To generalized the mimetic schemes proposed in [45], we consider a more general form of the diffusion coefficient, $\mathbb{K}\,k(p)$, where $\mathbb{K}$ is a discontinuous tensor independent of p and $k(p)$ is a discontinuous scalar function of p. The underlying mixed formulation is

$$\begin{aligned} \mathbf{u} &= -(\mathbb{K}\,\nabla)p, \\ \frac{\partial p}{\partial t} + \operatorname{div}(k\,\mathbf{u}) &= b. \end{aligned} \tag{34}$$

The combined operators $\operatorname{div}(k\cdot)$ and $\mathbb{K}\,\nabla(\cdot)$ are dual to each other with respect to the weighted L^2 inner products ($k\mathbb{K}^{-1}$ is the weight):

$$\int_\Omega (\operatorname{div} k\mathbf{u})\, q\, \mathrm{d}x = -\int_\Omega k\,\mathbb{K}^{-1}\mathbf{u}\cdot(\mathbb{K}\,\nabla)\, q\, \mathrm{d}x \qquad \forall \mathbf{u} \in H(div,\, \Omega),\ \forall q \in H_0^1(\Omega). \tag{35}$$

Consider again the three-step construction of a mimetic scheme. In the *first* step we need to specify the degrees of freedom. For the pressure variable, we consider the discrete space $\mathcal{C}_h$ of grid functions that consist of one value per cell and the discrete space Λ_h of grid functions that consist of one value per face. The discrete space $\mathcal{F}_h$ has the same dimension as in the linear case, but the discrete fluxes in $\mathcal{F}_h$ obey a different continuity condition:

$$k_f^{c_1}\, u_f^{c_1} = k_f^{c_2}\, u_f^{c_2} \tag{36}$$

on each interior face f shared by cells c_1 and c_2. Here, $k_f^{c_1}$ and $k_f^{c_2}$ are accurate one-side approximations of the diffusion coefficient k. For example, in regions where function k is continuous, we can take $k_f^{c_1} = k_f^{c_2} = k_f$ as a weighted average of the cell-centered values $k(p_{c_1})$ and $k(p_{c_2})$ calculated using the most recent approximation to solution p in the cells c_1 and c_2, respectively. The weights are the distances between the cell centers and face f.

The primary mimetic operator approximates the combined operator div$(k\cdot)$. It is defined locally on each mesh cell using a straightforward discretization of the divergence theorem [compare with formula (17)]:

$$(\mathcal{DIV}^k\,\mathbf{u}_h)\big|_c \equiv \mathcal{DIV}_c^k\,\mathbf{u}_c = \frac{1}{|c|}\sum_{f\in\partial c}\sigma_{c,f}\,|f|\,k_f^c u_f^c. \tag{37}$$

Since $\mathbf{u}_h$ is an algebraic vector, it is convenient to think about the discrete divergence operator $\mathcal{DIV}^k\colon \mathcal{F}_h \to \mathcal{C}_h$ as a matrix acting between two spaces. This matrix has full rank when $k_f^c > 0$. The proof is based on Lemma 2.5 in [24].

Let us introduce a cell-based diagonal matrix $\mathcal{K}_c$ formed by coefficients k_f^c, $f \in \partial c$. Then, the primary mimetic operators in (17) and (37) can be connected as follows:

$$\mathcal{DIV}_c^k\,\mathbf{u}_c = (\mathcal{DIV}_c\,\mathcal{K}_c)\mathbf{u}_c.$$

The *second* step is to define the inner products in spaces $\mathcal{C}_h$ and $\mathcal{F}_h$ that are accurate approximations of the integrals in (35). Such inner products can be defined again cell-by-cell. Moreover, the inner product in space $\mathcal{C}_h$ can be defined as in Sect. 3.2 by the relation $\big[p_c,\,q_c\big]_{\mathcal{C}_{h,c}} = |c|\,p_c\,q_c$.

The weight in the other L^2 inner product is given by $k\,\mathbb{K}^{-1}$. By our assumption, $\mathbb{K}$ is a piecewise constant tensor on mesh Ω_h. Instead, the scalar coefficient k can be a quite general non-negative function. An acceptable first-order error is committed when we replace k by the piecewise constant function with value $k_c = k(p_c)$ in cell c.

The consistency condition is expressed through the exactness property:

$$\big[\mathbf{u}_c^I,\,\mathbf{v}_c^I\big]_{\mathcal{F}_{h,c}} = \int_c k(p_c)\,\mathbb{K}_c^{-1}\mathbf{u}\cdot\mathbf{v}\,\mathrm{d}x \qquad \forall\mathbf{u} \in (\mathcal{P}^0(c))^d,\ \forall\mathbf{v} \in \mathcal{SF}_c.$$

Since $k_c\,\mathbb{K}_c^{-1}$ is a constant tensor in cell c, the derivation of the inner product matrix (still denoted by $\mathsf{M}_{\mathcal{F},c}$) proceeds as in Sect. 3.2.

The *third* step is to obtain the formula for the derived operator, which, in this case, is an approximation of the combined operator $\mathbb{K}\,\nabla(\cdot)$. The continuum Green formula for cell c is given by

$$\int_c \operatorname{div}(k\mathbf{u})q\,dx - \int_{\partial c} (k\mathbf{u}\cdot\mathbf{n})\,q\,dx = -\int_c k\,\mathbb{K}^{-1}(\mathbb{K}\,\nabla q)\cdot\mathbf{u}\,dx. \tag{38}$$

Now, the derived operator $\widetilde{\mathcal{GRAD}}_c\colon \mathcal{C}_{h,c} \times \Lambda_{h,c} \rightarrow \mathcal{F}_{h,c}$ satisfies the discrete integration by parts formula

$$\left[\mathcal{DIV}_c^k\,\mathbf{u}_c,\, q_c\right]_{\mathcal{C}_{h,c}} - \sum_{f\in\partial c} \sigma_{c,f}\,|f|\,k_f^c\,u_f^c\lambda_f = -\left[\mathbf{u}_c,\, \widetilde{\mathcal{GRAD}}_c \begin{pmatrix} q_c \\ \boldsymbol{\lambda}_c \end{pmatrix}\right]_{\mathcal{F}_{h,c}}$$

for all $\mathbf{u}_c \in \mathcal{F}_{h,c}$, $q_c \in \mathcal{C}_{h,c}$, and $\boldsymbol{\lambda}_c \in \Lambda_{h,c}$. This local mimetic formulation gives the following formula for the physical fluxes:

$$\mathcal{K}_c \begin{pmatrix} u_{f_1}^c \\ \vdots \\ u_{f_{n_c}}^c \end{pmatrix} = -\mathcal{K}_c\,\widetilde{\mathcal{GRAD}}_c \begin{pmatrix} p_c \\ \boldsymbol{\lambda}_c \end{pmatrix} = \mathcal{K}_c \mathsf{M}_{\mathcal{F},c}^{-1}\mathcal{K}_c \begin{pmatrix} \sigma_{c,f_1}\,|f_1|(p_c-\lambda_{f_1}) \\ \vdots \\ \sigma_{c,f_{n_c}}\,|f_{n_c}|(p_c-\lambda_{f_{n_c}}) \end{pmatrix}. \tag{39}$$

Note that this formula uses the symmetric matrix $\mathcal{K}_c\mathsf{M}_{\mathcal{F},c}^{-1}\mathcal{K}_c$. Our numerical experiments in [45] show that the resulting scheme is second-order accurate.

4.2.2 Marshak Heat Equation

Let us consider the modified Marshak heat equation [48, 52] in the rectangular domain $(0,3)\times(0,1)$ with zero source term, $\mathbb{K} = \mathsf{I}$, and $k(p) = p^3$. The initial value is $p(\mathbf{x},0) = 10^{-3}$. We set the time-dependent Dirichlet boundary condition $p = p^0(0,t) = 0.78\,t^{1/3}$ on the left side of Ω, the constant boundary condition $p = 10^{-3}$ on the right side, and the homogeneous Neumann boundary conditions on the remaining sides.

We solve the parabolic equation on a randomly perturbed quadrilateral mesh that have three times more cells in the x-direction than in the y-direction. We use the backward Euler time integration scheme and the weighted arithmetic average definition of the face-based diffusion coefficients k_f described above. To reduce the impact of the time integration error, we use small time steps. Comparison of pictures in Fig. 4 show that the new scheme fixes the deficiencies of the old MFD scheme and leads to the correct speed of propagation of the non-linear wave. Significant mesh refinement is needed to get a reasonable solution with the old MFD scheme.

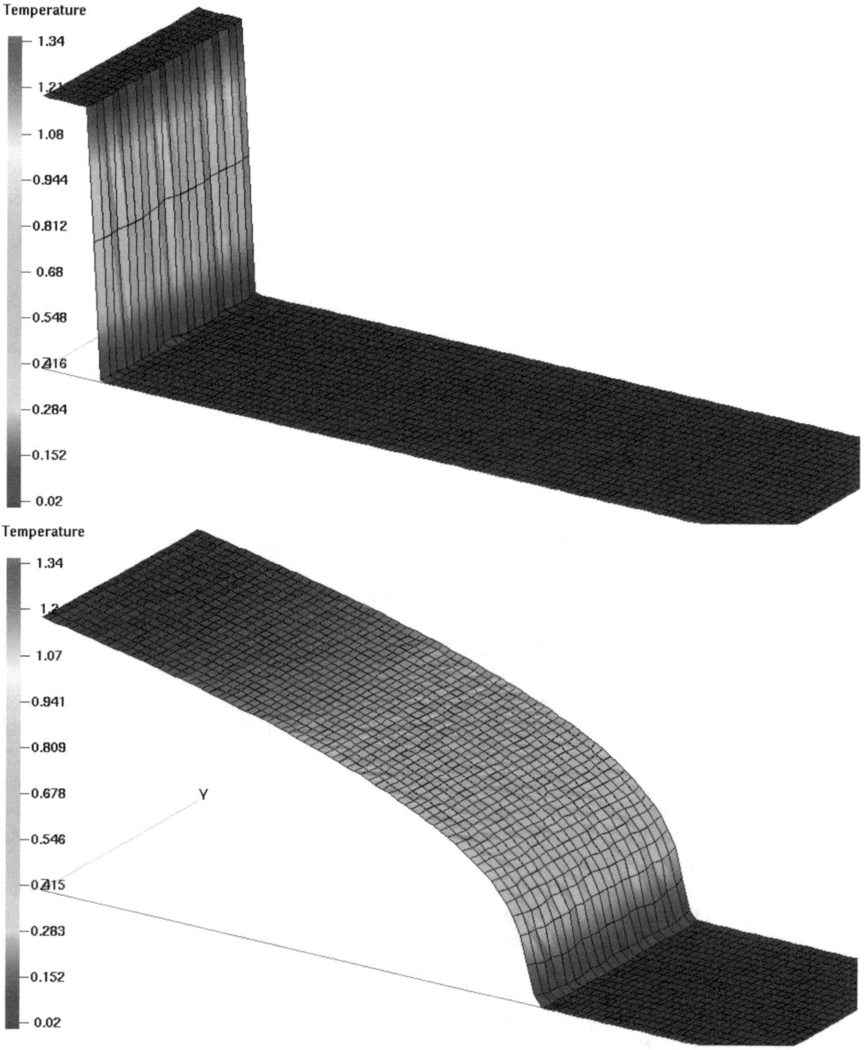

Fig. 4 Solution snapshots at time $t = 5.0$ for the standard (*top*) and new (*bottom*) MFD schemes. The solution in the *bottom panel* shows the correct and accurate position of the wave front at the chosen time

5 Conclusions

We described a few design principles used in the derivation of mimetic schemes for the numerical solution of PDEs. For diffusion equations, we established the bridges with a few FV and FE methods by showing that three popular discretization frameworks (MFD, FV and FE) use the equivalent design principles which leads to algebraically equivalent schemes. We illustrated the flexibility of the mimetic discretization framework to tackle challenging numerical issues in computer modeling of engineering problems with two examples: derivation of arbitrary-order schemes and convergent schemes for nonlinear problems with small diffusion coefficients.

Acknowledgements This work was carried out under the auspices of the National Nuclear Security Administration of the U.S. Department of Energy at Los Alamos National Laboratory under Contract No. DE-AC52-06NA25396. The authors acknowledge the support of the US Department of Energy Office of Science Advanced Scientific Computing Research (ASCR) Program in Applied Mathematics Research.

The meshes for the Marshak problem were created and managed using the mesh generation toolset MSTK (software.lanl.gov/MeshTools/trac) developed by Dr. Rao Garimella at Los Alamos National Laboratory.

References

1. P. Antonietti, L.B. da Veiga, M. Verani, A mimetic discretization of elliptic obstacle problems. Math. Comput. **82**, 1379–1400 (2013)
2. P.F. Antonietti, N. Bigoni, M. Verani, Mimetic discretizations of elliptic control problems. J. Sci. Comput. **56**(1), 14–27 (2013)
3. P.F. Antonietti, L.B. da Veiga, N. Bigoni, M. Verani, Mimetic finite differences for nonlinear and control problems. Math. Models Methods Appl. Sci. **24**(08), 1457–1493 (2014)
4. T. Arbogast, C. Dawson, P. Keenan, M. Wheeler, I. Yotov, Enhanced cell-centered finite differences for elliptic equations on general geometry. SIAM J. Sci. Comput. **19**(2), 404–425 (1998)
5. P. Bochev, J.M. Hyman, Principle of mimetic discretizations of differential operators, in *Compatible Discretizations. Proceedings of IMA Hot Topics Workshop on Compatible Discretizations*, ed. by D. Arnold, P. Bochev, R. Lehoucq, R. Nicolaides, M. Shashkov. IMA vol. 142 (Springer, Berlin, 2006)
6. J. Bonelle, Compatible discrete operator schemes on polyhedral meshes for elliptic and stokes equations. Ph.D. thesis. Technical report, University Paris-Est (2015)
7. J. Bonelle, A. Ern, Analysis of compatible discrete operator schemes for elliptic problems on polyhedral meshes. ESAIM Math. Model. Numer. Anal. **48**, 553–581 (2014)
8. F. Brezzi, K. Lipnikov, M. Shashkov, Convergence of the mimetic finite difference method for diffusion problems on polyhedral meshes. SIAM J. Numer. Anal. **43**(5), 1872–1896 (2005)
9. F. Brezzi, A. Buffa, K. Lipnikov, Mimetic finite differences for elliptic problems. Math. Model. Numer. Anal. **43**(2), 277–295 (2009)
10. F. Brezzi, A. Buffa, G. Manzini, Mimetic scalar products of discrete differential forms. J. Comput. Phys. **257**, Part B(0), 1228–1259 (2014)
11. F. Brezzi, R. Falk, L. Marini, Basic principle of mixed virtual element methods. ESAIM Math. Model. Numer. Anal. **48**(1), 1227–1240 (2014)
12. A. Cangiani, G. Manzini, Flux reconstruction and pressure post-processing in mimetic finite difference methods. Comput. Methods Appl. Mech. Eng. **197**(9–12), 933–945 (2008)
13. A. Cangiani, G. Manzini, A. Russo, Convergence analysis of the mimetic finite difference method for elliptic problems. SIAM J. Numer. Anal. **47**(4), 2612–2637 (2009)
14. A. Cangiani, F. Gardini, G. Manzini, Convergence of the mimetic finite difference method for eigenvalue problems in mixed form. Comput. Methods Appl. Mech. Eng. **200**(9–12), 1150–1160 (2011)

15. A. Cangiani, E. Georgoulis, P. Houston, *hp*-version discontinuous Galerkin methods on polygonal and polyhedral meshes. Math. Models Methods Appl. Sci. **24**(10), 2009–2041 (2014)
16. J. Castor, *Radiation Hydrodynamics* (Cambridge University Press, Cambridge, 2004)
17. B. Cockburn, J. Gopalakrishnan, R. Lazarov, Unified hybridization of discontinuous Galerkin, mixed, and continuous Galerkin methods for second order elliptic problems. SIAM J. Numer. Anal. **47**(2), 1319–1365 (2009)
18. Y. Coudière, G. Manzini, The discrete duality finite volume method for convection-diffusion problems. SIAM J. Numer. Anal. **47**(6), 4163–4192 (2010)
19. L.B. da Veiga, A mimetic finite difference method for linear elasticity. ESAIM Math. Model. Numer. Anal. **44**(2), 231–250 (2010)
20. L.B. da Veiga, D. Mora, A mimetic discretization of the Reissner–Mindlin plate bending problem. Numer. Math. **117**(3), 425–462 (2011)
21. L.B. da Veiga, V. Gyrya, K. Lipnikov, G. Manzini, Mimetic finite difference method for the Stokes problem on polygonal meshes. J. Comput. Phys. **228**(19), 7215–7232 (2009)
22. L.B. da Veiga, K. Lipnikov, G. Manzini, Error analysis for a mimetic discretization of the steady Stokes problem on polyhedral meshes. SIAM J. Numer. Anal. **48**(4), 1419–1443 (2010)
23. L.B. da Veiga, J. Droniou, G. Manzini, A unified approach to handle convection term in finite volumes and mimetic discretization methods for elliptic problems. IMA J. Numer. Anal. **31**(4), 1357–1401 (2011)
24. L.B. da Veiga, K. Lipnikov, G. Manzini, *The Mimetic Finite Difference Method*. Modeling, Simulations and Applications, vol. 11, 1st edn. (Springer, New York, 2014), 408 pp.
25. D.A. Di Pietro, A. Ern, *Mathematical Aspects of Discontinuous Galerkin Methods*. Mathématiques et Applications (Springer, Heidelberg, 2011)
26. D.A. Di Pietro, A. Ern, Hybrid high-order methods for variable diffusion problems on general meshes. C. R. Math. **353**, 31–34 (2014)
27. D. Di Pietro, A. Ern, S. Lemaire, An arbitrary-order and compact-stencil discretization of diffusion on general meshes based on local reconstruction operators. Comput. Methods Appl. Math. **14**(4), 461–472 (2014)
28. K. Domelevo, P. Omnes, A finite volume method for the Laplace equation on almost arbitrary two-dimensional grids. ESAIM Math. Model. Numer. Anal. **39**(6), 1203–1249 (2005)
29. J. Droniou, Finite volume schemes for diffusion equations: introduction to and review of modern methods. Math. Models Methods Appl. Sci. **24**(08), 1575–1619 (2014)
30. J. Droniou, R. Eymard, A mixed finite volume scheme for anisotropic diffusion problems on any grid. Numer. Math. **1**(105), 35–71 (2006)
31. J. Droniou, R. Eymard, Uniform-in-time convergence of numerical methods for non-linear degenerate parabolic equations. Numer. Math. **132**, 721–766 (2016)
32. J. Droniou, R. Eymard, T. Gallouët, R. Herbin, A unified approach to mimetic finite difference, hybrid finite volume and mixed finite volume methods. Math. Models Methods Appl. Sci. **20**(2), 265–295 (2010)
33. J. Droniou, R. Eymard, T. Gallouët, R. Herbin, Gradient schemes: a generic framework for the discretisation of linear, nonlinear and nonlocal elliptic and parabolic equations. Math. Models Methods Appl. Sci. **23**(13), 2395–2432 (2013)
34. G.M. Dusinberre, Heat transfer calculations by numerical methods. J. Am. Soc. Nav. Eng. **67**(4), 991–1002 (1955)
35. G.M. Dusinberre, *Heat-Transfer Calculation by Finite Differences* (International Textbook, Scranton, 1961)
36. R. Eymard, T. Gallouët, R. Herbin, Discretization of heterogeneous and anisotropic diffusion problems on general nonconforming meshes. SUSHI: a scheme using stabilization and hybrid interfaces. IMA J. Numer. Anal. **30**(4), 1009–1043 (2010)
37. V. Gyrya, K. Lipnikov, G. Manzini, The arbitrary order mixed mimetic finite difference method for the diffusion equation. Math. Model. Numer. Anal. (2015, accepted)
38. J. Hyman, M. Shashkov, The approximation of boundary conditions for mimetic finite difference methods. Comput. Math. Appl. **36**, 79–99 (1998)

39. J. Hyman, M. Shashkov, Mimetic discretizations for Maxwell's equations and the equations of magnetic diffusion. Prog. Electromagn. Res. **32**, 89–121 (2001)
40. J. Hyman, M. Shashkov, S. Steinberg, The numerical solution of diffusion problems in strongly heterogeneous non-isotropic materials. J. Comput. Phys. **132**(1), 130–148 (1997)
41. K. Lipnikov, J. Moulton, D. Svyatskiy, A multilevel multiscale mimetic (M^3) method for two-phase flows in porous media. J. Comput. Phys. **227**, 6727–6753 (2008)
42. K. Lipnikov, M. Shashkov, I. Yotov, Local flux mimetic finite difference methods. Numer. Math. **112**(1), 115–152 (2009)
43. K. Lipnikov, G. Manzini, F. Brezzi, A. Buffa, The mimetic finite difference method for 3D magnetostatics fields problems. J. Comput. Phys. **230**(2), 305–328 (2011)
44. K. Lipnikov, G. Manzini, M. Shashkov, Mimetic finite difference method. J. Comput. Phys. **257**, Part B(0), 1163–1227 (2014)
45. K. Lipnikov, G. Manzini, D. Moulton, M. Shashkov, The mimetic finite difference method for elliptic and parabolic problems with a staggered discretization of diffusion coefficient. J. Comput. Phys. **305**, 111–126 (2016)
46. G. Manzini, A. Russo, N. Sukumar, New perspectives on polygonal and polyhedral finite element methods. Math. Models Methods Appl. Sci. **24**(8), 1665–1699 (2014)
47. L. Margolin, M. Shashkov, P. Smolarkiewicz, A discrete operator calculus for finite difference approximations. Comput. Methods Appl. Mech. Eng. **187**(3–4), 365–383 (2000)
48. V.I. Maslyankin, Convergence of the iterative process for the quasilinear heat transfer equation. USSR Comput. Math. Math. Phys. **17**(1), 201–210 (1977)
49. A. Palha, P.P. Rebelo, R. Hiemstra, J. Kreeft, M. Gerritsma, Physics-compatible discretization techniques on single and dual grids, with application to the Poisson equation of volume forms. J. Comput. Phys. **257**, Part B(0), 1394–1422 (2014)
50. P. Raviart, J. Thomas, A mixed finite element method for second order elliptic problems, in *Mathematical Aspects of Finite Element Method*, ed. by I. Galligani, E. Magenes. Lecture Notes in Mathematics, vol. 606 (Springer, New York, 1977)
51. L. Richards, Capillary conduction of liquids through porous mediums. Physics 1 **5**, 318–333 (1931)
52. A. Samarskii, I. Sobol', Examples of the numerical calculation of temperature waves. USSR Comput. Math. Math. Phys. **3**(4), 945–970 (1963)
53. N. Sukumar, A. Tabarraei, Conforming polygonal finite elements. Int. J. Numer. Methods Eng. **61**(12), 2045–2066 (2004)
54. E. Wachspress, *A Rational Finite Element Basis* (Academic, New York, 1975)
55. J. Wang, X. Ye, A weak Galerkin mixed finite element method for second order elliptic problems. Math. Comput. **83**, 2101–2126 (2014)

Variational Multiscale Stabilization and the Exponential Decay of Fine-Scale Correctors

Daniel Peterseim

Abstract This paper reviews the variational multiscale stabilization of standard finite element methods for linear partial differential equations that exhibit multiscale features. The stabilization is of Petrov-Galerkin type with a standard finite element trial space and a problem-dependent test space based on pre-computed fine-scale correctors. The exponential decay of these correctors and their localisation to local cell problems is rigorously justified. The stabilization eliminates scale-dependent pre-asymptotic effects as they appear for standard finite element discretizations of highly oscillatory problems, e.g., the poor L^2 approximation in homogenization problems or the pollution effect in high-frequency acoustic scattering.

1 Introduction

In the past decades, the numerical analysis of partial differential equations (PDEs) was merely focused on the numerical approximation of sufficiently smooth solutions in the asymptotic regime of convergence. In the context of multiscale problems (and beyond), such results have only limited impact because the numerical approximation will hardly ever reach the asymptotic idealised regime under realistic conditions. Although a method performs well for sufficiently fine meshes it may fail completely on coarser (and feasible) scales of discretization. This happens for instance if the PDE exhibits rough and highly oscillatory solutions. Among the prominent applications are the numerical homogenization of elliptic boundary value problems with highly varying non-smooth diffusion coefficient, high-frequency time-harmonic acoustic wave propagation, and singularly perturbed problems such as convection-dominated flow.

The numerical approximation of such problems by finite element methods (FEMs) or related schemes is by no means straight-forward. The pure approximation

D. Peterseim (✉)
Institut für Numerische Simulation, Rheinische Friedrich-Wilhelms-Universität Bonn, Wegelerstr. 6, 53115 Bonn, Germany
e-mail: peterseim@ins.uni-bonn.de

G.R. Barrenechea et al. (eds.), *Building Bridges: Connections and Challenges in Modern Approaches to Numerical Partial Differential Equations*, Lecture Notes in Computational Science and Engineering 114, DOI 10.1007/978-3-319-41640-3_11

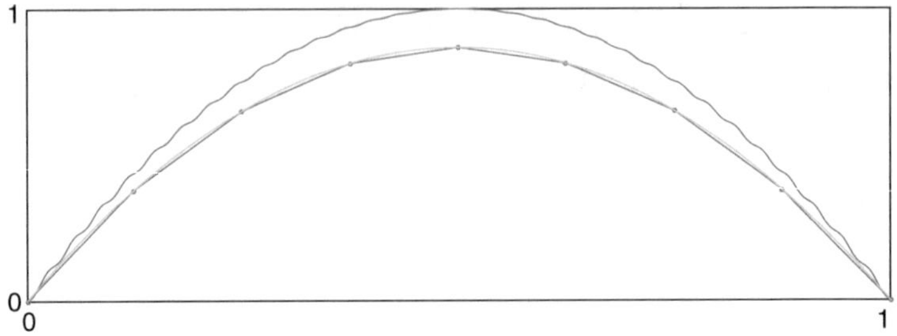

Fig. 1 Failure of FEM in homogenization problems: Consider the periodic problem $-\frac{d}{dx}A_\varepsilon(x)\frac{d}{dx}u_\varepsilon(x) = 1$ in the unit interval with homogeneous Dirichlet boundary condition, where $A_\varepsilon(x) := (2+\cos(2\pi x/\varepsilon))^{-1}$ for some small parameter $\varepsilon > 0$. The solution $u_\varepsilon = 4(x-x^2) - 4\varepsilon\left(\frac{1}{4\pi}\sin(2\pi\frac{x}{\varepsilon}) - \frac{1}{2\pi}x\sin(2\pi\frac{x}{\varepsilon}) - \frac{\varepsilon}{4\pi^2}\cos(2\pi\frac{x}{\varepsilon}) + \frac{\varepsilon}{4\pi^2}\right)$ is depicted in *blue* for $\varepsilon = 2^{-5}$. The P1-FE approximation (*red colored open circle*) on a uniform mesh of width h interpolates the curve $x \mapsto 2\sqrt{3}(x-x^2)$ whenever h is some multiple of the characteristic length scale ε and, hence, fails to approximate u_ε in any reasonable norm in the regime $h \geq \varepsilon$

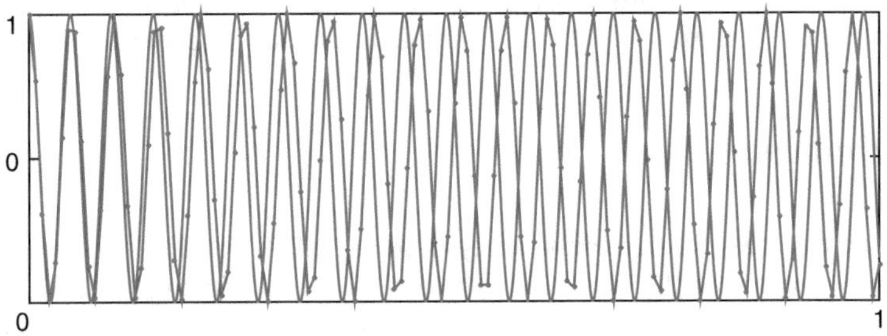

Fig. 2 Numerical dispersion in Helmholtz problems: Consider $-\frac{d^2}{dx^2}u_\varepsilon(x) - \kappa^2 u(x) = 0$ in the unit interval with $u(0) = 1$ and $\frac{d}{dx}u(1) = -i\kappa u(1)$ for some large parameter $\kappa > 0$. The solution $u_\kappa = \exp(-i\kappa x)$ is depicted in *blue* for $\kappa = 2^7$. The P1-FE approximation (*red colored open circle*) on a uniform mesh of width $h = 2^{-7} > 6 \cdot$ (wave length) fails to approximate u_κ due to the accumulation of phase errors

(e.g. interpolation) of the unknown solutions by finite elements already requires high spatial resolution to capture fast oscillations and heterogeneities on microscopic scales. When the function is described only implicitly as the solution of some partial differential equation, its approximation faces further scale-dependent pre-asymptotic effects caused by the under-resolution of relevant microscopic data. Examples are the poor L^2 approximation in homogenization problems (see Fig. 1) and the pollution effect [7] for Helmholtz problems with large wave numbers (see Fig. 2). We shall emphasise that, in the latter case, the existence and uniqueness of numerical approximations may not even be guaranteed in pre-asymptotic regimes.

Such situations require the stabilization of standard methods so that eventually a meaningful approximation on reasonably coarse scales of discretization becomes feasible. This paper aims to present a general framework for the stabilization of FEMs for multiscale problems with the aim to significantly reduce or even eliminate pre-asymptotic effects due to under-resolution. Our starting point will be the Variational Multiscale Method (VMS) originally introduced in [41, 43]. The method provides an abstract framework how to incorporate missing fine-scale effects into numerical problems governing coarse-scale behaviour [42]. One may interpret the VMS as a Petrov-Galerkin method using standard FE trial spaces and an operator-dependent test space that needs to be precomputed in general.

The construction of this operator-dependent test space is based on some stable projection onto the standard FE trial space and a corresponding scale decomposition of a function into its FE part given by the projection (the macroscopic/coarse-scale part) and a remainder that lies in the kernel of the projection operator (the microscopic/fine-scale part). The test functions are computed via a problem-dependent projection of the trial space into the space of fine-scale functions. This requires the solution of variational problems in the kernel of the projection—the fine-scale corrector problems. It has been observed empirically in certain applications that the Green's function associated with these fine-scale corrector problems—the so-called fine-scale Green's function [43]—may exhibit favourable exponential decay properties [43, 47] even though the decay of the classical full scale Green's function is only algebraic. It is this exponential decay property that allows one to turn the VMS into a feasible numerical method [44, 47].

Very recently, the exponential decay was rigorously proved for the first time in [50] in the context of multi-dimensional numerical homogenization. A key ingredient of the proof is the use of a (local) quasi-interpolation operator for the scale decomposition. Although the method presented in [50] still fits into the general framework of the VMS, it uses a different point of view on the method based on the orthogonalisation of coarse and fine scales with respect to the inner product associated with a symmetric and coercive model problem. This is why the method is now referred to as the Localized Orthogonal Decomposition (LOD) method. Subsequent work showed that these ideas can be generalised to other discretization techniques such as discontinuous Galerkin [21–23], Petrov-Galerkin formulations [25], mixed methods [33] and mesh-free methods [38]. Moreover, the method can also be reinterpreted in terms of the multiscale finite element method with special oversampling [35]. The class of problems that have been analysed by now includes semi-linear problems [36], high-contrast problems [9, 60], rough boundary conditions [34], problems on complicated geometries [24], linear and non-linear eigenvalue problems [37, 51, 52], parabolic problems [49], wave propagation [3, 10, 29, 59] and parametric problems [2].

This survey aims to reinterpret all those results in the abstract stabilization framework of the original VMS (Sect. 2) and aims to illustrate how the exponential decay of the fine-scale Green's function can be quantified (Sect. 3). We will show how these abstract results lead to super-localised numerical homogenization [35, 50] (Sect. 4) and pollution-free time-harmonic acoustic scattering (Sect. 5) [29, 59]. Section 6 contains some final remarks and also identifies methodological similarities and differences with some other numerical approaches that receive great attention these days, e.g., discontinuous Petrov-Galerkin methods (dPG) [18], Trefftz-type methods [31] and Isogeometric Analysis (IGA) [16].

2 Abstract Variational Multiscale Stabilization

This section is concerned with an abstract variational problem in a complex Hilbert space V as it appears for the weak formulation of second order PDEs. In this context, V is typically some closed subspace of the Sobolev space $W^{1,2}(\Omega;\mathbb{C}^m)$ for some bounded Lipschitz domain $\Omega \subset \mathbb{R}^d$. Let a denote a bounded sesquilinear form on $V \times V$ and let $F \in V'$ denote a bounded linear functional on V. We wish to find $u \in V$ satisfying the linear variational problem

$$\forall v \in V: \ a(u,v) = \overline{F(v)}. \tag{1}$$

We assume that the sesquilinear form a satisfies the inf-sup condition

$$\alpha := \inf_{0\neq v\in V} \sup_{0\neq w\in V} \frac{a(v,w)}{\|v\|_V \|w\|_V} = \inf_{0\neq w\in V} \sup_{0\neq v\in V} \frac{a(v,w)}{\|v\|_V \|w\|_V} > 0. \tag{2}$$

Under this condition, the abstract problem (1) is well-posed, i.e., for all $F \in V'$ there exists a unique solution $u \in V$ and the a priori bound

$$\|u\|_V \leq \alpha^{-1} \|F\|_{V'}$$

holds true; see, e.g., [4].

We wish to approximate the unknown solution u of (1) by some computable function. The standard procedure for approximation is the Galerkin method which simply chooses a finite-dimensional subspace $V_H \subset V$ (that contains simple functions such as piecewise polynomials) and restricts the variational problem (1) to this subspace. Usually, V_H belongs to some family of spaces parametrised by some abstract discretization parameter H, for instance the mesh size. This parameter (or set of parameters) provides some control on the approximation properties of V_H as $H \to 0$ at the price of an increasing computational cost in the sense of $\dim V_H \to \infty$. The Galerkin method seeks a function $G_H u \in V_H$ satisfying

$$\forall v_H \in V_H: \ a(G_H u, v_H) = \overline{F(v_H)} \ \big(= a(u, v_H) \big). \tag{3}$$

Recall that the well-posedness of the original problem (1) does not imply the well-posedness of the discrete variational problem (3) but needs to be checked for the particular application via discrete versions of the inf-sup condition (2). In many cases, such condition is only satisfied for H sufficiently small. This means that there is some threshold complexity for computing any Galerkin approximation and this threshold can be out of reach. Even if a Galerkin solution $G_H u$ exists and is computable, it might not provide the desired accuracy or does not reflect the relevant characteristic features of the solution, as we have seen in the introduction.

Therefore, we are interested in computing projections onto the discrete space V_H other than the Galerkin projection. Let $I_H : V \to V_H$ denote such a linear surjective projection operator and let us assume that it is bounded in the sense of the space $\mathcal{L}(V)$ of linear operators from V to V with finite operator norm

$$\|I_H\|_{\mathcal{L}(V)} := \sup_{0 \neq v \in V} \frac{\|I_H v\|_V}{\|v\|_V} < \infty.$$

Implicitly, we also assume that this operator norm does not depend on the discretization parameter H in a critical way. Possible choices of I_H include the orthogonal projection onto V_H with respect to the inner product of V or any Hilbert space $L \supset V$ containing V and mainly (local) quasi-interpolation operators of Clément or Scott-Zhang type as they are well-established in the finite element community in the context of a posteriori error estimation [11, 12, 14, 62].

2.1 *Petrov-Galerkin Characterisation of Finite Element Projections*

The Galerkin projection G_H is designed in such a way that its computation requires only the known data F associated with the unknown solution u. This section characterises the projection $I_H \in \mathcal{L}(V)$ as a Petrov-Galerkin discretization of (1) using V_H as the trial space and a non-standard test space $W_H \subset V$ that depends on the problem and the projection. The definition of W_H rests on the trivial observation that, for any $v \in V$,

$$a(I_H u, v) = \overline{F(v)} - a(u - I_H u, v). \tag{4}$$

The choice of a test function v in the subspace

$$W_H := \{w \in V \mid \forall z \in \operatorname{Ker} I_H : a(z, w) = 0\} \tag{5}$$

annihilates the second term on the right-hand side of (4) and, hence,

$$a(I_H u, w_H) = \overline{F(w_H)}$$

holds for all $w_H \in W_H$. This shows that $I_H u$ is a solution of the Petrov-Galerkin method: Find $u_H \in V_H$ such that

$$\forall w_H \in W_H : a(u_H, w_H) = \overline{F(w_H)}. \tag{6}$$

This characterisation of I_H is well known from the variational multiscale method as it is presented in [42].

The question whether or not (6) has a unique solution can not be answered under the general assumptions made so far. We need to assume the missing uniqueness to be able to proceed and one way of doing this is to assume that the dimensions of trial and test space are equal,

$$\dim W_H = \dim V_H. \tag{7}$$

In the present setting with a bounded operator I_H, this condition is equivalent to the well-posedness of the discrete variational problem (6), i.e., it admits a unique solution $u_H = I_H u \in V_H$ and

$$\|u_H\|_V \leq \|I_H\|_{\mathcal{L}(V)} \|u\|_V \leq \frac{\|I_H\|_{\mathcal{L}(V)}}{\alpha} \|F\|_{V'}.$$

The a priori estimate in turn implies a lower bound of the discrete inf-sup constant of the Petrov-Galerkin method by the quotient of the continuous inf-sup constant α and the continuity constant of I_H,

$$\inf_{0 \neq v_H \in V_H} \sup_{0 \neq w_H \in W_H} \frac{a(v_H, w_H)}{\|v_H\|_V \|w_H\|_V} \geq \frac{\alpha}{\|I_H\|_{\mathcal{L}(V)}} \leq \inf_{0 \neq w_H \in W_H} \sup_{0 \neq v_H \in V_H} \frac{a(v_H, w_H)}{\|v_H\|_V \|w_H\|_V}.$$

The test space W_H is the ideal test space for our purposes in the following sense. Assuming that we have access to it, the method (6) would enable us to compute $I_H u$ without the explicit knowledge of u. Although this will rarely be the case, we will see later that W_H can be approximated very efficiently in relevant cases. The discrete inf-sup conditions then indicate that the sufficiently accurate approximation of W_H will not harm the method, its stability properties or its subsequent error minimisation properties.

The continuity of the projection operator I_H readily implies the quasi-optimality of the Petrov-Galerkin method (6),

$$\|u - u_H\|_V = \|(1 - I_H)u\|_V \leq \|I_H\|_{\mathcal{L}(V)} \min_{v_H \in V_H} \|u - v_H\|_V. \tag{8}$$

Here, we have used that $\|I_H\|_{\mathcal{L}(V)} = \|1 - I_H\|_{\mathcal{L}(V)}$; see e.g. [63]. More importantly, the same arguments show that the Petrov-Galerkin method is quasi-optimal with respect to any other Hilbert space $L \supset V$ with norm $\|\cdot\|_L$ whenever $I_H \in \mathcal{L}(L)$,

$$\|u - u_H\|_L \leq \|I_H\|_{\mathcal{L}(L)} \min_{v_H \in V_H} \|u - v_H\|_L.$$

This quasi-optimality makes the ansatz very appealing and motivates its further investigation. Hence, in the remaining part of the paper, it is our aim to turn the method into a feasible numerical scheme while preserving these properties to a large extent. Although the discrete stability of the method depends on the stability properties of the original problem and, hence, on parameters such as the frequency in scattering problems, the quasi-optimality depends only on I_H and not necessarily on the problem.

2.2 Characterisation of the Ideal Test Space

A practical realisation of the Petrov-Galerkin method (6) requires a choice of bases in the discrete trial V_H and test space W_H. These choices will have big impact on the computational complexity. The underlying principle of finite elements is the locality of the bases which yields sparse linear systems and offers the possibility of linear computational complexity with respect to the number of degrees of freedom. Let $\{\lambda_j \mid j = 1, 2, \ldots, N_H = \dim V_H\}$ be such a local basis of V_H.

We shall derive a basis of the test space W_H defined in (5) by mapping the trial basis onto a test basis via some bijective operator $\mathcal{T}$, a so-called trial-to-test operator. Due to Assumption (7) such an operator exists, but there are many choices and we have to make a design decision. Our choice is that

$$I_H \circ \mathcal{T} = id \tag{9}$$

which is consistent with almost all existing practical realisations of the method but one might as well consider distance minimisation

$$\|(1 - \mathcal{T})v_H\|_V = \min_{w_H \in W_H} \|v_H - w_H\|_V.$$

The condition (9) fixes the (macroscopic) finite element part $I_H \mathcal{T} v_H = v_H$ of $\mathcal{T} v_H$ while the fine scale remainder $(1 - I_H)\mathcal{T} v_H$ is determined by the variational condition in the definition of W_H. Given $v_H \in V_H$, $(1 - I_H)\mathcal{T} v_H \in \operatorname{Ker} I_H$ satisfies

$$\forall z \in \operatorname{Ker} I_H : a(z, (1 - I_H)\mathcal{T} v_H) = -a(z, v_H). \tag{10}$$

This problem is referred to as the fine scale corrector problem for $v_H \in V_H$. Note that v_H can be replaced with any $v \in V$ so that $(1 - I_H)\mathcal{T}$ can be understood as an operator from V into $\operatorname{Ker} I_H$. We usually denote this operator the fine scale correction operator and write $\mathcal{C} := (1 - I_H)\mathcal{T}$. This operator is the Galerkin projection from V_H (or V) into $\operatorname{Ker} I_H$ related to the adjoint of the sesquilinear form a. It depends on the underlying variational problem and equips test functions with problem related features that are not present in V_H. In the context of elliptic PDEs, $\mathcal{C}$ is called the finescale Green's operator [41, 42].

For this construction to work we need to assume the well-posedness of the corrector problem (10), i.e., there is some constant $\beta > 0$ such that

$$\inf_{0 \neq v \in \operatorname{Ker} I_H} \sup_{0 \neq w \in \operatorname{Ker} I_H} \frac{a(v, w)}{\|v\|_V \|w\|_V} \geq \beta \leq \inf_{0 \neq w \in \operatorname{Ker} I_H} \sup_{0 \neq v \in \operatorname{Ker} I_H} \frac{a(v, w)}{\|v\|_V \|w\|_V}. \tag{11}$$

As for the Galerkin projection G_H onto V_H, these inf-sup conditions do not follow from their continuous counterparts (2) (unless a is coercive) and they might hold for sufficiently small H only. However, we were able to show in the context of the Helmholtz model problem of Sect. 5 that (11) holds in a much larger regime of the discretization parameter H than the corresponding conditions for the standard FEM do. In any case, condition (11) implies that the trial-to-test operator $\mathcal{T} = 1 + \mathcal{C}$ is a bounded linear projection operator from V to W_H with operator norm

$$\|\mathcal{T}\|_{\mathcal{L}(V)} = \|1 - \mathcal{T}\|_{\mathcal{L}(V)} = \|\mathcal{C}\|_{\mathcal{L}(V)} \leq \frac{C_a}{\beta},$$

where C_a denotes the continuity constant of the sesquilinear form a. Moreover, $\mathcal{T}|_{V_H} : V_H \to W_H$ is invertible with $(\mathcal{T}|_{V_H})^{-1} = I_H$ and $\{\mathcal{T}\lambda_j \mid j = 1, 2, \ldots, N_H\}$ defines a basis of W_H with

$$\frac{1}{\|I_H\|_{\mathcal{L}(V)}}\|\lambda_j\|_V \leq \|\mathcal{T}\lambda_j\|_V \leq \tfrac{C_a}{\beta}\|\lambda_j\|_V, \quad 1 \leq j \leq N_H.$$

In general, it cannot be expected (apart from one-dimensional exceptions where $\operatorname{Ker} I_H$ is a broken Sobolev space [42]) that the $\mathcal{T}\lambda_j$ have local support. On the contrary, their support will usually be global. However, we will show in the next section that they decay very fast in relevant applications; for illustrations see Sect. 4.

An important special case of the model problem (1) is the hermitian case. Note that hermiticity is preserved by the Petrov-Galerkin method in the following sense. For any $u_H, v_H \in V_H$, it holds that

$$a(u_H, \mathcal{T}v_H) = a(\mathcal{T}u_H, \mathcal{T}v_H) = \overline{a(\mathcal{T}v_H, \mathcal{T}u_H)} = \overline{a(v_H, \mathcal{T}u_H)}.$$

However, this hermiticity is typically lost once $\mathcal{T}$ is replaced with some approximation $\mathcal{T}_\ell$. In order to avoid a lack of hermiticity, previous papers such as [50] mainly used a variant of the method with W_H as the test and trial space. If hermiticity is important, one should follow this line. In this paper, we trade hermiticity for a cheaper method that avoids any costly communication between the fine-scale correctors that would be necessary in the hermitian version.

If the problem is non-hermitian, one might still consider a modified trial space based on the adjoint of $\mathcal{T}$ to improve approximation properties; see [45, 48, 59] for details. In a setting with a modified trial space, further generalisations are possible. Since V_H does not appear any more in the method, its conformity can be relaxed as it was recently proposed in [57] in the context of a multilevel solver for Poisson-type problems with L^∞ coefficients. This approach enables one to compute very general quantities of the solution such as piecewise mean values.

3 Exponential Decay of Fine-Scale Correctors

In many cases, the fine-scale correctors [i.e. the solutions of the fine-scale corrector problems (10)] have decay properties better than those of the Green's function associated with the underlying full-scale partial differential operator. To elaborate on this, we shall now assume that the space V is a closed subspace of $W^{1,2}(\Omega)$ with a local norm (the notation $\|\cdot\|_{V,\omega}$ means that the V-norm is restricted to some subdomain $\omega \subset \Omega$). Moreover, the sesquilinear form a is assumed to be local. This is the natural setting for scalar second order PDEs. The subsequent arguments can be easily generalised to vector-valued problems.

To be more precise regarding the locality of the basis mentioned above, we shall associate the basis functions of V_H with a set of geometric entities $\mathcal{N}_H$ called nodes (e.g. the vertices of a triangulation) and assume that these nodes are well distributed in the domain Ω in the sense of local quasi-uniformity. In this context, H refers to the maximal distance between nearest neighbours (the mesh size). Given some node $z \in \mathcal{N}_H$ and the corresponding basis function $\lambda_z \in V_H$, set the corrector $\phi_z = \mathcal{C}\lambda_z$ and recall from (10) that

$$a(w, \phi_z) = -a(w, \lambda_z)$$

for all $w \in \operatorname{Ker} I_H$.

We aim to show that there are constants $c > 0$ and $C > 0$ independent of H and R such that

$$\|\mathcal{C}\lambda_z\|_{V,\Omega\setminus B_R(z)} = \|\phi_z\|_{V,\Omega\setminus B_R(z)} \leq C\exp\left(-c\frac{R}{H}\right)\|\mathcal{C}\lambda_z\|_V, \tag{12}$$

where $B_R(z)$ denotes the ball of radius $R > 0$ centred at z.

We shall show how this result can be established and what kind of assumptions have to be made. Let $R > 2H$ and $r := R - H > H$ and let $\eta \in W^{1,\infty}(\Omega; [0, 1])$ be some cut-off function with $\eta = 0$ in $\Omega \setminus B_R(z)$, $\eta = 1$ in $B_r(z)$, and

$$\|\nabla\eta\|_{L^\infty(\Omega)} \leq C_\eta H^{-1} \tag{13}$$

for some generic constant C_η. In general, the fine-scale space $\operatorname{Ker} I_H$ is not closed under multiplication by a cut-off function and we will need to project the truncated function $\eta\phi_z$ back into $\operatorname{Ker} I_H$ by the operator $1 - I_H$. We assume that the concatenation of multiplication by η and $(1 - I_H)$ is stable and quasi-local in the sense that

$$\forall w \in \operatorname{Ker} I_H : \ \|(1 - I_H)(\eta w)\|_{V,B_R(z)\setminus B_r(z)} \leq C_{\eta,I_H}\|w\|_{V,B_{R'}(z)\setminus B_{r'}(z)} \tag{14}$$

holds with $r' := r - mH$ and $R' := R + mH$ and generic constants $C_{\eta,I_H} > 0$ and $m \in \mathbb{N}_0$ independent of H and z. Although the multiplication by η is not a stable operation in the full space V (think of a constant function), this result is possible in the space of fine scales for example if I_H enjoys quasi-local stability and approximation properties; see Sect. 4 below for an example. The quasi-locality of I_H is also used in the next argument.

Assuming that the inf-sup condition (11) holds, the corrector ϕ_z satisfies

$$\begin{aligned}\|\phi_z\|_{V,\Omega\setminus B_R(z)} &= \|(1-I_H)\phi_z\|_{V,\Omega\setminus B_R(z)}\\ &\leq \|(1-I_H)((1-\eta)\phi_z)\|_V\\ &\leq \beta^{-1}a(w,(1-I_H)((1-\eta)\phi_z))\\ &= \beta^{-1}\left(a(w,\phi_z)-a(w,(1-I_H)(\eta\phi_z))\right)\end{aligned}$$

for some $w \in \operatorname{Ker} I_H$ with $\|w\|_V = 1$. Since $\operatorname{supp}((1-I_H)((1-\eta)\phi_z)) \subset \Omega \setminus B_r(z)$ there is a good chance to actually find a function w with

$$\operatorname{supp} w \subset \operatorname{supp}((1-I_H)((1-\eta)\phi_z)) \subset \Omega \setminus B_r(z).$$

Of course, this is an assumption that needs to be verified in the particular application. Under this condition, the term $a(w,\phi_z) = a(w,\lambda_z)$ vanishes because the supports of w and λ_z have no overlap. This and (14) imply

$$\begin{aligned}\|\phi_z\|_{V,\Omega\setminus B_R(z)} &\leq \beta^{-1}C_aC_{\eta,I_H}\|\phi_z\|_{V,B_{R'}(z)\setminus B_{r'}(z)}\\ &= \beta^{-1}C_aC_{\eta,I_H}\left(\|\phi_z\|^2_{V,\Omega\setminus B_{r'}(z)} - \|\phi_z\|^2_{V,\Omega\setminus B_{R'}(z)}\right)^{1/2},\end{aligned}$$

where C_a denotes the continuity constant of the sesquilinear form a. Hence, the contraction

$$\|\phi_z\|^2_{V,\Omega\setminus B_{R'}(z)} \leq \frac{C'}{1+C'}\|\phi_z\|^2_{V,\Omega\setminus B_{R'-(2m+1)H}(z)}$$

holds with $C' := (\beta^{-1}C_aC_{\eta,I_H})^2$. The iterative application of this estimate with $R' \mapsto R' - (2m+1)H$ plus relabelling $R' \mapsto R$ leads to the conjectured decay result (12) with constants $C := (\frac{C'}{1+C'})^{-\frac{1}{2(2m+1)}}$ and $c := \left|\log(\frac{C'}{1+C'})\right| \frac{(1)}{2(2m+1)} > 0$.

The exponential decay motivates and justifies the localisation of the fine-scale corrector problems to local subdomains of diameter ℓH where $\ell \in \mathbb{N}$ is a new discretization parameter, the so-called oversampling parameter. It controls the perturbation with respect to the ideal global correctors. We will explain this localisation procedure on the basis of an example in Sect. 4 below. As a rule of thumb, the localisation to subdomains of diameter ℓH will introduce an error of order $\mathcal{O}(\exp(-\ell))$. As long as this error is small when compared with the inf-sup constant $\alpha\|I_H\|^{-1}_{\mathcal{L}(V)}$ of the ideal method, the stability and approximation properties of the method will be largely preserved.

4 Application to Numerical Homogenization of Elliptic PDEs

The first prototypical model problem concerns the diffusion problem $-\operatorname{div} A\nabla u = f$ in some bounded domain $\Omega \subset \mathbb{R}^d$ with homogeneous Dirichlet boundary condition. The difficulty is the strongly heterogeneous and highly varying (non-periodic) diffusion coefficient A. The heterogeneities and oscillations of the coefficient may appear on several non-separable scales. We assume that the diffusion matrix $A \in L^\infty\left(\Omega, \mathbb{R}^{d\times d}_{\mathrm{sym}}\right)$ is symmetric and uniformly elliptic with

$$0 < \alpha = \operatorname*{ess\,inf}_{x\in\Omega} \inf_{v\in\mathbb{R}^d\setminus\{0\}} \frac{(A(x)v)\cdot v}{v\cdot v}.$$

Given $f \in L^2(\Omega)$, we wish to find the unique weak solution $u \in V := H^1_0(\Omega)$ such that

$$a(u,v) := \int_\Omega (A\nabla u)\cdot\nabla v = \int_\Omega fv =: F(v) \quad \text{for all } v \in V. \tag{15}$$

It is well known that classical polynomial based FEMs can perform arbitrarily badly for such problems, see e.g. [6]. This is due to the fact that finite elements tend to average unresolved scales of the coefficient and the theory of homogenization shows that this way of averaging does not lead to meaningful macroscopic approximations. This is illustrated in the introduction. In the simple periodic example of Fig. 1, the averaging of the inverse of the diffusion coefficient A (harmonic averaging) would have lead to the correct macroscopic representation.

In computational homogenization, the impact of unresolved microstructures encoded in the rough coefficient A on the overall process is taken into account by the solution of local microscopic cell problems. While many approaches are empirically successful and robust for certain multiscale problems, the question whether such methods are stable and accurate beyond the strong assumptions of analytical homogenization regarding scale separation or even periodicity remained open for a long time. Only recently, the existence of an optimal approximation of the low-regularity solution space by some arbitrarily coarse generalised finite element space (that represents the homogenised problem) was shown in [5, 32]. However, the constructions therein include prohibitively expensive global solutions of the full fine scale problem or the solution of more involved eigenvalue problems. The first efficient and feasible construction, solely based on the solution of localised microscopic cell problems, was given and rigorously justified in [50] and later optimised and generalised in [35, 38]. A different approach with presumably similar properties was later suggested by Owhadi et al. [58] along with the notion of sparse super-localisation that reflects the locality of the discrete homogenised operator (similar to the sparsity of standard finite element matrices).

We shall now explain how the abstract theory of the previous sections is related to the LOD method [50] and its variants. Let $\mathcal{G}_H$ denote some regular (in the sense of

Ciarlet) finite element mesh into closed simplices and let $V_H := P_1(\mathcal{G}_H) \cap V$ denote the space of continuous functions that are affine when restricted to any element $T \in \mathcal{G}_H$. Let $I_H : V \to V_H$ be a quasi-interpolation operator that acts as a stable quasi-local projection in the sense that $I_H \circ I_H = I_H$ and that for any $T \in \mathcal{G}_H$ and all $v \in V$ there holds

$$H^{-1}\|v - I_H v\|_{L^2(T)} + \|I_H v\|_{V,T} \leq C_{I_H}\|\nabla v\|_{V,\Omega_T}, \tag{16}$$

where Ω_T refers to some neighbourhood of T (typically the union of T and the adjacent elements) and $\|\cdot\|_V := \|\nabla \cdot\|_{L^2(\Omega)}$. One possible choice (among many others) is to define $I_H := E_H \circ \Pi_H$, where Π_H is the piecewise L^2 projection onto $P^1(\mathcal{G}_H)$ and E_H is the averaging operator that maps $P_1(\mathcal{G}_H)$ to V_H by assigning to each interior vertex the arithmetic mean of the corresponding function values of the adjacent elements, that is, for any $v \in P_1(\mathcal{G}_H)$ and any free vertex $z \in \mathcal{N}_H$,

$$(E_H(v))(z) = \frac{1}{\operatorname{card}\{K \in \mathcal{G}_H \,:\, z \in K\}} \sum_{T \in \mathcal{G}_H : z \in T} v|_T(z).$$

For this choice, the proof of (16) follows from combining the well-established approximation and stability properties of Π_H and E_H, see for example [20]. The choice of I_H in [35, 50] was slightly different. Therein, the $L^2(\Omega)$-orthogonal projection onto V_H played the role of I_H. Since this a non-local operator, the analysis was based on the fact that the local quasi-interpolation operator in [12, Sect. 6] has the same kernel and, hence, induces the same method.

Following the recipe of Sect. 2.1 and taking into account the present setting with an inner product a, the ideal test space $W_H := (\operatorname{Ker} I_H)^{\perp_a}$ is simply the orthogonal complement (w.r.t. a) of the fine scale functions $\operatorname{Ker} I_H$.

Given the nodal basis of V_H, a basis of W_H is computed by means of the trial-to-test operator $\mathcal{T} = 1 + \mathcal{C}$, where

$$\forall w \in \operatorname{Ker} I_H : \; a(\mathcal{C}\lambda_z, w) = -a(\lambda_z, w). \tag{17}$$

It is easily checked that the assumptions made in Sect. 3 are satisfied in the present setting. In particular, formula (14) holds with $C_{\eta,I_H} = C_{I_H}(C_{I_H}C_\eta + 1)$ and $m = 2$. This follows from the product rule, (13), and the local approximation and stability properties (16) of I_H. This implies the exponential decay as it is stated in (12) with constants C and c independent of variations of the diffusion coefficient A. An example of a corrector and a test basis function are depicted in Fig. 3 to demonstrate the exponential decay.

We truncate the computational domain of the corrector problems to local subdomains of diameter ℓH roughly. We have not yet described how to do this in practice. The obvious way would be to simply replace Ω in (17) with suitable neighbourhoods of the nodes z. This procedure was used in [50]. However, it

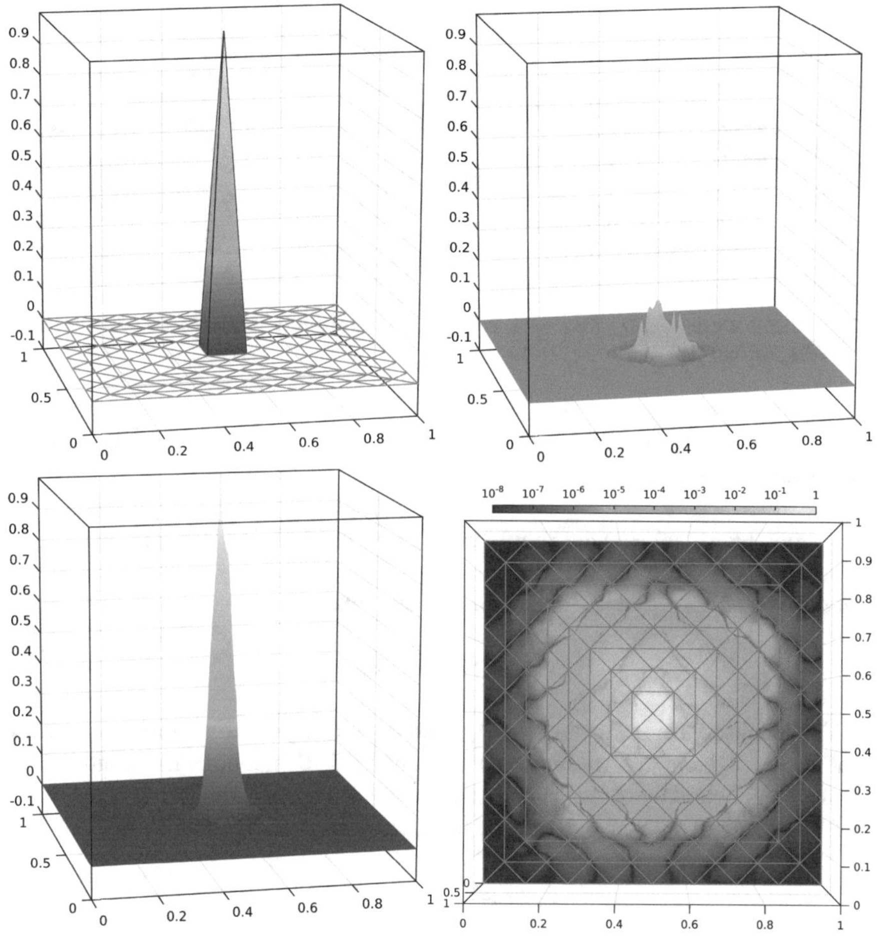

Fig. 3 Standard nodal basis function λ_z with respect to the coarse mesh $\mathcal{G}_H$ (*top left*), corresponding ideal corrector $\phi_z = \mathcal{C}\lambda_z$ (*top right*), and corresponding test basis function $\mathcal{T}\lambda_z = (1 + \mathcal{C})\lambda_z$ (*bottom left*). The *bottom right* figure shows a top view on the modulus of test basis function $\mathcal{T}\lambda_z = (1 + \mathcal{C})\lambda_z$ with logarithmic *color scale* to illustrate the exponential decay property. The underlying rough diffusion coefficient A is depicted in Fig. 6

turned out that it is advantageous to consider the following slightly more involved technique based on element correctors [35, 38].

We assign to any $T \in \mathcal{G}_H$ its ℓth order element patch $\Omega_{T,\ell}$ for a positive integer ℓ; see Fig. 4 for an illustration. Moreover, we define for all $v, w \in V$ and $\omega \subset \Omega$ the localised bilinear forms

$$a_\omega(v, w) := \int_\omega (A\nabla v) \cdot \nabla w.$$

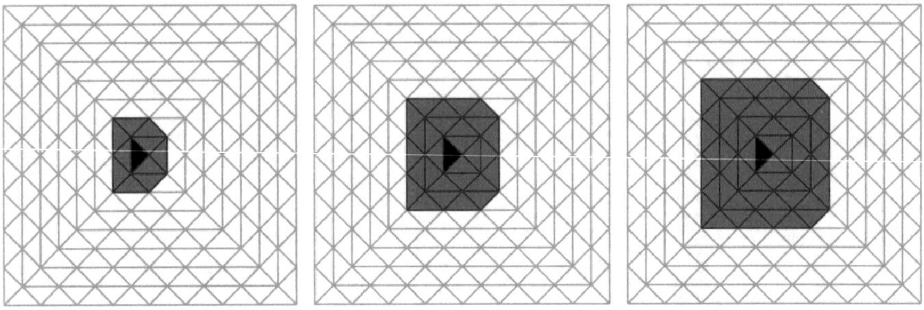

Fig. 4 Element patches $\Omega_{T,\ell}$ for $\ell = 1, 2, 3$ (from *left* to *right*) as they are used in the localised corrector problem (18)

Given any nodal basis function $\lambda_z \in V_H$, let $\phi_{z,\ell,T} \in \operatorname{Ker} I_H \cap W_0^{1,2}(\Omega_{T,\ell})$ solve the subscale corrector problem

$$a_{\Omega_{T,\ell}}(\phi_{z,\ell,T}, w) = -a_T(\lambda_z, w) \quad \text{for all } w \in \operatorname{Ker} I_H \cap W_0^{1,2}(\Omega_{T,\ell}). \tag{18}$$

Let $\phi_{z,\ell} := \sum_{T \in \mathcal{G}_H : z \in T} \phi_{z,\ell,T}$ and define the test function

$$\Lambda_{z,\ell} := \lambda_z + \phi_{z,\ell}.$$

The localised test basis function $\Lambda_{z,\ell}$ and the underlying correctors $\phi_{z,\ell,T}$ can be seen in Fig. 5. Note that we impose homogeneous Dirichlet boundary condition on the artificial boundary of the patch which is well justified by the fast decay.

More generally, we may define the localised correction operator $\mathcal{C}_\ell$ by

$$\mathcal{C}_\ell v_H := \sum_{z \in \mathcal{N}_H} v_H(z) \phi_{z,\ell}$$

as well as the localised trial-to-test operator

$$\mathcal{T}_\ell v_H := 1 + \mathcal{C}_\ell v_H = \sum_{z \in \mathcal{N}_H} v_H(z) \Lambda_{z,\ell}.$$

The space of test functions then reads

$$W_H^\ell := \mathcal{T}_\ell V_H = \operatorname{span}\{\Lambda_{z,\ell} \,:\, z \in \mathcal{N}_H\}$$

and the (localised) multiscale Petrov-Galerkin FEM seeks $u_{H,\ell} \in V_H$ such that

$$a(u_{H,\ell}, w_{H,\ell}) = (f, w_{H,\ell})_{L^2(\Omega)} \quad \text{for all } w_{H,\ell} \in W_{H,\ell}. \tag{19}$$

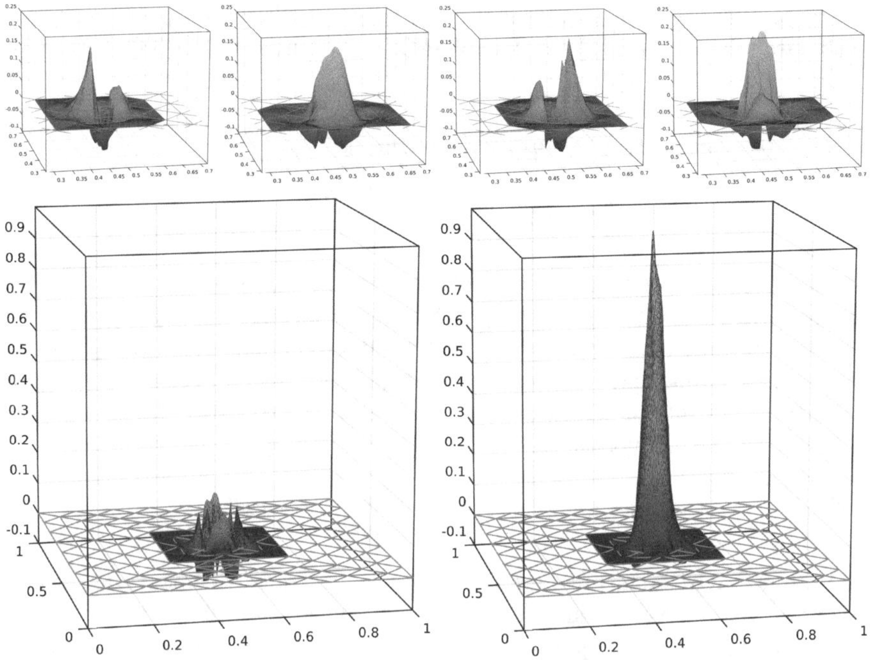

Fig. 5 Localised element correctors $\phi_{z,\ell,T}$ for $\ell = 2$ and all four elements T adjacent to the vertex $z = [0.5, 0.5]$ (*top*), localised nodal corrector $\phi_{z,\ell} = \mathcal{C}_\ell \lambda_z = \sum_{T \ni z} \phi_{z,\ell,T}$ (*bottom left*) and corresponding test basis function $\Lambda_{z,\ell} = \mathcal{T}_\ell \lambda_z = (1 + \mathcal{C}_\ell)\lambda_z$ (*bottom right*). The underlying rough diffusion coefficient is depicted in Fig. 6. The computations have been performed by standard linear finite elements on local fine meshes of with $h = 2^{-8}$. See Fig. 3 for a comparison with the ideal global corrector and basis

In previous papers [35, 38, 50] we have considered the symmetric version with $W_{H,\ell}$ as trial and test space and also the reverse version with $W_{H,\ell}$ as the trial space and V_H as test space [25]. All these methods are essentially equal in the ideal case and there are no major changes in the output after localisation (when only the V_H part of the discrete solution is considered). When it comes to implementation and computational complexity, the present Petrov-Galerkin version has the advantage that there is no communication between the correctors. This means that the fine-scale solutions of the corrector problems need not to be stored but only their interaction with the $\mathcal{O}(\ell^d)$ standard nodal basis functions in their patches; see also [25] for further discussions regarding those technical details.

The error analysis of the localised method follows similar arguments. Let $u_H \in V_H$ be the ideal Petrov-Galerkin approximation and let $e_H := u_H - u_{H,\ell} \in V_H$ denote the error with respect to the ideal method. Then there exists some $z_H \in W_H$ with $\|z_H\|_V = 1$ such that

$$\frac{\alpha}{\|I_H\|_{\mathcal{L}(V)}} \|e_H\|_V \leq a(e_H, z_H) = a(u_{H,\ell} - u, z_H - z_{H,\ell}),$$

where $z_{H,\ell} \in W_{H,\ell}$. The exponential decay property allows one to choose $z_{H,\ell}$ in such a way that $\|z_H - z_{H,\ell}\|_V \leq \tilde{C}\exp(-c\ell)$; see for instance [35, 38]. This shows that

$$\begin{aligned}\|u - u_{H,\ell}\|_V &\leq \|u - u_H\|_V + \|u_H - u_{H,\ell}\|_V\\ &\leq \|u - u_H\|_V + \frac{\|I_H\|_{\mathcal{L}(V)}C_a}{\alpha}\tilde{C}\exp(-c\ell)\|u - u_{H,\ell}\|_V.\end{aligned}$$

We shall emphasise that, in the present context, the constants $\tilde{C}$ and c are independent of variations of the rough diffusion tensor but they may depend on the contrast (the ratio between the global upper and lower bound of A). Using (8), this shows that the moderate choice $\ell \geq |\log(\alpha/(2\|I_H\|_{\mathcal{L}(V)}C_a\tilde{C}))|/c = \mathcal{O}(1)$ implies the quasi-optimality (and also the well-posedness) of the Petrov-Galerkin method with respect to the V-norm

$$\|u - u_{H,\ell}\|_V \leq 2\|I_H\|_{\mathcal{L}(V)} \min_{v_H \in V_H} \|u - v_H\|_V.$$

With regard to the fact that the V-best approximation may be poor and standard Galerkin would have provided us with an even better estimate at lower cost, this result is maybe not very impressive. Let us see if we can do something similar for the L^2-norm which appears to be the relevant measure in the context of homogenization problems. A standard duality argument shows that

$$\|e_H\|^2_{L^2(\Omega)} = a(e_H, z_H) = a(u - u_{H,\ell}, z_H - z_{H,\ell})$$

for some $z_H \in W_H$ with $\|z_H\|_V \leq C_3\alpha^{-1}\|I_H\|_{\mathcal{L}(V)}\|e_H\|_{L^2(\Omega)}$ and $z_{H,\ell} := \mathcal{T}_\ell I_H z_H \in W_{H,\ell}$. Similar arguments as before yield

$$\|u - u_{H,\ell}\|_{L^2(\Omega)} \leq C_1 \min_{v_H \in V_H} \|u - v_H\|_{L^2(\Omega)} + C_2\exp(-c\ell) \min_{v_H \in V_H} \|u - v_H\|_V,$$

where $C_1 := \|I_H\|_{\mathcal{L}(L^2(\Omega))}$ and $C_2 := C_a\tilde{C}C_3\alpha^{-1}\|I_H\|_{\mathcal{L}(V)}$. This shows that the method is accurate also in the L^2-norm regardless of the regularity of u. If the oversampling parameter is chosen such that $\ell \gtrsim \log H$, then the method is $\mathcal{O}(H)$ accurate in $L^2(\Omega)$ with no pre-asymptotic phenomena. This is the best worst-case rate one can expect for general $f \in V'$ and $A \in L^\infty$.

Note that the previous results hold true for general L^∞-coefficients and all constants are independent of the variations of the diffusion tensor as far as the contrast remains moderately bounded. In particular, the approach is by no means restricted to periodic coefficients or scale separation. For a more detailed discussion of high-contrast problems in this context we refer to [60].

The final step towards a fully practical method is the discretization of the fine-scale corrector problems. With regard to the possible low regularity of the solution, P_1 finite elements on a refined mesh $\mathcal{G}_h$ appears reasonable, but any other type of

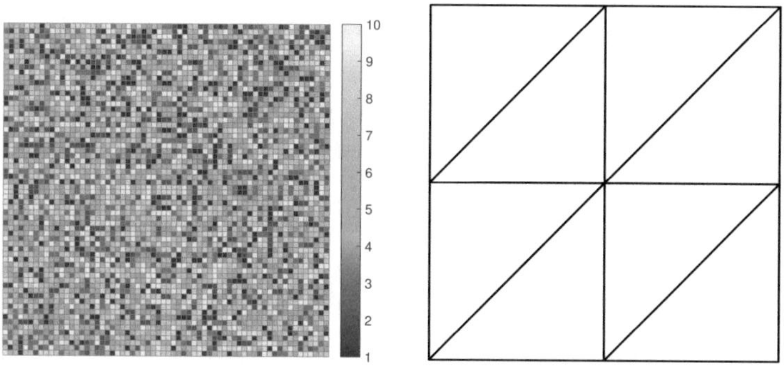

Fig. 6 Diffusion coefficient in the numerical experiment of Sect. 4 and coarsest mesh

discretisation is possible. Obviously, the fine-scale discretization parameter h has to be chosen fine enough to resolve all relevant features of the diffusion coefficient. The previous theory can be transferred to this case in a straight-forward way and we refer to [34, 35, 50] for the technical details.

To illustrate the previous estimates, we close this section with a numerical experiment. Let Ω be the unit square and the outer force $f \equiv 1$ in Ω. Consider the coefficient A that is piecewise constant with respect to a uniform Cartesian grid of width 2^{-6}. Its values are randomly chosen between 1 and 10; see Fig. 6. Consider uniform coarse meshes $\mathcal{G}_H$ of size $H = 2^{-1}, 2^{-2}, \ldots, 2^{-5}$ of Ω that certainly do not resolve the rough coefficient A appropriately. The reference mesh $\mathcal{G}_h$ has width $h = 2^{-9}$. Since no analytical solutions are available, the standard finite element approximation $u_h \in V_h$ on the reference mesh $\mathcal{G}_h$ serves as the reference solution. Doing this, we assume that u_h is sufficiently accurate and, necessarily, that $\mathcal{G}_h$ resolves the discontinuities of A. The corrector problems are also are also solved on this scale of numerical resolution.

The numerical results, i.e. errors with respect to the reference solution u_h are depicted in Fig. 7. The results are in agreement with the theoretical results. They are even better in the sense that $\ell = 1$ seems to be sufficient for quasi-optimality (with respect to u_h) in the present setup and parameter regime. We expect that the true errors with respect to u would behave similar in the beginning but level off at some point when the reference error starts to dominate the upscaling error. Still, the experiment clearly indicates that numerical homogenization is possible for very general L^∞-coefficients.

We refer to [3, 9, 23, 25, 34–38, 50, 51] for many more numerical experiments for several model problems including nonlinear stationary and non-stationary problems as well as eigenvalue problems.

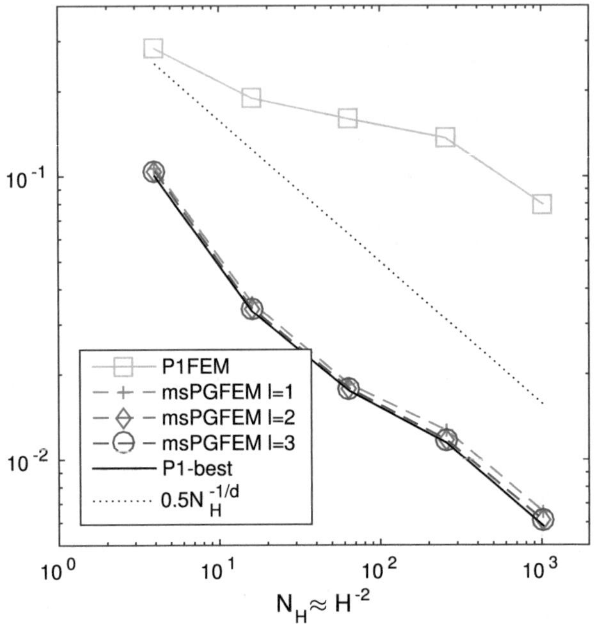

Fig. 7 Numerical experiment of Sect. 4. Relative L^2-errors of multiscale Petrov-Galerkin FEM (19) versus the number of degrees of freedom $N_H \approx H^{-2}$, where $H = 2^{-1}, \dots, 2^{-5}$ is the uniform coarse mesh size. The localisation parameter varies between $\ell = 1, \dots, 3$. The P_1-FE solution and the best-approximation in the P_1-FE space on the same coarse meshes are depicted for comparison

5 Application to High-Frequency Acoustic Scattering

This section will show that the abstract framework of Sects. 2–3 is indeed applicable beyond the coercive and symmetric model problem of the previous section. We consider the scattering of acoustic waves at a sound-soft scatterer modelled by the Helmholtz equation over a bounded Lipschitz domain $\Omega \subset \mathbb{R}^d$ ($d = 1, 2, 3$),

$$-\Delta u - \kappa^2 u = f \quad \text{in } \Omega, \tag{20a}$$

along with mixed boundary conditions of Dirichlet and Robin type

$$u = 0 \quad \text{on } \Gamma_D, \tag{20b}$$

$$\nabla u \cdot \nu - i\kappa u = 0 \quad \text{on } \Gamma_R. \tag{20c}$$

Here, the wave number $\kappa \gg 1$ is real and positive, i denotes the imaginary unit and $f \in L^2(\Omega, \mathbb{C})$. We assume that the boundary $\Gamma := \partial\Omega$ consists of two components

$$\partial\Omega = \overline{\Gamma_D \cup \Gamma_R}, \quad \overline{\Gamma}_D \cap \overline{\Gamma}_R = \emptyset$$

where Γ_D encloses the scatterer and Γ_R is an artificial truncation of the whole unbounded space. The vector ν denotes the unit normal vector that is outgoing from Ω.

Given $f \in L^2(\Omega, \mathbb{C})$, we wish to find $u \in V := \{v \in W^{1,2}(\Omega, \mathbb{C}) \mid v = 0 \text{ on } \Gamma_D\}$ such that, for all $v \in V$,

$$a(u, v) := \int_\Omega \nabla u \cdot \nabla \bar{v} - \kappa^2 \int_\Omega u\bar{v} - i\kappa \int_{\Gamma_R} u\bar{v} = \int_\Omega f\bar{v} =: \overline{F(w)}. \tag{21}$$

The space V is equipped with the usual κ-weighted norm $\|v\|_V^2 := \kappa^2 \|v\|_{L^2(\Omega)}^2 + \|\nabla v\|_{L^2(\Omega)}^2$. The presence of the Robin boundary condition (20c) ensures that this variational problem is well-posed in the sense of (2) with $\alpha = 1/C_{\mathrm{st}}(\kappa)$ for some κ-dependent stability constant $C_{\mathrm{st}}(\kappa)$; see e.g. [26]. The dependence on the wave number κ is not known in general. An exponential growth with respect to the wave number is possible [8] in non-generic domains. In most cases, the growth seems to be only polynomially, although this is an empirical rather than a theoretical statement, and sufficient geometric conditions for this to hold are rare [17, 26, 46, 53]. For the above scattering problem, we know that $C_{\mathrm{st}}(\kappa) \leq \mathcal{O}(\kappa)$ if Ω is convex and if the scatterer is star-shaped [39].

It is this κ-dependence in the stability of the problem that makes the numerical approximation by FEM or related schemes extremely difficult in the regime of large wave numbers. Any perturbation of the problem, e.g. by some discretization, can be amplified by $C_{\mathrm{st}}(\kappa)$. We have seen in the introduction that this is indeed observed in practice and causes a pre-asymptotic effect known as the pollution effect or numerical dispersion [7]. This effect puts very restrictive assumptions on the smallness of the underlying mesh that is much stronger than the minimal requirement for a meaningful representation of highly oscillatory functions from approximation theory, that is, to have at least 5–10 degrees of freedom per wave length and coordinate direction.

It is the aim of many newly developed methods to overcome or at least to reduce the pollution effect; see [19, 27, 28, 40, 64, 66] among many others. However, the only theoretical results regard high-order FEMs with the polynomial degree p coupled to the wave number κ via the relation $p \approx \log \kappa$ [26, 54–56]. Under this moderate assumption, those methods are stable and quasi-optimal in the regime $H\kappa/p \lesssim 1$ for certain model Helmholtz problems.

The multiscale method in [59] then showed that pollution in the numerical approximation of the Helmholtz problem can also be cured for a fairly large class of Helmholtz problems, including the acoustic scattering from convex non-smooth objects, by stabilization in the present framework. If the data of the problem (domain, boundary condition, force term) allows for polynomial-in-κ bounds of $C_{\mathrm{st}}(\kappa)$ and if the resolution condition $H\kappa \lesssim 1$ and the oversampling condition $\log(\kappa)/\ell \lesssim 1$ are satisfied, then the method is stable and quasi-optimal in the V-norm.

The recent paper [29] interprets the method of [59] in the present framework and we recall it here very briefly. Given the same discrete setup as in the previous section with some simplicial mesh $\mathcal{G}_H$, corresponding P_1 FE space $V_H := P_1(\mathcal{G}_H) \cap V$, and quasi-interpolation operator $I_H : V \to V_H$, the multiscale Petrov-Galerkin method is formally defined in the same way. We simply replace the inner product of Sect. 4 with the sesquilinear form a of this section.

Given any nodal basis function $\lambda_z \in V_H$, we construct a corresponding test basis function $\Lambda_{z,\ell}$ by the same procedure as in the previous section, $\Lambda_{z,\ell} := \lambda_z + \phi_{z,\ell}$, where $\phi_{z,\ell} := \sum_{T\in\mathcal{G}_H:z\in T} \phi_{z,T}$ and $\phi_{z,T}$ solves the cell problem

$$a_{\Omega_{T,\ell}}(w, \phi_{z,T}) = -a_T(w, \lambda_z) \quad \text{for all } w \in \operatorname{Ker} I_H \text{ with } \operatorname{supp} w \subset \bar{\Omega}_T.$$

Here,

$$a_\omega(u, v) := \int_{\Omega\cap\omega} \nabla u \cdot \nabla \bar{v} - \kappa^2 \int_{\Omega\cap\omega} u\bar{v} - i\kappa \int_{\Gamma_R\cap\partial\omega} u\bar{v}$$

for $\omega \in \{\Omega_{T,\ell}, T\}$. Note that the corrector problem inherits the boundary condition from the original problem is the patch boundary coincides with the boundary of Ω. On the part of the patch boundary that falls in the interior of Ω, we simply put the homogeneous Dirichlet condition. A major observation is that this corrector problem is well-posed and, in particular, coercive with $\beta = 1/3$ under the condition $H\kappa \le c_{\text{res}}$ for some given constant $0 < c_{\text{res}} = \mathcal{O}(1)$ that only depends on the constant in (16) but not on H or κ. This is because a satisfies a Gårding inequality and fine-scale functions satisfy $\|w\|_{L^2(\Omega)} \le C_{I_H} H \|\nabla w\|_{L^2(\Omega)}$. This coercivity also implies the desired exponential decay of the ideal correctors so that the choice $\Omega_{T,\ell}$ is well justified. This can also be observed in Fig. 8.

The space of localised test functions then reads $W_{H,\ell} := \operatorname{span}\{\Lambda_{z,\ell} : z \in \mathcal{N}_H\}$ and the multiscale Petrov-Galerkin FEM seeks $u_{H,\ell} \in V_H$ such that

$$a(u_{H,\ell}, w_{H,\ell}) = \overline{F(w_{H,\ell})} \quad \text{for all } w_{H,\ell} \in W_{H,\ell}. \tag{22}$$

The quasi-optimality result of the previous section is easily transferred to the present setting. The resolution condition $H\kappa \le c_{\text{res}}$ and the oversampling condition

$$\ell \ge |\log(\alpha/(2\|I_H\|_{\mathcal{L}(V)} C_a \tilde{C}))|/c = \mathcal{O}(\log C_{\text{st}}(\kappa))$$

imply the quasi-optimality (and stability) of the multiscale Petrov-Galerkin method with respect to the V-norm

$$\|u - u_{H,\ell}\|_V \le 2\|I_H\|_{\mathcal{L}(V)} \min_{v_H\in V_H} \|u - v_H\|_V.$$

Here, the constants c and $\tilde{C}$ are related to the exponential decay of the test basis [cf. (12)] and they are independent of κ under the resolution condition. We shall

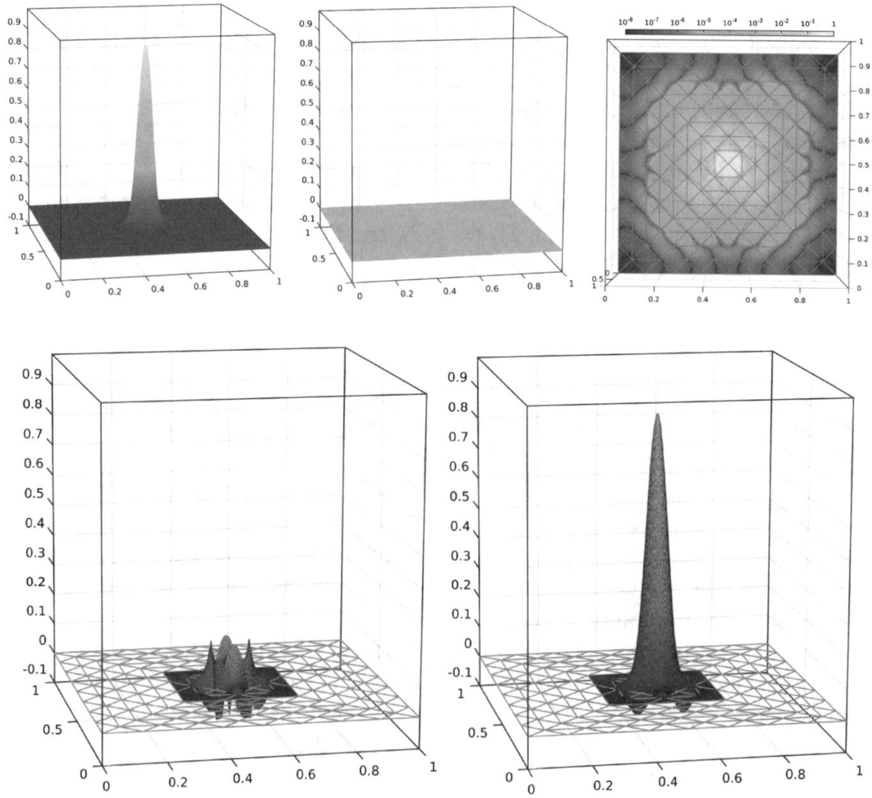

Fig. 8 Real and imaginary part of the ideal corrector $\mathcal{C}\lambda_z$ (*top left* and *middle*). The *top right* figure shows a top view on the modulus of test basis function $\mathcal{T}\lambda_z = (1 + \mathcal{C})\lambda_z$ with logarithmic *color scale* to illustrate the exponential decay property. The underlying computational domain is the unit square with a Robin boundary condition everywhere. The wave number $\kappa = 2^4$ is chosen such that the resolution condition on the coarse mesh is just satisfied. The localised nodal corrector $\phi_{z,\ell} = \mathcal{C}_\ell\lambda_z$ (*bottom left*) and corresponding test basis function $\Lambda_{z,\ell} = \mathcal{T}\lambda_z$ (*bottom right*) are real-valued because the patch boundary doesn't touch the domain boundary. The local fine meshes used in the computation have width $h = 2^{-8}$

emphasise that such a best-approximation property does not hold for standard FEMs which require e.g. $\kappa^2 H \lesssim 1$ for quasi-optimality [53] in the case of pure Robin boundary conditions on a convex planar domain. The FEM approximation is not even known to exist unless $\kappa^{3/2} H \lesssim 1$ in the simplest model problem without a scatterer [65].

For the multiscale Petrov-Galerkin method, the result means that pollution effects do not occur. Note that the resolution condition $H\kappa \leq c_{\text{res}}$ is somewhat minimal, because any meaningful approximation of the highly oscillatory solution of (20) requires at least 5–10 degrees of freedom per wave length and coordinate direction. Saying this, we assume that the fine scale corrector problems are solved sufficiently accurate; see [29, 59] for details.

We shall present a numerical experiment taken from [59] where this version of the method was already considered experimentally. Consider the scattering from sound-soft scatterer occupying the triangle Ω_D. The Sommerfeld radiation condition of the scattered wave is approximated by the Robin boundary condition on the boundary $\Gamma_R := \partial\Omega_R$ of the unit square so that $\Omega := (0,1)^2 \setminus \Omega_D$ is the computational domain; see Fig. 9. Given the wave number $\kappa = 2^7$, the incident wave $u_{inc}(x) := \exp(i\kappa\, x \cdot [\cos(0.5), \sin(0.5)]^T)$ is prescribed via an inhomogeneous Dirichlet boundary condition on $\Gamma_D := \partial\Omega_D$ and the scattered wave satisfies (20a) with $f \equiv 0$ and the boundary conditions

$$u = -u_{inc} \quad \text{on } \Gamma_D,$$

$$\nabla u \cdot \nu - i\kappa u = 0 \quad \text{on } \Gamma_R.$$

The error analysis extends to this setting in a straight-forward way.

We choose uniform coarse meshes of widths $H = 2^{-3}, \ldots, 2^{-7}$ as depicted in Fig. 9. The reference mesh $\mathcal{G}_h$ is derived by uniform mesh refinement of the coarse meshes and has mesh width $h = 2^{-9}$. The corresponding P_1 conforming finite element approximation on the reference mesh $\mathcal{G}_h$ is denoted by V_h. As in the previous section, we compare the coarse scale approximations $u_{H,\ell,h} \in V_H$ with some reference solution $u_h \in V_h$.

Figure 10 depicts the results for the multiscale Petrov-Galerkin method and shows that the pollution effect that is present in the P_1 FEM is eliminated when ℓ is moderately increased. For the present wave number $\ell = 2$ is sufficient.

Further numerical experiments are reported in [29, 59]. It is worth noting that the latter work also exploits the homogeneous structure of the PDE coefficients in the sense that only very few of the fine-scale corrector problems are actually solved due to translation invariance and symmetry. This makes the approach competitive.

A very natural and straight forward generalisation of the method would be the case of heterogeneous media. The previous section plus the analysis of this section strongly indicate the potential of the method to treat high oscillations or jumps in the PDE coefficients and the pollution effect in one stroke [10].

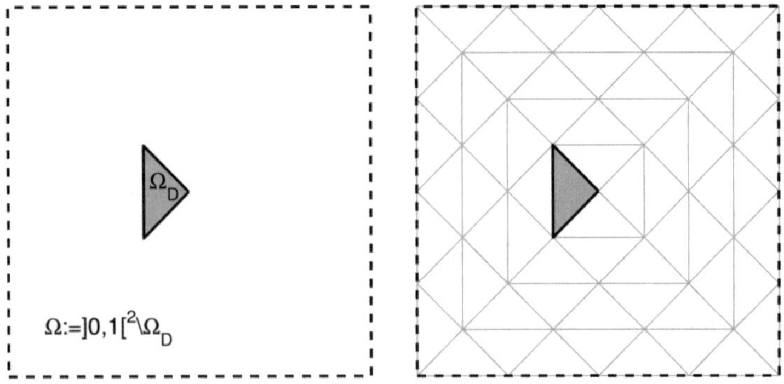

Fig. 9 Computational domain of the model problem of Sect. 5 and coarsest mesh

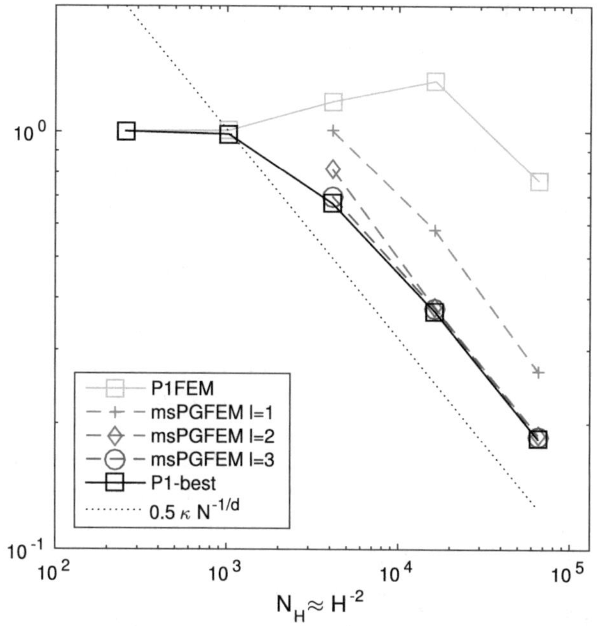

Fig. 10 Numerical experiment of Sect. 5: Relative V-norm errors of multiscale Petrov-Galerkin method (22) with wave number $\kappa = 2^7$ depending on the number of degrees of freedom $N_H \approx H^{-2}$, where $H = 2^{-5}, \ldots, 2^{-7}$ is the uniform coarse mesh size. The reference mesh size $h = 2^{-9}$ remains fixed. The oversampling parameter ℓ varies between 1 and 3. The P_1-FE solution and the best-approximation in the P_1-FE space on the same coarse meshes are depicted for comparison

6 Final Remarks

We have presented an abstract framework for the stabilization of numerical methods for multiscale partial differential equations with some focus on highly oscillatory problems. The methodology is based on the variational multiscale method and the more recent development of localised orthogonal decompositions. We have provided an abstract numerical analysis of the method which is applied to two representative model problems, a homogenization problem and a scattering problem. We have shown that the methodology can indeed eliminate critical scale-dependent pre-asymptotic effects in these cases. While the framework has already been applied successfully to other problem classes such as linear and non-linear eigenvalue problems, we expect that the framework will also be useful for convection-dominated flow, the problem that the variational method was initially designed for.

The multiscale method presented in this paper is shown to be stable and accurate under moderate assumptions on the discretization parameters relative to characteristic parameters and length scales of the problem. These valuable properties require the pre-computation of the test basis on subgrids. These pre-computations are both local and independent, but the worst-case (serial) complexity of the method can exceed the cost of a direct numerical simulation on a global sufficiently fine mesh. If the inherent parallelism of the local cell problems cannot be exploited during the computation, we still expect a significant gain with respect to computational complexity if the pre-computation can be reused several times in the context of parameter studies, coupled problems, optimal control problems or inverse problems. In many cases, there is also a lot of redundancy in the local problems which allows one to reduce the number of local problems drastically as it is shown in [29] in the context of acoustic scattering. We expect that this technique can be generalised to far more general situations using modern techniques of model order reduction [1, 61].

We may close the discussion with some rather philosophical remark regarding the stabilization of FEMs and their inter-element continuity properties. Presently, it is believed, e.g. in the context of time-harmonic wave propagation, that stability can be increased by relaxing inter-element continuity within a discontinuous Galerkin (DG) framework. The large number of variants includes the ultra weak variational formulation [13], Trefftz methods [40], dPG [19, 66], or the continuous interior penalty method [65]. There may be some truth in this but the general impression that relaxing continuity is the only way is certainly false as one can observe from the method presented in this paper. The multiscale Petrov-Galerkin does quite the opposite. The regularity of the test functions is increased compared to standard continuous finite elements, because they are solutions of second-order elliptic problems (at least in the ideal case). In general, test functions $w_H \in W_H$ have the property that $\operatorname{div} A\nabla w_H \in L^2(\Omega)$. In the context of the Helmholtz model problem of Sect. 5 where $A = 1$ this means that $\Delta W_H \subset L^2(\Omega)$. If Ω is convex and boundary conditions are appropriate, then $W_H \subset W^{2,2}(\Omega)$ (this can be observed for one basis function in Fig. 8). In this respect, our methodology clearly indicates that increased differentiability might as well lead to increased stability and accuracy. Similar effects have been observed for eigenvalue computations in IGA [15, 30] and also LOD [51]. This shows that breaking the inter-element continuity is not at all necessary for stability.

Acknowledgements The present work is the result of many fruitful collaborations over the past 4 years [9, 23, 29, 35–38, 50, 51, 60]. I would like to thank all my co-authors, in particular Axel Målqvist, Patrick Henning, and Dietmar Gallistl.

The author gratefully acknowledges support by the Hausdorff Center for Mathematics Bonn and by Deutsche Forschungsgemeinschaft in the Priority Program 1748 "Reliable simulation techniques in solid mechanics. Development of non-standard discretization methods, mechanical and mathematical analysis" under the project "Adaptive isogeometric modeling of propagating strong discontinuities in heterogeneous materials".

References

1. A. Abdulle, Y. Bai, Reduced-order modelling numerical homogenization. Philos. Trans. R. Soc. Lond. Ser. A Math. Phys. Eng. Sci. **372**(2021), 20130388, 23 (2014)
2. A. Abdulle, P. Henning, A reduced basis localized orthogonal decomposition. J. Comput. Phys. **295**, 379–401 (2015)

3. A. Abdulle, P. Henning, Localized orthogonal decomposition method for the wave equation with a continuum of scales. Math. Comput. (2015). doi:10.1090/mcom/3114
4. I. Babuška, Error-bounds for finite element method. Numer. Math. **16**, 322–333 (1970/1971)
5. I. Babuska, R. Lipton, Optimal local approximation spaces for generalized finite element methods with application to multiscale problems. Multiscale Model. Simul. **9**(1), 373–406 (2011)
6. I. Babuška, J.E. Osborn, Can a finite element method perform arbitrarily badly? Math. Comput. **69**(230), 443–462 (2000)
7. I.M. Babuška, S.A. Sauter, Is the pollution effect of the FEM avoidable for the Helmholtz equation considering high wave numbers? SIAM Rev. **42**(3), 451–484 (electronic) (2000)
8. T. Betcke, S.N. Chandler-Wilde, I.G. Graham, S. Langdon, M. Lindner, Condition number estimates for combined potential integral operators in acoustics and their boundary element discretisation. Numer. Methods Partial Differential Equations **27**(1), 31–69 (2011)
9. D. Brown, D. Peterseim, A multiscale method for porous microstructures (2014). ArXiv e-prints 1411.1944
10. D. Brown, D. Gallistl, D. Peterseim, Multiscale Petrov-Galerkin method for high-frequency heterogeneous Helmholtz equations (2015). ArXiv e-prints 1511.09244
11. C. Carstensen, Quasi-interpolation and a posteriori error analysis in finite element methods. M2AN Math. Model. Numer. Anal. **33**(6), 1187–1202 (1999)
12. C. Carstensen, R. Verfürth, Edge residuals dominate a posteriori error estimates for low order finite element methods. SIAM J. Numer. Anal. **36**(5), 1571–1587 (1999)
13. O. Cessenat, B. Despres, Application of an ultra weak variational formulation of elliptic PDEs to the two-dimensional Helmholtz problem. SIAM J. Numer. Anal. **35**(1), 255–299 (1998)
14. P. Clément, Approximation by finite element functions using local regularization. Rev. Française Automat. Informat. Recherche Opérationnelle Sér. RAIRO Analyse Numérique **9**(R-2), 77–84 (1975)
15. J.A. Cottrell, A. Reali, Y. Bazilevs, T.J.R. Hughes, Isogeometric analysis of structural vibrations. Comput. Methods Appl. Mech. Eng. **195**(41–43), 5257–5296 (2006)
16. J.A. Cottrell, T.J.R. Hughes, Y. Bazilevs, *Isogeometric Analysis: Toward Integration of CAD and FEA* (Wiley, New York, 2009)
17. P. Cummings, X. Feng, Sharp regularity coefficient estimates for complex-valued acoustic and elastic Helmholtz equations. Math. Models Methods Appl. Sci. **16**(1), 139–160 (2006)
18. L. Demkowicz, J. Gopalakrishnan, A class of discontinuous Petrov-Galerkin methods. II. Optimal test functions. Numer. Methods Partial Differential Equations **27**(1), 70–105 (2011)
19. L. Demkowicz, J. Gopalakrishnan, I. Muga, J. Zitelli, Wavenumber explicit analysis of a DPG method for the multidimensional Helmholtz equation. Comput. Methods Appl. Mech. Eng. **213/216**, 126–138 (2012)
20. D.A. Di Pietro, A. Ern, *Mathematical Aspects of Discontinuous Galerkin Methods*. Mathématiques & Applications, vol. 69 (Springer, Heidelberg, 2012)
21. D. Elfverson, A discontinuous Galerkin multiscale method for convection-diffusion problems (2015). ArXiv e-prints 1509.03523
22. D. Elfverson, E.H. Georgoulis, A. Målqvist, An adaptive discontinuous Galerkin multiscale method for elliptic problems. Multiscale Model. Simul. **11**(3), 747–765 (2013)
23. D. Elfverson, E.H. Georgoulis, A. Målqvist, D. Peterseim, Convergence of a discontinuous Galerkin multiscale method. SIAM J. Numer. Anal. **51**(6), 3351–3372 (2013)
24. D. Elfverson, M.G. Larson, A. Målqvist, Multiscale methods for problems with complex geometry (2015). ArXiv e-prints arXiv:1509.03991
25. D. Elfverson, V. Ginting, P. Henning, On multiscale methods in Petrov-Galerkin formulation. Numer. Math. **131**(4), 643–682 (2015). doi:10.1007/s00211-015-0703-z
26. S. Esterhazy, J.M. Melenk, On stability of discretizations of the Helmholtz equation, in *Numerical Analysis of Multiscale Problems*. Lecture Notes in Computer Science and Engineering, vol. 83 (Springer, Heidelberg, 2012), pp. 285–324
27. X. Feng, H. Wu, Discontinuous Galerkin methods for the Helmholtz equation with large wave number. SIAM J. Numer. Anal. **47**(4), 2872–2896 (2009)

28. X. Feng, H. Wu, *hp*-discontinuous Galerkin methods for the Helmholtz equation with large wave number. Math. Comput. **80**(276), 1997–2024 (2011)
29. D. Gallistl, D. Peterseim, Stable multiscale Petrov-Galerkin finite element method for high frequency acoustic scattering. Comput. Methods Appl. Mech. Eng. **295**, 1–17 (2015)
30. D. Gallistl, P. Huber, D. Peterseim, On the stability of the Rayleigh-Ritz method for eigenvalues (2015). INS Preprint 1527
31. C.J. Gittelson, R. Hiptmair, I. Perugia, Plane wave discontinuous galerkin methods: analysis of the h-version. ESAIM: Math. Model. Numer. Anal. **43**(2), 297–331, 2 (2009)
32. L. Grasedyck, I. Greff, S. Sauter, The AL basis for the solution of elliptic problems in heterogeneous media. Multiscale Model. Simul. **10**(1), 245–258 (2012)
33. F. Hellman, P. Henning, A. Målqvist, Multiscale mixed finite elements (2015). arXiv Preprint 1501.05526
34. P. Henning, A. Målqvist, Localized orthogonal decomposition techniques for boundary value problems. SIAM J. Sci. Comput. **36**(4), A1609–A1634 (2014)
35. P. Henning, D. Peterseim, Oversampling for the multiscale finite element method. Multiscale Model. Simul. **11**(4), 1149–1175 (2013)
36. P. Henning, A. Målqvist, D. Peterseim, A localized orthogonal decomposition method for semi-linear elliptic problems. ESAIM: Math. Model. Numer. Anal. **48**, 1331–1349 (2014)
37. P. Henning, A. Målqvist, D. Peterseim, Two-level discretization techniques for ground state computations of Bose-Einstein condensates. SIAM J. Numer. Anal. **52**(4), 1525–1550 (2014)
38. P. Henning, P. Morgenstern, D. Peterseim, Multiscale partition of unity, in *Meshfree Methods for Partial Differential Equations VII*, ed. by M. Griebel, M.A. Schweitzer. Lecture Notes in Computational Science and Engineering, vol. 100 (Springer International Publishing, Cham, 2015), pp. 185–204
39. U. Hetmaniuk, Stability estimates for a class of Helmholtz problems. Commun. Math. Sci. **5**(3), 665–678 (2007)
40. R. Hiptmair, A. Moiola, I. Perugia, Plane wave discontinuous Galerkin methods for the 2D Helmholtz equation: analysis of the *p*-version. SIAM J. Numer. Anal. **49**(1), 264–284 (2011)
41. T.J.R. Hughes, Multiscale phenomena: Green's functions, the Dirichlet-to-Neumann formulation, subgrid scale models, bubbles and the origins of stabilized methods. Comput. Methods Appl. Mech. Eng. **127**(1–4), 387–401 (1995)
42. T. Hughes, G. Sangalli, Variational multiscale analysis: the fine-scale Green's function, projection, optimization, localization, and stabilized methods. SIAM J. Numer. Anal. **45**(2), 539–557 (2007)
43. T.J.R. Hughes, G.R. Feijóo, L. Mazzei, J.-B. Quincy, The variational multiscale method—a paradigm for computational mechanics. Comput. Methods Appl. Mech. Eng. **166**(1–2), 3–24 (1998)
44. M.G. Larson, A. Målqvist, Adaptive variational multiscale methods based on a posteriori error estimation: energy norm estimates for elliptic problems. Comput. Methods Appl. Mech. Eng. **196**(21–24), 2313–2324 (2007)
45. M.G. Larson, A. Målqvist, A mixed adaptive variational multiscale method with applications in oil reservoir simulation. Math. Models Methods Appl. Sci. **19**(07), 1017–1042 (2009)
46. C. Makridakis, F. Ihlenburg, I. Babuška, Analysis and finite element methods for a fluid-solid interaction problem in one dimension. Math. Models Methods Appl. Sci. **06**(08), 1119–141 (1996)
47. A. Målqvist, Adaptive variational multiscale methods. Ph.D. Thesis, Chalmers Tekniska Högskola, Sweden (2005)
48. A. Målqvist, Multiscale methods for elliptic problems. Multiscale Model. Simul. **9**, 1064–1086 (2011)
49. A. Målqvist, A. Persson, Multiscale techniques for parabolic equations. ArXiv e-prints, 1504.08140 (2015)
50. A. Målqvist, D. Peterseim, Localization of elliptic multiscale problems. Math. Comput. **83**(290), 2583–2603 (2014)

51. A. Målqvist, D. Peterseim, Computation of eigenvalues by numerical upscaling. Numer. Math. **130**(2), 337–361 (2015)
52. A. Målqvist, D. Peterseim, Generalized finite element methods for quadratic eigenvalue problems (2016). ESAIM: M2AN. doi:10.1051/m2an/2016019
53. J.M. Melenk, *On Generalized Finite-Element Methods* (ProQuest LLC, Ann Arbor, MI, 1995). Ph.D. thesis, University of Maryland, College Park
54. J.M. Melenk, S.A. Sauter, Convergence analysis for finite element discretizations of the Helmholtz equation with Dirichlet-to-Neumann boundary conditions. Math. Comput. **79**(272), 1871–1914 (2010)
55. J.M. Melenk, S. Sauter, Wavenumber explicit convergence analysis for Galerkin discretizations of the Helmholtz equation. SIAM J. Numer. Anal. **49**(3), 1210–1243 (2011)
56. J.M. Melenk, A. Parsania, S. Sauter, General DG-methods for highly indefinite Helmholtz problems. J. Sci. Comput. **57**(3), 536–581 (2013)
57. H. Owhadi, Multigrid with rough coefficients and multiresolution operator decomposition from hierarchical information games (2015). ArXiv e-prints, 1503.03467
58. H. Owhadi, L. Zhang, L. Berlyand, Polyharmonic homogenization, rough polyharmonic splines and sparse super-localization. ESAIM: Math. Model. Numer. Anal. **48**(2), 517–552 (2014)
59. D. Peterseim, Eliminating the pollution effect in Helmholtz problems by local subscale correction. ArXiv e-prints, 1411.1944 (2014)
60. D. Peterseim, R. Scheichl, Rigorous numerical upscaling at high contrast. Comput. Methods Appl. Math. doi:10.1515/mcom-2016-0022 (2016)
61. G. Rozza, D.B.P. Huynh, A.T. Patera, Reduced basis approximation and a posteriori error estimation for affinely parametrized elliptic coercive partial differential equations: application to transport and continuum mechanics. Arch. Comput. Meth. Eng. **15**(3), 229–275 (2008)
62. L.R. Scott, S. Zhang, Finite element interpolation of nonsmooth functions satisfying boundary conditions. Math. Comput. **54**(190), 483–493 (1990)
63. D.B. Szyld, The many proofs of an identity on the norm of oblique projections. Numer. Algorithms **42**(3–4), 309–323 (2006)
64. R. Tezaur, C. Farhat, Three-dimensional discontinuous Galerkin elements with plane waves and Lagrange multipliers for the solution of mid-frequency Helmholtz problems. Int. J. Numer. Methods Eng. **66**(5), 796–815 (2006)
65. H. Wu, Pre-asymptotic error analysis of CIP-FEM and FEM for the Helmholtz equation with high wave number. Part I: linear version. IMA J. Numer. Anal. **34**(3), 1266–1288 (2014)
66. J. Zitelli, I. Muga, L. Demkowicz, J. Gopalakrishnan, D. Pardo, V.M. Calo, A class of discontinuous PetrovGalerkin methods. part IV: the optimal test norm and time-harmonic wave propagation in 1D. J. Comput. Phys. **230**(7), 2406–2432 (2011)

Discontinuous Galerkin Methods for Time-Dependent Convection Dominated Problems: Basics, Recent Developments and Comparison with Other Methods

Chi-Wang Shu

Abstract In this survey article, we will give a short summary of the basic algorithm issues of discontinuous Galerkin methods for time-dependent convection dominated problems. We will then give a few representative examples of recent developments of discontinuous Galerkin methods for such problems, and provide comparisons with several other types of numerical methods commonly used for similar or related problems. For the comparison, we concentrate mainly on the methods presented in the London Mathematical Society EPSRC Durham Symposium on Building Bridges: Connections and Challenges in Modern Approaches to Numerical Partial Differential Equations.

1 Introduction

This survey article is based on and expanded from the lectures given by the author in the London Mathematical Society EPSRC Durham Symposium on Building Bridges: Connections and Challenges in Modern Approaches to Numerical Partial Differential Equations, which was held on July 8–16, 2014 at the University of Durham. The central topic of these lectures given by the author is the discontinuous Galerkin (DG) method for time-dependent convection dominated problems.

We will start with a concise summary of the basic algorithm issues of discontinuous Galerkin methods for time-dependent convection dominated problems in Sect. 2. In Sect. 3, we will give a few representative examples of recent developments of discontinuous Galerkin methods for such problems. Since the main theme of the Durham Symposium was on "building bridges" between different numerical methods, we will attempt to give comparison remarks between the discontinuous Galerkin method and several other numerical methods commonly used to solve

C.-W. Shu (✉)
Division of Applied Mathematics, Brown University, Providence, RI 02912, USA
e-mail: shu@dam.brown.edu

G.R. Barrenechea et al. (eds.), *Building Bridges: Connections and Challenges in Modern Approaches to Numerical Partial Differential Equations*,
Lecture Notes in Computational Science and Engineering 114,
DOI 10.1007/978-3-319-41640-3_12

similar or related problems. For the purpose of comparison, we will concentrate mainly on the methods presented in the Durham Symposium.

2 Discontinuous Galerkin Methods for Time-Dependent Convection Dominated Problems

We are concerned with time-dependent convection dominated partial differential equations (PDEs). These include hyperbolic conservation laws, convection dominated convection-diffusion equations, convection dominated convection-dispersion equations, etc. For such problems, generic solutions contain shocks or rapid changing regions in the solutions of the PDEs, calling for special care in the design of numerical methods in order to achieve stability and accuracy. Our discussions will mainly concentrate on the spatial discretization. The time variable is either left undiscretized (method of lines) or discretized by standard explicit Runge-Kutta or multi-step methods, for example the total-variation-diminishing (TVD) or strong-stability-preserving (SSP) time discretizations [41, 94]. This is adequate as long as the problem remains truly convection dominated, hence the time step restriction for explicit time discretization will not be too severe.

Discontinuous Galerkin methods are a class of finite element methods using discontinuous basis functions, which are usually chosen as piecewise polynomials, but could also be chosen as other types of functions to suit specific needs (e.g. in [119, 133]). The DG methods are especially suitable for time-dependent convection dominated PDEs with explicit time discretizations. Since the basis functions are completely discontinuous, the DG methods have the flexibility which is not shared by typical continuous finite element methods, such as the allowance of arbitrary triangulation with hanging nodes, complete freedom in changing the polynomial degrees or even the type of basis functions in each element independent of that in the neighbors (*p* adaptivity), and extremely local data structure (elements only communicate with immediate neighbors regardless of the order of accuracy of the scheme) and the resulting embarrassingly high parallel efficiency (usually more than 99 % for a fixed mesh, and more than 80 % for a dynamic load balancing with adaptive meshes which change often during time evolution), see, e.g. [7, 86, 95]. The DG method is also very friendly to the GPU environment [58].

The first discontinuous Galerkin method was introduced in 1973 by Reed and Hill [85], in the framework of neutron transport, i.e. a time independent linear hyperbolic equation. A major development of the DG method is carried out by Cockburn et al. in a series of papers [20, 21, 23, 27, 28], in which they have established a framework to easily solve *nonlinear* time-dependent problems, such as the Euler equations of gas dynamics, using explicit, nonlinearly stable high order Runge-Kutta time discretizations [94] and DG discretization in space with exact or approximate Riemann solvers as interface fluxes and total variation bounded (TVB) nonlinear limiters [92] to achieve non-oscillatory properties for strong shocks.

The DG method has found rapid applications in diverse areas. For more details, we refer to the survey paper [29], and other papers in that Springer volume, which contains the conference proceedings of the First International Symposium on Discontinuous Galerkin Methods held at Newport, Rhode Island in 1999. The lecture notes [18] is a good reference for many details, as well as the extensive review paper [24]. The review paper [113] covers the local DG method for partial differential equations (PDEs) containing higher order spatial derivatives, such as Navier-Stokes equations. There are three special issues devoted to the discontinuous Galerkin methods [25, 26, 32], which contain many interesting papers in the development of the method in all aspects including algorithm design, analysis, implementation and applications. There are also a few recent books and lecture notes [35, 46, 57, 65, 88, 93] on DG methods.

2.1 Discontinuous Galerkin Method for Conservation Laws

The discontinuous Galerkin method was first designed as an effective numerical method for solving hyperbolic conservation laws, which may have discontinuous solutions. It remains to be the focal application area of DG methods and the area where the advantage of DG method is most clearly demonstrated.

We start our discussion with the one dimensional conservation law

$$u_t + f(u)_x = 0. \tag{1}$$

We assume the following mesh to cover the computational domain [0, 1], consisting of cells $I_i = [x_{i-\frac{1}{2}}, x_{i+\frac{1}{2}}]$, for $1 \le i \le N$, where

$$0 = x_{\frac{1}{2}} < x_{\frac{3}{2}} < \cdots < x_{N+\frac{1}{2}} = 1.$$

We denote

$$\Delta x_i = x_{i+\frac{1}{2}} - x_{i-\frac{1}{2}}, \quad 1 \le i \le N; \qquad h = \max_{1 \le i \le N} \Delta x_i.$$

We define a finite element space consisting of piecewise polynomials

$$V_h^k = \left\{ v : v|_{I_i} \in P^k(I_i);\ 1 \le i \le N \right\}, \tag{2}$$

where $P^k(I_i)$ denotes the set of polynomials of degree up to k defined on the cell I_i. The semi-discrete DG method for solving (1) is defined as follows: find the unique function $u_h = u_h(t) \in V_h^k$ such that, for all test functions $v_h \in V_h^k$ and all $1 \le i \le N$, we have

$$\int_{I_i} (u_h)_t (v_h) dx - \int_{I_i} f(u_h)(v_h)_x dx + \hat{f}_{i+\frac{1}{2}} v_h(x^-_{i+\frac{1}{2}}) - \hat{f}_{i-\frac{1}{2}} v_h(x^+_{i-\frac{1}{2}}) = 0. \tag{3}$$

Here, $\hat{f}_{i+\frac{1}{2}}$ is the numerical flux, which is a single valued function defined at the cell interface $x_{i+\frac{1}{2}}$ and in general depends on the values of the numerical solution u_h from both sides of the interface

$$\hat{f}_{i+\frac{1}{2}} = \hat{f}(u_h(x^-_{i+\frac{1}{2}}, t), u_h(x^+_{i+\frac{1}{2}}, t)).$$

For scalar equations, we use the so-called monotone fluxes from finite difference and finite volume methodology for solving conservation laws. We refer to, e.g., [63] for more details and examples of monotone fluxes.

There are several key issues of DG methods for solving conservation laws (1) which are worth mentioning:

Cell Entropy Inequality and Energy Stability It is well known that weak solutions of (1) may not be unique and the unique, physically relevant weak solution (the so-called entropy solution) satisfies the following entropy inequality

$$U(u)_t + F(u)_x \leq 0 \tag{4}$$

in distribution sense, for any convex entropy $U(u)$ satisfying $U''(u) \geq 0$ and the corresponding entropy flux $F(u) = \int^u U'(u)f'(u)du$. It is usually quite difficult to prove a discrete entropy inequality for finite difference or finite volume schemes, especially for high order schemes and when the flux function $f(u)$ in (1) is not convex or concave, see, e.g. [66, 77]. However, it turns out that it is easy to prove that the DG scheme (3) satisfies a cell entropy inequality [54].

Proposition 2.1 ([54]) *The solution u_h to the semi-discrete DG scheme (3) satisfies the following cell entropy inequality*

$$\frac{d}{dt}\int_{I_i} U(u_h)\,dx + \hat{F}_{i+\frac{1}{2}} - \hat{F}_{i-\frac{1}{2}} \leq 0 \tag{5}$$

for the square entropy $U(u) = \frac{u^2}{2}$, for a consistent entropy flux

$$\hat{F}_{i+\frac{1}{2}} = \hat{F}(u_h(x^-_{i+\frac{1}{2}}, t), u_h(x^+_{i+\frac{1}{2}}, t))$$

satisfying $\hat{F}(u, u) = F(u)$.

We remark that the result holds for the piecewise polynomial space (2) with any degree k. Also, the same result holds for the multi-dimensional DG scheme on any triangulation [54], for symmetric hyperbolic systems [47], as well as for the fully discrete Runge-Kutta DG scheme for linear equations [126]. Such cell entropy inequalities (which is essentially a local stability result), which hold even when the exact solution of the conservation law (1) is discontinuous, and the trivial consequence of global L^2 stability of the numerical solution stated below, are the main strong points of discontinuous Galerkin methods for solving convection dominated problems. A major line of research for DG methods is to prove such stability results for various DG methods designed for different nonlinear convection dominated PDEs. The stability result is more important than error estimates. This is because stability holds for general solutions, including discontinuous solutions, while most error estimates only hold for smooth solutions.

Proposition 2.2 ([54]) *For periodic or compactly supported boundary conditions, the solution u_h to the semi-discrete DG scheme (3) satisfies the following L^2 stability*

$$\frac{d}{dt}\int_0^1 (u_h)^2 dx \le 0, \tag{6}$$

or

$$\|u_h(\cdot,t)\| \le \|u_h(\cdot,0)\|. \tag{7}$$

Here and below, an unmarked norm is the usual L^2 norm.

Limiters and Total Variation Stability For discontinuous solutions, the cell entropy inequality (5) and the L^2 stability (6), although helpful, are not enough to control spurious numerical oscillations near discontinuities. In practice, especially for problems containing strong discontinuities, we often need to apply nonlinear limiters to control these oscillations and to obtain provable total variation stability. Most of the limiters studied in the literature come from the methodologies of finite volume and finite difference high resolution schemes.

A limiter can be considered as a post-processor for the DG solution. If a cell is suspected to contain a possible discontinuity (the so-called *troubled cells*), the DG polynomial in this cell is replaced by a new polynomial, of the same degree and with the same cell average (for conservation), which is hopefully less oscillatory than the old one. The difficulty is that one would hope also that, if the solution in this *troubled cells* happens to be smooth, then the new polynomial should still be as high order accurate as the old one. Some limiters are applied to all cells, while they should take effect (actually change the polynomial) only in the cells near the discontinuities. The total variation diminishing (TVD) limiters [44] belong to this class. Unfortunately, such limiters tend to take effect also in some cells in which the solution is smooth, for example in cells near smooth extrema of the exact solution. Accuracy is therefore lost in such cells. The total variation bounded (TVB) limiters [92], applied to Runge-Kutta DG (RKDG) schemes in [20, 23, 27, 28], attempt to remove this difficulty and to ensure that the limiter takes effect only in cells near the discontinuities. The TVB limiters are widely used in applications, because of their simplicity in implementation. However, they involve a parameter, related to the value of the second derivative of the exact solution near smooth extrema, which must be chosen by the user. The moment-based limiter [7] and the improved moment limiter [9] also belong to this class, and they are specifically designed for DG methods and limit the moments of the polynomial sequentially, from the highest order moment downwards. Unfortunately, the moment-based limiters may also take effect in certain smooth cells, thereby destroying accuracy in these cells. There are other types of limiters discussed in the literature which we will not review in this paper. We will however discuss some recent developments on the design of limiters in Sect. 3.

Error Estimates for Smooth Solutions If we assume the exact solution of (1) is smooth, we can obtain optimal L^2 error estimates. That is, for piecewise polynomial of degree k, we can obtain $(k+1)$-th order L^2 error estimates for the DG solution. Such error estimates can be obtained for the general nonlinear conservation law (1) and for fully discretized RKDG methods, for both scalar problems and for symmetrizable hyperbolic systems, as long as the purely upwind numerical fluxes are used and tensor product meshes and polynomial spaces are used, see, e.g. [10, 72, 121, 122, 126]. Optimal error estimates can also be obtained for DG methods on certain special types of triangulations for non-tensor product cases with purely upwind fluxes, e.g. in [30, 87]. For other cases, namely fluxes which are not purely upwind, or general unstructured meshes, sub-optimal L^2 error estimates of order $(k+1/2)$ can be proved, e.g. [56]. More recently, optimal $(k+1)$-th order L^2 error estimates are obtained for certain upwind-biased but non-purely upwind fluxes in [74]. We will discuss some recent developments on the error estimates of DG methods for non-smooth solutions in Sect. 3.

2.2 *Discontinuous Galerkin Method for Convection Diffusion Equations*

The DG method can be extended to solve time dependent convection diffusion equations

$$u_t + \sum_{i=1}^{d} f_i(u)_{x_i} - \sum_{i=1}^{d}\sum_{j=1}^{d}(a_{ij}(u)u_{x_j})_{x_i} = 0, \tag{8}$$

where $(a_{ij}(u))$ is a symmetric, semi-positive definite matrix. We still assume the PDE (8) is convection dominated, namely $(a_{ij}(u))$ is small (e.g. has a small coefficient in front) or could even vanish (degenerate) in certain parts of the computational domain.

For equations containing higher order spatial derivatives, such as the convection diffusion equation (8), care must be taken when designing discontinuous Galerkin methods. A naive and careless application of the discontinuous Galerkin method directly to the heat equation containing second derivatives could yield a method which behaves nicely in the computation but is "inconsistent" with the original equation and has $O(1)$ errors to the exact solution [24, 120].

There are however several different ways to correctly generalize DG schemes to solve the convection diffusion equations (8).

Local Discontinuous Galerkin Method The idea of the local discontinuous Galerkin (LDG) method for time-dependent partial differential equations with higher derivatives, such as the convection diffusion equation (8), is to rewrite the equation into a first order system, then apply the discontinuous Galerkin method on the system. A key ingredient for the success of such methods is the correct design of interface numerical fluxes. These fluxes must be designed to guarantee stability and local solvability of all the auxiliary variables introduced to approximate the derivatives of the solution. The local solvability of all the auxiliary variables is why the method is called a "local" discontinuous Galerkin method in [22].

The first local discontinuous Galerkin method was developed by Cockburn and Shu [22], for the convection diffusion equation (8) containing second derivatives. Their work was motivated by the successful numerical experiments of Bassi and Rebay [4] for the compressible Navier-Stokes equations.

We consider the one dimensional convection diffusion equation

$$u_t + f(u)_x = (a(u)u_x)_x \tag{9}$$

with $a(u) \geq 0$, as an example. We rewrite this equation as the following system

$$u_t + f(u)_x = (b(u)q)_x, \qquad q - B(u)_x = 0, \tag{10}$$

where

$$b(u) = \sqrt{a(u)}, \qquad B(u) = \int^u b(u)du. \tag{11}$$

The finite element space is still given by (2). The semi-discrete LDG scheme is defined as follows. Find $u_h, q_h \in V_h^k$ such that, for all test functions $v_h, p_h \in V_h^k$ and all $1 \leq i \leq N$, we have

$$\begin{aligned} &\textstyle\int_{I_i}(u_h)_t(v_h)dx - \int_{I_i}(f(u_h) - b(u_h)q_h)(v_h)_x dx \\ &+(\hat{f} - \hat{b}\hat{q})_{i+\frac{1}{2}}(v_h)^-_{i+\frac{1}{2}} - (\hat{f} - \hat{b}\hat{q})_{i-\frac{1}{2}}(v_h)^+_{i-\frac{1}{2}} = 0, \\ &\textstyle\int_{I_i} q_h p_h dx + \int_{I_i} B(u_h)(p_h)_x dx - \hat{B}_{i+\frac{1}{2}}(p_h)^-_{i+\frac{1}{2}} + \hat{B}_{i-\frac{1}{2}}(p_h)^+_{i-\frac{1}{2}} = 0. \end{aligned} \tag{12}$$

Here, all the "hat" terms are the numerical fluxes, namely single valued functions defined at the cell interface $x_{i+\frac{1}{2}}$ which typically depend on the discontinuous numerical solution from both sides of the interface. We already know from Sect. 2.1 that the convection flux $\hat{f}$ should be chosen as a monotone flux. However, the upwinding principle is no longer a valid guiding principle for the design of the diffusion fluxes $\hat{b}$, $\hat{q}$ and $\hat{B}$. In [22], sufficient conditions for the choices of these diffusion fluxes to guarantee the stability of the scheme (12) are given. Here, we will discuss a particularly attractive choice, called "alternating fluxes", defined as

$$\hat{b} = \frac{B(u_h^+) - B(u_h^-)}{u_h^+ - u_h^-}, \qquad \hat{q} = q_h^+, \qquad \hat{B} = B(u_h^-). \tag{13}$$

The important point is that $\hat{q}$ and $\hat{B}$ should be chosen from different directions. Thus, the choice

$$\hat{b} = \frac{B(u_h^+) - B(u_h^-)}{u_h^+ - u_h^-}, \qquad \hat{q} = q_h^-, \qquad \hat{B} = B(u_h^+)$$

is also fine.

Notice that, from the second equation in the scheme (12), we can solve q_h explicitly and locally (in cell I_i) in terms of u_h, by inverting the small mass matrix inside the cell I_i. This is why the method is referred to as the "local" discontinuous Galerkin method.

Similar to the case for hyperbolic conservation laws, we also have the following "cell entropy inequality" for the LDG method (12).

Proposition 2.3 ([22]) *The solution u_h, q_h to the semi-discrete LDG scheme (12) satisfies the following "cell entropy inequality"*

$$\frac{1}{2}\frac{d}{dt}\int_{I_i}(u_h)^2\,dx + \int_{I_i}(q_h)^2 dx + \hat{F}_{i+\frac{1}{2}} - \hat{F}_{i-\frac{1}{2}} \leq 0 \tag{14}$$

for some consistent entropy flux

$$\hat{F}_{i+\frac{1}{2}} = \hat{F}(u_h(x^-_{i+\frac{1}{2}},t), q_h(x^-_{i+\frac{1}{2}},t); u_h(x^+_{i+\frac{1}{2}},t), q_h(x^+_{i+\frac{1}{2}}))$$

satisfying $\hat{F}(u,u) = F(u) - ub(u)q$ where, as before, $F(u) = \int^u uf'(u)du$.

The proof does not depend on the accuracy of the scheme, namely it holds for the piecewise polynomial space (2) with any degree k. Also, the same proof can be given for multi-dimensional LDG schemes on any triangulation. As before, the cell entropy inequality trivially implies an L^2 stability of the numerical solution.

Proposition 2.4 *For periodic or compactly supported boundary conditions, the solution u_h, q_h to the semi-discrete LDG scheme (12) satisfies the following L^2 stability*

$$\frac{d}{dt}\int_0^1 (u_h)^2 dx + 2\int_0^1 (q_h)^2 dx \leq 0, \tag{15}$$

or

$$\|u_h(\cdot,t)\| + 2\int_0^t \|q_h(\cdot,\tau)\| d\tau \leq \|u_h(\cdot,0)\|. \tag{16}$$

Both the cell entropy inequality (14) and the L^2 stability (15) are valid regardless of whether the convection diffusion equation (9) is convection dominated or diffusion dominated and regardless of whether the exact solution is smooth or not. The diffusion coefficient $a(u)$ can be degenerate (equal to zero) in any part of the domain. The stability analysis is also valid for multi-dimensional cases (8) on arbitrary triangulations. The LDG method is particularly attractive for convection dominated convection diffusion equations, when traditional continuous finite element methods are less stable.

If we assume the exact solution of (9) is smooth, we can obtain optimal $O(h^{k+1})$ L^2 error estimates. Such error estimates can be obtained for the general nonlinear convection diffusion equation (9), see, e.g. [110]. For multi-dimensional problems, the optimal $O(h^{k+1})$ error estimates can be obtained on tensor product meshes and polynomial spaces. For general triangulations and piecewise polynomials of degree k, a sub-optimal error estimate of $O(h^k)$ can be obtained, see [22, 110].

Internal Penalty Discontinuous Galerkin Methods Another important class of DG methods for solving diffusion equations is the class of internal penalty discontinuous Galerkin methods. We will use the simple heat equation

$$u_t = u_{xx} \tag{17}$$

to demonstrate the idea. If we multiply both sides of (17) by a test function v and integrate over the cell I_i, and integrate by parts for the right-hand-side, we obtain the equality

$$\int_{I_i} u_t v dx = -\int_{I_i} u_x v_x dx + (u_x)_{i+\frac{1}{2}} v^-_{i+\frac{1}{2}} - (u_x)_{i-\frac{1}{2}} v^+_{i-\frac{1}{2}} \tag{18}$$

where we have used superscripts $\pm$ on v at cell boundaries to prepare for numerical schemes involving functions which are discontinuous at those cell boundaries. Summing over i, we obtain, with periodic boundary conditions for simplicity, the following equality

$$\int_a^b u_t v dx = -\sum_{i=1}^{N} \int_{I_i} u_x v_x dx - \sum_{i=1}^{N} (u_x)_{i+\frac{1}{2}} [v]_{i+\frac{1}{2}} \tag{19}$$

where $[w] \equiv w^+ - w^-$ denotes the jump of w at the cell interface. If we attempt to convert the equality (19) into a numerical scheme, we could try the following. Find $u_h \in V_h^k$ such that, for all test functions $v_h \in V_h^k$, we have

$$\int_a^b (u_h)_t (v_h) dx = -\sum_{i=1}^{N} \int_{I_i} (u_h)_x (v_h)_x dx - \sum_{i=1}^{N} \{(u_h)_x\}_{i+\frac{1}{2}} [v_h]_{i+\frac{1}{2}} \tag{20}$$

where $\{w\} \equiv \frac{1}{2}(w^+ + w^-)$ denotes the average of w at the cell interface. This scheme is actually the "bad" scheme considered in [24, 120], which is shown to be unstable. Notice that the right-hand-side of (20) is not symmetric with respect to u_h and v_h. We can therefore add another term to symmetrize it, obtaining the following scheme. Find $u_h \in V_h^k$ such that, for all test functions $v_h \in V_h^k$, we have

$$\begin{aligned}\int_a^b (u_h)_t(v_h)dx = &-\sum_{i=1}^N \int_{I_i}(u_h)_x(v_h)_x dx\\ &-\sum_{i=1}^N \{(u_h)_x\}_{i+\frac{1}{2}}[v_h]_{i+\frac{1}{2}} - \sum_{i=1}^N \{(v_h)_x\}_{i+\frac{1}{2}}[u_h]_{i+\frac{1}{2}}. \end{aligned} \tag{21}$$

Notice that, since the exact solution is continuous, the additional term $-\sum_{i=1}^N \{(v_h)_x\}_{i+\frac{1}{2}}[u_h]_{i+\frac{1}{2}}$ is zero if the numerical solution u_h is replaced by the exact solution u, hence the scheme is consistent. Scheme (21) is symmetric, unfortunately it is still unconditionally unstable. In order to stabilize the scheme, a further penalty term must be added, resulting in the following symmetric internal penalty discontinuous Galerkin (SIPG) method [2, 101]

$$\begin{aligned}\int_a^b (u_h)_t(v_h)dx = &-\sum_{i=1}^N \int_{I_i}(u_h)_x(v_h)_x dx - \sum_{i=1}^N \{(u_h)_x\}_{i+\frac{1}{2}}[v_h]_{i+\frac{1}{2}}\\ &-\sum_{i=1}^N \{(v_h)_x\}_{i+\frac{1}{2}}[u_h]_{i+\frac{1}{2}} - \sum_{i=1}^N \frac{\alpha}{h}[u_h]_{i+\frac{1}{2}}[v_h]_{i+\frac{1}{2}}. \end{aligned} \tag{22}$$

Clearly, the scheme (22) is still symmetric, and it can be proved [2, 101] that, for sufficiently large α, it is stable and has optimal $O(h^{k+1})$ order convergence in L^2. The disadvantage of this scheme is that it involves a parameter α which has to be chosen adequately to ensure stability. Another possible way to obtain a stable scheme is to change the sign of the last term in the unstable scheme (21), resulting in the following non-symmetric internal penalty discontinuous Galerkin (NIPG) method [5, 76] of Baumann and Oden

$$\begin{aligned}\int_a^b (u_h)_t(v_h)dx = &-\sum_{i=1}^N \int_{I_i}(u_h)_x(v_h)_x dx\\ &-\sum_{i=1}^N \{(u_h)_x\}_{i+\frac{1}{2}}[v_h]_{i+\frac{1}{2}} + \sum_{i=1}^N \{(v_h)_x\}_{i+\frac{1}{2}}[u_h]_{i+\frac{1}{2}}. \end{aligned} \tag{23}$$

This scheme is not symmetric, however it is L^2 stable and convergent, although it has a sub-optimal $O(h^k)$ order of L^2 errors for even k [5, 76, 120].

There are other types of DG methods involving the internal penalty methodology, for example the direct discontinuous Galerkin (DDG) methods [68, 69]. We refer to [3] for a comprehensive review of DG methods for elliptic problems, most of which also apply to the second derivative part of time-dependent convection diffusion equations as well.

Ultra Weak Discontinuous Galerkin Methods Ultra weak discontinuous Galerkin methods are designed in [14]. Let us again use the simple heat equation (17) to demonstrate the idea. If we multiply both sides of (17) by a test function v and integrate over the cell I_i, and integrate by parts twice for the right-hand-side, we obtain the equality

$$\begin{aligned}\int_{I_i} u_t v dx = & \int_{I_i} u v_{xx} dx + (u_x)_{i+\frac{1}{2}} v_{i+\frac{1}{2}} - (u_x)_{i-\frac{1}{2}} v_{i-\frac{1}{2}} \\ & - u_{i+\frac{1}{2}} (v_x)_{i+\frac{1}{2}} + u_{i-\frac{1}{2}} (v_x)_{i-\frac{1}{2}}.\end{aligned} \tag{24}$$

We can then follow the general principle of designing DG schemes, namely converting the solution u and its derivatives at the cell boundary into numerical fluxes, and taking values of the test function v and its derivatives at the cell boundary by values inside the cell I_i, to obtain the following scheme. Find $u_h \in V_h^k$ such that, for all test functions $v_h \in V_h^k$ and all $1 \le i \le N$, we have

$$\begin{aligned}\int_{I_i} (u_h)_t \, v_h dx = & \int_{I_i} u_h \, (v_h)_{xx} dx + \widehat{u_x}_{i+\frac{1}{2}} (v_h)^-_{i+\frac{1}{2}} - \widehat{u_x}_{i-\frac{1}{2}} (v_h)^+_{i-\frac{1}{2}} \\ & - \hat{u}_{i+\frac{1}{2}} ((v_h)_x)^-_{i+\frac{1}{2}} + \hat{u}_{i-\frac{1}{2}} ((v_h)_x)^+_{i-\frac{1}{2}}.\end{aligned} \tag{25}$$

The crucial ingredient for the stability of the scheme (25) is still the choice of numerical fluxes. It is proved in [14] that the following choice of numerical fluxes

$$\hat{u}_{i+\frac{1}{2}} = (u_h)^-_{i+\frac{1}{2}}, \qquad \widehat{u_x}_{i+\frac{1}{2}} = ((u_h)_x)^+_{i+\frac{1}{2}} + \frac{\alpha}{h}[u_h]_{i+\frac{1}{2}} \tag{26}$$

would yield a stable DG scheme if the constant $\alpha > 0$ is sufficiently large. Notice that the choice in (26) is a combination of alternating fluxes and internal penalty. The following choice of alternating fluxes would also work

$$\hat{u}_{i+\frac{1}{2}} = (u_h)^+_{i+\frac{1}{2}}, \qquad \widehat{u_x}_{i+\frac{1}{2}} = ((u_h)_x)^-_{i+\frac{1}{2}} + \frac{\alpha}{h}[u_h]_{i+\frac{1}{2}}.$$

Sub-optimal L^2 error estimates are given in [14] for the scheme (25) with the fluxes (26) for $k \ge 1$. In numerical experiments, optimal L^2 convergence rate of $O(h^{k+1})$ is observed for all $k \ge 1$. The scheme can be easily generalized to the general nonlinear convection-diffusion equation (9) with the same stability property [14].

2.3 Discontinuous Galerkin Method for PDEs Containing Higher Order Spatial Derivatives

We now consider the DG methods for solving PDEs containing higher order spatial derivatives. We consider two types of PDEs: dispersive wave equations and diffusion equations.

Dispersive Wave Equations We start with the following general KdV type equations

$$u_t + \sum_{i=1}^{d} f_i(u)_{x_i} + \sum_{i=1}^{d} \left(r_i'(u) \sum_{j=1}^{d} g_{ij}(r_i(u)_{x_i})_{x_j} \right)_{x_i} = 0, \tag{27}$$

where $f_i(u)$, $r_i(u)$ and $g_{ij}(q)$ are arbitrary (smooth) nonlinear functions. The one-dimensional KdV equation

$$u_t + (\alpha u + \beta u^2)_x + \sigma u_{xxx} = 0, \tag{28}$$

where α, β and σ are constants, is a special case of the general class (27).

Stable LDG schemes for solving (27) are first designed in [115]. We will concentrate our discussion for the one-dimensional case. For the one-dimensional generalized KdV type equations

$$u_t + f(u)_x + (r'(u)g(r(u)_x)_x)_x = 0, \tag{29}$$

where $f(u)$, $r(u)$ and $g(q)$ are arbitrary (smooth) nonlinear functions, the LDG method is based on rewriting it as the following system

$$u_t + (f(u) + r'(u)p)_x = 0, \qquad p - g(q)_x = 0, \qquad q - r(u)_x = 0. \tag{30}$$

The finite element space is still given by (2). The semi-discrete LDG scheme is defined as follows. Find $u_h, p_h, q_h \in V_h^k$ such that, for all test functions $v_h, w_h, z_h \in V_h^k$ and all $1 \le i \le N$, we have

$$\begin{aligned}
&\int_{I_i} (u_h)_t (v_h) dx - \int_{I_i} (f(u_h) + r'(u_h)p_h)(v_h)_x dx \\
&\quad + (\hat{f} + \widehat{r'}\hat{p})_{i+\frac{1}{2}} (v_h)^-_{i+\frac{1}{2}} - (\hat{f} + \widehat{r'}\hat{p})_{i-\frac{1}{2}} (v_h)^+_{i-\frac{1}{2}} = 0, \qquad (31)\\
&\int_{I_i} p_h w_h dx + \int_{I_i} g(q_h)(w_h)_x dx - \hat{g}_{i+\frac{1}{2}} (w_h)^-_{i+\frac{1}{2}} + \hat{g}_{i-\frac{1}{2}} (w_h)^+_{i-\frac{1}{2}} = 0, \\
&\int_{I_i} q_h z_h dx + \int_{I_i} r(u_h)(z_h)_x dx - \hat{r}_{i+\frac{1}{2}} (z_h)^-_{i+\frac{1}{2}} + \hat{r}_{i-\frac{1}{2}} (z_h)^+_{i-\frac{1}{2}} = 0.
\end{aligned}$$

Here again, all the "hat" terms are the numerical fluxes, namely single valued functions defined at the cell interfaces which typically depend on the discontinuous numerical solution from both sides of the interface. We already know from Sect. 2.1 that the convection flux $\hat{f}$ should be chosen as a monotone flux. It is important to design the other fluxes suitably in order to guarantee stability of the resulting LDG scheme. In fact, the upwinding principle is still a valid guiding principle here, since the KdV type equation (29) is a dispersive wave equation for which waves are propagating with a direction. For example, the simple linear equation

$$u_t + u_{xxx} = 0,$$

which corresponds to (29) with $f(u) = 0$, $r(u) = u$ and $g(q) = q$, admits the following simple wave solution

$$u(x,t) = \sin(x+t),$$

that is, information propagates from right to left. This motivates the following choice of numerical fluxes, discovered in [115]:

$$\widehat{r'} = \frac{r(u_h^+) - r(u_h^-)}{u_h^+ - u_h^-}, \qquad \hat{p} = p_h^+, \qquad \hat{g} = \hat{g}(q_h^-, q_h^+), \qquad \hat{r} = r(u_h^-). \tag{32}$$

Here, $-\hat{g}(q_h^-, q_h^+)$ is a monotone flux for $-g(q)$, namely $\hat{g}$ is a non-increasing function in the first argument and a non-decreasing function in the second argument. The important point is again the "alternating fluxes", namely $\hat{p}$ and $\hat{r}$ should come from opposite sides. Thus

$$\widehat{r'} = \frac{r(u_h^+) - r(u_h^-)}{u_h^+ - u_h^-}, \qquad \hat{p} = p_h^-, \qquad \hat{g} = \hat{g}(q_h^-, q_h^+), \qquad \hat{r} = r(u_h^+)$$

would also work.

It is quite interesting to observe that monotone fluxes, which are originally designed for hyperbolic conservation laws, can be used also for nonlinear dispersive equations to obtain stability. Also notice that, from the third equation in the scheme (31), we can solve q_h explicitly and locally (in cell I_i) in terms of u_h, by inverting the small mass matrix inside the cell I_i. Then, from the second equation in the scheme (31), we can solve p_h explicitly and locally (in cell I_i) in terms of q_h. Thus only u_h is the global unknown and the auxiliary variables q_h and p_h can be solved in terms of u_h locally. This is why the method is referred to as the "local" discontinuous Galerkin method.

Similar to the case for hyperbolic conservation laws and convection diffusion equations, we have the following "cell entropy inequality" for the LDG method (31) with the flux choice (32).

Proposition 2.5 ([115]) *The solution u_h to the semi-discrete LDG scheme (31) with the fluxes (32) satisfies the following "cell entropy inequality"*

$$\frac{1}{2}\frac{d}{dt}\int_{I_i}(u_h)^2\,dx + \hat{F}_{i+\frac{1}{2}} - \hat{F}_{i-\frac{1}{2}} \le 0 \tag{33}$$

for some consistent entropy flux

$$\hat{F}_{i+\frac{1}{2}} = \hat{F}(u_h(x^-_{i+\frac{1}{2}},t), p_h(x^-_{i+\frac{1}{2}},t), q_h(x^-_{i+\frac{1}{2}},t); u_h(x^+_{i+\frac{1}{2}},t), p_h(x^+_{i+\frac{1}{2}},t), q_h(x^+_{i+\frac{1}{2}}))$$

satisfying $\hat{F}(u,u) = F(u) + ur'(u)p - G(q)$ where $F(u) = \int^u uf'(u)du$ and $G(q) = \int^q qg(q)dq$.

The proof of this proposition does not depend on the accuracy of the scheme, namely it holds for the piecewise polynomial space (2) with any degree k. Also, the same proof can be given for the multi-dimensional LDG scheme solving (27) on any triangulation.

As before, the cell entropy inequality trivially implies an L^2 stability of the numerical solution.

Proposition 2.6 ([115]) *For periodic or compactly supported boundary conditions, the solution u_h to the semi-discrete LDG scheme (31) with the fluxes (32) satisfies the following L^2 stability*

$$\frac{d}{dt}\int_0^1 (u_h)^2 dx \le 0, \tag{34}$$

or

$$\|u_h(\cdot,t)\| \le \|u_h(\cdot,0)\|. \tag{35}$$

Again, both the cell entropy inequality (33) and the L^2 stability (34) are valid regardless of whether the KdV type equation (29) is convection dominated or dispersion dominated and regardless of whether the exact solution is smooth or not. The dispersion flux $r'(u)g(r(u)_x)_x$ can be degenerate (equal to zero) in any part of the domain. The LDG method is particularly attractive for convection dominated convection dispersion equations, when traditional continuous finite element methods may be less stable. In [115], this LDG method is used to study the dispersion limit of the Burgers equation, for which the third derivative dispersion term in (29) has a small coefficient which tends to zero.

Sub-optimal L^2 error estimates for this scheme, for both linear and nonlinear problems, are obtained in [110, 115]. In [114], Xu and Shu proved optimal L^2 error estimate of order $O(h^{k+1})$ for this scheme, when applied to linear PDEs. This proof involves new techniques, because of the wave nature of the third order dispersive wave equation and hence a lack of control of the derivatives. The approach in [114] is to establish stability not only for u_h as in (33), but also for q_h and p_h approximating u_x and u_{xx}.

Ultra weak discontinuous Galerkin methods for such dispersive wave equations can also be designed, see [14]. The choice of the numerical fluxes is the same as that for the LDG scheme, see (32).

Both the LDG method and the ultra weak DG method can be designed for other dispersive wave equations containing higher order (usually odd order) derivatives. Examples include equations with fifth order spatial derivatives [116], the so-called $K(m, n)$ equation [64], the fifth-order fully nonlinear $K(n, n, n)$ equations [106], the nonlinear Schrödinger (NLS) equation [107], the two dimensional Kadomtsev-Petviashvili (KP) equations [108], and the Camassa-Holm (CH) equation [111].

Dissipative Equations DG methods have also been designed for other dissipative equations containing higher even order derivatives. The list includes the time dependent convection bi-harmonic equation [116], where the numerical fluxes for the LDG scheme are chosen following the same "alternating fluxes" principle similar to the second order convection-diffusion equation (8), namely the flux pairs corresponding to u and u_{xxx}, and the flux pairs corresponding to u_x and u_{xx}, should be chosen in an alternating fashion within each pair. A cell entropy inequality and the L^2 stability of the LDG scheme for the nonlinear bi-harmonic equation can be proved [116], which do not depend on the smoothness of the solution, the order of accuracy of the scheme, or the triangulation. Optimal L^2 error estimates can be proved for the linear biharmonic equation, for both structured and unstructured meshes, see [36]. The list of equations for which LDG (and for some cases also ultra weak DG) methods have been designed also includes the Kuramoto-Sivashinsky type equations [109], the Cahn-Hilliard equation and the Cahn-Hilliard system [102, 103], and the surface diffusion equation and the Willmore flow equation [52, 53, 112].

3 A Few Recent Developments

In this section we give a few examples of recent developments on DG methods for convection dominated PDEs.

3.1 Nonlinear Limiters

As we mentioned in the previous section, even though the DG method is usually energy stable, this stability is not strong enough to prevent spurious oscillations or even blow-ups of the numerical solution in the presence of strong discontinuities. Therefore, nonlinear limiters are often needed to control such spurious oscillations. Another situation that nonlinear limiters are needed is to enforce physical bounds (e.g. maximum principle, positivity-preserving, etc.) while maintaining high order accuracy.

WENO Limiters The limiters based on the weighted essentially non-oscillatory (WENO) methodology are designed with the objective of maintaining the original high order accuracy even if the limiters take effect in smooth cells. These limiters are based on the WENO methodology for finite volume schemes [55, 70], and involve nonlinear reconstructions of the polynomials in troubled cells using the information of neighboring cells. The WENO reconstructed polynomials have the same high order of accuracy as the original polynomials when the solution is smooth, and they are (essentially) non-oscillatory near discontinuities. Qiu and Shu [83] and Zhu et al. [135] designed WENO limiters using the usual WENO reconstruction based on cell averages of neighboring cells as in [48, 55, 91], to reconstruct the values of the solutions at certain Gaussian quadrature points in the target cells, and then rebuild the solution polynomials from the original cell average and the reconstructed values at the Gaussian quadrature points through a numerical integration for the moments. This limiter needs to use the information from not only the immediate neighboring cells but also neighbors' neighbors, making it complicated to implement in multi-dimensions, especially for unstructured meshes [48, 123, 135]. It also destroys the local data structure of the base DG scheme (which needs only to communicate with immediate neighbors). The effort in [80, 82] attempts to construct Hermite type WENO approximations, which use the information of not only the cell averages but also the lower order moments such as slopes, to reduce the spread of reconstruction stencils. However for higher order methods the information of neighbors' neighbors is still needed.

More recently, Zhong and Shu [134] developed a new WENO limiting procedure for RKDG methods on structured meshes. The idea is to reconstruct the entire polynomial, instead of reconstructing point values or moments in the classical WENO reconstructions. That is, the entire reconstruction polynomial on the target cell is a convex combination of polynomials on this cell and its immediate neighboring cells, with suitable adjustments for conservation and with the nonlinear weights of the convex combination following the classical WENO procedure. The main advantage of this limiter is its simplicity in implementation, as it uses only the information from immediate neighbors and the linear weights are always positive. This simplicity is more prominent for multi-dimensional unstructured meshes, which is studied in [136] for two-dimensional unstructured triangular meshes. Further improvements of this limiter are carried out in [137]. In the next section, we will discuss the adaptation of this WENO limiter to another class of numerical methods, the so-called CPR schemes.

The WENO limiters are typically applied only in designated "troubled cells", in order to save computational cost and to minimize the influence to accuracy in smooth regions. Therefore, a troubled cell indicator is needed, to correctly identify cells near discontinuities in which the limiters should be applied. Qiu and Shu in [81] have compared several troubled cell indicators. In practice, the TVB indicator [92] and the KXRCF indicator [62] are often the best choices.

Bound-Preserving Limiters In many convection dominated problems, the physical quantities have desired bounds which are satisfied by the exact solutions of the

PDEs. For example, for two-dimensional incompressible Euler or Navier-Stokes equations written in a vorticity-streamfunction formulation, the vorticity satisfies a maximum principle. For Euler equations of compressible gas dynamics, density and pressure remain positive (non-negative) when their initial values are positive. It would certainly be desirable if numerical solutions obey the same bounds. If the numerical solution goes out of the bounds because of spurious oscillations, it would either be non-physical (e.g. negative density, negative internal energy, a percentage of a component which goes below zero or above one), or worse still, it could lead to nonlinear instability and blowups of the code because the PDE becomes ill-posed (e.g. the Euler equations of compressible gas dynamics become ill-posed for negative density or pressure).

Not all limiters designed for controlling spurious oscillations can enforce the bound-preserving property. When they do, they often degenerate the order of accuracy of the original scheme in smooth regions.

Recently, a general framework is established to preserve strict bounds (maximum principle for scalar problems and positivity of relevant quantities for scalar problems or systems) for DG and finite volume schemes, while maintaining provable high order accuracy of the original schemes. These techniques apply to multi-dimensions in general unstructured triangulations as well. See [124, 125, 131].

We will not repeat here the details of this general framework and refer the readers to the references. We will summarize here the main steps in this framework:

1. We first find a first order base DG scheme, using piecewise polynomials of degree zero (piecewise constants), which can be proved to be bound-preserving under certain CFL conditions for Euler forward time discretization. Notice that a first order DG scheme is the same as a first order finite volume scheme.
   For scalar hyperbolic conservation laws (1), the first order DG scheme using any monotone numerical flux would satisfy a maximum principle. For Euler equations of compressible gas dynamics, several first order schemes, including the Godunov scheme [39], the Lax-Friedrichs scheme [79, 125], the Harten-Lax-van Leer (HLLE) scheme [45], and the Boltzmann type kinetic scheme [78], among others, are positivity-preserving for density and pressure.
2. We then apply a simple scaling limiter to the high order DG solution at time level n. If the DG solution at time level n in cell I_i is a polynomial $p_i(x)$, we replace it by the limited polynomial $\tilde{p}_i(x)$ defined by

$$\tilde{p}_i(x) = \theta_i(p_i(x) - \bar{u}_i^n) + \bar{u}_i^n$$

   where $\bar{u}_i^n$ is the cell average of $p_i(x)$, and

$$\theta_i = \min\left\{ \left| \frac{M - \bar{u}_i^n}{M_i - \bar{u}_i^n} \right|, \left| \frac{m - \bar{u}_i^n}{m_i - \bar{u}_i^n} \right|, 1 \right\},$$

with

$$M_i = \max_{x \in S_i} p_i(x), \qquad m_i = \min_{x \in S_i} p_i(x)$$

here M and m are the desired global upper and lower bounds to be preserved, and S_i is the set of certain Legendre Gauss-Lobatto quadrature points of the cell I_i. Clearly, this limiter is just a simple scaling of the original polynomial around its average.

3. We then evolve the solution by Euler forward time discretization, or by TVD or SSP Runge-Kutta time discretization [41, 94].

We can see that this procedure is very simple and inexpensive to implement. The scaling limiter is completely local inside the cell I_i and involves only evaluation of the DG polynomial at pre-determined quadrature points. The procedure can be applied in arbitrary triangular meshes. Amazingly, this simple process leads to mathematically provable bound-preserving property without degenerating the high order accuracy of the base DG scheme.

For scalar nonlinear conservation laws, passive convection in a divergence-free velocity field, and 2D incompressible Euler equations in the vorticity-streamfunction formulation, high order DG schemes maintaining maximum principle have been designed in Zhang and Shu [124] and in Zhang et al. [131].

For scalar nonlinear convection diffusion equations, second order DG schemes on unstructured triangulations maintaining maximum principle have been designed in Zhang et al. [132].

For Euler equations of gas dynamics, high order DG schemes maintaining positivity of density and pressure (or internal energy) have been designed in Zhang and Shu [125, 127–129] and in Zhang et al. [131].

For shallow water equations, high order DG schemes maintaining non-negativity of water height have been designed in Xing et al. [104].

Positivity-preserving semi-Lagrangian DG schemes have been designed in Qiu and Shu [84] and in Rossmanith and Seal [90].

3.2 DG Method for Hyperbolic Equations Involving δ-Functions

In a hyperbolic conservation law

$$\begin{aligned} &u_t + f(u)_x = g(x,t), && (x,t) \in R \times (0,T], \\ &u(x,0) = u_0(x), && x \in R, \end{aligned} \tag{36}$$

the initial condition u_0, or the source term $g(x,t)$, or the solution $u(x,t)$ may contain δ-singularities. Such singularities are more difficult to handle than discontinuities in the solutions. Many high order schemes would easily blow up in the presence of δ-function singularities, because of the severe oscillations leading to non-physical regimes (e.g. negative density). On the other hand, if one applies traditional limiters such as various slope limiters to enforce stability, the resolution of the δ-function singularities could be seriously deteriorated. Resolution is also seriously affected by other commonly used strategies such as modifications by an approximate Gaussian to smear out the δ-function.

Since DG methods are based on weak formulations, they can be designed directly to handle δ-function singularities. Recently, we have designed and analyzed DG schemes for solving linear and nonlinear PDE models with δ-function singularities [117, 118]. For linear problems, we prove stability and high order error estimates in negative norms when the DG method is applied, and propose post-processing techniques to recover high order accuracy in strong norms away from these δ-function singularities. For nonlinear problems, such as Krause's consensus models [11] and pressureless Euler equations [13], an adequate design of bound preserving limiter, within the framework described in Sect. 3.1, to enforce the physical bounds without compromising resolution of δ-function singularities is shown to be crucial. With such limiters, high resolution and highly stable results can be obtained for such difficult nonlinear models.

On a related issue, a difficult but important problem is to prove error estimates for DG methods applied to hyperbolic equations with discontinuous solutions. In [130], optimal error estimate for the explicit Runge-Kutta discontinuous Galerkin method to solve a linear hyperbolic equation in one dimension with discontinuous but piecewise smooth initial data is provided, for arbitrary polynomial degree $k \geq 1$, and the third order explicit TVD Runge-Kutta time discretization under the standard CFL condition. The error estimate is obtained for a L^2-norm excluding a pollution region of the size $O(\sqrt{T}h^{1/2}\log(1/h))$, where T is the final time and h is the maximum cell length. Numerical experiments confirm the sharpness of the size of the pollution region. For earlier work in this direction, we refer to [19].

3.3 DG Method for Second Order Wave Equations

Second order wave equations arise frequently in applications. The simplest example is

$$u_{tt} = u_{xx}, \tag{37}$$

with suitable initial and boundary conditions. One way to solve such equations is to rewrite them into first order hyperbolic systems, for example, (37) can be rewritten as

$$u_t + v_x = 0, \qquad v_t + u_x = 0, \tag{38}$$

with suitable initial and boundary conditions, by introducing the new variable v which satisfies $v_x = -u_t$. Therefore, standard DG method with upwind fluxes can be used to solve the hyperbolic system (38), with stability proofs and optimal error estimates. The DG schemes with upwind fluxes for solving (38) is dissipative, that is, energy is not conserved but is dissipated during time evolution. This is not a concern, in fact it is even desirable, when the solution of the PDE is discontinuous, especially for nonlinear wave systems. However, when the solution of the PDE is smooth and one would like to solve the wave equations over long time, there is an advantage in using an energy-conserving scheme. A DG scheme for solving the first order system (38) can be made energy-conserving by using the central fluxes, however this would reduce the order of accuracy of the scheme for odd degree polynomials. In [105], an LDG method directly approximating the second order wave equation (37), using alternating fluxes, is studied. It is proved that this LDG scheme is energy conserving and is optimally convergent in L^2. Numerical examples indicate that such a method has an advantage for long time simulation. Generalizations to multiple dimensions and to waves in heterogeneous (including discontinuous) media are carried out in [15]. For studies of DG methods for second order wave equations, we refer also to [1, 16, 17, 40, 42, 89].

4 Comparison with Other Methods

In this section we provide comparisons with several other types of numerical methods which are often used for similar or related problems. For the comparison, we concentrate mainly on the methods presented in the London Mathematical Society EPSRC Durham Symposium on Building Bridges: Connections and Challenges in Modern Approaches to Numerical Partial Differential Equations.

4.1 The Correction Procedure via Reconstruction (CPR) Schemes

Recently, a new correction procedure via reconstruction (CPR) scheme framework [43, 50, 51, 97] was developed to solve hyperbolic conservation laws. This method was originally developed in [50] to solve conservation laws on structured meshes, under the name of flux reconstruction. In [97], the CPR framework was extended to 2D triangular and mixed grids under lifting collocation penalty. In [43], CPR was further extended to 3D hybrid meshes. The idea of CPR is to choose the degrees of freedom as approximations to point values of the solution at certain "solution points" in the cell, and compute the residue as the derivatives of a certain flux approximation polynomial evaluated at those solution points. As such, the method resembles a finite difference scheme, yet by special choice of the solution points and the flux approximation polynomial, the method can be made conservative, accurate and stable. For linear equations, by choosing suitable correction functions which are used in building the flux approximation polynomial, the CPR framework can unify several well-known methods such as the DG, staggered-grid (SG) multi-domain spectral method [59–61], the spectral volume (SV) [96, 98–100] and spectral difference (SD) methods [71, 73]. The CPR framework is based on a nodal differential form, with an element-wise discontinuous piecewise polynomial solution space, thus it can be considered as the DG method with a suitable numerical quadrature for the integration of the nonlinear terms. The advantage is that it solves the conservation laws in a differential form, similar to a finite difference scheme. In the SG or SD method, two groups of grid points are needed, i.e., the solution points and the flux points. However, the CPR framework involves only one group of grid points, namely, the solution points. Hence, the CPR framework is easier to understand and more efficient to implement. Because of the relationships to DG methods, many strategies designed for DG schemes, for example the compact WENO limiters and bound-preserving limiters reviewed in Sect. 3.1, can be generalized to CPR schemes, see [37, 38].

4.2 The HDG Methods

Hybrid DG (HDG) methods, as surveyed by Cockburn in this Durham Symposium, form a special class of DG methods, particularly suitable for solving steady state problems and implicit time discretized versions of time-dependent problems. The idea of HDG methods is to introduce the fluxes (traces) at element interfaces as independent solution variables, rather than computing them from the two limits of the solution polynomials from the two adjacent cells. It would appear that such a strategy would increase the memory and computational cost for the degrees of freedoms, however it turns out that the solution polynomials inside cells can be obtained from the interface variables via solving local problems, hence the global system to be solved involves only the degrees of freedom from element interfaces, thus reducing the size of the system and its computational cost. Stability properties and error estimates are also usually stronger for HDG than for regular DG. While HDG methods have advantages over other DG methods for many steady state problems or implicitly discretized time dependent problems, for time dependent problems with explicit time marching, the original DG method still has its advantage because of the local nature of the time evolution. We refer to, e.g. [12, 31, 75] and the lectures of Cockburn in this Durham Symposium for more details of HDG methods.

4.3 The Isogeometric Methods

The isogeometric methods, as surveyed by Hughes et al. in this Durham Symposium, form a class of finite element methods with a very nice integration between geometry and discretization. The approximation spaces are usually based on specially defined spline functions, with good regularity and excellent representation of geometry. Since the finite element discretization is based on the same approximation spaces, the solution of complicated PDEs on complicated surfaces becomes an integrated process, saving a lot of computational resources and forming nice theoretical properties of accuracy and stability. Comparing with DG methods, which are based on discontinuous approximation spaces, the isogeometric methods are based on much smoother approximation spaces with C^1 or smoother splines. Such smoother function spaces certainly allow for a much smaller set of degrees of freedom to reach the same order of accuracy, and they also allow for a better representation of the geometry. On the other hand, the DG method has more flexibility in approximating rapidly changing and discontinuous problems, which would be difficult for globally defined methods. Perhaps the DG methodology can be combined with the isogeometric methodology to deal with problems of isolated singularities, such as sharp edges in the geometry. We refer to, e.g. [49] and the lectures of Hughes et al. in this Durham Symposium for more details of the isogeometric methods.

4.4 The DPG Method

The discontinuous Petrov Galerkin (DPG) method, as surveyed by Demkowicz in this Durham Symposium, is similar to DG in the solution space consisting of discontinuous, piecewise polynomials. However, unlike the DG method which uses the same function space for the test functions, the DPG method uses another function space of the same dimension. This extra freedom is used to choose the test function space in order to minimize the error in a specific norm. For certain simple linear problems, this task can be accomplished exactly, while for other problems the test functions themselves must be constructed numerically, from solving local problems using polynomials of higher degree. With such special construction, the DPG method can usually achieve much smaller error using the same number of degrees of freedom, comparing with regular DG methods. It would be interesting to generalize such methods to solve nonlinear time dependent problems with shocked solutions, and compare them with regular DG methods. We refer to, e.g. [33, 34] and the lectures of Demkowicz in this Durham Symposium for more details of the DPG methods.

4.5 The Virtual Element Methods

The virtual element methods, as surveyed by Beirão da Veiga et al. in this Durham Symposium, form a new class of continuous finite element methods which are particularly suitable for solving problems on arbitrary polygon meshes. Traditional polynomial based continuous finite element methods are difficult to design on such meshes. If non-polynomial basis functions are used, their efficient computational implementation is a major issue. The virtual element methods rely on a very clever idea in using such non-polynomial basis functions in a "virtual" fashion, namely one would not need to know their formulas or even use them explicitly in the implementation. Only the existence of these basis functions and their certain properties are used in the design and analysis of the virtual element methods, while their implementation is similar to that of the regular polynomial based continuous finite elements. It is well known that one advantage of DG versus continuous finite elements is that DG is very flexible to be used on meshes of any polygon shape. The disadvantage of continuous finite elements over DG on this aspect is largely overcome by the new virtual element methods. We refer to, e.g. [6] and the lectures of Beirão da Veiga et al. in this Durham Symposium for more details of the virtual element methods.

4.6 The Mimetic Finite Difference Methods

The mimetic finite difference methods, as surveyed by Lipnikov in this Durham Symposium, form a class of finite difference methods which share many good properties of finite element methods in stability and accuracy, on arbitrary unstructured meshes. These nice properties are achieved through introducing discrete operators which satisfy fundamental identities of vector and tensor calculus. There is some similarity with the CPR schemes surveyed in Sect. 4.1, in that a finite difference method is designed on arbitrary meshes with good physical property such as conservation, stability and accuracy, although the approaches in the design of the two algorithms are apparently different. The mimetic finite difference methods have relationships to the virtual element methods discussed in the previous subsection and can be considered as one of the predecessors of the virtual element methods. We refer to, e.g. [8, 67] and the lectures of Lipnikov in this Durham Symposium for more details of the mimetic finite difference methods.

Acknowledgements The research of the author is supported partially by NSF grant DMS-1418750 and DOE grant DE-FG02-08ER25863.

References

1. S. Adjerid, H. Temimi, A discontinuous Galerkin method for the wave equation. Comput. Methods Appl. Mech. Eng. **200**, 837–849 (2011)
2. D.N. Arnold, An interior penalty finite element method with discontinuous elements. SIAM J. Numer. Anal. **19**, 742–760 (1982)
3. D.N. Arnold, F. Brezzi, B. Cockburn, L.D. Marini, Unified analysis of discontinuous Galerkin methods for elliptic problems. SIAM J. Numer. Anal. **39**, 1749–1779 (2002)
4. F. Bassi, S. Rebay, A high-order accurate discontinuous finite element method for the numerical solution of the compressible Navier-Stokes equations. J. Comput. Phys. **131**, 267–279 (1997)
5. C.E. Baumann, J.T. Oden, A discontinuous *hp* finite element method for convection-diffusion problems. Comput. Methods Appl. Mech. Eng. **175**, 311–341 (1999)
6. L. Beirão da Veiga, F. Brezzi, A. Cangiani, G. Manzini, L.D. Marini, A. Russo, Basic principles of virtual element methods. Math. Models Methods Appl. Sci. **23**, 199–214 (2013)
7. R. Biswas, K.D. Devine, J. Flaherty, Parallel, adaptive finite element methods for conservation laws. Appl. Numer. Math. **14**, 255–283 (1994)
8. F. Brezzi, K. Lipnikov, V. Simoncini, A family of mimetic finite difference methods on polygonal and polyhedral meshes. Math. Models Methods Appl. Sci. **15**, 1533–1551 (2005)
9. A. Burbeau, P. Sagaut, Ch.H. Bruneau, A problem-independent limiter for high-order Runge-Kutta discontinuous Galerkin methods. J. Comput. Phys. **169**, 111–150 (2001)
10. E. Burman, A. Ern, M.A. Fernández, Explicit Runge-Kutta schemes and finite elements with symmetric stabilization for first-order linear PDE systems. SIAM J. Numer. Anal. **48**, 2019–2042 (2010)
11. C. Canuto, F. Fagnani, P. Tilli, An Eulerian approach to the analysis of Krause's consensus models. SIAM J. Control Optim. **50**, 243–265 (2012)
12. F. Celiker, B. Cockburn, K. Shi, A projection-based error analysis of HDG methods for Timoshenko beams. Math. Comput. **81**, 131–151 (2012)
13. G.-Q. Chen, H. Liu, Formation of δ-shocks and vacuum states in the vanishing pressure limit of solutions to the Euler equations for isentropic fluids. SIAM J. Math. Anal. **34**, 925–938 (2003)
14. Y. Cheng, C.-W. Shu, A discontinuous Galerkin finite element method for time dependent partial differential equations with higher order derivatives. Math. Comput. **77**, 699–730 (2008)
15. C.-S. Chou, C.-W. Shu, Y. Xing, Optimal energy conserving local discontinuous Galerkin methods for second-order wave equation in heterogeneous media. J. Comput. Phys. **272**, 88–107 (2014)
16. E.T. Chung, B. Engquist, Optimal discontinuous Galerkin methods for wave propagation. SIAM J. Numer. Anal. **44**, 2131–2158 (2006)
17. E.T. Chung, B. Engquist, Optimal discontinuous Galerkin methods for the acoustic wave equation in higher dimensions. SIAM J. Numer. Anal. **47**, 3820–3848 (2009)
18. B. Cockburn, Discontinuous Galerkin methods for convection-dominated problems, in *High-Order Methods for Computational Physics*, ed. by T.J. Barth, H. Deconinck. Lecture Notes in Computational Science and Engineering, vol. 9 (Springer, Berlin, 1999), pp. 69–224
19. B. Cockburn, J. Guzmán, Error estimates for the Runge-Kutta discontinuous Galerkin method for the transport equation with discontinuous initial data. SIAM J. Numer. Anal. **46**, 1364–1398 (2008)
20. B. Cockburn, C.-W. Shu, TVB Runge-Kutta local projection discontinuous Galerkin finite element method for conservation laws II: general framework. Math. Comput. **52**, 411–435 (1989)
21. B. Cockburn, C.-W. Shu, The Runge-Kutta local projection P^1-discontinuous-Galerkin finite element method for scalar conservation laws. Math. Model. Numer. Anal. **25**, 337–361 (1991)

22. B. Cockburn, C.-W. Shu, The local discontinuous Galerkin method for time-dependent convection diffusion systems. SIAM J. Numer. Anal. **35**, 2440–2463 (1998)
23. B. Cockburn, C.-W. Shu, The Runge-Kutta discontinuous Galerkin method for conservation laws V: multidimensional systems. J. Comput. Phys. **141**, 199–224 (1998)
24. B. Cockburn, C.-W. Shu, Runge-Kutta Discontinuous Galerkin methods for convection-dominated problems. J. Sci. Comput. **16**, 173–261 (2001)
25. B. Cockburn, C.-W. Shu, Foreword for the special issue on discontinuous Galerkin method. J. Sci. Comput. **22–23**, 1–3 (2005)
26. B. Cockburn, C.-W. Shu, Foreword for the special issue on discontinuous Galerkin method. J. Sci. Comput. **40**, 1–3 (2009)
27. B. Cockburn, S.-Y. Lin, C.-W. Shu, TVB Runge-Kutta local projection discontinuous Galerkin finite element method for conservation laws III: one dimensional systems. J. Comput. Phys. **84**, 90–113 (1989)
28. B. Cockburn, S. Hou, C.-W. Shu, The Runge-Kutta local projection discontinuous Galerkin finite element method for conservation laws IV: the multidimensional case. Math. Comput. **54**, 545–581 (1990)
29. B. Cockburn, G. Karniadakis, C.-W. Shu, The development of discontinuous Galerkin methods, in *Discontinuous Galerkin Methods: Theory, Computation and Applications*, ed. by B. Cockburn, G. Karniadakis, C.-W. Shu. Lecture Notes in Computational Science and Engineering, Part I: Overview, vol. 11 (Springer, Berlin, 2000), pp. 3–50
30. B. Cockburn, B. Dong, J. Guzmán, Optimal convergence of the original DG method for the transport-reaction equation on special meshes. SIAM J. Sci. Comput. **46**, 1250–1265 (2008)
31. B. Cockburn, B. Dong, J. Guzmán, M. Restelli, R. Sacco, A hybridizable discontinuous Galerkin method for steady-state convection-diffusion-reaction problems. SIAM J. Sci. Comput. **31**, 3827–3846 (2009)
32. C. Dawson, Foreword for the special issue on discontinuous Galerkin method. Comput. Methods Appl. Mech. Eng. **195**, 3183 (2006)
33. L. Demkowicz, J. Gopalakrishnan, A class of discontinuous Petrov-Galerkin methods. Part I: the transport equation. Comput. Methods Appl. Mech. Eng. **199**, 1558–1572 (2010)
34. L. Demkowicz, J. Gopalakrishnan, A class of discontinuous Petrov-Galerkin methods. II. Optimal test functions. Numer. Methods Partial Differ. Equ. **27**, 70–105 (2011)
35. D.A. Di Pietro, A. Ern, *Mathematical Aspects of Discontinuous Galerkin Methods* (Springer, Berlin, Heidelberg, 2012)
36. B. Dong, C.-W. Shu, Analysis of a local discontinuous Galerkin method for linear time-dependent fourth-order problems. SIAM J. Numer. Anal. **47**, 3240–3268 (2009)
37. J. Du, C.-W. Shu, M. Zhang, A simple weighted essentially non-oscillatory limiter for the correction procedure via reconstruction (CPR) framework. Appl. Numer. Math. **95**, 173–198 (2015)
38. J. Du, C.-W. Shu, M. Zhang, A simple weighted essentially non-oscillatory limiter for the correction procedure via reconstruction (CPR) framework on unstructured meshes. Appl. Numer. Math. **90**, 146–167 (2015)
39. B. Einfeldt, C.D. Munz, P.L. Roe, B. Sjögreen, On Godunov-Type methods near low densities. J. Comput. Phys. **92**, 273–295 (1991)
40. L. Fezoui, S. Lanteri, S. Lohrengel, S. Piperno, Convergence and stability of a discontinuous Galerkin time-domain method for the 3D heterogeneous Maxwell equations on unstructured meshes. Math. Model. Numer. Anal. **39**, 1149–1176 (2005)
41. S. Gottlieb, D. Ketcheson, C.-W. Shu, *Strong Stability Preserving Runge-Kutta and Multistep Time Discretizations* (World Scientific, Singapore, 2011)
42. M.J. Grote, A. Schneebeli, D. Schötzau, Discontinuous Galerkin finite element method for the wave equation. SIAM J. Numer. Anal. **44**, 2408–2431 (2006)
43. T. Haga, H. Gao, Z.J. Wang, A high-order unifying discontinuous formulation for the Navier-Stokes equations on 3D mixed grids. Math. Model. Nat. Phenom. **6**, 28–56 (2011)
44. A. Harten, High resolution schemes for hyperbolic conservation laws. J. Comput. Phys. **49**, 357–393 (1983)

45. A. Harten, P.D. Lax, B. van Leer, On upstream differencing and Godunov type schemes for hyperbolic conservation laws. SIAM Rev. **25**, 35–61 (1983)
46. J. Hesthaven, T. Warburton, *Nodal Discontinuous Galerkin Methods* (Springer, New York, 2008)
47. S. Hou, X.-D. Liu, Solutions of multi-dimensional hyperbolic systems of conservation laws by square entropy condition satisfying discontinuous Galerkin method. J. Sci. Comput. **31**, 127–151 (2007)
48. C. Hu, C.-W. Shu, Weighted essentially non-oscillatory schemes on triangular meshes. J. Comput. Phys. **150**, 97–127 (1999)
49. T.J.R. Hughes, J.A. Cottrell, Y. Bazilevs, Isogeometric analysis: CAD, finite elements, NURBS, exact geometry and mesh refinement. Comput. Methods Appl. Mech. Eng. **194**, 4135–4195 (2005)
50. H.T. Huynh, A flux reconstruction approach to high-order schemes including discontinuous Galerkin methods. AIAA Paper 2007–4079 (2007)
51. H.T. Huynh, A reconstruction approach to high-order schemes including discontinuous Galerkin for diffusion. AIAA Paper 2009–403 (2009)
52. L. Ji, Y. Xu, Optimal error estimates of the local discontinuous Galerkin method for Willmore flow of graphs on Cartesian meshes. Int. J. Numer. Anal. Model. **8**, 252–283 (2011)
53. L. Ji, Y. Xu, Optimal error estimates of the local discontinuous Galerkin method for surface diffusion of graphs on Cartesian meshes. J. Sci. Comput. **51**, 1–27 (2012)
54. G.-S. Jiang, C.-W. Shu, On cell entropy inequality for discontinuous Galerkin methods. Math. Comput. **62**, 531–538 (1994)
55. G.-S. Jiang, C.-W. Shu, Efficient implementation of weighted ENO schemes. J. Comput. Phys. **126**, 202–228 (1996)
56. C. Johnson, J. Pitkäranta, An analysis of the discontinuous Galerkin method for a scalar hyperbolic equation. Math. Comput. **46**, 1–26 (1986)
57. G. Kanschat, *Discontinuous Galerkin Methods for Viscous Flow* (Deutscher Universitäts Verlag, Wiesbaden, 2007)
58. A. Klockner, T. Warburton, J. Bridge, J. Hesthaven, Nodal discontinuous Galerkin methods on graphics processors. J. Comput. Phys. **228**, 7863–7882 (2010)
59. D.A. Kopriva, A conservative staggered-grid Chebyshev multidomain method for compressible flows. II: a semi-structured method. J. Comput. Phys. **128**, 475–488 (1996)
60. D.A. Kopriva, A staggered-grid multidomain spectral method for the compressible Navier Stokes equations. J. Comput. Phys. **143**, 125–158 (1998)
61. D.A. Kopriva, J.H. Kolias, A conservative staggered-grid Chebyshev multidomain method for compressible flows. J. Comput. Phys. **125**, 244–261 (1996)
62. L. Krivodonova, J. Xin, J.-F. Remacle, N. Chevaugeon, J.E. Flaherty, Shock detection and limiting with discontinuous Galerkin methods for hyperbolic conservation laws. Appl. Numer. Math. **48**, 323–338 (2004)
63. R.J. LeVeque, *Numerical Methods for Conservation Laws* (Birkhauser, Basel, 1990)
64. D. Levy, C.-W. Shu, J. Yan, Local discontinuous Galerkin methods for nonlinear dispersive equations. J. Comput. Phys. **196**, 751–772 (2004)
65. B. Li, *Discontinuous Finite Elements in Fluid Dynamics and Heat Transfer* (Birkhauser, Basel, 2006)
66. P.-L. Lions, P.E. Souganidis, Convergence of MUSCL and filtered schemes for scalar conservation law and Hamilton-Jacobi equations. Numer. Math. **69**, 441–470 (1995)
67. K. Lipnikov, G. Manzini, M. Shashkov, Mimetic finite difference method. J. Comput. Phys. **257**, 1163–1227 (2014)
68. H. Liu, J. Yan, The direct discontinuous Galerkin (DDG) methods for diffusion problems. SIAM J. Numer. Anal. **47**, 675–698 (2009)
69. H. Liu, J. Yan, The direct discontinuous Galerkin (DDG) methods for diffusion with interface corrections. Commun. Comput. Phys. **8**, 541–564 (2010)
70. X. Liu, S. Osher, T. Chan, Weighted essentially non-oscillatory schemes. J. Comput. Phys. **115**, 200–212 (1994)

71. Y. Liu, M. Vinokur, Z.J. Wang, Spectral difference method for unstructured grids I: basic formulation. J. Comput. Phys. **216**, 780–801 (2006)
72. J. Luo, C.-W. Shu, Q. Zhang, A priori error estimates to smooth solutions of the third order Runge-Kutta discontinuous Galerkin method for symmetrizable systems of conservation laws. ESAIM: Math. Model. Numer. Anal. **49**, 991–1018 (2015)
73. G. May, A. Jameson, A spectral difference method for the Euler and Navier-Stokes equations on unstructured meshes. AIAA Paper 2006–304 (2006)
74. X. Meng, C.-W. Shu, B. Wu, Optimal error estimates for discontinuous Galerkin methods based on upwind-biased fluxes for linear hyperbolic equations. Math. Comput. **85**, 1225–1261 (2016). doi:http://dx.doi.org/10.1090/mcom/3022
75. N.C. Nguyen, J. Peraire, B. Cockburn, An implicit high-order hybridizable discontinuous Galerkin method for nonlinear convection-diffusion equations. J. Comput. Phys. **228**, 8841–8855 (2009)
76. J.T. Oden, I. Babuvska, C.E. Baumann, A discontinuous *hp* finite element method for diffusion problems. J. Comput. Phys. **146**, 491–519 (1998)
77. S. Osher, E. Tadmor, On the convergence of the difference approximations to scalar conservation laws. Math. Comput. **50**, 19–51 (1988)
78. B. Perthame, Second-order Boltzmann schemes for compressible Euler equations in one and two space dimensions. SIAM J. Numer. Anal. **29**, 1–19 (1992)
79. B. Perthame, C.-W. Shu, On positivity preserving finite volume schemes for Euler equations. Numer. Math. **73**, 119–130 (1996)
80. J.-X. Qiu, C.-W. Shu, Hermite WENO schemes and their application as limiters for Runge-Kutta discontinuous Galerkin method: one-dimensional case. J. Comput. Phys. **193**, 115–135 (2003)
81. J.-X. Qiu, C.-W. Shu, A comparison of troubled-cell indicators for Runge-Kutta discontinuous Galerkin methods using weighted essentially nonoscillatory limiters. SIAM J. Sci. Comput. **27**, 995–1013 (2005)
82. J.-X. Qiu, C.-W. Shu, Hermite WENO schemes and their application as limiters for Runge-Kutta discontinuous Galerkin method II: two dimensional case. Comput. Fluids **34**, 642–663 (2005)
83. J.-X. Qiu, C.-W. Shu, Runge-Kutta discontinuous Galerkin method using WENO limiters. SIAM J. Sci. Comput. **26**, 907–929 (2005)
84. J.-M. Qiu, C.-W. Shu, Positivity preserving semi-Lagrangian discontinuous Galerkin formulation: theoretical analysis and application to the Vlasov-Poisson system. J. Comput. Phys. **230**, 8386–8409 (2011)
85. W.H. Reed, T.R. Hill, Triangular mesh methods for the neutron transport equation. Technical Report LA-UR-73-479, Los Alamos Scientific Laboratory (1973)
86. J.-F. Remacle, J. Flaherty, M. Shephard, An adaptive discontinuous Galerkin technique with an orthogonal basis applied to Rayleigh-Taylor flow instabilities. SIAM Rev. **45**, 53–72 (2003)
87. G.R. Richter, An optimal-order error estimate for the discontinuous Galerkin method. Math. Comput. **50**, 75–88 (1988)
88. B. Rivière, *Discontinuous Galerkin Methods for Solving Elliptic and Parabolic Equations. Theory and Implementation* (SIAM, Philadelphia, 2008)
89. B. Rivière, M.F. Wheeler, Discontinuous finite element methods for acoustic and elastic wave problems. Contemp. Math. **329**, 271–282 (2003)
90. J.A. Rossmanith, D.C. Seal, A positivity-preserving high-order semi-Lagrangian discontinuous Galerkin scheme for the Vlasov-Poisson equations. J. Comput. Phys. **230**, 6203–6232 (2011)
91. J. Shi, C. Hu, C.-W. Shu, A technique of treating negative weights in WENO schemes. J. Comput. Phys. **175**, 108–127 (2002)
92. C.-W. Shu, TVB uniformly high-order schemes for conservation laws. Math. Comput. **49**, 105–121 (1987)
93. C.-W. Shu, Discontinuous Galerkin methods: general approach and stability, in *Numerical Solutions of Partial Differential Equations*, ed. by S. Bertoluzza, S. Falletta, G. Russo,

C.-W. Shu. Advanced Courses in Mathematics CRM Barcelona (Birkhäuser, Basel, 2009), pp.149–201
94. C.-W. Shu, S. Osher, Efficient implementation of essentially non-oscillatory shock-capturing schemes. J. Comput. Phys. **77**, 439–471 (1988)
95. A. Stock, J. Neudorfer, M. Riedlinger, G. Pirrung, G. Gassner, R. Schneider, S. Roller, C.-D. Munz, Three-dimensional numerical simulation of a 30-GHz gyrotron resonator with an explicit high-order discontinuous-Galerkin-based parallel particle-in-cell method. IEEE Trans. Plasma Sci. **40**, 1860–1870 (2012)
96. Z.J. Wang, Spectral (finite) volume method for conservation laws on unstructured grids: basic formulation. J. Comput. Phys. **178**, 210–251 (2002)
97. Z.J. Wang, H. Gao, A unifying lifting collocation penalty formulation including the discontinuous Galerkin, spectral volume/difference methods for conservation laws on mixed grids. J. Comput. Phys. **228**, 8161–8186 (2009)
98. Z.J. Wang, Y. Liu, Spectral (finite) volume method for conservation laws on unstructured grids II: extension to two-dimensional scalar equation. J. Comput. Phys. **179**, 665–697 (2002)
99. Z.J. Wang, Y. Liu, Spectral (finite) volume method for conservation laws on unstructured grids III: one-dimensional systems and partition optimization. J. Sci. Comput. **20**, 137–157 (2004)
100. Z.J. Wang, L. Zhang, Y. Liu, Spectral (finite) volume method for conservation laws on unstructured grids IV: extension to two-dimensional Euler equations. J. Comput. Phys. **194**, 716–741 (2004)
101. M.F. Wheeler, An elliptic collocation-finite element method with interior penalties. SIAM J. Numer. Anal. **15**, 152–161 (1978)
102. Y. Xia, Y. Xu, C.-W. Shu, Local discontinuous Galerkin methods for the Cahn-Hilliard type equations. J. Comput. Phys. **227**, 472–491 (2007)
103. Y. Xia, Y. Xu, C.-W. Shu, Application of the local discontinuous Galerkin method for the Allen-Cahn/Cahn-Hilliard system. Commun. Comput. Phys. **5**, 821–835 (2009)
104. Y. Xing, X. Zhang, C.-W. Shu, Positivity preserving high order well balanced discontinuous Galerkin methods for the shallow water equations. Adv. Water Resour. **33**, 1476–1493 (2010)
105. Y. Xing, C.-S. Chou, C.-W. Shu, Energy conserving local discontinuous Galerkin methods for wave propagation problems. Inverse Problems Imaging **7**, 967–986 (2013)
106. Y. Xu, C.-W. Shu, Local discontinuous Galerkin methods for three classes of nonlinear wave equations. J. Comput. Math. **22**, 250–274 (2004)
107. Y. Xu, C.-W. Shu, Local discontinuous Galerkin methods for nonlinear Schrödinger equations. J. Comput. Phys. **205**, 72–97 (2005)
108. Y. Xu, C.-W. Shu, Local discontinuous Galerkin methods for two classes of two dimensional nonlinear wave equations. Physica D **208**, 21–58 (2005)
109. Y. Xu, C.-W. Shu, Local discontinuous Galerkin methods for the Kuramoto-Sivashinsky equations and the Ito-type coupled KdV equations. Comput. Methods Appl. Mech. Eng. **195**, 3430–3447 (2006)
110. Y. Xu, C.-W. Shu, Error estimates of the semi-discrete local discontinuous Galerkin method for nonlinear convection-diffusion and KdV equations. Comput. Methods Appl. Mech. Eng. **196**, 3805–3822 (2007)
111. Y. Xu, C.-W. Shu, A local discontinuous Galerkin method for the Camassa-Holm equation. SIAM J. Numer. Anal. **46**, 1998–2021 (2008)
112. Y. Xu, C.-W. Shu, Local discontinuous Galerkin method for surface diffusion and Willmore flow of graphs. J. Sci. Comput. **40**, 375–390 (2009)
113. Y. Xu, C.-W. Shu, Local discontinuous Galerkin methods for high-order time-dependent partial differential equations. Commun. Comput. Phys. **7**, 1–46 (2010)
114. Y. Xu, C.-W. Shu, Optimal error estimates of the semi-discrete local discontinuous Galerkin methods for high order wave equations. SIAM J. Numer. Anal. **50**, 79–104 (2012)
115. J. Yan, C.-W. Shu, A local discontinuous Galerkin method for KdV type equations. SIAM J. Numer. Anal. **40**, 769–791 (2002)

116. J. Yan, C.-W. Shu, Local discontinuous Galerkin methods for partial differential equations with higher order derivatives. J. Sci. Comput. **17**, 27–47 (2002)
117. Y. Yang, C.-W. Shu, Discontinuous Galerkin method for hyperbolic equations involving δ-singularities: negative-order norm error estimates and applications. Numer. Math. **124**, 753–781 (2013)
118. Y. Yang, D. Wei, C.-W. Shu, Discontinuous Galerkin method for Krause's consensus models and pressureless Euler equations. J. Comput. Phys. **252**, 109–127 (2013)
119. L. Yuan, C.-W. Shu, Discontinuous Galerkin method based on non-polynomial approximation spaces. J. Comput. Phys. **218**, 295–323 (2006)
120. M. Zhang, C.-W. Shu, An analysis of three different formulations of the discontinuous Galerkin method for diffusion equations. Math. Models Methods Appl. Sci. **13**, 395–413 (2003)
121. Q. Zhang, C.-W. Shu, Error estimates to smooth solutions of Runge-Kutta discontinuous Galerkin methods for scalar conservation laws. SIAM J. Numer. Anal. **42**, 641–666 (2004)
122. Q. Zhang, C.-W. Shu, Error estimates to smooth solutions of Runge-Kutta discontinuous Galerkin method for symmetrizable systems of conservation laws. SIAM J. Numer. Anal. **44**, 1703–1720 (2006)
123. Y.-T. Zhang, C.-W. Shu, Third order WENO scheme on three dimensional tetrahedral meshes. Commun. Comput. Phys. **5**, 836–848 (2009)
124. X. Zhang, C.-W. Shu, On maximum-principle-satisfying high order schemes for scalar conservation laws. J. Comput. Phys. **229**, 3091–3120 (2010)
125. X. Zhang, C.-W. Shu, On positivity preserving high order discontinuous Galerkin schemes for compressible Euler equations on rectangular meshes. J. Comput. Phys. **229**, 8918–8934 (2010)
126. Q. Zhang, C.-W. Shu, Stability analysis and a priori error estimates to the third order explicit Runge-Kutta discontinuous Galerkin method for scalar conservation laws. SIAM J. Numer. Anal. **48**, 1038–1063 (2010)
127. X. Zhang, C.-W. Shu, Maximum-principle-satisfying and positivity-preserving high order schemes for conservation laws: survey and new developments. Proc. R. Soc. A **467**, 2752–2776 (2011)
128. X. Zhang, C.-W. Shu, Positivity-preserving high order discontinuous Galerkin schemes for compressible Euler equations with source terms. J. Comput. Phys. **230**, 1238–1248 (2011)
129. X. Zhang, C.-W. Shu, A minimum entropy principle of high order schemes for gas dynamics equations. Numer. Math. **121**, 545–563 (2012)
130. Q. Zhang, C.-W. Shu, Error estimates for the third order explicit Runge-Kutta discontinuous Galerkin method for linear hyperbolic equation in one-dimension with discontinuous initial data. Numer. Math. **126**, 703–740 (2014)
131. X. Zhang, Y. Xia, C.-W. Shu, Maximum-principle-satisfying and positivity-preserving high order discontinuous Galerkin schemes for conservation laws on triangular meshes. J. Sci. Comput. **50**, 29–62 (2012)
132. Y. Zhang, X. Zhang, C.-W. Shu, Maximum-principle-satisfying second order discontinuous Galerkin schemes for convection-diffusion equations on triangular meshes. J. Comput. Phys. **234**, 295–316 (2013)
133. Y. Zhang, W. Wang, J. Guzmán, C.-W. Shu, Multi-scale discontinuous Galerkin method for solving elliptic problems with curvilinear unidirectional rough coefficients. J. Sci. Comput. **61**, 42–60 (2014)
134. X. Zhong, C.-W. Shu, A simple weighted essentially nonoscillatory limiter for Runge-Kutta discontinuous Galerkin methods. J. Comput. Phys. **232**, 397–415 (2012)
135. J. Zhu, J.-X. Qiu, C.-W. Shu, M. Dumbser, Runge-Kutta discontinuous Galerkin method using WENO limiters II: unstructured meshes. J. Comput. Phys. **227**, 4330–4353 (2008)
136. J. Zhu, X. Zhong, C.-W. Shu, J.-X. Qiu, Runge-Kutta discontinuous Galerkin method using a new type of WENO limiters on unstructured meshes. J. Comput. Phys. **248**, 200–220 (2013)
137. J. Zhu, X. Zhong, C.-W. Shu, J.-X. Qiu, Runge-Kutta discontinuous Galerkin method with a simple and compact Hermite WENO limiter. Commun. Comput. Phys. **19**, 944–969 (2016)

Foundations of the MHM Method

Christopher Harder and Frédéric Valentin

Abstract An abstract setting for the construction and analysis of the Multiscale Hybrid-Mixed (MHM for short) method is proposed. We review some of the most recent developments from this standpoint, and establish relationships with the classical lowest-order Raviart-Thomas element and the primal hybrid method, as well as with some recent multiscale methods. We demonstrate the reach of the approach by revisiting the wellposedness and error analysis of the MHM method applied to the Laplace problem. In the process, we devise new theoretical results for this model.

1 Introduction

In the last decade, there has been an extensive development of massive parallel computer architectures. With single processors limited by technical issues such as heating, computers have instead been built to leverage a large number of processors (grouped in cores) of mild speed and storage capacities. This new paradigm has led to a revision of what is expected from simulators from the viewpoint of numerical algorithms. Although precision and robustness remain fundamental properties of numerical methods for extreme-scale computational science, the underlying algorithms must be naturally shaped to take advantage of the new massively parallel architectures and built-in fault tolerance (see [40] for an interesting overview). In fact, failure is a certainty in long-time simulations on such new generation parallel architectures since at least one processor is certain to break down in such a scenario

C. Harder
Metropolitan State University of Denver, P.O. Box 173362, Campus Box 38, Denver, CO 80217-3362, USA
e-mail: harderc@msudenver.edu

F. Valentin (✉)
National Laboratory for Scientific Computing - LNCC, Av. Getúlio Vargas, 333, 25651-070 Petrópolis, RJ, Brazil
e-mail: valentin@lncc.br

G.R. Barrenechea et al. (eds.), *Building Bridges: Connections and Challenges in Modern Approaches to Numerical Partial Differential Equations*,
Lecture Notes in Computational Science and Engineering 114,
DOI 10.1007/978-3-319-41640-3_13

[14]. Consequently, numerical methods for extreme-scale computing should induce asynchronous and communication-avoiding algorithms to prevent loss of long-term simulations entirely.

Numerical methods built upon the "divide-and-conquer" philosophy satisfy the architectural imperatives of high-performance computers better than classical methods operating only on the finest scale of the discretization. Indeed, splitting the computation of extreme simulations into a set of independent problems of smaller size turns out to be a way to circumvent faults and to allow spatial and time data locality while taking full advantage (in terms of performance) of the granularity of the new generation of computer architectures. There is a panoply of possibilities in the literature, starting in the eighties with the domain decomposition technique [24] (see [42] for a survey). Also, the residual-free bubbles (RFB) [8, 11–13], the variational multiscale method [29, 30], and the enriched finite element methods [1, 2, 9, 17, 23] may be seen from this perspective, although they were not originally presented this way; being devised for precision on coarse meshes at the price of solving element-wise independent problems, these methods carry in their construction the desired approach. Hybridization has been also used recently in the context of Discontinuous Galerkin method in [15, 37] leading to methods that share this philosophy.

In this context, multiscale numerical methods appeared as an attractive "divide-and-conquer" option to handle heterogeneous problems (see [20, 43, 44], just to cite a few). The approach started with the pioneering work by Babuška and Osborn [7] and was further extended to higher dimensions by Hou and Wu [28]. The latter has been proved to be closely related to the RFB method proposed in [41]. Overall, the idea relies on basis functions specially designed to upscale submesh scales to an overlying coarse mesh. As a result, such numerical methods become precise on coarse meshes. Particularly interesting is the fact that the multiscale basis functions can be locally computed through completely independent problems.

Recently, a new family of multiscale finite element methods, named Multiscale Hybrid-Mixed (MHM) method, was introduced in [26] and further analyzed in [3]. The MHM method is devised from the primal hybridization of the original formulation as proposed in [39] and allowed to localize computations. This is made possible by the characterization of the exact solution in terms of the solution of a global formulation posed on the skeleton of a (coarse) partition of the domain, and the solution of independent local problems. The Lagrange multipliers play the role of Neumann boundary conditions for the local problems. Such a decomposition drives discretization, decouples the global and local problems and gives rise to the following staggered algorithm: given a coarse partition of the domain, compute

- the multiscale basis functions from independent element-wise problems, and
- the degrees of freedom on faces from the global face-based formulation.

The MHM method has a notably general formulation that recovers some well-established finite element methods, such as the ones proposed in [16, 38, 39], under appropriate hypotheses. It also shares the same goals of the multiscale mortar mixed finite element method [6] and the spectral multiscale hybridizable discontinuous

Galerkin method [21], and the multiscale method in [32] with a different viewpoint which induces a different algorithm. The method requires neither scale separation nor periodicity of the media when used for highly heterogenous coefficient problems. Moreover, it produces precise numerical primal and dual variables, with respect to the characteristic size of the mesh (c.f. [3]) and is shown to be robust with respect to small physical coefficients (c.f. [35]).

To be more precise, let us illustrate the idea of the MHM formulation for the Laplace problem. Let $\Omega \subset \mathbb{R}^d$, $d \in \{2, 3\}$, with a polygonal boundary $\partial\Omega$. The standard weak formulation consists of finding $u \in H_0^1(\Omega)$ such that

$$\int_\Omega \kappa\, \nabla u \cdot \nabla v\, d\mathbf{x} = \int_\Omega f\, v\, d\mathbf{x} \quad \text{for all } v \in H_0^1(\Omega)\,, \tag{1}$$

where κ is a positive definite second-order tensor which is assumed to be uniformly bounded and $f \in L^2(\Omega)$. In this setting, Raviart and Thomas [39] considered (taking κ equal to the identity tensor) the primal hybrid version of (1) on a family of regular partitions $\{\mathcal{T}_H\}_{H>0}$ of Ω composed of elements K with boundary ∂K. With the space of Lagrange multipliers M given by (the spaces having their usual meaning, see [33])

$$M := \left\{\sigma \cdot \boldsymbol{n}_K|_{\partial K} \in H^{-1/2}(\partial K),\ \forall K \in \mathcal{T}_H\ :\ \sigma \in H(\text{div};\Omega)\right\}\,,$$

where $\boldsymbol{n}_K$ stands for the unit outward normal vector on ∂K, the primal hybrid formulation of the problem is to find $u \in H^1(\mathcal{T}_H)$ and $\lambda \in M$ such that

$$\begin{cases} \displaystyle\sum_{K\in\mathcal{T}_H}\left[\int_K \kappa\, \nabla u \cdot \nabla v\, d\mathbf{x} + (\lambda\,, v)_{\partial K}\right] = \sum_{K\in\mathcal{T}_H}\int_K f\, v\, d\mathbf{x} \quad \text{for all } v \in H^1(\mathcal{T}_H)\,, \\ \displaystyle\sum_{K\in\mathcal{T}_H}(\mu\,, u)_{\partial K} = 0 \quad \text{for all } \mu \in M\,. \end{cases} \tag{2}$$

Here $H^1(\mathcal{T}_H)$ stands for the functions in $L^2(\Omega)$ such that their restriction to element $K \in \mathcal{T}_H$ belongs to $H^1(K)$ and $(\cdot,\cdot)_{\partial K}$ is the $H^{-1/2}(\partial K) \times H^{1/2}(\partial K)$ duality product. It has been proved (see [39] for instance) that Problem (2) has a unique solution. Moreover, $u \in H_0^1(\Omega)$ is the solution to problem (1) and $\lambda = -\kappa\, \nabla u \cdot \boldsymbol{n}_K$ on ∂K for each $K \in \mathcal{T}_H$.

In [3] it is shown that problem (2) can be stated in an equivalent global-local formulation. In a broad sense, we observe that $H^1(\mathcal{T}_H)$ decomposes into the space of piecewise constants V_0 and its orthogonal complement, i.e.,

$$H^1(\mathcal{T}_H) = V_0 \oplus V_0^\perp\,,$$

where $V_0^\perp := H^1(\mathcal{T}_H) \cap L_0^2(\mathcal{T}_H)$ ($L_0^2(\mathcal{T}_H)$ stands for the functions in $L^2(\Omega)$ with mean value equal to zero in each $K \in \mathcal{T}_H$). Testing the first equation of (2) against $V_0^\perp$ shows that $u_0^\perp$, which is the part of the solution u belonging to $V_0^\perp$, may be characterized as

$$u_0^\perp = T\lambda + \hat{T}f.$$

Here, the bounded linear operators T, $\hat{T}$ with image in $V_0^\perp$ are defined via local problems. Specifically, given $\mu \in M$ and $q \in L^2(\Omega)$, for each $K \in \mathcal{T}_H$, $T\mu|_K$ and $\hat{T}q|_K$ satisfy,

$$\int_K \kappa \nabla T\mu \, \nabla w \, d\mathbf{x} = -(\mu, w)_{\partial K} \quad \text{for all } w \in H^1(K) \cap L_0^2(K), \tag{3}$$

$$\int_K \kappa \, \nabla \hat{T} q \, \nabla w \, d\mathbf{x} = \int_K q \, w \, d\mathbf{x} \quad \text{for all } w \in H^1(K) \cap L_0^2(K). \tag{4}$$

Hence, the exact solution u of (2) may be decomposed as follows

$$u = u_0 + T\lambda + \hat{T}f,$$

where $u_0 \in V_0$. Now, testing (2) against $M \times V_0$ shows that $(\lambda, u_0) \in M \times V_0$ satisfy,

$$\begin{cases} \displaystyle\sum_{K\in\mathcal{T}_H} (\lambda, v_0)_{\partial K} = \sum_{K\in\mathcal{T}_H} \int_K f \, v_0 \, d\mathbf{x} \quad \text{for all } v_0 \in V_0, \\ \displaystyle\sum_{K\in\mathcal{T}_H} [(\mu, T\lambda)_{\partial K} + (\mu, u_0)_{\partial K}] = -\sum_{K\in\mathcal{T}_H} (\mu, \hat{T}f)_{\partial K} \quad \text{for all } \mu \in M, \end{cases} \tag{5}$$

which is the global part of the formulation.

The MHM method stems from the coupled problems (3)–(5). Selecting a finite dimensional subspace $M_H \subset M$ allows for finding solutions to (3) in terms of basis functions and gives rise to a one-level MHM method in the form of (5). In such a case, one assumes that the corresponding local problems are computed exactly, i.e, a closed formula for the multiscale basis functions is available. The wellposedness and *a priori* and *a posteriori* error estimates of the one-level MHM method were addressed in [3, 35].

Although particular cases exist where a closed formula for solutions to local problems (3) and (4) are known (c.f. [26]), the solutions must generally be approximated. This yields the two-level MHM method. The strategy is to select a finite dimensional subspace $V_h(K)$ of $H^1(K) \cap L_0^2(K)$ (which may be finite element spaces which are different in each K) and then set up a numerical method T_h and $\hat{T}_h$ at the second level. These choices are general and "only" require approximation properties for T_h and $\hat{T}_h$. The underlying MHM method reads: Find $(\lambda_H, u_0^{H,h}) \in$

$M_H \times V_0$ such that

$$\begin{cases} \displaystyle\sum_{K\in\mathcal{T}_H}\int_{\partial K}\lambda_H\, v_0 = \sum_{K\in\mathcal{T}_H}\int_K f\, v_0 \quad \text{for all } v_0 \in V_0\,, \\ \displaystyle\sum_{K\in\mathcal{T}_H}\left[\int_{\partial K}\mu_H\, T_h\,\lambda_H + \int_{\partial K}\mu_H\, u_0^{H,h}\right] = -\sum_{K\in\mathcal{T}_H}\int_{\partial K}\mu_H\,\hat{T}_h f \quad \text{for all } \mu_H \in M_H\,. \end{cases} \tag{6}$$

Some comments are necessary at this point. First, observe that the first equation in (6) assures the discrete local conservation with respect to external forces, and the second equation is responsible, through the action of T_h and $\hat{T}_h$ on the basis functions of M_H and f, for upscaling information "lost" by the mesh. High order of convergence (as well as super-convergence) is achieved by increasing the quality of approximation of the Lagrange λ_H on faces. Interestingly, this can be done independently on each face which makes the MHM method particularly attractive to be used within space adaptive algorithms (see [3, 27] for instance).

It is also worth pointing out how the MHM method matches the modern massively parallel architectures. Observe that global formulation (6) is responsible for coupling the degrees of freedoms, and as such, it could be the source of the standard difficulties with respect to parallelization. But, the computational effort in solving such a global problem is drastically decreased as it overlies on top of a coarse mesh skeleton with only face-based degrees of freedom involved for λ_H and a degree of freedom for each $K \in \mathcal{T}_H$. Also, owing to this feature, the MHM method (6) can undertake a second level of hybridization which leads to a positive definite global problem to be solved in place of the mixed one in (6). In conclusion, the computational cost involved in obtaining the degrees of freedom in (6) is completely overshadowed by the local basis computations. The good news is that, although there are many local problems, they are entirely local and independent to one another, and thus, they match perfectly to the architecture provided by the modern extreme-scale computers. Scalability of the MHM method is currently under investigation.

Extensions to the linear elasticity and the advective-reactive dominated models have been proposed in [25, 27], respectively. For the former, an analysis of the two-level MHM method was also addressed. The analysis of the various operators, although performed case-by-case, shares a number of common points. Thereby, the primary goal of this work is to set up an abstract setting that accounts for the analysis of the one- and two-level MHM methods in a broad sense. To this end, we uncover the mathematical mechanism from which the MHM global-local formulation derives from the classical primal hybrid formulation [39]. The wellposedness of the MHM formulation is thus shown to be a consequence of the wellposedness of the its classical primal hybrid counterpart. This interconnection is paramount to establish existence and uniqueness of solutions as well as optimality of the errors for the MHM method in the context of subspaces. On the other hand, the classical discrete primal hybrid method in [39] may be recovered within the

MHM formalism, yielding a new characterization of its numerical solution. Also, the relationship of the MHM method with some classical (the Raviart-Thomas element) and well-established recent methods (the MsFEM [16] and the subgrid upscaling method, UpFEM, [4]) is highlighted, and some comments with respect to the similarities/differences between the MHM method and the DPG [19] and the HDG method [21] are given. Finally, a new analysis of an existing MHM method is connected to the framework, with particular emphasis on the general properties of the approximations generated by solving the local problems approximatively. The interested reader can also find a related work on this direction in [31].

The rest of this work is outlined as follows. The abstract problem is posed and analyzed in Sect. 2. Section 3 is dedicated to wellposedness and best approximation properties within this general context. Previous sections are then connected to the MHM method applied to the Laplace problem in Sect. 4. Particularly, this section introduces the one-level and a new two-level analysis of the MHM method for the Laplace problem in the context of subspaces. Sect. 5 is devoted to establishing relationships with other approaches. Conclusions are stated in Sect. 6.

2 General Setting of the MHM Methods

Let W and X denote Hilbert spaces with associated inner products $(.,.)_W$ and $(.,.)_X$, respectively, and Λ and M denote reflexive Banach spaces with norms $\|.\|_\Lambda$ and $\|.\|_M$. As usual, Λ' and M' mean the dual spaces of Λ and M, and $\langle\cdot,\cdot\rangle_\Lambda$ and $\langle\cdot,\cdot\rangle_M$ their respective duality products. Furthermore, we suppose $\mathcal{A} : W \to X$, $\mathcal{B} : \Lambda \to X$, and $\mathcal{C} : M \to W$ are bounded linear operators.

Consider the problem, for $f \in X$ and $g \in M'$: Find $w \in W$ and $\lambda \in \Lambda$ such that,

$$\begin{cases} (\mathcal{A}w, x)_X + (\mathcal{B}\lambda, x)_X = (f, x)_X & \text{for all } x \in X, \\ (\mathcal{C}\mu, w)_W = \langle g, \mu\rangle_M & \text{for all } \mu \in M. \end{cases} \tag{7}$$

We recognize the standard abstract form of mixed or hybrid formulations [10]. In this work, it will be explored in the context of primal hybrid formulations, with the Laplace model serving as an example in Sect. 4.

Since the primal hybrid formulation (2) fits (7), we look next for an equivalent form of (7) that includes the MHM formulation. The nullspace $N(\mathcal{A})$ and the range $R(\mathcal{A})$ of $\mathcal{A}$ play an integral part in the development of such an equivalence. Recall the decompositions (see, e.g., [22]),

$$W = N(\mathcal{A}) \oplus \overline{R(\mathcal{A}^T)} \quad \text{and} \quad X = N\left(\mathcal{A}^T\right) \oplus \overline{R(\mathcal{A})}. \tag{8}$$

As usual, $\mathcal{A}^T$ denotes the adjoint operator of $\mathcal{A}$, i.e., $\mathcal{A}^T : X \to W$ given by

$$(w, \mathcal{A}^T x)_W := (\mathcal{A}w, x)_X .$$

The analysis will also make use of the adjoint operators $\mathcal{B}^T : X \to \Lambda'$ and $\mathcal{C}^T : W \to M'$ given by

$$\langle \mathcal{B}^T x, \nu \rangle_\Lambda := (\mathcal{B}\nu, x)_X\,, \quad \text{and} \quad \langle \mathcal{C}^T w, \mu \rangle_M := (\mathcal{C}\mu, w)_W\,.$$

Also central to the definition of the equivalent form of (7) is existence of the generalized inverse of operator $\mathcal{A}$, denoted by $\mathcal{A}^\dagger : X \to W$. In order to ensure the generalized inverse is bounded, $\mathcal{A}$ in (7) is assumed to have closed range. In this context, the definition of generalized inverse and some properties are recalled (see, e.g., [34, Definition 1.1]).

Lemma 1 *Suppose $\mathcal{A} : W \to X$ is a bounded operator with closed range. Then, the generalized inverse operator $\mathcal{A}^\dagger$ is uniquely defined by*

$$\mathcal{A}^\dagger := \begin{cases} \mathcal{A}_1^{-1} & \text{on } R(\mathcal{A})\,, \\ 0 & \text{on } R(\mathcal{A})^\perp\,, \end{cases} \tag{9}$$

where $\mathcal{A}_1^{-1} : R(\mathcal{A}) \to R(\mathcal{A}^T)$ stands for the bounded inverse of $\mathcal{A}_1 := \mathcal{A}|_{R(\mathcal{A}^T)}$. Moreover, the following properties hold

$$\mathcal{A}\mathcal{A}^\dagger\mathcal{A} = \mathcal{A}, \quad \mathcal{A}^\dagger\mathcal{A}\mathcal{A}^\dagger = \mathcal{A}^\dagger, \quad \mathcal{A}^\dagger\mathcal{A} = \mathcal{P}_{R(\mathcal{A}^T)}, \quad \mathcal{A}\mathcal{A}^\dagger = \mathcal{P}_{R(\mathcal{A})}, \tag{10}$$

where $\mathcal{P}_{R(\mathcal{A}^T)}$ and $\mathcal{P}_{R(\mathcal{A})}$ are orthogonal projections onto $R(\mathcal{A}^T)$ and $R(\mathcal{A})$ with respect to the inner-products in W and X, respectively.

Proof See [34, page 317] and [36, Theorem 1].

Under the assumptions of Lemma 1, an abstract version of problem (7) emerges: Find $w_0 \in N(\mathcal{A})$ and $\lambda \in \Lambda$ such that

$$\begin{cases} -(\mathcal{C}\mu, \mathcal{A}^\dagger\mathcal{B}\lambda)_W + (\mathcal{C}\mu, w_0)_W = \langle g, \mu \rangle_M - (\mathcal{C}\mu, \mathcal{A}^\dagger f)_W & \text{for all } \mu \in M, \\ (\mathcal{B}\lambda, x_0)_X = (f, x_0)_X & \text{for all } x_0 \in N(\mathcal{A}^T)\,. \end{cases} \tag{11}$$

This formulation is justified in Theorem 1 below. It will be shown in Sect. 4 that the global formulation (5) (which induces the MHM method and uses operators defined locally via (3) and (4)) fits (11). First, we establish the equivalence between problems (7) and (11). The following technical result, which is a consequence of Lemma 1 is needed.

Lemma 2 *Under the conditions of Lemma 1, $f \in X$, $u \in W$, and $\nu \in \Lambda$ satisfy $\mathcal{A}u + \mathcal{B}\nu = f$ if and only if*

(i) $\mathcal{P}_{R(\mathcal{A}^T)}u = \mathcal{A}^\dagger f - \mathcal{A}^\dagger\mathcal{B}\nu\,,$
(ii) $\mathcal{P}_{N(\mathcal{A}^T)}(f - \mathcal{B}\nu) = 0\,,$

where $\mathcal{P}_{N(\mathcal{A}^T)} := I - \mathcal{P}_{R(\mathcal{A})}$, and I is the identity operator.

Proof Assume $\mathcal{A}u = f - \mathcal{B}\nu$. The first condition is a result of applying $\mathcal{A}^\dagger$ on both sides and using the properties of the generalized inverse in Lemma 1. Likewise, the second condition follows upon applying $\mathcal{P}_{N(\mathcal{A}^T)}$ on both sides and observing that $\mathcal{P}_{N(\mathcal{A}^T)}\mathcal{A}u = 0$. Now, assume conditions (*i*) and (*ii*) hold. Using Lemma 1 and the first condition,

$$\begin{aligned}\mathcal{A}u = \mathcal{A}\mathcal{A}^\dagger\mathcal{A}u &= \mathcal{A}\mathcal{P}_{R(\mathcal{A}^T)}u \\ &= \mathcal{A}\mathcal{A}^\dagger(f - \mathcal{B}\nu) \\ &= \mathcal{P}_{R(\mathcal{A})}(f - \mathcal{B}\nu).\end{aligned}$$

This together with the second condition implies $\mathcal{A}u = f - \mathcal{B}\nu$.

Remark 1 In the case $\mathcal{A}u = f - \mathcal{B}\nu$ is a problem statement for given $f \in X$ and $\nu \in \Lambda$, condition (*i*) in Lemma 2 discusses solvability in the factor space $W/N(\mathcal{A})$ and condition (*ii*) indicates a compatibility requirement on the data f and ν. The prototypical example for these two conditions is the Laplace problem posed with Neumann boundary conditions.

Now, we consider the main result of this section establishing the equivalence between (7) and (11).

Theorem 1 *Consider $f \in X$ and $g \in M'$. Under the conditions of Lemma 1, problem* (7) *admits a unique solution $(w, \lambda) \in W \times \Lambda$ if and only if problem* (11) *admits a unique solution $(w_0, \lambda) \in N(\mathcal{A}) \times \Lambda$. Moreover, the following characterization holds*

$$w = w_0 - \mathcal{A}^\dagger\mathcal{B}\lambda + \mathcal{A}^\dagger f. \tag{12}$$

Proof Denote by $w \in W$ and $\lambda \in \Lambda$ the unique solution to problem (7), which in operator form is,

$$\begin{cases}\mathcal{A}w + \mathcal{B}\lambda = f, \\ \mathcal{C}^T w = g.\end{cases} \tag{13}$$

By condition (*i*) in Lemma 2, observe w decomposes as,

$$w = w_0 - \mathcal{A}^\dagger\mathcal{B}\lambda + \mathcal{A}^\dagger f, \tag{14}$$

where $w_0 := (\boldsymbol{I} - \mathcal{P}_{R(\mathcal{A}^T)})w := \mathcal{P}_{N(\mathcal{A})}w$. Substituting (14) into the second equation of (13) and using condition (*ii*) in Lemma 2 yields,

$$\begin{cases}\mathcal{P}_{N(\mathcal{A}^T)}\mathcal{B}\lambda = \mathcal{P}_{N(\mathcal{A}^T)}f, \\ -\mathcal{C}^T\mathcal{A}^\dagger\mathcal{B}\lambda + \mathcal{C}^T w_0 = g - \mathcal{C}^T\mathcal{A}^\dagger f.\end{cases} \tag{15}$$

So (15), which is the operator form of (11), is satisfied by w_0 and λ.

Now, suppose $u_0 \in N(\mathcal{A})$ and $\nu \in \Lambda$ also satisfy (15) and define

$$u := u_0 - \mathcal{A}^{\dagger}\mathcal{B}\nu + \mathcal{A}^{\dagger}f . \tag{16}$$

It follows immediately from the second equation in (15) that

$$\mathcal{C}^T u = g.$$

Moreover, using properties (10) and the definition of u,

$$\mathcal{A}u = -\mathcal{A}\mathcal{A}^{\dagger}\mathcal{B}\nu + \mathcal{A}\mathcal{A}^{\dagger}f = \mathcal{P}_{R(\mathcal{A})}\left(-\mathcal{B}\nu + f\right) ,$$

from which the first equation in (15) implies,

$$\mathcal{A}u = \mathcal{P}_{R(\mathcal{A})}\left(-\mathcal{B}\nu + f\right) + \mathcal{P}_{N\left(\mathcal{A}^T\right)}\left(-\mathcal{B}\nu + f\right) = -\mathcal{B}\nu + f .$$

It follows that u and ν satisfy (13). Therefore, $u = w$ and $\nu = \lambda$ by the assumed uniqueness of solution of (13), which yields $w_0 = u_0$. This establishes that (15) has a unique solution if (13) has a unique solution.

Now, suppose (15) has a unique solution $u_0 \in N(\mathcal{A})$ and $\nu \in \Lambda$. By the second paragraph above, $u \in W$ defined in (16) and ν satisfy (13). If w and λ also satisfy (13), it follows from the first paragraph that w_0 and λ also satisfy (15). By uniqueness of the solution of (15), $w_0 = u_0$ and $\lambda = \nu$ so that $u = w$. Therefore, (13) has a unique solution if (15) has a unique solution.

2.1 The Subspace Case

Problems (7) and (11) are now reformulated in terms of closed subspaces $M_s \subset M$, $\Lambda_s \subset \Lambda$, $W_s \subset W$, and $X_s \subset X$. Suppose $\mathcal{A}_s : W_s \to X_s$ is a bounded linear map with closed range, and denote its generalized inverse by $\mathcal{A}_s^{\dagger} : X_s \to W_s$ defined through Lemma 1 (with $\mathcal{A}_s$ in place of $\mathcal{A}$). Let $\mathcal{B}_s : \Lambda_s \to X_s$ and $\mathcal{C}_s : M_s \to W_s$ be the bounded linear maps induced by $\mathcal{B}$ and $\mathcal{C}$, i.e.,

$$(\mathcal{B}_s\lambda_s, x_s)_X = (\mathcal{B}\lambda_s, x_s)_X \quad \text{and} \quad (\mathcal{C}_s\mu_s, w_s)_W = (\mathcal{C}\mu_s, w_s)_W . \tag{17}$$

In this setting, problem (7) may be recast as: Given $f \in X$ and $g \in M'$, find $w_s \in W_s$ and $\lambda_s \in \Lambda_s$ such that

$$\begin{cases} (\mathcal{A}_s\, w_s, x_s)_X + (\mathcal{B}_s\, \lambda_s, x_s)_X = (f, x_s)_X & \text{for all } x_s \in X_s, \\ (\mathcal{C}_s\, \mu_s, w_s)_W = \langle g, \mu_s \rangle_M & \text{for all } \mu_s \in M_s. \end{cases} \tag{18}$$

This problem, as established in Corollary 1 below, is equivalent to: Find $w_0 \in N(\mathcal{A}_s)$ and $\lambda_s \in \Lambda_s$ such that

$$\begin{cases} -(\mathcal{C}_s\,\mu_s, \mathcal{A}_s^\dagger\,\mathcal{B}_s\,\lambda_s)_W + (\mathcal{C}_s\,\mu_s, w_0)_W = \langle g, \mu_s\rangle_M - (\mathcal{C}_s\,\mu_s, \mathcal{A}_s^\dagger\mathcal{P}_{X_s}f)_W & \text{for all } \mu_s \in M_s\,, \\ (\mathcal{B}_s\,\lambda_s, x_0)_X = (f, x_0)_X \quad \text{for all } x_0 \in N\left(\mathcal{A}_s^T\right)\,, \end{cases} \tag{19}$$

where $\mathcal{P}_{X_s}$ is the X-orthogonal projection onto X_s.

It is important to establish connections between (18) and (19) in anticipation of an error analysis in the next section.The following result is a straight-forward application of Theorem 1.

Corollary 1 *Let operator* $\mathcal{A}_s : W_s \to X_s$ *be bounded with closed range. Problem* (18) *admits a unique solution* $(w_s, \lambda_s) \in W_s \times \Lambda_s$ *if and only if problem* (19) *admits a unique solution* $(w_0, \lambda_s) \in N(\mathcal{A}_s) \times \Lambda_s$. *Furthermore, it holds*

$$w_s = w_0 - \mathcal{A}_s^\dagger\mathcal{B}_s\,\lambda_s + \mathcal{A}_s^\dagger\mathcal{P}_{X_s}f\,. \tag{20}$$

Proof Problem (18) may be expressed in operator form as

$$\begin{cases} \mathcal{A}_s w_s + \mathcal{B}_s\,\lambda_s = \mathcal{P}_{X_s}f, \\ \mathcal{C}_s^T\,w_s = g_s\,, \end{cases} \tag{21}$$

where g_s stands for the functional g acting on the space M_s. The result follows analogously the proof of Theorem 1, using the operators $\mathcal{A}_s$, $\mathcal{B}_s$, and $\mathcal{C}_s$ in place of $\mathcal{A}$, $\mathcal{B}$, and $\mathcal{C}$, respectively.

Remark 2 There exists the possibility to make use of subspaces Λ_s and M_s but insist that the entire spaces W and X, as well as operator $\mathcal{A}$, are used in (11). Such a statement is not merely academic, and it will be in the genesis of the one-level MHM method. The related version of (11) reads: Find $w_0 \in N(\mathcal{A})$ and $\lambda_s \in \Lambda_s$ such that

$$\begin{cases} -(\mathcal{C}\mu_s, \mathcal{A}^\dagger\mathcal{B}\,\lambda_s)_W + (\mathcal{C}\,\mu_s, w_0)_W = \langle g, \mu_s\rangle_M - (\mathcal{C}\,\mu_s, \mathcal{A}^\dagger f)_W & \text{for all } \mu_s \in M_s\,, \\ (\mathcal{B}\,\lambda_s, x_0)_X = (f, x_0)_X \quad \text{for all } x_0 \in N\left(\mathcal{A}^T\right)\,. \end{cases} \tag{22}$$

Remark 3 Result (20) is remarkable since it indicates that the solution w_s of (11) may be formed by searching for an element of the *subspace* of W_s generated by the image of operator $\mathcal{A}_s^\dagger\mathcal{B}_s$ restricted to Λ_s, an element of the kernel of $\mathcal{A}_s$, and the image of f through operator $\mathcal{A}_s^\dagger\mathcal{P}_{X_s}$. This structure will be explored more deeply in the construction of the MHM methods.

The next section provides the analysis of problem (19).

3 Wellposedness and Best Approximation

In the last section, wellposedness of (18) (defined or not in terms of proper subspaces) was shown, in the context of operator $\mathcal{A}_s$ which is bounded with closed range, to be equivalent to the wellposedness of problem (19). In this section, best approximation results are provided for problem (19). First, we recall the standard necessary and sufficient conditions for the abstract problem (18) to be well-posed (see, e.g., [10]). These induce the conditions of Corollary 1 for (19) to be well-posed.

It is well known that necessary and sufficient conditions for problem (18) (and (7)) to be well-posed are the existence of positive constants $\alpha_{s,1}$, $\alpha_{s,2}$ and $\alpha_{s,3}$ such that

$$
\begin{aligned}
\alpha_{s,1}\|w\|_W &\leqslant \sup_{x\in N\left(\mathcal{B}_s^T\right)} \frac{(\mathcal{A}_s w, x)_X}{\|x\|_X} \quad \text{for all } w \in N\left(\mathcal{C}_s^T\right), \\
\{x \in N\left(\mathcal{B}_s^T\right) &: (\mathcal{A}_s w, x)_X = 0 \text{ for all } w \in N\left(\mathcal{C}_s^T\right)\} = \{0\}, \\
\alpha_{s,2}\|\nu\|_\Lambda &\leqslant \sup_{x\in X_s} \frac{(\mathcal{B}_s \nu, x)_X}{\|x\|_X} \quad \text{for all } \nu \in \Lambda_s, \\
\alpha_{s,3}\|\mu\|_M &\leqslant \sup_{w\in W_s} \frac{(\mathcal{C}_s \mu, w)_W}{\|w\|_W} \quad \text{for all } \mu \in M_s.
\end{aligned}
\tag{23}
$$

Above and hereafter we lighten notation and understand the supremum to be taken over sets excluding the zero function, even though this is not specifically indicated. The next result establishes similar necessary and sufficient conditions for the wellposedness of problem (19), and is a direct consequence of Corollary 1. First, consider the following inequalities, which are a direct consequence of the properties of the generalized inverse operator presented in Lemma 1,

$$
\begin{aligned}
\|\mathcal{A}_s^\dagger x\|_W &\leqslant \|\mathcal{A}_s^\dagger\|\,\|x\|_X \quad \text{for all } x \in X_s, \\
\frac{1}{\|\mathcal{A}_s^\dagger\|}\|w\|_W &\leqslant \|\mathcal{A}_s w\|_X \quad \text{for all } w \in R\left(\mathcal{A}_s^T\right).
\end{aligned}
\tag{24}
$$

Lemma 3 *Let $\mathcal{N}(\mathcal{B}_s)$ and $\mathcal{N}(\mathcal{C}_s)$ be the following null spaces*

$$
\begin{aligned}
\mathcal{N}(\mathcal{B}_s) &:= \left\{\mu_s \in \Lambda_s : (\mathcal{B}_s\,\mu_s, x_s)_X = 0 \quad \textit{for all } x_s \in N\left(\mathcal{A}_s^T\right)\right\}, \\
\mathcal{N}(\mathcal{C}_s) &:= \{\mu_s \in M_s : (\mathcal{C}_s\,\mu_s, w_s)_W = 0 \quad \textit{for all } w_s \in N\left(\mathcal{A}_s\right)\}.
\end{aligned}
$$

Conditions (23) *hold if and only if the following conditions are satisfied:*

1. $\exists\, \beta_{s,1} > 0$ *such that,*

i. $\beta_{s,1}\|\nu\|_\Lambda \leqslant \sup_{\mu\in\mathcal{N}(\mathcal{C}_s)} \dfrac{(\mathcal{C}_s\mu, \mathcal{A}_s^\dagger \mathcal{B}_s\, \nu)_W}{\|\mu\|_M}$ *for all* $\nu \in \mathcal{N}(\mathcal{B}_s)$,

ii. $\{\mu \in \mathcal{N}(\mathcal{C}_s) \,:\, (\mathcal{C}_s\mu, \mathcal{A}_s^\dagger \mathcal{B}_s\, \nu)_W = 0\, \textit{for all}\ \nu \in \mathcal{N}(\mathcal{B}_s)\} = \{0\}$.

2. $\exists\, \beta_{s,2} > 0$ *such that* $\beta_{s,2}\|x_0\|_X \leqslant \sup_{\nu\in\Lambda_s} \dfrac{(\mathcal{B}_s\nu, x_0)_X}{\|\nu\|_\Lambda}$ *for all* $x_0 \in N\left(\mathcal{A}_s^T\right)$.

3. $\exists\, \beta_{s,3} > 0$ *such that* $\beta_{s,3}\|w_0\|_W \leqslant \sup_{\mu\in M_s} \dfrac{(\mathcal{C}_s\mu, w_0)_W}{\|\mu\|_M}$ *for all* $w_0 \in N(\mathcal{A}_s)$.

Moreover, problem (19) *has an unique solution* $(\lambda_s, w_0) \in \Lambda_s \times V_0$ *if only if the conditions above hold, and we get*

$$\|\lambda_s\|_\Lambda \leqslant \frac{1}{\beta_{s,1}}\|g\|_{M'} + \left(\frac{\|\mathcal{A}_s^\dagger\|\|\mathcal{C}_s\|}{\beta_{s,1}}\left(1+\frac{\|\mathcal{B}_s\|}{\beta_{s,2}}\right)+\frac{1}{\beta_{s,2}}\right)\|f\|_X,$$

$$\|w_0\|_W \leqslant \frac{1}{\beta_{s,3}}\left(1+\frac{\|\mathcal{A}_s^\dagger\|\|\mathcal{B}_s\|\|\mathcal{C}_s\|}{\beta_{s,1}}\right)\|g\|_M$$
$$+ \frac{\|\mathcal{A}_s^\dagger\|\|\mathcal{C}_s\|}{\beta_{s,3}}\left(1+\frac{\|\mathcal{A}_s^\dagger\|\|\mathcal{B}_s\|\|\mathcal{C}_s\|}{\beta_{s,1}}\left(1+\frac{\|\mathcal{B}_s\|}{\beta_{s,2}}\right)+\frac{\|\mathcal{B}_s\|}{\beta_{s,2}}\right)\|f\|_X.$$

Proof The result follows from the standard condition for mixed/hybrid formulations [10] and from Corollary 1.

With the above theoretical structure, approximation results may be presented.

Lemma 4 *Let* $\mathcal{A} : W \to X$ *and* $\mathcal{A}_s : W_s \to X_s$ *be bounded operators with closed range. Assume* $N(\mathcal{A}_s) \subset N(\mathcal{A})$ *and* $N\left(\mathcal{A}_s^T\right) \subset N\left(\mathcal{A}^T\right)$, *and the conditions of Lemma 3 hold for problems* (11) *and* (19). *If* $(\lambda, w_0) \in \Lambda \times N(\mathcal{A})$ *and* $(\lambda_s, w_0^s) \in \Lambda_s \times N(\mathcal{A}_s)$ *are the solutions of* (11) *and* (19), *respectively, then*

$$\|\lambda-\lambda_s\|_\Lambda \leqslant \inf_{z_s\in\Lambda_s^\star}\left\{\left(1+\frac{\|\mathcal{C}_s\|\|\mathcal{B}\|\|\mathcal{A}^\dagger\|}{\beta_{s,1}}\right)\|\lambda-z_s\|_\Lambda + \frac{\|\mathcal{C}_s\|}{\beta_{s,1}}\|(\mathcal{A}^\dagger-\mathcal{A}_s^\dagger\mathcal{P}_{X_s})(f-\mathcal{B}z_s)\|_W\right\},$$

where $\Lambda_s^\star := \left\{\mu_s \in \Lambda_s : (\mathcal{B}_s\mu_s, x_0)_X = (f, x_0)_X \quad \textit{for all}\ x_0 \in N\left(\mathcal{A}_s^T\right)\right\}$ *and*

$$\|w_0-w_0^s\|_W \leqslant \frac{\|\mathcal{C}_s\|\|\mathcal{A}^\dagger\|\|\mathcal{B}\|}{\beta_{s,3}}\|\lambda-\lambda_s\|_\Lambda + \frac{\|\mathcal{C}_s\|}{\beta_{s,3}}\|(\mathcal{A}^\dagger-\mathcal{A}_s^\dagger\mathcal{P}_{X_s})(f-\mathcal{B}\lambda_s)\|_W$$
$$+\left(1+\frac{\|\mathcal{C}_s\|}{\beta_{s,3}}\right)\inf_{y_0\in N(\mathcal{A}_s)}\|w_0-y_0\|_W.$$

Moreover, for w_s *given in* (20) *and* w *given in* (12)*, it holds*

$$\|\boldsymbol{w}-\boldsymbol{w}_s\|_W \leqslant \left(\frac{\|\mathcal{C}_s\| \|\mathcal{A}^\dagger\| \|\mathcal{B}\|}{\beta_{s,3}} + \|\mathcal{A}^\dagger\| \|\mathcal{B}\| \right) \|\lambda - \lambda_s\|_\Lambda + \left(1 + \frac{\|\mathcal{C}\|}{\beta_{s,3}}\right) \inf_{y_0 \in N(\mathcal{A}_s)} \|w_0 - y_0\|_W$$
$$+ \left(1 + \frac{\|\mathcal{C}_s\|}{\beta_{s,3}}\right) \|(\mathcal{A}^\dagger - \mathcal{A}_s^\dagger \mathcal{P}_{X_s})(f - \mathcal{B}\lambda_s)\|_W \,.$$

Proof Let $z_s \in \Lambda_s^\star$. As a result $\lambda_s - z_s \in \mathcal{N}(\mathcal{B}_s)$ and from item (*i*) in Lemma 3, problems (11) and (19), and observing that $\mathcal{B}_s = \mathcal{P}_{X_s}\mathcal{B}$ from (17), it holds

$$\begin{aligned}
\beta_{s,1}\|\lambda_s - z_s\|_\Lambda &\leqslant \sup_{\mu_s \in \mathcal{N}(\mathcal{C}_s)} \frac{(\mathcal{C}_s\,\mu_s, \mathcal{A}_s^\dagger \mathcal{B}_s(\lambda_s - z_s))_W}{\|\mu_s\|_M} \\
&= \sup_{\mu_s \in \mathcal{N}(\mathcal{C}_s)} \frac{(\mathcal{C}_s\,\mu_s, \mathcal{A}^\dagger \mathcal{B}\lambda - \mathcal{A}_s^\dagger \mathcal{B}_s z_s)_W - (\mathcal{C}_s\,\mu_s, \mathcal{A}^\dagger \mathcal{B}\lambda - \mathcal{A}_s^\dagger \mathcal{B}_s \lambda_s)_W}{\|\mu_s\|_M} \\
&= \sup_{\mu_s \in \mathcal{N}(\mathcal{C}_s)} \frac{(\mathcal{C}_s\,\mu_s, \mathcal{A}^\dagger \mathcal{B}\lambda - \mathcal{A}_s^\dagger \mathcal{B}_s z_s)_W - (\mathcal{C}_s\,\mu_s, (\mathcal{A}^\dagger - \mathcal{A}_s^\dagger \mathcal{P}_{X_s})f)_W}{\|\mu_s\|_M} \\
&= \sup_{\mu_s \in \mathcal{N}(\mathcal{C}_s)} \frac{(\mathcal{C}_s\,\mu_s, \mathcal{A}^\dagger \mathcal{B}(\lambda - z_s))_W - (\mathcal{C}_s\,\mu_s, (\mathcal{A}^\dagger - \mathcal{A}_s^\dagger \mathcal{P}_{X_s})(\mathcal{B}z_s - f)_W}{\|\mu_s\|_M} \\
&\leqslant \|\mathcal{C}_s\| \|\mathcal{B}\| \|\mathcal{A}^\dagger\| \|\lambda - z_s\|_\Lambda + \|\mathcal{C}_s\| \|(\mathcal{A}^\dagger - \mathcal{A}_s^\dagger \mathcal{P}_{X_s})(\mathcal{B}z_s - f)\|_W \,,
\end{aligned}$$

where the continuity of $\mathcal{A}_s$, $\mathcal{B}_s$, and $\mathcal{C}_s$ are used. The first result follows from the triangle inequality. Now, suppose $y_0 \in N(\mathcal{A}_s)$. From item (3) in Lemma 3, problems (11) and (19), and $\mathcal{B}_s = \mathcal{P}_{X_s}\mathcal{B}$ from (17), it holds

$$\begin{aligned}
\beta_{s,3}\|w_0^s - y_0\|_W &\leqslant \sup_{\mu_s \in M_s} \frac{(\mathcal{C}_s\mu_s, w_0^s - y_0)_W}{\|\mu_s\|_W} \\
&= \sup_{\mu_s \in M_s} \frac{(\mathcal{C}_s\mu_s, w_0 - y_0)_W + (\mathcal{C}_s\mu_s, w_0^s - w_0)_W}{\|\mu_s\|_W} \\
&= \sup_{\mu_s \in M_s} \frac{(\mathcal{C}_s\mu_s, w_0 - y_0)_W + (\mathcal{C}_s\mu_s, \mathcal{A}_s^\dagger \mathcal{B}_s\lambda_s - \mathcal{A}^\dagger \mathcal{B}\lambda + (\mathcal{A}^\dagger - \mathcal{A}_s^\dagger \mathcal{P}_{X_s})f)_W}{\|\mu_s\|_W} \\
&= \sup_{\mu_s \in M_s} \frac{(\mathcal{C}_s\mu_s, w_0 - y_0)_W + (\mathcal{C}_s\mu_s, \mathcal{A}^\dagger \mathcal{B}(\lambda_s - \lambda))_W + (\mathcal{C}_s\mu_s, (\mathcal{A}^\dagger - \mathcal{A}_s^\dagger \mathcal{P}_{X_s})(f - \mathcal{B}\lambda_s))_W}{\|\mu_s\|_W} \\
&\leqslant \|\mathcal{C}_s\| \|w_0 - y_0\|_W + \|\mathcal{C}_s\| \|\mathcal{A}^\dagger\| \|\mathcal{B}\| \|\lambda - \lambda_s\|_\Lambda + \|\mathcal{C}_s\| \|(\mathcal{A}^\dagger - \mathcal{A}_s^\dagger \mathcal{P}_{X_s})(f - \mathcal{B}\lambda_s)\|_W \,.
\end{aligned}$$

The second result follows by the triangle inequality.

For the final result, decompositions (12) and (20) of w and w_s and $\mathcal{B}_s = \mathcal{P}_{X_s}\mathcal{B}$ from (17) are used to get

$$\begin{aligned}\|w - w_s\|_W &\leqslant \|w_0 - w_0^s\|_W + \|\mathcal{A}^\dagger\mathcal{B}\lambda - \mathcal{A}_s^\dagger\mathcal{B}_s\lambda_s + (\mathcal{A}^\dagger - \mathcal{A}_s^\dagger\mathcal{P}_{X_s})f\|_W \\ &\leqslant \|w_0 - w_0^s\|_W + \|\mathcal{A}^\dagger\mathcal{B}(\lambda - \lambda_s)\|_W + \|(\mathcal{A}^\dagger - \mathcal{A}_s^\dagger\mathcal{P}_{X_s})(f - \mathcal{B}\lambda_s)\|_W \\ &\leqslant \|w_0 - w_0^s\|_W + \|\mathcal{A}^\dagger\|\|\mathcal{B}\|\|\lambda - \lambda_s\|_\Lambda + \|(\mathcal{A}^\dagger - \mathcal{A}_s^\dagger\mathcal{P}_{X_s})(f - \mathcal{B}\lambda_s)\|_W\end{aligned}$$

and the result follows.

Remark 4 In the particular case formulation (22) is adopted, the estimates in Lemma 4 simplify to

$$\begin{aligned}\|\lambda - \lambda_s\|_\Lambda &\leqslant \inf_{z\in\Lambda_s^\star}\left(1 + \frac{\|\mathcal{C}\|\|\mathcal{A}^\dagger\|\|\mathcal{B}\|}{\beta_{s,1}}\right)\|\lambda - z\|_\Lambda\,, \\ \|w_0 - w_0^s\|_W &\leqslant \frac{\|\mathcal{C}\|\|\mathcal{A}^\dagger\|\|\mathcal{B}\|}{\beta_{s,3}}\|\lambda - \lambda_s\|_\Lambda\,.\end{aligned}$$

It is common to define $\mathcal{A}_s$ (although other choices would be also possible) through the action of $\mathcal{A}$ in the following standard way

$$(\mathcal{A}_s w_s, x_s)_X = (\mathcal{A}w_s, x_s)_X \quad \text{for all } w_s \in W_s,\ x_s \in X_s\,. \tag{25}$$

Due to the consistency that results from such a choice, an estimate for the terms $\|(\mathcal{A}^\dagger - \mathcal{A}_s^\dagger\mathcal{P}_{X_s})(f - \mathcal{B}\lambda_s)\|_W$ in Lemma 4 may be given. Consider the improved best approximation result of Xu and Zikatanov [45, Theorem 2] extended to the present setting.

Lemma 5 *Let operator $\mathcal{A} : W \to X$ be bounded with closed range and suppose $\mathcal{A}_s : W_s \to X_s$ defined by* (25) *has closed range. If $N(\mathcal{A}) = N(\mathcal{A}_s) \subset W_s$ and $N(\mathcal{A}^T) = N(\mathcal{A}_s^T) \subset X_s$, then*

$$\|\mathcal{A}^\dagger x - \mathcal{A}_s^\dagger\mathcal{P}_{X_s}x\|_W \leqslant \|\mathcal{A}_s^\dagger\|\|\mathcal{A}\| \inf_{v_s\in W_s}\|\mathcal{A}^\dagger x - v_s\|_W \quad \textit{for all } x \in X\,.$$

Proof Define $\Pi_s : W \to R(\mathcal{A}_s^T)$ (which is a projection according to (25)) by

$$(\mathcal{A}_s\,\Pi_s\,z, x_s)_X = (\mathcal{A}\,z, x_s)_X \quad \text{for all } x_s \in R(\mathcal{A}_s) \text{ and } z \in W\,. \tag{26}$$

Given $v_s \in R(\mathcal{A}_s^T)$, Eqs. (24) and (25) imply $(I - \Pi_s)v_s = 0$. So, by [45, Lemma 5]

$$\|z - \Pi_s z\|_W = \|(I - \Pi_s)(z - v_s)\|_W \leqslant \|\Pi_s\|\|z - v_s\|_W \ \text{ for all } z \in W, \tag{27}$$

where $\|\Pi_s\| := \sup_{z\in W} \frac{\|\Pi_s z\|_W}{\|z\|_W}$. Since $\Pi_s z \in R\left(\mathcal{A}_s^T\right)$ and $\mathcal{A}_s^\dagger$ is bounded, Lemma 1 and (26) imply,

$$\begin{aligned}
\|\Pi_s z\|_W &= \|\mathcal{A}_s^\dagger \mathcal{A}_s \Pi_s z\|_W \\
&\leqslant \|\mathcal{A}_s^\dagger\| \frac{(\mathcal{A}_s \Pi_s z, \mathcal{A}_s \Pi_s z)_X}{\|\mathcal{A}_s \Pi_s z\|_X} \\
&= \|\mathcal{A}_s^\dagger\| \frac{(\mathcal{A} z, \mathcal{A}_s \Pi_s z)_X}{\|\mathcal{A}_s \Pi_s z\|_X} \\
&\leqslant \|\mathcal{A}_s^\dagger\| \|\mathcal{A}\| \|z\|_W .
\end{aligned}$$

Therefore, using the inequality above in (27), we obtain

$$\|z - \Pi_s z\|_W \leqslant \|\mathcal{A}_s^\dagger\| \|\mathcal{A}\| \inf_{v_s \in R\left(\mathcal{A}_s^T\right)} \|z - v_s\|_W \quad \text{for all } z \in W. \tag{28}$$

If $z \in R\left(\mathcal{A}^T\right)$, this estimate may be improved. Choose arbitrary $v_0 \in N(\mathcal{A}_s)$ and $v_s \in R\left(\mathcal{A}_s^T\right)$. Then, $(z - v_s, v_0)_W = 0$ since $N(\mathcal{A}) = N(\mathcal{A}_s) \subset W_s$. Under these assumptions,

$$\|z - v_s\|_W^2 \leqslant \|z - v_s\|_W^2 + \|v_0\|_W^2 = \|z - (v_s + v_0)\|_W^2 .$$

Therefore, using (28) and the decomposition $W_s = R\left(\mathcal{A}_s^T\right) \oplus N(\mathcal{A}_s)$,

$$\|z - \Pi_s z\|_W \leqslant \|\mathcal{A}_s^\dagger\| \|\mathcal{A}\| \inf_{v_s \in W_s} \|z - v_s\|_W \quad \text{for all } z \in R\left(\mathcal{A}^T\right).$$

For arbitrary $x \in X$, define $z = \mathcal{A}^\dagger x$ and $z_s = \mathcal{A}_s^\dagger \mathcal{P}_{X_s} x$. The proof is complete if $z_s = \Pi_s z$. From (26), the definitions of z and z_s, Lemma 1, and the assumption $N\left(\mathcal{A}^T\right) = N\left(\mathcal{A}_s^T\right) \subset X_s$, it holds

$$\begin{aligned}
\mathcal{P}_{X_s} \mathcal{A}_s \Pi_s z &= \mathcal{P}_{X_s} \mathcal{A} z \\
&= \mathcal{P}_{X_s} \mathcal{P}_{R(\mathcal{A})} x \\
&= \mathcal{P}_{X_s} (I - \mathcal{P}_{N(\mathcal{A}^T)}) x \\
&= \mathcal{P}_{X_s} (I - \mathcal{P}_{N(\mathcal{A}_s^T)}) x \\
&= \mathcal{P}_{X_s} (I - \mathcal{P}_{N(\mathcal{A}_s^T)}) \mathcal{P}_{X_s} x \\
&= \mathcal{P}_{X_s} \mathcal{P}_{R(\mathcal{A}_s)} \mathcal{P}_{X_s} x \\
&= \mathcal{P}_{X_s} \mathcal{A}_s z_s .
\end{aligned}$$

By (24), $z_s = \Pi_s z$ and the result follows.

4 Application to the Laplace Problem

In this section, the abstract theory developed in Sect. 2 is leveraged to develop and analyze MHM methods for the Laplace problem (1). The first step is to show that the primal hybrid formulation (2) fits the form of problem (7).

Recall that $\{\mathcal{T}_H\}_{H>0}$ is a family of regular partitions of Ω composed of elements K with boundary ∂K. A partition $\mathcal{T}_H$ is parameterized by $H = \max_{K\in\mathcal{T}_H} H_K$, where $H_K = \operatorname{diam} K$. As presented in the introduction, the test and trial spaces involved in the definition of problem (2) coincide and can be recognized as

$$X,\, W \;=\; H^1(\mathcal{T}_H)\,, \tag{29}$$

$$M,\, \Lambda = \{\sigma\cdot\boldsymbol{n}_K\,|_{\partial K} \in H^{-1/2}(\partial K),\ \forall K\in\mathcal{T}_H \;:\; \sigma \in H(\text{div};\Omega)\}\,. \tag{30}$$

Here, $H^1(\mathcal{T}_H)$ is a Hilbert space equipped with the inner product, for $u, v \in H^1(\mathcal{T}_H)$,

$$(u,v)_X = \sum_{K\in\mathcal{T}_H} \frac{1}{d_\Omega^2}(u,v)_K + (\nabla\, u, \nabla\, v)_K, \tag{31}$$

where $(.,.)_K$ is the standard $L^2(K)$ inner product. The induced norm is denoted by

$$\|v\|_X^2 := \sum_{K\in\mathcal{T}_H}\left(\frac{1}{d_\Omega^2}\|v\|_{0,K}^2 + \|\nabla v\|_{0,K}^2\right).$$

We shall use extensively the following notation, for $u, v \in X$ and $\mu \in M$,

$$(u,v)_{\mathcal{T}_H} := \sum_{K\in\mathcal{T}_H}(u,v)_K \quad \text{and} \quad (\mu,v)_{\partial\mathcal{T}_H} := \sum_{K\in\mathcal{T}_H}(u,v)_{\partial K}\,,$$

where that $(\mu, v)_{\partial K}$ represents the duality product $H^{-1/2}(\partial K)$ and $H^{1/2}(\partial K)$. Also, the broken H^1 semi-norm is denoted by

$$|u|_{1,h} := \left[\sum_{K\in\mathcal{T}_H}|u|_{1,K}^2\right]^{1/2}.$$

Furthermore, the space of Lagrange multipliers M is normed by,

$$\|\mu\|_M = \sup_{v\in H^1(\mathcal{T}_H)} \frac{(\mu,v)_{\partial\mathcal{T}_H}}{\|v\|_X}\,. \tag{32}$$

Denoting the quotient norm on M by

$$|||\mu|||_M := \inf_{\substack{\sigma\in H(\text{div};\Omega)\\ \sigma\cdot\boldsymbol{n}_K=\mu \text{ on } \partial K,\, K\in\mathcal{T}_H}} \|\boldsymbol{\sigma}\|_{\text{div}}, \tag{33}$$

where

$$\|\sigma\|_{\text{div}}^2 = \sum_{K \in \mathcal{T}_H} \left(\|\sigma\|_{0,K}^2 + d_\Omega^2 \| \operatorname{div} \sigma \|_{0,K}^2 \right) ,$$

the following equivalence holds (see [3])

$$\frac{\sqrt{2}}{2} |||\mu|||_M \leqslant \|\mu\|_M \leqslant |||\mu|||_M . \tag{34}$$

Consequently the space M (e.g. Λ) is also a reflexive Banach space with respect to the norm $||| \cdot |||_M$.

In order to realize (2) in the form (7), operators $\mathcal{A}$, $\mathcal{B}$, and $\mathcal{C}$ must be defined. First, the self-adjoint, bounded linear operator $\mathcal{A} : X \to X$ is defined by

$$(\mathcal{A}v, w)_X = (\kappa \nabla v, \nabla w)_{\mathcal{T}_H} \quad \text{for all } v, w \in X. \tag{35}$$

According to the definition of $\mathcal{A}$, the null space corresponds to piecewise constant functions, i.e. $N(\mathcal{A}) = V_0$, and then $R(\mathcal{A})$ is closed and coincides with $X \cap L_0^2(\mathcal{T}_H)$. Next, let $\mathcal{C} = \mathcal{B} = j_1$, where $j_1 : M \to X$ is a bounded linear operator defined by

$$(j_1 \mu, x)_X = (\mu, x)_{\partial \mathcal{T}_H} \qquad \text{for all } \mu \in M, \ x \in X . \tag{36}$$

By the next lemma, j_1 is an isometry onto the subspace $H_0^1(\Omega)^\perp \subset X$, i.e., the spaces $H_0^1(\Omega)^\perp$ and M are isomorphic so that $R(j_1)$ is closed.

Lemma 6 *Let L be a bounded linear functional acting on X. The following conditions are equivalent:*

1. *$L(x) = 0$ for all $x \in H_0^1(\Omega)$;*
2. *$\exists ! \, \mu \in M$ such that $L(x) = (\mu, x)_{\partial \mathcal{T}_H}$ for all $x \in X$;*
3. *$\exists ! \, v \in (H_0^1(\Omega))^\perp$ such that $L(x) = (v, x)_X$ for all $x \in X$.*

Moreover, $\|L\|_{X'} = \|\mu\|_M = \|v\|_X$.

Proof The equivalence of (1) and (2) is established in [39, Lemma 1] (modified to the present context). Moreover, the equivalence of (1) and (3) stems from the Riesz Representation Theorem. Finally, the equality of norms follows by standard results.

With these operators, problem (2) is recast as follows: Find $u \in X$ and $\lambda \in M$ such that

$$\begin{cases} (\mathcal{A}u, v)_X + (j_1 \lambda, v)_X = (j_2 f, v)_X & \text{for all } v \in X, \\ (j_1 \mu, u)_X = 0 \quad \text{for all } \mu \in M , \end{cases} \tag{37}$$

where $j_2 : L^2(\Omega) \to X$ is defined by,

$$(j_2 q, v)_X = (q, v)_\Omega \quad \text{for all } q \in L^2(\Omega)\,. \tag{38}$$

Thereby, formulation (37) is equivalent to (2) and, then following closely [39], conditions 1 and 2 of Lemma 6 establish that problem (2) is well-posed and forms the solution to problem (1). This result is summarized in the next lemma [39, Theorem 1].

Lemma 7 *Problem* (2) *(e.g.* (37)*) has a unique solution* $(u, \lambda) \in X \times M$. *Moreover,* $u \in H_0^1(\Omega)$ *is the solution to problem* (1) *and* $\lambda = -\kappa \nabla u \cdot \boldsymbol{n}_K |_{\partial K}$ *for each* $K \in \mathcal{T}_H$.

Since operator $\mathcal{A}$ defined in (35) is bounded with closed range, the following problem is well-posed by Lemma 7 and Theorem 1: Find $(\lambda, u_0) \in M \times V_0$ such that

$$\begin{cases} -(j_1\mu, \mathcal{A}^\dagger j_1 \lambda)_X + (j_1\mu, u_0)_X = -(j_1\mu, \mathcal{A}^\dagger j_2 f)_X & \text{for all } \mu \in M\,, \\ (j_1\lambda, v_0)_X = (j_2 f, v_0)_X & \text{for all } v_0 \in V_0\,. \end{cases} \tag{39}$$

Note that $j_2 f$ is the representative in X of $f \in L^2(\Omega)$, so that $j_2 f$ here plays the role of the corresponding right-hand side in the abstract setting.

The next result demonstrates that problem (5) is well-posed. Such a result was originally shown in [3] by establishing conditions (1)–(3) in Lemma 3. Here, we propose an alternative (and indirect) proof by showing that the generalized inverse $\mathcal{A}^\dagger$ in (39) may be characterized *locally* in terms of operators T and $\hat{T}$ defined in (3) and (4), respectively. As a result, the wellposedness of (5) follows from Lemma 7 and Theorem 1. This is the subject of the next lemma.

Theorem 2 *There exists a unique solution* (u_0, λ) *to problem* (5). *Moreover, the solution* u *of* (37) *reads*

$$u = u_0 + T\lambda + \hat{T} f\,. \tag{40}$$

Proof First note that the generalized inverse $\mathcal{A}^\dagger$ has the property that

$$\mathcal{A}^\dagger j_2\, q = \hat{T} q \quad \text{for all } q \in L^2(\Omega)\,. \tag{41}$$

Indeed, by the definition of $\mathcal{A}$ in (35), it follows that for all $w \in R(\mathcal{A})$

$$(\mathcal{A}\hat{T}q, w)_X = (\kappa\, \nabla \hat{T}q, \nabla w)_{\mathcal{T}_H} = (q, w)_{\mathcal{T}_H} = (j_2\, q, w)_X = (\mathcal{A}\mathcal{A}^\dagger j_2\, q, w)_X,$$

where $\mathcal{A}\mathcal{A}^\dagger = \mathcal{P}_{R(\mathcal{A})}$ from Lemma 1. Therefore,

$$(\mathcal{A}(\hat{T} - \mathcal{A}^\dagger j_2)\, q, w)_X = 0 \quad \text{for all } w \in R(\mathcal{A})\,,$$

and then $\mathcal{A}(\hat{T} - \mathcal{A}^{\dagger} j_2)\, q = 0$. Since $R(\mathcal{A})$ is closed, this implies that

$$(\hat{T} - \mathcal{A}^{\dagger} j_2)\, q \in N(\mathcal{A}) \cap R\left(\mathcal{A}^{T}\right) \quad \text{for all } q \in L^2(\Omega)$$

and (41) follows. Next, the generalized inverse has also the property that

$$\mathcal{A}^{\dagger} j_1 \mu = -T\mu \quad \text{for all } \mu \in M\,. \tag{42}$$

This is achieved by employing the same argument used to show (41), i.e., from Lemma 6 it holds that for all $w \in R(\mathcal{A})$

$$(\mathcal{A}T\mu, w)_X = -(\mu, w)_{\partial\mathcal{T}_H} = -(j_1\mu, w)_X = -(\mathcal{A}\mathcal{A}^{\dagger} j_1\, \mu, w)_X\,,$$

and then (42) follows. Problem (5) is well-posed from (39), Theorem 1 and using the characterizations (41) and (42). Decomposition (14) follows from (12) in Theorem 1, and (41) and (42) with $\mathcal{B} = j_1$.

Remark 5 An interesting characteristic of the formulation (39) (e.g. (5)) is that it may be written in an alternative form involving divergence. Define for an arbitrary $\mu \in M$

$$\sigma^{\mu} := \kappa\nabla\mathcal{A}^{\dagger} j_1\mu = -\kappa\nabla T\mu\,.$$

Now, since $\mathcal{A}^{T} = \mathcal{A}$ and ${\mathcal{A}^{\dagger}}^{T} = \mathcal{A}^{\dagger}$, the definition of $\mathcal{A}$ and Lemma 1 imply

$$\begin{aligned}(j_1\mu, \mathcal{A}^{\dagger} j_1\lambda)_X &= (j_1\mu, \mathcal{A}^{\dagger}\mathcal{A}\mathcal{A}^{\dagger} j_1\lambda)_X\\ &= (\mathcal{A}\mathcal{A}^{\dagger} j_1\mu, \mathcal{A}^{\dagger} j_1\lambda)_X\\ &= (\sigma^{\mu}, \kappa^{-1}\sigma^{\lambda})_{\mathcal{T}_H}\,.\end{aligned}$$

Next, note from Lemma 1 that $(\mathcal{A}\mathcal{A}^{\dagger} j_1\mu, z_0)_X = 0$ for each $z_0 \in V_0$. Let $z \in R(\mathcal{A})$ be arbitrary. From the definition (10) of generalized inverse $\mathcal{A}^{\dagger}$, it holds

$$\begin{aligned}(\mu, z)_{\partial\mathcal{T}_H} &= (j_1\mu, z)_X\\ &= (\mathcal{A}\mathcal{A}^{\dagger} j_1\mu, z)_X\\ &= -(\nabla\cdot\sigma^{\mu}, z)_{\mathcal{T}_H} + \sum_{K\in\mathcal{T}_H} (\sigma^{\mu}\cdot\boldsymbol{n}_K, z)_{\partial K}\,,\end{aligned}$$

and it follows that $\sigma^{\mu}|_K$, for all $K \in \mathcal{T}_H$, satisfies (in a weak sense)

$$-\nabla\cdot\sigma^{\mu} = x_0 \quad \text{in } K \quad \text{and} \quad \sigma^{\mu}\cdot\boldsymbol{n}_K = \mu \quad \text{on } \partial K\,, \tag{43}$$

where $x_0 \in V_0$ is such that $x_0 = -\frac{1}{|K|}\int_{\partial K}\mu$. Observe that σ^μ belongs to $H(\text{div};\Omega)$ since $\sigma^\mu \cdot \boldsymbol{n}_K|_{\partial K} \in M$ for all $K \in \mathcal{T}_H$. Therefore,

$$(j_1\mu, z_0)_X = (\mu, z_0)_{\partial\mathcal{T}_H} = \sum_{K\in\mathcal{T}_H}(\sigma^\mu \cdot \boldsymbol{n}_K, z_0)_{\partial K} = (\nabla\cdot\sigma^\mu, z_0)_{\mathcal{T}_H}\,.$$

So, problem (39) may be written in the form: Find $(\lambda, u_0) \in M \times V_0$ such that

$$\begin{cases} (\sigma^\mu, \kappa^{-1}\sigma^\lambda)_{\mathcal{T}_H} + (\nabla\cdot\sigma^\mu, u_0)_{\mathcal{T}_H} = (j_1\mu, \mathcal{A}^\dagger j_2 f)_X \quad \text{for all } \mu \in M\,, \\ (\nabla\cdot\sigma^\lambda, v_0)_{\mathcal{T}_H} = (f, v_0)_{\mathcal{T}_H} \quad \text{for all } v_0 \in V_0\,. \end{cases} \tag{44}$$

4.1 The Subspace Case

Let $X_s \subset X$ and $M_s \subset M$ be closed subspaces (possibly infinite-dimensional) with M given in (30), and consider the primal hybrid problem: Find $u_s \in X_s$ and $\lambda_s \in M_s$ such that

$$\begin{cases} (\kappa\nabla u_s, \nabla x_s)_{\mathcal{T}_H} + (\lambda_s, x_s)_{\partial\mathcal{T}_H} = (f, x_s)_{\mathcal{T}_H} \quad \text{for all } x_s \in X_s\,, \\ (\mu_s, u_s)_{\partial\mathcal{T}_H} = 0 \quad \text{for all } \mu_s \in M_s\,. \end{cases} \tag{45}$$

This problem may be expressed in the form (18) using $\mathcal{A}_s : X_s \to X_s$ defined by,

$$(\mathcal{A}_s w_s, x_s)_X = (\mathcal{A} w_s, x_s)_X = (\kappa\nabla w_s, \nabla x_s)_{\mathcal{T}_H} \quad \text{for all } w_s, x_s \in X_s\,, \tag{46}$$

and setting $\mathcal{B}_s = \mathcal{C}_s = \mathcal{P}_{X_s} j_1$, i.e.,

$$(\mathcal{P}_{X_s} j_1\mu_s, w_s)_X = (j_1\mu_s, w_s)_X = (\mu_s, w_s)_{\partial\mathcal{T}_H} \quad \text{for all } w_s \in X_s\,, \tag{47}$$

where (36) was used.

By the next result, (45) is well-posed under an equivalent condition on the compatibility of spaces X_s and M_s and the assumption that M_s contains the piecewise constant space on faces M_0. Such a result is adapted from [39, Theorem 2] to the present case.

Lemma 8 *Assume $M_0 \subset M_s$ and $\mathcal{C}_s$ has closed range. Problem* (45) *is well-posed if and only if*

$$\{\mu_s \in M_s : (\mathcal{C}_s\mu_s, x_s)_X = 0 \quad \textit{for all } x_s \in X_s\} = \{0\}\,. \tag{48}$$

Proof Following closely [39, Theorem 2], the condition $M_0 \subset M_s$ implies that

$$C\,\|x_s\|_X^2 \leqslant (\mathcal{A}_s x_s, x_s)_X \quad \text{for all } x_s \in N\left(\mathcal{C}_s^T\right)\,, \tag{49}$$

where C is independent of H. The coercivity result (49) implies the first and second conditions in (23). Also, condition (48), along the closure of $R(\mathcal{C}_s)$, leads to the surjectivity of $\mathcal{C}_s^T$. This is equivalent to the third (and fourth) condition in (23), and then the result follows.

We are ready to establish the wellposedness of the following global-local problem: Find $(\lambda_s, u_0) \in M_s \times V_0$ such that

$$\begin{aligned} (\lambda_s, v_0)_{\partial\mathcal{T}_H} &= (f, v_0)_{\mathcal{T}_H} \quad \text{for all } v_0 \in V_0\,, \\ (\mu_s, T_s\,\lambda_s)_{\partial\mathcal{T}_H} + (\mu_s, u_0)_{\partial\mathcal{T}_H} &= -(\mu_s, \hat{T}_s f)_{\partial\mathcal{T}_H} \quad \text{for all } \mu_s \in M_s\,, \end{aligned} \tag{50}$$

where operators $T_s : M_s \to X_s \cap L_0^2(\mathcal{T}_H)$ and $\hat{T}_s : L^2(\Omega) \to X_s \cap L_0^2(\mathcal{T}_H)$ are defined piecewise in $K \in \mathcal{T}_H$ by

$$\mu_s \in M_s, \quad (\kappa\,\nabla\, T_s\,\mu_s, \nabla\, w_s)_K = -(\mu_s, w_s)_{\partial K} \quad \text{for all } w_s \in X_s \cap L_0^2(\mathcal{T}_H)\,, \tag{51}$$

$$q \in L^2(\Omega), \quad (\kappa\ \nabla\,\hat{T}_s\,q, \nabla\, w_s)_K = (q, w_s)_K \quad \text{for all } w_s \in X_s \cap L_0^2(\mathcal{T}_H)\,. \tag{52}$$

Theorem 3 *Assume that $M_0 \subset M_s$ and* (48) *hold. Then, problem* (50) *is well-posed. Moreover, if (λ, u_0) and (λ_s, u_0^s) are the solutions of* (2) *and* (50)*, respectively, then the following estimates hold*

$$\|\lambda - \lambda_s\|_M \leqslant \inf_{z \in M_s^\star} \left\{ \left(1 + \frac{\|\mathcal{A}^\dagger\|}{\beta_{s,1}}\right) \|\lambda - z\|_M + \frac{1}{\beta_{s,1}} \| \left(\hat{T} - \hat{T}_s\right) f + (T - T_s)\, z\|_X \right\}\,,$$

$$\|u_0 - u_0^s\|_X \leqslant \frac{\|\mathcal{A}_s^\dagger\|}{\beta_{s,3}} \|\lambda - \lambda_s\|_M + \frac{1}{\beta_{s,3}} \| \left(\hat{T} - \hat{T}_s\right) f + (T - T_s)\,\lambda_s\|_X\,,$$

where $M_s^\star := \{\mu_s \in M_s : (\mu_s, x_0)_{\partial\mathcal{T}_H} = (f, x_0)_\Omega \quad \text{for all } x_0 \in V_0\}$.

Proof First note that, from (46) the finite dimensional null space $N(\mathcal{A}_s) = V_0 = N(\mathcal{A})$ and the image space $R(\mathcal{A}_s) = X_s \cap L_0^2(\mathcal{T}_H)$ is closed. As a result, the operator $\mathcal{A}_s$ has a generalized inverse $\mathcal{A}_s^\dagger$ from Lemma 1 and satisfies the properties (10). Next, following closely the proof in Theorem 2, it follows

$$T_s = -\mathcal{A}_s^\dagger \mathcal{P}_{X_s} j_1 \quad \text{and} \quad \hat{T}_s = \mathcal{A}_s^\dagger \mathcal{P}_{X_s} j_2\,,$$

and, thus, problem (50) can be rewritten in the form (19). Also, notice that operators $\mathcal{C}_s$ and $\mathcal{B}_s$ have closed range from (47). The result follows using Corollary 1 and Lemma 8. The estimates follow from Lemma 4, with $\|\mathcal{C}_s\| = \|\mathcal{B}_s\| = \|\mathcal{B}\| = \|\mathcal{C}\| = 1$.

We next present best approximation results for the full solution $u_0^s + T_s\,\lambda_s + \hat{T}_s f$ in the X and L^2 norms. Unlike in classical results, the estimate in the L^2 norm does not require extra regularity.

Lemma 9 *Assume that $M_0 \subset M_s$ and* (48) *hold. If (u_s, λ_s) and (u_0^s, λ_s) solve* (45) *and* (50)*, respectively, then*

$$u_s = u_0^s + T_s \lambda_s + \hat{T}_s f ,$$

and the following estimate holds

$$\|\boldsymbol{u} - u_s\|_X \leqslant \left(\frac{\|\mathcal{A}_s^\dagger\|}{\beta_{s,3}} + \|\mathcal{A}^\dagger\| \right) \|\lambda - \lambda_s\|_M + \left(1 + \frac{1}{\beta_{s,3}}\right) \| \left(\hat{T} - \hat{T}_s\right) f + (T - T_s)\, \lambda_s \|_X .$$

Furthermore, it holds

$$\|\boldsymbol{u} - u_s\|_{0,\Omega} \leqslant C H \,|\boldsymbol{u} - u_s|_{1,h} .$$

Proof The first estimate follows from Lemma 4, observing that $\|\mathcal{C}\| = \|\mathcal{C}_s\| = \|\mathcal{B}_s\| = 1$. The second result follows closely the proof proposed first in [35]. We revisit it here in a more general form. From the definition of $\boldsymbol{u}$ and $\boldsymbol{u}_s$, we get

$$\begin{aligned} \|\boldsymbol{u} - u_s\|_{0,\Omega} &\leqslant \|T\,\lambda + \hat{T} f + u_0 - T_s\,\lambda_s - \hat{T}_s f - u_0^s\|_{0,\Omega} \\ &\leqslant \|u_0 - u_0^s\|_{0,\Omega} + C H \,|T\,\lambda + \hat{T} f - T_s\,\lambda_s - \hat{T}_s f|_{1,h} \end{aligned} \tag{53}$$

where we used the triangle inequality, the Poincaré inequality, and the assumption on the regularity of the mesh. Next, we estimate $\|u_0 - u_0^s\|_{0,\Omega}$. Without losing generality, we assume that $u_0 - u_0^s \in V_0$ does not vanish in $K \in \mathcal{T}_H$. It is known that there exists a vector-valued function $\boldsymbol{\sigma}^\star$ belonging to the lowest-order Raviart-Thomas space such that $\nabla \cdot \boldsymbol{\sigma}^\star = u_0 - u_0^s$ and $\|\boldsymbol{\sigma}^\star\|_{0,K} \leqslant C H_K \|\nabla \cdot \boldsymbol{\sigma}^\star\|_{0,K}$ in each $K \in \mathcal{T}_H$. We recall that $\boldsymbol{\sigma}^\star \cdot \boldsymbol{n}_K\,|_{\partial K}$ is piecewise constant for all K in $\mathcal{T}_H$. Now, from (50), the fact that $T\lambda\,|_K$, $\hat{T} f\,|_K$ and $T_s\lambda_s,|_K$, $\hat{T}_s f\,|_K$ belong to $L_0^2(K)$ for all $K \in \mathcal{T}_H$, the Cauchy-Schwarz inequality, and the regularity of the mesh, we get

$$\begin{aligned} \|u_0 - u_0^s\|_{0,\Omega}^2 &= (\nabla \cdot \boldsymbol{\sigma}^\star, u_0 - u_0^s)_{\mathcal{T}_H} \\ &= \sum_{K \in \mathcal{T}_H} (\boldsymbol{\sigma}^\star \cdot \boldsymbol{n}_K, u_0 - u_0^s)_{\partial K} \\ &= -\sum_{K \in \mathcal{T}_H} (\boldsymbol{\sigma}^\star \cdot \boldsymbol{n}_K, T\lambda - T_s\lambda_s + \hat{T} f - \hat{T}_s f)_{\partial K} \\ &= -(\boldsymbol{\sigma}^\star, \nabla(T\lambda - T_s\lambda_s + \hat{T} f - \hat{T}_s f))_{\mathcal{T}_H} + (\nabla \cdot \boldsymbol{\sigma}^\star, T\lambda - T_s\lambda_s + \hat{T} f - \hat{T}_s f)_{\mathcal{T}_H} \\ &= -(\boldsymbol{\sigma}^\star, \nabla(T\lambda - T_s\lambda_s + \hat{T} f - \hat{T}_s f))_{\mathcal{T}_H} \\ &\leqslant \sum_{K \in \mathcal{T}_H} \|\boldsymbol{\sigma}^\star\|_{0,K} \|\nabla(T\lambda - T_s\lambda_s + \hat{T} f - \hat{T}_s f)\|_{0,K} \end{aligned}$$

$$\leqslant C \sum_{K \in \mathcal{T}_H} H_K \|\nabla \cdot \boldsymbol{\sigma}^\star\|_{0,K} \|\nabla(T\lambda - T_s\lambda_s + \hat{T}f - \hat{T}_s f)\|_{0,K}$$

$$\leqslant C H \|u_0 - u_0^s\|_{0,\Omega} |T\lambda - T_s\lambda_s + \hat{T}f - \hat{T}_s f|_{1,h} .$$

Collecting the previous results, we get from (53) the existence of C such that

$$\|\boldsymbol{u} - u_s\|_{0,\Omega} \leqslant C H |\boldsymbol{u} - u_s|_{1,h} .$$

Remark 6 The result in Lemma 9 is the first to address the relationship between the classical hybrid formulation and the global-local MHM formulation in the case of subspaces. Also new are the error estimates in Theorem 3 and Lemma 9 measuring the impact of replacing T and $\hat{T}$ by T_s and $\hat{T}_s$. This is important in preparation for the next section in which particular choices are made for the approximation spaces.

5 Relationship with the MHM Method and Beyond

The abstract framework introduced in the last sections are now used to recover some of the classical and recent numerical methods in the literature. As such, the wellposedness and best approximation results for them follow naturally from the last sections. Particular subspaces are chosen with respect to $\mathcal{T}_H$ being a triangulation of $\Omega \subset \mathbb{R}^2$ into simplex elements $K \in \mathcal{T}_H$. In all cases we pick

$$M_s = M_l := \{\mu \in M \,:\, \mu|_F \in \mathbb{P}_l(F), \text{ for all } F \subset \partial K \text{ and } K \in \mathcal{T}_H\} , \tag{54}$$

where $\mathbb{P}_l(F)$ stands for the space of polynomial functions with degree less or equal to $l \geqslant 0$ on F. This space has a well-known interpolation result [39, Lemma 9], namely, given a function $\phi \in H^{m+1}(\mathcal{T}_H)$, with $1 \leqslant m \leqslant l+1$, there is $\mu_l \in M_l$ (taken as the L^2 projection of μ onto space $\mathbb{P}_l(F)$) and positive constants independent of H, such that

$$\|\mu_l\|_M \leqslant C \|\phi\|_{m+1,\Omega}, \quad \text{and} \quad \|\psi - \mu_l\|_M \leqslant C H^m |\phi|_{m+1,\Omega} , \tag{55}$$

where $\psi \in M$ is such that $\psi = \nabla\phi \cdot \boldsymbol{n}_K$ when restricted to ∂K for all $K \in \mathcal{T}_H$.

5.1 The One-Level MHM Method

The one-level MHM method is recovered by taking $X_s := H^1(\mathcal{T}_H)$ and letting $T = T_s$ and $\hat{T} = \hat{T}_s$ defined in (3) and (4), respectively. The resulting method corresponds to find $(\lambda_l, u_0^H) \in M_l \times V_0$ such that

$$\begin{cases} (\mu_l, T\,\lambda_l)_{\partial\mathcal{T}_H} + (\mu_l, u_0^H)_{\partial\mathcal{T}_H} = -(\mu_l, \hat{T} f)_{\partial\mathcal{T}_H} & \text{for all } \mu_l \in M_l\,, \\ (\lambda_l, v_0)_{\partial\mathcal{T}_H} = (f,\, v_0)_{\mathcal{T}_H} & \text{for all } v_0 \in V_0\,. \end{cases} \tag{56}$$

Observe that problems (3) and (4) can be written in the following (equivalent) strong form

$$-\nabla\cdot(\kappa\nabla T\,\mu) = c_K^\mu \quad \text{in } K, \quad -\kappa\nabla T\,\mu\cdot\boldsymbol{n}_K = \mu \quad \text{on } F\subset\partial K\,, \tag{57}$$

and

$$-\nabla\cdot\left(\kappa\nabla\hat{T}\,q\right) = q - \bar{q}_K \quad \text{in } K, \quad \kappa\nabla\hat{T}\,q\cdot\boldsymbol{n}_K = 0 \quad \text{on } F\subset\partial K\,, \tag{58}$$

where

$$\bar{q}_K := \frac{1}{|K|}\int_K q\, d\mathbf{x} \quad \text{and} \quad c_K^\mu := \frac{1}{|K|}\int_{\partial K}\mu\, ds\,. \tag{59}$$

It is easy to see from (54) and the definition of the space X_s that the conditions in Lemma 8 hold so that problem (56) is well-posed from Theorem 3. In addition, using again Theorem 3, Lemma 9, and (55), we get the following error estimates

$$\begin{aligned} \|u_0 - u_0^H\|_X + H\,\|\lambda - \lambda_l\|_M &\leqslant C\,H^{m+1}\,\|u\|_{m+1}\,, \\ \|u - u_H\|_{0,\Omega} + H\,|u - u_H|_{1,h} &\leqslant C\,H^{m+1}\,\|u\|_{m+1}\,, \\ \|\sigma - \sigma_H\|_{\text{div}} &\leqslant C\,H^{m}\,\|u\|_{m+1}\,, \end{aligned} \tag{60}$$

where $1 \leqslant m \leqslant l+1$, and $\sigma := \kappa\,\nabla(T\,\lambda + \hat{T} f)$, and

$$u_H = u_0^H + T\,\lambda_l + \hat{T} f \quad \text{and} \quad \sigma_H := \kappa\,\nabla(T\,\lambda_l + \hat{T} f)\,.$$

For the third error estimate in (60), we used $\sigma_H \in H(\text{div};\Omega)$ and $\nabla\cdot(\sigma - \sigma_H) = 0$ in each $K \in \mathcal{T}_H$ as a result of problem (56)–(58). The independence of the constant in (60) with respect to H follows from the independence of the inf-sup constant in (23) in the context of the discrete primal hybrid method for the Laplace problem (see [39]). The convergence results (60) were first established in [3] proving conditions in Lemma 3 directly.

Remark 7 The one-level MHM method has been analyzed (see [35] for details) assuming highly oscillatory coefficient and periodicity. Convergence with respect to the (small) characteristic length of oscillations ε was then addressed under the assumption $\frac{\varepsilon}{H} < C = O(1)$ and without any oversampling technique. One of the main result is that the numerical solution is *resonance-free*, and the following estimates

$$|u - u_H|_{1,h} \leqslant C\,(H + \varepsilon^{1/2})\|f\|_{0,\Omega} \quad \text{and} \quad \|u - u_H\|_{0,\Omega} \leqslant C\,(H^2 + H\,\varepsilon^{1/2})\|f\|_{0,\Omega}\,,$$

hold under mild regularity assumptions.

It is instructive to describe the algorithm underlying the one-level MHM method (56). Discretization decouples the local problems (3) and (4) from the global one (56). Thereby, a staggered algorithm can be used to solve the system. To see this more clearly, consider $T\,\lambda_l$ in more detail. Suppose $\{\psi_i\}_{i=1}^{\dim M_l}$ is a basis for M_l, and define the set $\{\eta_i\}_{i=1}^{\dim M_l} \subset H^1(\mathcal{T}_H) \cap L^2_0(\mathcal{T}_H)$ such that,

$$-\nabla\cdot(\kappa\nabla\eta_i) \;=\; c_K^{\psi_i} \quad \text{in } K, \qquad -\kappa\nabla\eta_i\cdot\boldsymbol{n}_K \;=\; \psi_i \quad \text{on } F\subset\partial K\,, \tag{61}$$

i.e., $\eta_i = T\,\psi_i$, where ψ_i changes its sign in (61) according to the sign of $\boldsymbol{n}\cdot\boldsymbol{n}_K\,|_F$. Now, giving $\lambda_l = \sum_{i=1}^{\dim M_l} c_i\,\psi_i$ in M_l, $c_i \in \mathbb{R}$, the linearity of problem (57) implies we may uniquely write

$$T\,\lambda_l \;=\; \sum_{i=1}^{\dim M_l} c_i\,T\,\psi_i \;=\; \sum_{i=1}^{\dim M_l} c_i\,\eta_i\,.$$

Therefore, the degrees of freedom c_i's of λ_l are "inherited" by $T\,\lambda_l$. It then follows that

$$u_H = u_0^H + \sum_{i=1}^{\dim M_l} c_i\,\eta_i + \hat{T}f\,. \tag{62}$$

As a result, the global formulation (56) is responsible for computing the degrees of freedom of u_0^H (one per element) and the c_i's in (62), once the multiscale basis functions η_i's and $\hat{T}f$ are available from the local problems. Also, it is interesting to note that heterogeneous and/or high-contrast aspects of the media automatically impact the design of the basis functions η_i's as well as $\hat{T}f$ as they are driven by (61) and (4), respectively. In Fig. 1 we depict an example of such a basis function in the case the coefficient κ oscillates and $M_s = M_0$.

Although the one-level MHM method is impractical (in general) since basis $\{\eta_i\}_{i=1}^{\dim M_l}$ and $\hat{T}f$ are not readily available, an interesting case arises where a closed formula is indeed available. From these, we recover the classical lowest-order Raviart-Thomas finite element method. Also, we can recognize from this perspective the well-known MsFEM [16] and the UpFEM [4]. These cases are the subject of the next sections.

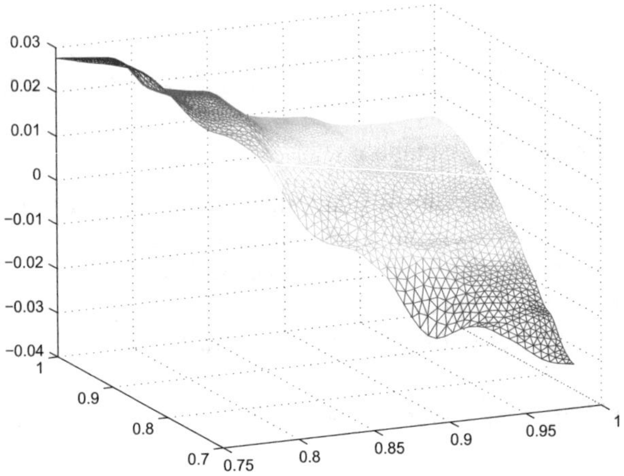

Fig. 1 Illustration of a typical multiscale function η_i on a triangle associated to a piecewise constant interpolation function on faces

5.1.1 The Lowest-Order Raviart-Thomas Element

Assume that $\kappa = I$ (where I is the 2×2 identity tensor) and let $M_s = M_0$. Next, consider the following function defined on Ω with respect to face F of the partition $\mathcal{T}_H$

$$\eta_F^K(\mathbf{x}) = \begin{cases} \pm \dfrac{H_F}{2\,|K|} \left(\dfrac{\mathbf{x}\cdot\mathbf{x}}{2} - \mathbf{x}\cdot\mathbf{x}_F + C \right) & \mathbf{x} \in K \text{ and } F \subset \partial K,\ K \in \mathcal{T}_H\,, \\ 0 & \text{otherwise.} \end{cases} \tag{63}$$

Here, $\mathbf{x}_F$ represents the vertex opposite to the face F, and for the two $K \in \mathcal{T}_H$ sharing face F, one function is taken with the positive sign and the other with the negative sign. Furthermore, constant C is chosen so that $\eta_F^K \in L_0^2(K)$, and $|K|$ is the area of K. It is easily verified that $\eta_F^K = T\,\mu_F$ where $\mu_F \in M_0$.

Define the set $\{\sigma^{\mu_F}\}$, for all $F \subset \partial K$ and $K \in \mathcal{T}_H$, such that

$$\sigma^{\mu_F}|_K = -\nabla T\,\mu_F\,|_K = \pm \frac{H_F}{2\,|K|}\,(\mathbf{x} - \mathbf{x}_F)\,.$$

It is clear upon taking $\{\mu_F\}$ as a basis for M_0 that the set $\{\sigma^{\mu_F}\}$ is a basis for the local lowest-order Raviart-Thomas finite element space. Moreover, in light of (44), we recognize the mixed-method version of the MHM global method as a modified version of the lowest-order Raviart-Thomas method for the mixed form of the

Laplace problem, namely: Find $\lambda_F \in M_0$ and $u_0 \in V_0$ such that

$$\begin{cases} (\sigma^{\mu_F}, \sigma^{\lambda_F})_{\mathcal{T}_H} + (\nabla\cdot\sigma^{\mu_F}, u_0)_{\mathcal{T}_H} = -(\mu_F, \hat{T} f)_{\partial\mathcal{T}_H} & \text{for all } \mu_F \in M_0\,, \\ (\nabla\cdot\sigma^{\lambda_F}, v_0)_{\mathcal{T}_H} = (f, v_0)_{\mathcal{T}_H} & \text{for all } v_0 \in V_0\,. \end{cases} \tag{64}$$

As such, if f is piecewise constant on $\mathcal{T}_H$, so, from (58) we get that $\hat{T}f = 0$ and then we recover the lowest-order Raviart-Thomas method for the mixed form of the Laplace problem *exactly*. The well-known error estimates for the lowest-order Raviart-Thomas method are then obtained from (60). In addition, we observe a super-convergence of the error in the L^2 norm for the "updated" u_H. A similar result was also proved in [4].

5.1.2 Multiscale Finite Element Methods

In the case the Laplace problem involves a more complex coefficient κ, the one-level MHM method (56) might been seen as a generalization of the multiscale finite element method (MsFEM) by Chen and Hou [16]. Indeed, the case when κ highly oscillates inside $K \in \mathcal{T}_H$ has been handled in [16] using the solution of the local problem (57) directly in formulation (64) (assuming M_0 to discretize flux on faces). However, the second local problem (58) has not been considered, meaning the multiscale method in [16] does not include the term $-(\mu_F, \hat{T} f)_{\partial\mathcal{T}_H}$. As mentioned in the previous section, if f is piecewise constant on $\mathcal{T}_H$, then $\hat{T}f = 0$ and we recover exactly the MsFEM.

The subgrid upscaling method (UpFEM) was introduced in [4] and recovered the method in [16] inside an enhanced space framework which includes the term $(f, T\mu_l)_{\mathcal{T}_H}$ on the right-hand side. After some algebraic manipulations, it may be seen as the present method (in its lower order version) using the space M_1 to approach the flux across faces. In fact, observe that the term $-(\mu_l, \hat{T} f)_{\partial\mathcal{T}_H}$ may written equivalently as

$$-(\mu_l, \hat{T} f)_{\partial\mathcal{T}_H} = (\kappa\, \nabla T\mu_l, \nabla\hat{T} f)_{\mathcal{T}_H} = (T\mu_l, f)_{\mathcal{T}_H}\,,$$

where we used the definition of operators T and $\hat{T}$ in (3) and (4), respectively. Nevertheless, the way the UpFEM method is built is fundamentally different than the present work. This prohibits establishing such a relationship for higher-order interpolation spaces M_l, $l \geqslant 2$. Finally, and unlike the present approach, the cited works start with the mixed Laplace problem. As a consequence, the local problems are also of mixed form, and so, a two-level stable finite element pair of spaces is necessarily adopted.

5.2 *The Two-Level MHM Method*

As we mentioned, in the case the Laplace problem involves a more complex coefficient κ, either due to multiscale or high-contrast aspects for instance, or if one adopts higher-order interpolation on faces, a two-level method must be employed to find an approximate local solution of (51) and (52), respectively. This is the subject in this section. First, we notice that the two-level MHM method is quite general and can embed a large variety of two-level numerical methods. Here, the choice made will allow us to establish a relationship between the two-level MHM method and the discrete primal hybrid formulation proposed in [39]. To this end, we shall strongly use the abstract results proved in the previous sections. Also, some remarks with respect to the recent proposed Discontinuous Petrov-Galerkin (DPG) method [19] and the Hybrid Discontinuous Galerkin (HDG) method [21] close this section.

5.2.1 The Primal Hybrid Numerical Method

Define $X_s = V_h$ as follows

$$V_h := \oplus_{K\in\mathcal{T}_H}\mathbb{P}_k(K) \subset H^1(\mathcal{T}_H)\,, \tag{65}$$

where $P_k(K)$ is the space of polynomial functions of degree equal or less k on K. The discrete version of the primal hybrid formulation in [39] reads: Find $u_h \in V_h$ and $\lambda_l \in M_l$ such that

$$\begin{cases} (\kappa\nabla u_h, \nabla v_h)_{\mathcal{T}_H} + (\lambda_l, v_h)_{\partial\mathcal{T}_H} = (f, v_h)_{\mathcal{T}_H} \quad \text{for all } v_h \in V_h\,, \\ (\mu_l, u_h)_{\partial\mathcal{T}_H} = 0 \quad \text{for all } \mu_l \in M_l\,. \end{cases} \tag{66}$$

We recall that the following result [39, Lemma 4 and Theorem 2] provides the necessary and sufficient condition to (66) be well-posed.

Lemma 10 *Let the spaces M_l and V_h be defined in* (54) *and* (65)*, respectively. Then, the compatibility condition* (48) *holds if and only if*

$$k \geqslant \begin{cases} l+1 & \text{when } l \text{ is even}\,, \\ l+2 & \text{when } l \text{ is odd}\,. \end{cases}$$

Now, we choose to approximate local problems $T\,\lambda_l\,|_K$ and $\hat{T}f\,|_K$ in each $K \in \mathcal{T}_H$ as follows

$$\begin{aligned} (\kappa\nabla T_h\,\lambda_l, \nabla v_h)_K &= -(\lambda_l, v_h)_{\partial K} \quad \text{for all } v_h \in V_h \cap L^2_0(K)\,, \\ (\kappa\nabla \hat{T}_h\,\lambda_l, \nabla v_h)_K &= (f, v_h)_K \quad \text{for all } v_h \in V_h \cap L^2_0(K)\,, \end{aligned} \tag{67}$$

and we take $T_s = T_h$ and $\hat{T}_s = \hat{T}_h$ in (50). Note that local problems are discretized by the classical Galerkin method based on a single element, and require no further discretization of each element. As such, it can be seen as a p-method at the second level. The two-level MHM method reads: Find $(\lambda_l, u_0^{H,h}) \in M_l \times V_0$ such that

$$\left\{\begin{aligned} (\mu_l, T_h\,\lambda_l)_{\partial\mathcal{T}_H} + (\mu_l, u_0^{H,h})_{\partial\mathcal{T}_H} &= -(\mu_l, \hat{T}_h f)_{\partial\mathcal{T}_H} \quad \text{for all } \mu_l \in M_l\,, \\ (\lambda_l, v_0)_{\partial\mathcal{T}_H} &= (f, v_0)_{\mathcal{T}_H} \quad \text{for all } v_0 \in V_0\,. \end{aligned}\right. \tag{68}$$

Next, we establish the wellposedness and the optimality of the method (68). It is important to mention that the two-level method (68) and its analysis for the Laplace problem are *new*. Such results are a consequence of Theorem 3, Lemma 9 and Lemma 10, and the following standard interpolation result (see [22] for instance):

$$\inf_{v_h \in \mathbb{P}_k(K)} \|v - v_h\|_{1,K} \leqslant C H_K^m\,|v|_{m+1,K} \quad \text{for all } v \in H^{m+1}(K),\ 1 \leqslant m \leqslant k\,. \tag{69}$$

Theorem 4 *Assume the compatibility condition in Lemma 10 holds. Then, MHM method* (68) *is well-posed and*

$$u_h = u_0^{H,h} + T_h\,\lambda_l + \hat{T}_h f\,, \tag{70}$$

where (u_h, λ_l) *solves* (66) *and* $(u_0^{H,h}, \lambda_l)$ *solves* (68). *Furthermore, if* $u \in H^{m+1}(\Omega)$, *for* $1 \leqslant m \leqslant l + 1$, *then the following estimates hold*

$$\begin{aligned} \|\lambda - \lambda_l\|_M + \|u_0 - u_0^{H,h}\|_X &\leqslant C\,H^m |u|_{m+1,\Omega}\,, \\ \|u - u_h\|_{0,\Omega} + H\,|u - u_h|_{1,h} &\leqslant C\,H^{m+1}\,|u|_{m+1,\Omega}\,. \end{aligned} \tag{71}$$

Proof From Lemma 10, and since $M_0 \subset M_l$ from definition (54), Lemma 8 holds. As such, method (68) is well-posed from Theorem 3. Characterization (70) is a consequence of Lemma 9.

Now, from Theorem 3 we get

$$\|\lambda - \lambda_l\|_M \leqslant \inf_{z \in M_s^\star} \left\{ \left(1 + \frac{\|\mathcal{A}^\dagger\|}{\beta_{s,1}}\right) \|\lambda - z\|_M + \frac{1}{\beta_{s,1}} \|\left(\hat{T} - \hat{T}_h\right) f + (T - T_h)\,z\|_X \right\}\,.$$

First, the estimate of the first term in the right hand side above results from (55) used with the L^2 projection of $\lambda \in M$ onto the space $\mathbb{P}_l(F)$, here denoted ν_l. Observing that $\nu_l \in M_s^\star$ (from the first equation in (5)), we get

$$\inf_{z \in M_s^\star} \|\lambda - z\|_M \leqslant \|\lambda - \nu_l\|_M \leqslant C\,H^m |u|_{m+1,\Omega}\,.$$

The second term is dealt with using the standard interpolation estimate (69) (together with $k \geqslant l+1 \geqslant m \geqslant 1$) and Lemma 5 to get

$$\inf_{z \in M_s^\star} \| \left(\hat{T} - \hat{T}_h\right) f + (T - T_h)\, z \|_X \leqslant \| \left(\hat{T} - \hat{T}_h\right) f + (T - T_h)\, \nu_l \|_X \leqslant C H^m |u|_{m+1,\Omega} \,,$$

and the first estimate in (71) follows from Theorem 3. The second estimate in (71) is a straightforward consequence of the first estimate and Lemma 9. Note that since the constants in (60) are independent of H, the constants $\beta_{s,1}$ and $\beta_{s,2} = \beta_{s,3}$ in Lemma 3 are independent of H. Therefore, the constants C here are also independent of H.

Remark 8 As mentioned, the second level discretization can be quite general, i.e., the MHM method could have worked with different interpolation spaces V_h and discrete methods T_h and $\hat{T}_h$ locally. The one adopted in this section is only an example in order to establish a relationship with the classical primal hybrid method [39]. Discontinuous Galerkin method, stabilized method or even mixed methods, just to cite a few, may be adopted as a second-level solver, and induce a large variety of MHM methods. It would be interesting, for instance, to establish some relationship with some members of the family of DG and/or HDG methods, or with some stabilized or enriched finite element methods within the current framework.

Remark 9 Recently the Discontinuous Petrov-Galerkin (DPG) method, first proposed in [18] for the first-order system of the Poisson equation, has been revisited and reformulated for the original second-order Poisson problem in [19]. The DPG methodology starts at the continuous level, looking for the solution in the $H_0^1(\Omega)$ space and relaxing the continuity of the test functions. Specifically, the idea is to find $(u_{DPG}, \lambda_{DPG}) \in H_0^1(\Omega) \times \Pi_{K \in \mathcal{T}_H} H^{-1/2}(\partial K)$ such that

$$(\nabla u_{DPG}, \nabla v)_{\mathcal{T}_H} + (\lambda_{DPG}, v)_{\partial \mathcal{T}_H} = (f, v)_{\mathcal{T}_H}, \quad \text{for all } v \in H^1(\mathcal{T}_H)\,, \tag{72}$$

where λ_{DPG} is the numerical flux from the DPG terminology. Observe that, although not directly stated in [19], Problem (72) is nothing but the primal hybrid formulation of the Poisson problem, i.e., (u_{DPG}, λ_{DPG}) coincides with (u, λ) solution of (2) (with $\kappa = I$) as a result of Lemma 7. Therefore, the MHM and DPG methods share the same starting point, but ultimately differ in the way the finite dimensional subspaces are selected to approximate the exact solutions. For the latter, $H_0^1(\Omega)$-conformity is imposed for the discrete trial space, and local solvers are employed to approximate optimal test functions driven by the choice of trial space. Such a strategy differs from that of the MHM, thereby yielding distinct numerical methods.

Remark 10 The Hybridizable Discontinuous Galerkin (HDG) methods share similarities with the current setting. In particular, the solutions may be eliminated with respect to the Lagrange multiplier and the right-hand side f to yield a face-based global method in a fewer number of degrees of freedom. A multiscale HDG method was proposed recently [21] on top of a multiscale trace space for the Lagrange multipliers, an idea also pursued in [5]. Nonetheless, differences appear as well.

Importantly, the form of the local and global problems are different. This owes to the fact that the HDG methods take as their starting point the dual hybrid formulation of the Laplace problem (i.e., the hybridization of the first-order mixed formulation) which has already been discretized through a discontinuous Galerkin method. As such, the formulations are modified so as to introduce numerical fluxes (which are prescribed in advance) and are balanced by a stabilization parameter. Such a definition of the numerical flux impacts the final form of the global problem. Also, as a result of the hybridization strategy, the local problems are of mixed type with prescribed Dirichlet boundary conditions. This last feature seems to make the HDG method more sensitive to the choice of interpolations for the Lagrange multipliers on the boundary elements than the MHM method. Such a flexibility in avoiding more involved discrete spaces for the Lagrange multipliers is particularly attractive in the case of highly heterogenous problems and/or singularly perturbed problems with crossing face interfaces [27] (see also [35] for a view of the accuracy of the MHM method using polynomial interpolations for a highly oscillatory case). These characteristics make the MHM and the HDG methods intrinsically different.

6 Conclusion

The abstract framework built around the MHM method opened a new perspective for their analysis and construction. Indeed, it led to an alternative proof of known results as well as new error estimates for the two-level methods. In the process, discrete solutions of the classical primal hybrid formulation in [38] were characterized using fewer degrees of freedom. Also, the MHM methods are shown to be closely related to the lowest-order Raviart-Thomas element, the MsFEM and the UpFEM under appropriate conditions. Particularly, the latter can be recovered within the MHM strategy depending on the choice of approximation spaces.

The Laplace model was used as a proof of concept of the proposed abstract setting. It is worth mentioning that the present scope also includes the version of the MHM method applied to the elasticity and to the reactive-advective-diffusive equations presented in [25] and [27], respectively. Specifically, the essential results of the error analysis in [25] can be entirely recovered within this new perspective. Also, the wellposedness and the best approximation properties of the MHM method for the reactive-advective dominant model in [27] fit the present framework. However, the important study of the dependence of constants in the error estimates with respect to the physical coefficients deserves further investigation. As such, these new results will be addressed in forthcoming works.

This work investigated the mathematical structures and theoretical aspect of the MHM methods. Interested readers can find a large set of numerical validations for the MHM method in [3, 26, 35]. Algorithmic aspects of the MHM method were left out of the scope on purpose. This is, indeed, a vast subject on its own which must be deeply investigated. Although the MHM method is of the "divide and conquer" form, the method deserves intensive validation in terms of computational performance and memory allocation within new massively parallel architectures. Behind these aspects is the question of the capacity of the numerical algorithm, which underlies the MHM method, to be optimized on such computers in comparison with other domain decomposition methods. This work is currently in progress on the new petaflop computer recently acquired at LNCC.

References

1. A. Allendes, G.R. Barrenechea, E. Hernandez, F. Valentin, A two-level enriched finite element method for a mixed problem. Math. Comput. **80**, 11–41 (2011)
2. R. Araya, G.R. Barrenechea, L.P. Franca, F. Valentin, Stabilization arising from PGEM: a review and further developments. Appl. Numer. Math. **59**, 2065–2081 (2009)
3. R. Araya, C. Harder, D. Paredes, F. Valentin, Multiscale hybrid-mixed method. SIAM J. Numer. Anal. **51**, 3505–3531 (2013)
4. T. Arbogast, K. Boyd, Subgrid upscaling and mixed multiscale finite elements. SIAM J. Numer. Anal. **44**, 1150–1171 (2006)
5. T. Arbogast, H. Xiao, A multiscale mortar mixed space based on homogenization for heterogeneous elliptic problems. SIAM J. Numer. Anal. **51**, 377–399 (2013)
6. T. Arbogast, G. Pencheva, M.F. Wheeler, I. Yotov, A multiscale mortar mixed finite element method. SIAM Multiscale Model. Simul. **6**, 319–346 (2007)
7. I. Babuška, E. Osborn, Generalized finite element methods: their performance and their relation to mixed methods. SIAM J. Numer. Anal. **20**, 510–536 (1983)
8. C. Baiocchi, F. Brezzi, L.P. Franca, Virtual bubbles and Galerkin-Least-Squares type methods (Ga.L.S.). Comput. Methods Appl. Mech. Eng. **105**, 125–141 (1993)
9. G.R. Barrenechea, L.P. Franca, F. Valentin, A symmetric nodal conservative finite element method for the Darcy equation. SIAM J. Numer. Anal. **47**, 3652–3677 (2009)
10. F. Brezzi, M. Fortin, *Mixed and Hybrid Finite Element Methods*. Springer Series in Computational Mathematics, vol. 15 (Springer, Berlin/New York, 1991)
11. F. Brezzi, A. Russo, Choosing bubbles for advection-diffusion problems. Math. Models Methods Appl. Sci. **4**, 571–587 (1994)
12. F. Brezzi, M.O. Bristeau, L.P. Franca, M. Mallet, G. Rogé, A relationship between stabilized finite element methods and the Galerkin method with bubble functions. Comput. Methods Appl. Mech. Eng. **96**, 117–129 (1992)
13. F. Brezzi, L.P. Franca, T.J.R. Hughes, A. Russo, $b = \int g$. Comput. Methods Appl. Mech. Eng. **145**, 329–339 (1997)
14. F. Cappelo, A. Geis, B. Gropp, L. Kale, B. Kramer, M. Sinir, Toward exascale resilience. Int. J. High Perform. Comput. Appl. **23**, 374–388 (2009)
15. B. Chabaud, B. Cockburn, Uniform-in-time superconvergence of hdg methods for the heat equation. Math. Comput. **81**, 107–129 (2011)
16. Z. Chen, T. Hou, A mixed multiscale finite element method for elliptic problems with oscillating coefficients. Math. Comput. **72**, 541–576 (2002)
17. A. Coutinho, L.P. Franca, F. Valentin, Numerical multiscale methods. Int. J. Numer. Meth. Fluids **70**, 403–419 (2012)
18. L. Demkowicz, J. Gopalakrishnan, Analysis of the dpg method for the Poisson equation. SIAM J. Numer. Anal. **49**, 1788–1809 (2011)
19. L. Demkowicz, J. Gopalakrishnan, A primal dpg method without a first order reformulation. Comput. Math. Appl. **66**, 1058–1064 (2013)
20. Y. Efendiev, T. Hou, X. Wu, Convergence of a nonconforming multiscale finite element method. SIAM J. Numer. Anal. **37**, 888–910 (2000) (electronic)
21. Y. Efendiev, R. Lazarov, M. Moon, K. Shi, A spectral multiscale hybridizable discontinuous galerkin method for second order elliptic problems. Comput. Meth. Appl. Mech. Eng. **292**, 243–256 (2015)

22. A. Ern, J.-L. Guermond, *Theory and Practice of Finite Elements* (Springer, Berlin/New York, 2004)
23. C. Farhat, I. Harari, L.P. Franca, The discontinuous enrichment method. Comput. Methods Appl. Mech. Eng. **190**, 6455–6479 (2001)
24. R. Glowinski, M.F. Wheeler, Domain decomposition and mixed finite element methods for elliptic problems, in *First International Symposium on Domain Decomposition Methods for Partial Differential Equations*, ed. by R. Glowinski, G.H. Golub, G.A. Meurant, J.Périaux (Society for Industrial and Applied Mathematics, Philadelphia, 1988), pp. 144–172
25. C. Harder, A.L. Madureira, F. Valentin, A hybrid-mixed method for elasticity. ESAIM Math. Model. Numer. Anal. **5**(2), 311–336 (2016)
26. C. Harder, D. Paredes, F. Valentin, A family of multiscale hybrid-mixed finite element methods for the Darcy equation with rough coefficients. J. Comput. Phys. **245**, 107–130 (2013)
27. C. Harder, D. Paredes, F. Valentin, On a multiscale hybrid-mixed method for advective-reactive dominated problems with heterogenous coefficients. SIAM Multiscale Model. Simul. **13**, 491–518 (2015)
28. T.Y. Hou, X. Wu, A multiscale finite element method for elliptic problems in composite materials and porous media. J. Comput. Phys. **134**, 169–189 (1997)
29. T. Hughes, G. Sangalli, Variational multiscale analysis: the fine-scale green's function, projection, optimization, localization, and stabilized methods. SIAM J. Numer. Anal. **45**, 539–557 (2007)
30. T.J.R. Hughes, G.R. Feijoo, L. Mazzei, J. Quincy, The variational multiscale method - a paradigm for computational mechanics. Comput. Methods Appl. Mech. Eng. **166**, 3–24 (1998)
31. A. Madureira, Abstract multiscale-hybrid-mixed methods. Calcolo **52**(4), 543–557 (2015)
32. A. Malqvist, D. Peterseim, Localization of elliptic multiscale problems. Math. Comput. **83**, 2583–2603 (2014)
33. W. McLean, *Strongly Elliptic Systems and Boundary Integral Equations* (Cambridge University Press, Cambridge, 2000)
34. M.Z. Nashed, Generalized inverses, normal solvability, and iteration for singular operator equations, in *Nonlinear Functional Analysis and Applications*. Proceedings of an Advanced Seminar Conducted by the Mathematics Research Center, the University of Wisconsin, Madison, Wisconsin, 1970 (Academic, New York, 1971), pp. 311–359
35. D. Paredes, F. Valentin, H.M. Versieux, On the robusteness of multiscale hybrid-mixed methods. Math. Comput. (2016). doi:dx.doi.org/10.1090/mcom/3108 (to appear)
36. W.V. Petryshyn, On generalized inverses and on the uniform convergence of $(i - pk)^n$ with application to iterative methods. J. Math. Anal. Appl. **18**, 417–439 (1967)
37. D.D.A. Pietro, A. Ern, S. Lemaire, A review of hybrid high-order methods: formulations, computational aspects, comparison with other methods, hal-01163569, INRIA (2015)
38. P. Raviart, J. Thomas, A mixed finite element method for 2nd order elliptic problems, in *Mathematical Aspect of Finite Element Methods*. Lecture Notes in Mathematics, vol. 606 (Springer, New York, 1977), pp. 292–315
39. P. Raviart, J. Thomas, Primal hybrid finite element methods for 2nd order elliptic equations. Math. Comput. **31**, 391–413 (1977)
40. U. Rude, New mathematics for extremescale computational science. SIAM News, June 1st 2014
41. G. Sangalli, Capturing small scales in elliptic problems using a Residual-Free Bubbles finite element method. SIAM Multiscale Model. Simul. **1**, 485–503 (2003)
42. A. Toselli, O. Widlund, *Domain Decomposition Methods-Algorithms and Theory*. Springer Series in Computational Mathematics, vol. 34 (Springer, Berlin, 2005)
43. W. E, B. Engquist, The heterogeneous multiscale methods. Commun. Math. Sci. **1**(1), 87–132 (2003)
44. M.F. Wheeler, G. Xue, I. Yotov, A multiscale mortar multipoint flux mixed finite element method. Math. Models Methods Appl. Sci. **46**, 759–796 (2012)
45. J. Xu, L. Zikatanov, Some observations on Babuska and Brezzi theories. Math. Comput. **94**, 195–202 (2003)

Erratum to: A Survey of Trefftz Methods for the Helmholtz Equation

Ralf Hiptmair, Andrea Moiola, and Ilaria Perugia

Erratum to:
Chapter 8 in: G.R. Barrenechea et al. (eds.), *Building Bridges: Connections and Challenges in Modern Approaches to Numerical Partial Differential Equations*, Lecture Notes in Computational Science and Engineering 114,
DOI 10.1007/978-3-319-41640-3_8

During production process incorrect manuscript for chapter 8 was erroneously used which has now been replaced after original publication.

The updated online version of the original chapter can be found under
http://dx.doi.org/10.1007/978-3-319-41640-3_8

R. Hiptmair (✉)
Seminar for Applied Mathematics, ETH Zürich, 8092 Zürich, Switzerland
email: hiptmair@sam.math.ethz.ch; ralfh@ethz.ch

A. Moiola
Department of Mathematics and Statistics, University of Reading,
Whiteknights, PO Box 220, Reading RG6 6AX, UK
email: a.moiola@reading.ac.uk

I. Perugia
Faculty of Mathematics, University of Vienna, 1090 Vienna, Austria

Department of Mathematics, University of Pavia, 27100 Pavia, Italy
email: ilaria.perugia@univie.ac.at

G.R. Barrenechea et al. (eds.), *Building Bridges: Connections and Challenges in Modern Approaches to Numerical Partial Differential Equations*,
Lecture Notes in Computational Science and Engineering 114,
DOI 10.1007/978-3-319-41640-3_14

Editorial Policy

1. Volumes in the following three categories will be published in LNCSE:

i) Research monographs
ii) Tutorials
iii) Conference proceedings

Those considering a book which might be suitable for the series are strongly advised to contact the publisher or the series editors at an early stage.

2. Categories i) and ii). Tutorials are lecture notes typically arising via summer schools or similar events, which are used to teach graduate students. These categories will be emphasized by Lecture Notes in Computational Science and Engineering. **Submissions by interdisciplinary teams of authors are encouraged.** The goal is to report new developments – quickly, informally, and in a way that will make them accessible to non-specialists. In the evaluation of submissions timeliness of the work is an important criterion. Texts should be well-rounded, well-written and reasonably self-contained. In most cases the work will contain results of others as well as those of the author(s). In each case the author(s) should provide sufficient motivation, examples, and applications. In this respect, Ph.D. theses will usually be deemed unsuitable for the Lecture Notes series. Proposals for volumes in these categories should be submitted either to one of the series editors or to Springer-Verlag, Heidelberg, and will be refereed. A provisional judgement on the acceptability of a project can be based on partial information about the work: a detailed outline describing the contents of each chapter, the estimated length, a bibliography, and one or two sample chapters – or a first draft. A final decision whether to accept will rest on an evaluation of the completed work which should include

- at least 100 pages of text;
- a table of contents;
- an informative introduction perhaps with some historical remarks which should be accessible to readers unfamiliar with the topic treated;
- a subject index.

3. Category iii). Conference proceedings will be considered for publication provided that they are both of exceptional interest and devoted to a single topic. One (or more) expert participants will act as the scientific editor(s) of the volume. They select the papers which are suitable for inclusion and have them individually refereed as for a journal. Papers not closely related to the central topic are to be excluded. Organizers should contact the Editor for CSE at Springer at the planning stage, see *Addresses* below.

In exceptional cases some other multi-author-volumes may be considered in this category.

4. Only works in English will be considered. For evaluation purposes, manuscripts may be submitted in print or electronic form, in the latter case, preferably as pdf- or zipped ps-files. Authors are requested to use the LaTeX style files available from Springer at http://www.springer.com/gp/authors-editors/book-authors-editors/manuscript-preparation/5636 (Click on LaTeX Template → monographs or contributed books).

For categories ii) and iii) we strongly recommend that all contributions in a volume be written in the same LaTeX version, preferably LaTeX2e. Electronic material can be included if appropriate. Please contact the publisher.

Careful preparation of the manuscripts will help keep production time short besides ensuring satisfactory appearance of the finished book in print and online.

5. The following terms and conditions hold. Categories i), ii) and iii):

Authors receive 50 free copies of their book. No royalty is paid.
Volume editors receive a total of 50 free copies of their volume to be shared with authors, but no royalties.

Authors and volume editors are entitled to a discount of 33.3 % on the price of Springer books purchased for their personal use, if ordering directly from Springer.

6. Springer secures the copyright for each volume.

Addresses:

Timothy J. Barth
NASA Ames Research Center
NAS Division
Moffett Field, CA 94035, USA
barth@nas.nasa.gov

Michael Griebel
Institut für Numerische Simulation
der Universität Bonn
Wegelerstr. 6
53115 Bonn, Germany
griebel@ins.uni-bonn.de

David E. Keyes
Mathematical and Computer Sciences
and Engineering
King Abdullah University of Science
and Technology
P.O. Box 55455
Jeddah 21534, Saudi Arabia
david.keyes@kaust.edu.sa

and

Department of Applied Physics
and Applied Mathematics
Columbia University
500 W. 120 th Street
New York, NY 10027, USA
kd2112@columbia.edu

Risto M. Nieminen
Department of Applied Physics
Aalto University School of Science
and Technology
00076 Aalto, Finland
risto.nieminen@aalto.fi

Dirk Roose
Department of Computer Science
Katholieke Universiteit Leuven
Celestijnenlaan 200A
3001 Leuven-Heverlee, Belgium
dirk.roose@cs.kuleuven.be

Tamar Schlick
Department of Chemistry
and Courant Institute
of Mathematical Sciences
New York University
251 Mercer Street
New York, NY 10012, USA
schlick@nyu.edu

Editor for Computational Science
and Engineering at Springer:
Martin Peters
Springer-Verlag
Mathematics Editorial IV
Tiergartenstrasse 17
69121 Heidelberg, Germany
martin.peters@springer.com

Lecture Notes in Computational Science and Engineering

1. D. Funaro, *Spectral Elements for Transport-Dominated Equations.*
2. H.P. Langtangen, *Computational Partial Differential Equations.* Numerical Methods and Diffpack Programming.
3. W. Hackbusch, G. Wittum (eds.), *Multigrid Methods V.*
4. P. Deuflhard, J. Hermans, B. Leimkuhler, A.E. Mark, S. Reich, R.D. Skeel (eds.), *Computational Molecular Dynamics: Challenges, Methods, Ideas.*
5. D. Kröner, M. Ohlberger, C. Rohde (eds.), *An Introduction to Recent Developments in Theory and Numerics for Conservation Laws.*
6. S. Turek, *Efficient Solvers for Incompressible Flow Problems.* An Algorithmic and Computational Approach.
7. R. von Schwerin, ***Multi Body System SIM****ulation.* Numerical Methods, Algorithms, and Software.
8. H.-J. Bungartz, F. Durst, C. Zenger (eds.), *High Performance Scientific and Engineering Computing.*
9. T.J. Barth, H. Deconinck (eds.), *High-Order Methods for Computational Physics.*
10. H.P. Langtangen, A.M. Bruaset, E. Quak (eds.), *Advances in Software Tools for Scientific Computing.*
11. B. Cockburn, G.E. Karniadakis, C.-W. Shu (eds.), *Discontinuous Galerkin Methods.* Theory, Computation and Applications.
12. U. van Rienen, *Numerical Methods in Computational Electrodynamics.* Linear Systems in Practical Applications.
13. B. Engquist, L. Johnsson, M. Hammill, F. Short (eds.), *Simulation and Visualization on the Grid.*
14. E. Dick, K. Riemslagh, J. Vierendeels (eds.), *Multigrid Methods VI.*
15. A. Frommer, T. Lippert, B. Medeke, K. Schilling (eds.), *Numerical Challenges in Lattice Quantum Chromodynamics.*
16. J. Lang, *Adaptive Multilevel Solution of Nonlinear Parabolic PDE Systems.* Theory, Algorithm, and Applications.
17. B.I. Wohlmuth, *Discretization Methods and Iterative Solvers Based on Domain Decomposition.*
18. U. van Rienen, M. Günther, D. Hecht (eds.), *Scientific Computing in Electrical Engineering.*
19. I. Babuška, P.G. Ciarlet, T. Miyoshi (eds.), *Mathematical Modeling and Numerical Simulation in Continuum Mechanics.*
20. T.J. Barth, T. Chan, R. Haimes (eds.), *Multiscale and Multiresolution Methods.* Theory and Applications.
21. M. Breuer, F. Durst, C. Zenger (eds.), *High Performance Scientific and Engineering Computing.*
22. K. Urban, *Wavelets in Numerical Simulation.* Problem Adapted Construction and Applications.
23. L.F. Pavarino, A. Toselli (eds.), *Recent Developments in Domain Decomposition Methods.*

24. T. Schlick, H.H. Gan (eds.), *Computational Methods for Macromolecules: Challenges and Applications.*

25. T.J. Barth, H. Deconinck (eds.), *Error Estimation and Adaptive Discretization Methods in Computational Fluid Dynamics.*

26. M. Griebel, M.A. Schweitzer (eds.), *Meshfree Methods for Partial Differential Equations.*

27. S. Müller, *Adaptive Multiscale Schemes for Conservation Laws.*

28. C. Carstensen, S. Funken, W. Hackbusch, R.H.W. Hoppe, P. Monk (eds.), *Computational Electromagnetics.*

29. M.A. Schweitzer, *A Parallel Multilevel Partition of Unity Method for Elliptic Partial Differential Equations.*

30. T. Biegler, O. Ghattas, M. Heinkenschloss, B. van Bloemen Waanders (eds.), *Large-Scale PDE-Constrained Optimization.*

31. M. Ainsworth, P. Davies, D. Duncan, P. Martin, B. Rynne (eds.), *Topics in Computational Wave Propagation.* Direct and Inverse Problems.

32. H. Emmerich, B. Nestler, M. Schreckenberg (eds.), *Interface and Transport Dynamics.* Computational Modelling.

33. H.P. Langtangen, A. Tveito (eds.), *Advanced Topics in Computational Partial Differential Equations.* Numerical Methods and Diffpack Programming.

34. V. John, *Large Eddy Simulation of Turbulent Incompressible Flows.* Analytical and Numerical Results for a Class of LES Models.

35. E. Bänsch (ed.), *Challenges in Scientific Computing - CISC 2002.*

36. B.N. Khoromskij, G. Wittum, *Numerical Solution of Elliptic Differential Equations by Reduction to the Interface.*

37. A. Iske, *Multiresolution Methods in Scattered Data Modelling.*

38. S.-I. Niculescu, K. Gu (eds.), *Advances in Time-Delay Systems.*

39. S. Attinger, P. Koumoutsakos (eds.), *Multiscale Modelling and Simulation.*

40. R. Kornhuber, R. Hoppe, J. Périaux, O. Pironneau, O. Wildlund, J. Xu (eds.), *Domain Decomposition Methods in Science and Engineering.*

41. T. Plewa, T. Linde, V.G. Weirs (eds.), *Adaptive Mesh Refinement – Theory and Applications.*

42. A. Schmidt, K.G. Siebert, *Design of Adaptive Finite Element Software.* The Finite Element Toolbox ALBERTA.

43. M. Griebel, M.A. Schweitzer (eds.), *Meshfree Methods for Partial Differential Equations II.*

44. B. Engquist, P. Lötstedt, O. Runborg (eds.), *Multiscale Methods in Science and Engineering.*

45. P. Benner, V. Mehrmann, D.C. Sorensen (eds.), *Dimension Reduction of Large-Scale Systems.*

46. D. Kressner, *Numerical Methods for General and Structured Eigenvalue Problems.*

47. A. Boriçi, A. Frommer, B. Joó, A. Kennedy, B. Pendleton (eds.), *QCD and Numerical Analysis III.*

48. F. Graziani (ed.), *Computational Methods in Transport.*

49. B. Leimkuhler, C. Chipot, R. Elber, A. Laaksonen, A. Mark, T. Schlick, C. Schütte, R. Skeel (eds.), *New Algorithms for Macromolecular Simulation.*

50. M. Bücker, G. Corliss, P. Hovland, U. Naumann, B. Norris (eds.), *Automatic Differentiation: Applications, Theory, and Implementations.*

51. A.M. Bruaset, A. Tveito (eds.), *Numerical Solution of Partial Differential Equations on Parallel Computers.*

52. K.H. Hoffmann, A. Meyer (eds.), *Parallel Algorithms and Cluster Computing.*

53. H.-J. Bungartz, M. Schäfer (eds.), *Fluid-Structure Interaction.*

54. J. Behrens, *Adaptive Atmospheric Modeling.*

55. O. Widlund, D. Keyes (eds.), *Domain Decomposition Methods in Science and Engineering XVI.*

56. S. Kassinos, C. Langer, G. Iaccarino, P. Moin (eds.), *Complex Effects in Large Eddy Simulations.*

57. M. Griebel, M.A Schweitzer (eds.), *Meshfree Methods for Partial Differential Equations III.*

58. A.N. Gorban, B. Kégl, D.C. Wunsch, A. Zinovyev (eds.), *Principal Manifolds for Data Visualization and Dimension Reduction.*

59. H. Ammari (ed.), *Modeling and Computations in Electromagnetics: A Volume Dedicated to Jean-Claude Nédélec.*

60. U. Langer, M. Discacciati, D. Keyes, O. Widlund, W. Zulehner (eds.), *Domain Decomposition Methods in Science and Engineering XVII.*

61. T. Mathew, *Domain Decomposition Methods for the Numerical Solution of Partial Differential Equations.*

62. F. Graziani (ed.), *Computational Methods in Transport: Verification and Validation.*

63. M. Bebendorf, *Hierarchical Matrices.* A Means to Efficiently Solve Elliptic Boundary Value Problems.

64. C.H. Bischof, H.M. Bücker, P. Hovland, U. Naumann, J. Utke (eds.), *Advances in Automatic Differentiation.*

65. M. Griebel, M.A. Schweitzer (eds.), *Meshfree Methods for Partial Differential Equations IV.*

66. B. Engquist, P. Lötstedt, O. Runborg (eds.), *Multiscale Modeling and Simulation in Science.*

67. I.H. Tuncer, Ü. Gülcat, D.R. Emerson, K. Matsuno (eds.), *Parallel Computational Fluid Dynamics 2007.*

68. S. Yip, T. Diaz de la Rubia (eds.), *Scientific Modeling and Simulations.*

69. A. Hegarty, N. Kopteva, E. O'Riordan, M. Stynes (eds.), *BAIL* 2008 – *Boundary and Interior Layers.*

70. M. Bercovier, M.J. Gander, R. Kornhuber, O. Widlund (eds.), *Domain Decomposition Methods in Science and Engineering XVIII.*

71. B. Koren, C. Vuik (eds.), *Advanced Computational Methods in Science and Engineering.*

72. M. Peters (ed.), *Computational Fluid Dynamics for Sport Simulation.*

73. H.-J. Bungartz, M. Mehl, M. Schäfer (eds.), *Fluid Structure Interaction II - Modelling, Simulation, Optimization.*

74. D. Tromeur-Dervout, G. Brenner, D.R. Emerson, J. Erhel (eds.), *Parallel Computational Fluid Dynamics 2008.*

75. A.N. Gorban, D. Roose (eds.), *Coping with Complexity: Model Reduction and Data Analysis.*

76. J.S. Hesthaven, E.M. Rønquist (eds.), *Spectral and High Order Methods for Partial Differential Equations.*

77. M. Holtz, *Sparse Grid Quadrature in High Dimensions with Applications in Finance and Insurance.*

78. Y. Huang, R. Kornhuber, O.Widlund, J. Xu (eds.), *Domain Decomposition Methods in Science and Engineering XIX.*

79. M. Griebel, M.A. Schweitzer (eds.), *Meshfree Methods for Partial Differential Equations V.*

80. P.H. Lauritzen, C. Jablonowski, M.A. Taylor, R.D. Nair (eds.), *Numerical Techniques for Global Atmospheric Models.*

81. C. Clavero, J.L. Gracia, F.J. Lisbona (eds.), *BAIL 2010 – Boundary and Interior Layers, Computational and Asymptotic Methods.*

82. B. Engquist, O. Runborg, Y.R. Tsai (eds.), *Numerical Analysis and Multiscale Computations.*

83. I.G. Graham, T.Y. Hou, O. Lakkis, R. Scheichl (eds.), *Numerical Analysis of Multiscale Problems.*

84. A. Logg, K.-A. Mardal, G. Wells (eds.), *Automated Solution of Differential Equations by the Finite Element Method.*

85. J. Blowey, M. Jensen (eds.), *Frontiers in Numerical Analysis - Durham 2010.*

86. O. Kolditz, U.-J. Gorke, H. Shao, W. Wang (eds.), *Thermo-Hydro-Mechanical-Chemical Processes in Fractured Porous Media - Benchmarks and Examples.*

87. S. Forth, P. Hovland, E. Phipps, J. Utke, A. Walther (eds.), *Recent Advances in Algorithmic Differentiation.*

88. J. Garcke, M. Griebel (eds.), *Sparse Grids and Applications.*

89. M. Griebel, M.A. Schweitzer (eds.), *Meshfree Methods for Partial Differential Equations VI.*

90. C. Pechstein, *Finite and Boundary Element Tearing and Interconnecting Solvers for Multiscale Problems.*

91. R. Bank, M. Holst, O. Widlund, J. Xu (eds.), *Domain Decomposition Methods in Science and Engineering XX.*

92. H. Bijl, D. Lucor, S. Mishra, C. Schwab (eds.), *Uncertainty Quantification in Computational Fluid Dynamics.*

93. M. Bader, H.-J. Bungartz, T. Weinzierl (eds.), *Advanced Computing.*

94. M. Ehrhardt, T. Koprucki (eds.), *Advanced Mathematical Models and Numerical Techniques for Multi-Band Effective Mass Approximations.*

95. M. Azaïez, H. El Fekih, J.S. Hesthaven (eds.), *Spectral and High Order Methods for Partial Differential Equations ICOSAHOM 2012.*

96. F. Graziani, M.P. Desjarlais, R. Redmer, S.B. Trickey (eds.), *Frontiers and Challenges in Warm Dense Matter.*

97. J. Garcke, D. Pflüger (eds.), *Sparse Grids and Applications – Munich 2012.*

98. J. Erhel, M. Gander, L. Halpern, G. Pichot, T. Sassi, O. Widlund (eds.), *Domain Decomposition Methods in Science and Engineering XXI.*

99. R. Abgrall, H. Beaugendre, P.M. Congedo, C. Dobrzynski, V. Perrier, M. Ricchiuto (eds.), *High Order Nonlinear Numerical Methods for Evolutionary PDEs - HONOM 2013.*

100. M. Griebel, M.A. Schweitzer (eds.), *Meshfree Methods for Partial Differential Equations VII.*

101. R. Hoppe (ed.), *Optimization with PDE Constraints - OPTPDE 2014.*

102. S. Dahlke, W. Dahmen, M. Griebel, W. Hackbusch, K. Ritter, R. Schneider, C. Schwab, H. Yserentant (eds.), *Extraction of Quantifiable Information from Complex Systems.*

103. A. Abdulle, S. Deparis, D. Kressner, F. Nobile, M. Picasso (eds.), *Numerical Mathematics and Advanced Applications - ENUMATH 2013.*

104. T. Dickopf, M.J. Gander, L. Halpern, R. Krause, L.F. Pavarino (eds.), *Domain Decomposition Methods in Science and Engineering XXII.*

105. M. Mehl, M. Bischoff, M. Schäfer (eds.), *Recent Trends in Computational Engineering - CE2014.* Optimization, Uncertainty, Parallel Algorithms, Coupled and Complex Problems.

106. R.M. Kirby, M. Berzins, J.S. Hesthaven (eds.), *Spectral and High Order Methods for Partial Differential Equations - ICOSAHOM'14.*

107. B. Jüttler, B. Simeon (eds.), *Isogeometric Analysis and Applications 2014.*

108. P. Knobloch (ed.), *Boundary and Interior Layers, Computational and Asymptotic Methods – BAIL 2014.*

109. J. Garcke, D. Pflüger (eds.), *Sparse Grids and Applications – Stuttgart 2014.*

110. H. P. Langtangen, *Finite Difference Computing with Exponential Decay Models.*

111. A. Tveito, G.T. Lines, *Computing Characterizations of Drugs for Ion Channels and Receptors Using Markov Models.*

112. B. Karazösen, M. Manguoğlu, M. Tezer-Sezgin, S. Göktepe, Ö. Uğur (eds.), *Numerical Mathematics and Advanced Applications - ENUMATH 2015.*

113. H.-J. Bungartz, P. Neumann, W.E. Nagel (eds.), *Software for Exascale Computing - SPPEXA 2013-2015.*

114. G.R. Barrenechea, F. Brezzi, A. Cangiani, E.H. Georgoulis (eds.), *Building Bridges: Connections and Challenges in Modern Approaches to Numerical Partial Differential Equations.*

For further information on these books please have a look at our mathematics catalogue at the following URL: www.springer.com/series/3527

Monographs in Computational Science and Engineering

1. J. Sundnes, G.T. Lines, X. Cai, B.F. Nielsen, K.-A. Mardal, A. Tveito, *Computing the Electrical Activity in the Heart.*

For further information on this book, please have a look at our mathematics catalogue at the following URL: www.springer.com/series/7417

Texts in Computational Science and Engineering

1. H. P. Langtangen, *Computational Partial Differential Equations.* Numerical Methods and Diffpack Programming. 2nd Edition
2. A. Quarteroni, F. Saleri, P. Gervasio, *Scientific Computing with MATLAB and Octave.* 4th Edition
3. H. P. Langtangen, *Python Scripting for Computational Science.* 3rd Edition
4. H. Gardner, G. Manduchi, *Design Patterns for e-Science.*
5. M. Griebel, S. Knapek, G. Zumbusch, *Numerical Simulation in Molecular Dynamics.*
6. H. P. Langtangen, *A Primer on Scientific Programming with Python.* 5th Edition
7. A. Tveito, H. P. Langtangen, B. F. Nielsen, X. Cai, *Elements of Scientific Computing.*
8. B. Gustafsson, *Fundamentals of Scientific Computing.*
9. M. Bader, *Space-Filling Curves.*
10. M. Larson, F. Bengzon, *The Finite Element Method: Theory, Implementation and Applications.*
11. W. Gander, M. Gander, F. Kwok, *Scientific Computing: An Introduction using Maple and MATLAB.*
12. P. Deuflhard, S. Röblitz, *A Guide to Numerical Modelling in Systems Biology.*
13. M. H. Holmes, *Introduction to Scientific Computing and Data Analysis.*
14. S. Linge, H. P. Langtangen, *Programming for Computations* - A Gentle Introduction to Numerical Simulations with MATLAB/Octave.
15. S. Linge, H. P. Langtangen, *Programming for Computations* - A Gentle Introduction to Numerical Simulations with Python.

For further information on these books please have a look at our mathematics catalogue at the following URL: www.springer.com/series/5151

Printed in the United States
By Bookmasters